AMATEUR ROCKET MOTOR CONSTRUCTION

AMATEUR ROCKET MOTOR CONSTRUCTION

A COMPLETE GUIDE TO THE CONSTRUCTION OF HOMEMADE SOLID FUEL ROCKET MOTORS

by

DAVID SLEETER

THE TELEFLITE CORPORATION
Moreno Valley, CA

Copyright © 2004 by David G. Sleeter
Photos & illustrations copyright © 1992-2004 by David G. Sleeter
With the exception of the cover and the 3 photographs in the dedication, all photos and illustrations are by the author.

Cover Design by Rose Freeland of Big Concepts Inc., South Pasadena, CA

All rights reserved. No part of this book may be reproduced or transmitted in any form or by any means, electronic or mechanical, including photocopying, recording, or by any information storage and retrieval system, without permission in writing from the publisher.

Published by The Teleflite Corporation
11620 Kitching St.
Moreno Valley, CA 92557
www.teleflite.com
www.amateur-rocketry.com

ISBN #0-930387-04-X
Library of Congress catalogue card #95-090904
Printed in the United States of America

A Good Book on Rocket Engineering:
Sutton, George P. *Rocket Propulsion Elements*. John Wiley & Sons (New York, Chichester, Brisbane, Toronto, Singapore; numerous editions from 1949 to the present).

A Good Book on Chemistry:
Conkling, John A. *Chemistry of Pyrotechnics*. Marcel Dekker, Inc. (New York, 1985).

A Good Movie:
October Sky. Universal Pictures (1999).

Historical Sources:

Baker, David. *The History of Manned Space Flight*. Crown Publishers (New York, 1985).

Cernan, Eugene A. & Arnold, H.J.P. et al. *Man in Space, An Illustrated History of Spaceflight*. CLB Publishing (England 1993).

Davis, Tenney L., Ph.D. *The Chemistry of Powder and Explosives*. John Wiley & Sons (New York & London, 1943) plus numerous reprints by other publishers to the present.

Lampton, Christopher. *Wernher von Braun*. Franklin Watts (New York, 1988).

Lewis, Richard S. *From Vinland to Mars, A Thousand Years of Exploration*. Quadrangle/The New York Times Book Co. (New York, 1976)

Ley, Willy. *Rockets, Missiles, and Space Travel*. The Viking Press (New York, 1951).

von Braun, Wernher & Ordway, Frederick. *History of Rocketry & Space Travel*. Crowell. (New York, 1969).

Walter, William. *Space Age*. Random House (New York, 1992).

DEDICATION

To the late **Mr. B.J. Humphreys** for his lifetime of contribution to amateur rocketry, and for his friendship. From the days of his youth to the end of his life, Jim loved rockets and everything about them. During the latter half of his life he developed muscular dystrophy, and in 1973, just to prove that he could do it, became the first person in history to drive a rocket powered wheel chair.

For the next 20 years he remained active in amateur rocketry, and always generous with his time, continued to lend his help and support whenever and wherever he could. In October of 1993 Jim embarked on the *next* great adventure. His enthusiasm, his good humor, and his refined and intelligent company are missed by all who knew him.

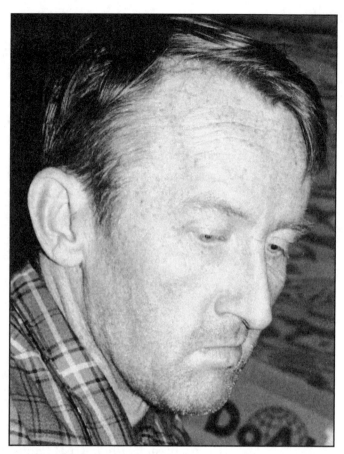

B.J. Humphreys in 1952. Photo courtesy of Nicky Humphreys. Photographer unknown.

Rocket builders B.J. Humphreys, Jim Medsker, and Ed Parker with one of their two stage, solid fuel rockets. It was launched from the Reaction Research Society's Mojave Desert test site in September of 1955. It held the distinction of being the largest two stage rocket built by a civilian rocket society as of that date. Photo courtesy of Nicky Humphreys. Photographer unknown.

Jim, sitting in his rocket powered wheel chair, is interviewed by reporters outside the Anaheim Convention center in 1973. Photo courtesy of Nicky Humphreys. Photo by Jack McConnell.

Contents

1. IMPORTANT INFORMATION & SAFETY PROCEDURES

Some Interesting History 1
    The Years 1232 to 1770 1
    Haidar and Tippu Ali 2
    William Congreve 2
    William Hale 2
    Konstantin Tsiolkovskii 3
    Robert Goddard 4
    Pedro Paulet 5
    Hermann Oberth and The Germans 6
The Attempted Destruction of Amateur Rocketry 9
Accidents During the Fifites, and the Events that Followed 11
The Tripoli Rocketry Association 12
The Reaction Research Society 12
About Black Powder 12
    Sensitivity To Impact 13
    Sensitivity to Sparks and Static Electricity 14
    Sensitivity to Friction 14
    The Piezoelectric Effect 15
    Storage of Black Powder 16
    Disposal of Black Powder 16
    Handling and Storage of Finished Motors 16
    Safety During Propellant Preparation and Motor Construction 16
    Aren't You Afraid that a Terrorist Will Use a Homemade Rocket? 18
    Won't Someone Use Homemade Black Powder to Make a Bomb? 18
Can I Sell the Rocket Motors that I Make? 18
What is the Safety Record so Far? 18
What if a Serious Accident Happens? 19

2. BASIC CONCEPTS

Basic Rocket Function 20
Cardboard Motor Casings 20
Clay Nozzles 20
The De Laval Rocket Nozzle 20
Basic Terminology 21
End Burners 23
Cylindrical Core Burners 24
Time Delays 25
Single Stage Rockets 29
Two Stage Rockets 34
Motor Classification 34

3. TOOLS YOU BUY

Weighing Equipment 37
    A Triple Beam Balance 37
    Reloading Scales 40
    Other Scales 40
    Scale Maintenance 40

Powder Mills 40
    The Manufacture of Commercial Black Powder 40
    Can I use a Mortar and Pestle? 41
    Can I use a Blender? 41
    Ball Mills and Rock Tumblers 42
    Vacuum Pumps 46
Disposable Syringes 48
    How to use a Syringe 48
Mixing Cups 51
Other Handy Things 52

4. TOOLS YOU MAKE

General Comments 53
Nozzle Molds 54
Improvised Centering Rings 59
Casing Retainers 60
An Improvised Plastic Casing Retainer 65
Core Spindles 66
An Improvised Core Spindle 72
Tamps 72
Wooden Tamps - A Homemade Boring Jig 77
An Improvised Wooden Tamp 79
A Homemade Vacuum Chamber 81
Homemade Sieves 85
A Homemade Coal Crusher 86

5. TOOL DRAWINGS

Casing Retainer Notes 88
Nozzle Mold Notes 88
Core Spindle Notes 89
Tamp Notes 89
Casing Retainer Drawings Begin on 90
Nozzle Mold Drawings Begin on 95
Core Spindle Drawings Begin on 121
Tamp Drawings Begin on 145
    Plug and Socket Detail Drawings Begin on 170
De Laval Nozzle Tool Drawings Begin on 172

6. THE CHEMICALS

Chemical Grades 174
Oxidizers 174
    Potassium Nitrate Fertilizer 175
    Potassium Nitrate Stump Remover 175
    Potassium Nitrate from a Chemical Dealer 175
    Extracted Potassium Nitrate 176
    Sodium Nitrate 181
    Homemade Sodium Nitrate 181
    Extracted Sodium Nitrate 185
Fuels 187
    Coal 187
    Charcoal 188
    Carbon Black 191
    Powdered Sugar 192
Sulfur 192
    Soil Sulfur 193
    Dusting Sulfur 193

       Flowers of Sulfur ... 193
       Powdered Sulfur ... 193
    Binders ... 194
       Red Gum ... 194
       Goma Laca Shellac ... 194
       Vinsol Resin ... 195
    Burn Rate Modifiers ... 195
       Baking Soda ... 195
    Solvents ... 196
       Acetone ... 196
       Denatured Alcohol ... 196
       Evaporation Rates ... 196
    Nozzle Clay Materials ... 197
       Powdered Clay ... 197
       Grog ... 198
       Sand ... 198
       Paraffin Wax ... 199
 Additional Notes on Oxidizers ... 199
 Additional Notes on Fuels ... 199

7. CHEMICAL PREPARATION

Drying ... 200
Oxidizers ... 200
    Milled Potassium Nitrate and Milled Sodium Nitrate (4 Hour) ... 200
    Precipitated Potassium Nitrate ... 201
Fuels ... 206
    Milled Anthracite (12 Hour) ... 206
    Milled Mesquite Charcoal (6 Hour) ... 207
    Milled Garden Charcoal (6 Hour) ... 208
    Airfloat Charcoal ... 209
    Carbon Black ... 209
    Powdered Sugar ... 209
Sulfur ... 209
Homemade Nozzle Clay ... 209

8. MOTOR CASINGS

How to Buy Convolute Tubes ... 211
Small Homemade Tubes ... 213
How to Buy Paper ... 221
    Kraft Linerboard ... 222
    Tagboard ... 224
    A Caveat and a Suggestion ... 224
    Red Rosin Paper ... 224
Large Homemade Tubes ... 225
Drying ... 232
Glue Coating ... 232

9. WORKING with PROPELLANTS

Propellant Names ... 234
The Propellant Mixing Procedure ... 235
Propellant Burn Rate Control ... 237
Adding the Binder Solvent ... 238
    The Pellet Test ... 240
The Logic Behind the Propellant Formulas ... 245
The KS Propellant ... 252

10. THE MOTOR BUILDING PROCESS

- An Illustrated Outline 253
- Your First Rocket Motor 274
    - Getting Started 274
    - Forming the Nozzle 278
    - Loading the Propellant 281
    - Adding a Time Delay 285
    - Forming a Porthole 287
- Air-Drying 290
- Vacuum-Drying 290
    - Vacuum Pump Maintenance 291
- The Effect of Baking Soda on Motor Performance 292
- The Effect of Core Length on Motor Performance 292
- The Time Delay Burn Rate Test 293
- Solving Problems 296
    - A Tilted Core Spindle 296
    - A Slipped Casing Retainer 299
    - Rapid Solvent Evaporation 299
    - Using up Leftover Propellant 300
- Dedicating your Tools 301
- Multistage Ignition 301
    - Preparing a Booster for Pyrotechnic Staging 301
    - Preparing a Second Stage Motor for Pyrotechnic Staging 303
    - Preparing for a Two Stage Rocket Flight 307
- Parachute Ejection 308
- End Burners 310
    - Construction Technique and Tooling 310
- Improving Performance with De Laval Nozzles 319
- An Important and Helpful Cleanup Note 322
- De Laval Nozzle Drawings Begin on 323

11. KG1 PROPELLANT and MOTOR DESIGNS

- The KG1 Propellant Formula 328
- Improving Motor Performance 329
- KG1 Motor Designs Begin on 330

12. KG3 PROPELLANT and MOTOR DESIGNS

- The KG3 Propellant Formula 348
- Improving Motor Performance 349
- KG3 Motor Designs Begin on 350

13. KS PROPELLANT and MOTOR DESIGNS

- The KS Propellant Formula 368
- KS Motor Designs Begin on 370

14. NG6 PROPELLANT and MOTOR DESIGNS

- The NG6 Propellant Formula 382
- Improving Motor Performance 382
- NG6 Motor Designs Begin on 384

15. NV6 PROPELLANT and MOTOR DESIGNS

- The NV6 Propellant Formula 418
- Improving Motor Performance 418
- NV6 Motor Designs Begin on 420

16. THE ORIGINAL 1983 MOTOR BUILDING PROCESS

Propellant Formulas and Propellant Preparation 446
Preparing the Casings 447
    The Original Method 447
    A Better Method 447
The Motor Building Process 452
Drying the Motors 454
General Advice 455

17. ELECTRIC IGNITERS and a HOMEMADE IGNITION SYSTEM

A Paper-Match Igniter 456
A High-Heat Igniter 459
    A Fast-Drying Igniter Composition 460
A Mini-Igniter 460
General Advice 462
A Homemade Ignition System 462
    How to Use the Ignition System 465

18. TEST EQUIPMENT

My Own Test Equipment 468
A Homemade Rocket Motor Test Stand 470
    The Test for Maximum Thrust 475
A Homemade Chart Recorder 478
    A Thrust-Time Curve 482
    Chart Speed 483
    Maximum Thrust 484
    Total Burn Time 486
    Total Impulse 487
    Average Thrust 488
    Propellant Mass 489
    Specific Impulse 490
    Equipment Limitations 492
    Safety During Motor Testing 492

19. FLIGHT

A Basic Stick Rocket 496
    Launching a Single Stage Stick Rocket 500
A Two Stage Stick Rocket 501
    Launching a Two Stage Stick Rocket 504
Practicing with a Time Delay 504
Fitting a Homemade Motor to a Commercially Made Rocket 504
Aerodynamic Balance 504
Flight Safety 507

INDEX 508

Preface

In 1982, while walking the dogs in a nearby field, I met a man flying a rocket. Powered by small, solid fuel rocket motors, it took to the air with a startling "whoosh", and I was immediately impressed by its power. Curious to know how the little motors were made, I bought three of them from the man. Then I went looking for a book on the subject. When a month of considerable effort turned up *nothing*, I decided to figure it out for myself. I found an article on skyrockets in an old encyclopedia, cleared off a table in the garage, and in my spare time I went to work. Eighteen months later I published *my own* book on rocket motors, and in the years since, my company, The Teleflite Corporation, has sold more than 10,000 copies of that book to rocket amateurs throughout the world.

My original building method required that you add water to the propellant prior to loading. Before the motors could be fired, they had to be dried, and this could take up to a month. In 1989, in an effort to remedy that situation, I began to experiment with acetone-soluble gums. Over the next 7 years I perfected a series of *fast* drying propellants and a dozen other improvements, including larger motors, the use of wax to improve the quality of the nozzle clay, the use of carbon black to cure combustion instability, and a simple way to control a propellant's burn rate by adding or subtracting baking soda. This new and much larger book is the result of that work.

By reading it, you will learn how to buy or make the chemicals, how to process the chemicals, how to mix your own propellants, how to assemble and use your own test equipment, and how to build and fly 54 *proven and tested* solid fuel rocket motors using simple hand tools and easy-to-find materials. Depending on the formula and where you buy the chemicals, the homemade rocket fuel will cost from 25 cents to 2 dollars per pound. Depending on the *size*, a finished rocket motor will cost from 15 cents to 3 dollars, **about 1/10 the cost of a commercially made motor.** The motors vary in size from a **C-6**, which you can fly in an area the size of a school yard, to an **I-65** that will take a three foot rocket to an altitude of 7,000 feet. Motor burn times vary from 1/2 second to 6.3 seconds, and motor thrusts range from 4 pounds to 58 pounds.

The smallest motors air-dry in 12 hours, and the largest in 72, but with a vacuum pump and a homemade vacuum chamber (plans provided herein), you can dry them in an hour. Though metal working equipment is not needed to *build* the motors, a metal lathe and a milling machine are needed to make some of the hand tools. For readers who are new to machine work, I've provided complete instructions on how to make each tool. **For readers *without* machine tools, I've included instructions on how to improvise with things like plastic pipe, hose clamps, and wooden dowels.** These improvised tools are primitive, but they work just fine, and they'll provide you with enough experience to know if you want to make a further investment in the hobby.

Though this book may *appear* complex, please be assured that the motor building process is *simple*. It takes just eight hand tools to make one of these rocket motors, and you can finish a **D** or an **E** motor in 10 minutes. The large volume of material presented is due to the fact that there are 5 propellant formulas and 54 motor designs to choose from, plus copious amounts of information on what to buy, how to shop, and even how to make some of the chemicals yourself, or extract them from common garden products.

Red gum is the glue or "binder" that holds four of these propellants together. In 1994 there was a worldwide shortage of red gum when the Australian government declared the Yacca trees, from which it is harvested, a protected species. Australia reversed that declaration in 1995. They resumed shipments of red gum, and the future supply is now secure. Their action, however, shook the confidence of the fireworks industry, and during the shortage, the industry developed two substitutes. One, called vinsol resin, is a by-product of America's logging industry. The other, called goma laca shellac, is a man made synthetic gum manufactured in Spain. With the recent introduction of these two alternatives, the availability of red gum will never again be an issue, and I discuss both at length in **Chapter 6**.

I am aware that readers experienced in rocketry and machine work might think the number of illustrations in this book to be excessive. But I've written this book for beginners as well. When you're doing something for the very first time, it is helpful to compare what you've just done, made, or bought to what you see in a *picture*. The lack of a picture can lead to hesitation and mistakes. It is therefore better to have too many pictures than not enough.

The motors in this book, though simple in design, are *amazingly* powerful, and when you see how well they work, *I think you will be impressed*. But before you proceed, **please read this book from cover to cover. Please pay special attention to Chapter 1, and please pay attention to all notices regarding safety**. Rocket motors derive their power and thrust from high energy chemical mixtures. They need not be feared, but they should *always* be treated with care and respect.

● My Thanks to Contributors

> My thanks to Rick Loehr for his help with several technical problems, and my thanks to Randy Thompson for his excellent motor testing software. My thanks also to the people who preserved the history of rocketry, and to the authors and publishers of the books listed under **Historical Sources** for making the myriad details accessible to the rest of us.
>
> David Sleeter, Moreno Valley, CA 2004

● About the Microphotographs

One of the tools in my shop is a scanning electron microscope that I use for high resolution microphotography. Five of the photos in this book were taken with this instrument, and you will immediately notice that things like charcoal and carbon black, which normally appear black, are *not* black in the photos. This is because a scanning electron microscope illuminates the specimen with an electron beam. The electron beam causes the specimen to emit a glow of *secondary* electrons, and it's *this glow* that the microscope scans and photographs. Therefore, regardless of how they appear to the naked eye, unless the images have been artificially colored, objects viewed with a scanning electron microscope appear in varying shades of *gray*.

1. IMPORTANT INFORMATION & SAFETY PROCEDURES.

FIRST and MOST IMPORTANT! Do *not* add or substitute other chemicals either to-or-for the chemicals described in this book! Specifically do *NOT* use POTASSIUM CHLORATE, SODIUM CHLORATE, POTASSIUM PERCHLORATE, SODIUM PERCHLORATE, POTASSIUM PERMANGANATE, AMMONIUM PERCHLORATE, PHOSPHOROUS, MATCH HEADS, ALUMINUM, MAGNESIUM, or ANY METALS AT ALL! *All* **of these substances will form DANGEROUS and EXPLOSIVE mixtures when combined with the other ingredients described herein,** *AND OTHER CHEMICALS WILL FORM DANGEROUS MIXTURES AS WELL.* **When mixing or handling propellants, be sure to ground yourself and all containers and utensils according to the instructions on page 14.**

If you decide to experiment on your own outside the bounds of the instructions provided in this book, consult with a professional chemist before you begin.

Some Interesting History

Like all of the things that exist in nature, the most primitive form of jet propulsion was not invented by anyone. It was *observed* by many people in many cultures at many different times. Toss a handful of chestnuts into a campfire, and if you are lucky, you'll see what I mean. As the moisture inside the nuts turns to steam, the ones that are tightly sealed burst open with a "*pop*". If one of the nuts has a tiny hole in its shell, it wiggles and dances among the hot coals with an audible "*hiss*". A careful observer sees a jet of steam coming out of the hole, and notices that the nut moves in a direction *opposite* the jet of steam. The efforts of many people over thousands of years to duplicate, improve, and enlarge upon this, and similar natural phenomena, resulted in the *gradual* development of the jet propelled vehicles that we build and use today.

A rocket is a *special* jet propelled vehicle that, unlike other jet propelled devices, carries *within* itself everything that it needs to operate, and this *includes* all of the necessary chemicals. The rocket *motor* is the device that burns the chemicals, produces the jet, and thereby generates propulsion. The jet engines used in aircraft sustain combustion by taking oxygen from the surrounding air, but if oxygen is needed, a rocket motor carries *its own* oxygen, and can therefore operate in the vacuum of Space.

● The Years 1232 to 1770

A simple steam jet like the chestnut described above is of little practical use because it ceases to function the moment it leaves the heat of the fire. With the invention of *gunpowder*, it first became possible to produce a self-sustaining, gas-generating reaction *outside* the fire, and this allowed the construction of the first jet propelled device that carried *its own* chemicals and made *its own* fire, the first *true* rocket motor. The logical step of attaching the motor to an arrow, which was designed to fly straight, resulted in a free-flying stick rocket, and history first records the use of these rockets in a battle at Kai-Fung-Fu in the year 1232 A.D. The Chinese used them to set fire to their Mongol attackers' camps, and they fired them in great numbers, so it is obvious that workers already experienced in their construction had made them and stockpiled them before the battle began. Unfortunately, if any notes about those people ever existed, those notes were long ago discarded, and if a record ever existed of who made the *first* rocket, that record has been lost in the mists of time.

The realization that rockets were of use to the military spurred a heightened interest in making them work better, but a poor understanding of science and engineering prevented early designers from turning them into truly accurate weapons. What these primitive rockets lacked in accuracy they often made up for in size, and ongoing efforts to make them larger culminated in 1688 when an officer in the German army designed a 132-pound stick-rocket with a guide pole made of wood, and a *huge* motor casing made of sail cloth saturated with dried and hardened glue.

An interesting footnote in the history of rocketry is the story of Wan Hoo. Wan Hoo was a wealthy Chinese nobleman who lived around the year 1500, and reliable sources say that Wan Hoo was *fascinated* by rockets. In a poorly conceived effort to become the world's first rocket pilot, he mounted two giant kites and 47 large skyrockets on a chair of his own design. At the appointed hour he strapped himself into the chair, and ordered his assistants to light the fuses. From that moment on, the descriptions of what happened and where he actually went are confusing. But I think it appropriate that astronomers have named a *crater* after Wan Hoo, and that the one they chose is on the far side of The Moon.

● Haidar and Tippu Ali

After the battle of Kai-Fung-Fu, rockets were used in military engagements throughout the world. But because of their small size and their inherent inaccuracy, they had little effect on their intended targets. So their use was limited until the 1770s when Prince Haidar Ali, the Muslim ruler of the Mysore region in India, developed a series of rockets weighing 6 to 12 pounds with motor tubes made of iron. Until that time, motor tubes were made of things like paper or bamboo. The iron tubes had a *much* greater burst strength, and this allowed Haidar's rockets to operate at a higher chamber pressure. The elevated pressure gave his rockets as *much* higher thrust and a *far* greater range, and the heavy iron tubes had a greater impact when they hit something.

These rockets were so successful that by 1780 Haidar had established a specially-trained rocket corps of 1,200 men, and following his death in 1782, his son, Tippu, increased the size of the corps to 5,000. Firing these rockets in great numbers, Tippu's troops used them very effectively in battles against the British at Srirangapatnam in 1792 and 1799. In one engagement, it was reported that a single rocket killed three British soldiers, and wounded another four.

● William Congreve

Back in England at the Woolwich Arsenal the success of Haidar's rockets caught the attention of a young Colonel named William Congreve. Congreve realized their potential, and decided to improve and enlarge upon Haidar's design. He began by developing a standard propellant formula and a standard production technique. Then he designed and built a large, iron stick-rocket that weighed 32 pounds, carried 7 pounds of incendiary chemicals, and flew 3,000 yards.

In 1803 the French Revolution and the "Storming of the Bastille" were just 14 years past. France was sympathetic with the recent *American* revolution, and ongoing hostilities between the old British monarchy and the new French republic flared into open warfare. By 1805, in preparation for an invasion of England, Napoleon had amassed a large store of troops and attack-supplies in the city of Boulogne. Boulogne was on the east coast of the English Channel, just 28 miles from England, and British anxiety over the situation was considerable. When Congreve suggested that they try a rocket attack on Boulogne, his proposal was quickly approved.

On October 18th ten English boats loaded with Congreve rockets assembled on the waters off Boulogne, but a violent storm rocked the boats, and made it impossible to aim the rockets. On November 21st they tried again. The seas were calmer, and they fired the rockets successfully, but hit nothing of value, and were *greatly* embarrassed when their intended French targets shouted defiant obscenities, and strutted up and down the shore waiving the burned-out rocket casings. During the Summer of 1806 there was a brief try at a peace agreement, but the accumulated hatred was too great, and in the Fall of that year the fighting began anew. In October of 1806 the British launched a third attack on Boulogne, and this time they hit their targets with devastating effect. Exact accounts of the number of rockets used vary from 200 to 2,000, but in any case, most of Boulogne was set on fire, Napoleon's invasion supplies were destroyed, and William Congreve's place in history was assured.

In 1807, still at war with France, and angered by Dutch collaboration with the French, the British launched *25,000* rockets against the Danish city of Copenhagen. The resulting *and spectacular* rain of fire almost destroyed Copenhagen. They used the rockets extensively in the U.S. vs. British "War of 1812", and in 1814, *still* at war with the United States, they used them in an unsuccessful attack on Baltimore's Fort McHenry. The black powder rocket propellant burned with a pink-colored flame, and a young lawyer named Francis Scott Key, who witnessed the event, described the "red glare" of the Congreve rockets in a poem titled *Defence of Fort M'Henry*. Liking the poem considerably, his brother-in-law (a local judge) had copies printed, and passed them out to his friends. On September 20th the *Defence of Fort M'Henry* was published in the *Baltimore Patriot*, and then in newspapers throughout the country. Shortly thereafter someone added Key's lyrics to the music from a drinking song written by John Stafford Smith. Together they became *The Star Spangled Banner*, and in 1931 the U.S. Congress declared *The Star Spangled Banner* to be the national anthem of the United States.

Congreve eventually built rockets weighing up to 300 pounds, and in 1815 he improved their accuracy by moving the guide stick from the side of the rocket to its central axis, and surrounding it with 5 equally-spaced exhaust jets. William Congreve approached the design of his rockets with logic and good scientific thinking, and should probably be considered the world's first, true, rocket engineer.

● William Hale

Congreve rockets and the ones that preceded them used long sticks or poles to establish aerodynamic stability, but the poles were heavy and flexible. They made the rockets sensitive to cross winds, and when they whipped around in flight, they diverted the rockets from their intended path. If you could eliminate the pole, and stabilize the rocket in some other manner, you could make it more accurate. As any child knows, a gyroscope, once set in motion, maintains its orientation in space until its spinning rotor stops. In 1846, with a clear understanding of the gyroscopic principle, a British inventor named William Hale got rid of the pole, and built a *spin*-stabilized rocket. In his first experiments he used multiple jets like Congreve, but he angled them around the rocket's central axis. In later designs he used multiple *straight* jets, but arranged them so that the exhaust from each jet passed over an angled vane. Impressed by the accuracy of Hale's rockets, the U.S. Army bought his patent, and used them to a limited extent in the war with Mexico.

In the 1860s both Hale rockets *and* Congreve rockets were occasionally used in the Civil War, and thereafter in minor conflicts around the world. From 1864 through 1881 the Russians used a small version of a Congreve rocket in a long series of battles with local tribes in the mountains of Turkestan. But as these events occurred, work was progressing on other types of weaponry, and by 1890 remarkable improvements in the range and accuracy of artillery (i.e. "big guns") had rendered even the best of these primitive rockets obsolete.

● Konstantin Tsiolkovskii

To the people of the 19th Century the rocket was a firework and a weapon of war, but that limited perception would soon change. In 1866 in the Russian province of Ryasan a 9 year old boy was losing his hearing from the aftereffects of scarlet fever. He had *seventeen* brothers and sisters. His name was Konstantin Eduardovich Tsiolkovskii, and though he would never actually build a rocket, he would eventually be acknowledged as the father of modern spaceflight.

By the age of 10 he was deaf, and unable to attend school, he developed an impressive appetite for reading. His father was a forestry worker and an amateur inventor who nurtured his son's interest in science. As a teenager young Konstantin moved to Moscow where he educated himself by haunting the city's libraries. At one of those libraries he met and befriended the great Russian philosopher, Nikolai Fedorov, who took him on as a student. For a young man who was deaf, this patient and Socratic form of learning was ideal, and his private lessons with Federov were an acceptable substitute for the University lectures that his deafness prevented him from hearing. Fedorov advocated the attainment of immortality through technology, and the ultimate expansion of humanity into the cosmos; a kind of galactic manifest destiny that came to be known as "Cosmism". Federov's ideas plus Tsiolkovskii's love for the novels of Jules Verne solidified his direction in life. In 1874 at the age of 17 he became *fascinated* with the idea of spaceflight.

In 1878 he moved back to Ryasan, where he earned a teaching certificate. In 1879 he moved 50 miles south of Moscow to the city of Borovsk, where he took a job teaching math. While in Borovsk he met and married his wife, Barbara, started a family, and in 1881 began a lifetime of *theoretical* research in aeronautics, rocketry, and space travel. In 1883 in his first serious paper he described the effects of weightlessness and the basic requirements for building of a space station. In 1885 he suggested that workmen in Earth orbit could build machines for turning solar energy into electric power. In a subsequent paper he explained that spinning a space station would generate artificial gravity. In 1898 he developed a mathematical formula that determined a rocket's speed at any moment, its rate of gas outflow, its mass, and the amount of fuel needed to put it in space. In a 1903 paper titled *Investigating Space with Reaction Devices* he proved mathematically that rockets would be most efficient if they were powered by the chemical reaction between liquid hydrogen and liquid oxygen. As of this writing, these are the *exact* propellants used by the main engines on the Space Shuttle.

During his lifetime he wrote 500 papers on rocketry and related topics. He suggested the use of gyroscopes for directional control, double wall living quarters for protection from meteors, air locks for entering and exiting a space ship, reclined seats to protect astronauts from G forces, space suits for working in the vacuum of space, and even self-contained space colonies. In a 1926 paper titled *Plan of Space Exploration* he expressed his *own* cosmic philosophy, and divided his plan for humanity's future into the 16 steps that follow. **1**-the creation of rocket airplanes with wings, **2**-increasing the speed and altitude of these planes, **3**-the creation of rockets *without* wings, **4**-the ability to land them on the ocean, **5**-reaching escape velocity, and the first flight into Earth orbit, **6**-lengthening rocket flight times in space, **7**-the experimental use of plants to make an artificial atmosphere, **8**-the use of pressurized space suits for working outside a spaceship, **9**-the construction of orbiting greenhouses for plants, **10**-the construction of large, orbital habitats around the Earth, **11**-the use of solar energy to grow food, heat crew quarters, and for transportation throughout the solar system, **12**-the colonization of the asteroid belt, **13**-the colonization of the entire solar system, **14**-the achievement of individual and social perfection, **15**-overcrowding of the solar system followed by colonization of the galaxy, and **16**-the slow death of the Sun, and the migration of the remaining population to the stars.

Steps **1** through **8** and much of step **11** have already happened. I anxiously await number **14**, and for the rest we'll *all* have to wait and see. In his later years Tsiolkovskii moved south to the city of Kaluga where he continued to think and write. In 1919 he was elected to the Socialist Academy, the predecessor of the U.S.S.R. Academy of Science. In the 1920s the Soviet government granted him the well-earned honor and security of a pension. Though he never built a rocket himself, his work was an inspiration to the people who *did*. Among them was a young man named Sergei Korolev, who would eventually become the chief architect of the modern Soviet space program, and one of the people responsible for making Yuri Gagarin *the first man in space*. Konstantin Tsiolkovskii died in 1935 at the age of 78.

● Robert Goddard

Throughout the 19th Century the hearts of Americans were captivated by the drama of westward expansion. But by 1901 the great adventure was over, and the old wagon trails had given up their souls to the railroads. *Adventurous* young men were dreaming of *new* frontiers, and a *very* smart high school student named Robert Goddard was fascinated by *Space*. Like Tsiolkovskii, he loved the writings of Wells and Verne, but *unlike* Tsiolkovskii, he was mechanically gifted. Throughout his career, he was remarkably successful at getting money for his research, and he would eventually build *real, working models* of the things he designed.

Robert Goddard was born in Worcester, Massachusetts on October 5th, 1882. As a teenager he attended South High School. Then he then went on to college at Worcester Polytechnic Institute. In 1908 he started graduate school at Worcester's Clark University, and in 1909 concluded, like Tsiolkovskii, that a rocket operating on liquid hydrogen and liquid oxygen would be the most efficient. He earned a Ph D in 1911, and was i*mmediately* hired by Clark as a physics professor. From 1912 to 1913 he studied at Princeton. Then he returned to Clark in 1914, where he split his time between teaching and making both black powder and smokeless powder rockets, which he tested in a meadow near Coes Pond. Clark let him use the physics department's machine shop, but he had to pay for the materials himself. As the rockets grew larger they cost more to build, and by 1916 he was badly in need of additional funds. Privately he dreamed of space flight, but he had the wisdom not to discuss the subject with the conservative and *unimaginative* scientific community of his time. Instead he wrote a proposal to develop rockets for high altitude *atmospheric* research. *Entirely* avoiding the subject of "Space ", he *wisely* titled it A Method of Attaining Extreme Altitudes, and submitted it to The Smithsonian Institution in September of 1916. To his great and wonderful surprise, he won a grant of $5,000, an amount that allowed him to continue his work with a new level of focus and commitment.

In 1917 America entered World War I. The Army Signal Corps, having heard of Goddard's work, offered *even more* financial support if he would help them develop rocket powered weapons. When Goddard agreed, the Army sent him to the *then - secluded* Mount Wilson, California, and gave him an initial budget of $20,000. On the mile-high terrain above Pasadena, with the vacant land around the Solar Observatory as a test area, he quickly developed a hand-held, shoulder-mounted rocket launcher, the forerunner of the modern "Bazooka". Immensely impressed, on November 7th, 1918, the Army asked him to start work on a rocket that could be launched from an airplane. But four days later Germany surrendered, World War I ended, and so did the Army's interest in rocket research. Goddard returned to Worcester a few weeks later.

Back at Clark in an effort to attract new financial support, his old mentor, Dr. Arthur Webster, suggested that he let The Smithsonian actually publish A Method of Attaining Extreme Altitudes. Embarrassed by his lack of progress on the subject of this paper, Goddard refused, then reluctantly changed his mind when Webster, with *benevolent* intentions, threatened to do it without his permission. The first copies rolled off the press around New Years of 1920, and then, to Goddard's *great* consternation and surprise, on January 12th a large headline appeared on the front page of the *Boston American*. It said "**Modern Jules Verne Invents Rocket to Reach Moon**", and within days, newspapers across the country were calling Dr. Goddard "the moon man".

The triggers for this publicity were a chapter titled Calculation of Minimum Mass Required to Raise One Pound to an Infinite Altitude, and the *casual* suggestion that if a rocket actually hit the moon, a sufficient amount of flash powder might make an explosion bright enough to be seen with a telescope. The media blew Goddard's statements all out of proportion, and his efforts to clarify what he'd said were ignored. *The New York Times* criticized him for not knowing that a rocket wouldn't work in the vacuum of space (*The Times* was wrong!). More than 100 people volunteered to go on the first trip. The publicity agent of the famed actress, Mary Pickford, asked him to "*deliver a message from Miss Pickford to the Moon*", and worst of all, Dr. Goddard's media-created moon-man-persona became the subject of many jokes. In disgust, Goddard returned to the University, where his spirits were marginally improved by the good news that The Smithsonian had granted him another $3,500. From 1920 to 1923 he worked at the Navy's Indian Head Powder Factory in Maryland, and in 1922, came to the *important* conclusion that only a *liquid*-fuel rocket would have sufficient power to reach the altitudes that he wanted to explore.

In 1924, and back at Clark, he built and tested his first liquid fuel rocket motor, but it was too small and underpowered to lift its own weight. In December of 1925 he tested a motor with sufficient power to rise off the ground, and in January of 1926 he built a rocket that rose to the top of the test stand and pulled at the restraints. Knowing that *this* one would fly, he took it out to his Aunt Effie Ward's farm near Auburn, Massachusetts, and on March 16th, 1926, launched it from a pipe fitting stand. The rocket rose just 41 feet, and was airborne for only 2-1/2 seconds. *But it was the first time in history* that a *liquid*-fuel rocket had lifted free of the ground, and flown under its own power. Goddard's wife filmed the event, but sadly missed the flight itself, because the home movie cameras of that time had just 8 seconds of film. She started the camera too soon, and ran out of film before the rocket took off.

From 1926 to 1929 Goddard improved his motors' performance, and developed new equipment for measuring their chamber pressure and thrust. His flights at Auburn continued until July of 1929 when a large and *loud* rocket frightened the neighbors, who immediately complained, and had the firemarshal shut him down. This rocket carried scientific instruments, and shortly thereafter the publicity generated by the flight drew the attention of the famous aviator, Charles Lindbergh. Lindbergh visited Goddard, and *greatly* impressed, got him a $50,000 grant from the Guggenheim Foundation. With this kind of money, Goddard could *greatly* increase the size and power of his rockets, *but he couldn't fly them in Massachusetts*. In search of a suitable site, he took a trip out West, and in July of 1930 set up shop on a ranch near Roswell, New Mexico. While at Roswell, Goddard worked diligently on pumps, nozzles, combustion chambers, gimballed steering, and guidance systems, but the maximum altitude achieved was only 8,500 feet. Despite public perception to the contrary, the Roswell years were Goddard's most productive. With his wife, his assistants, a remote location, and better facilities, he could work at a scale not possible before. With the exception of a two-year break from 1932 to 1933, he stayed at Roswell until 1941.

While the public awaited the achievement of a great altitude, Goddard understood that it was far more important to perfect the performance and reliability of his rocket motors' *components*. In a 1937 letter to his old friend and onetime colleague, Dr. Clarence N. Hickman, he said, "*It is, as you can imagine, a fascinating life. The drawback is, that until there has been a great and spectacular height reached, no laymen, and not many scientists, will concede that you have accomplished anything, and of course there is a vast amount of spade work of much importance, that must be done first.*" To the disappointment of the world-at-large, the rest of Goddard's life would be characterized by this unromantic "spade work", and it wasn't until after his death that the importance of what he'd done was appreciated.

In 1940 Goddard and "The Guggenheim" proposed that the military explore the use of rockets for the jet-assisted takeoff of airplanes, but the idea was rejected. In 1942, having entered the war with Japan, the military *predictably* changed its mind. From 1942 to 1945 Goddard worked on jet-assisted takeoff and variable-thrust rocket motors at the Naval Engineering Experiment Station at Annapolis, Maryland. On June 18th, 1945 his doctor discovered a tumor in his throat. Surgeons removed it the following day, but the effort was to no avail, and he died on August 10th, 1945.

Throughout his life Goddard was secretive about his work, and unwilling to share his findings with other scientists. His early experience with the news media enhanced his reclusiveness, and from that time on he kept his thoughts about actual space flight to himself. Nevertheless, he did keep records on the subject, and in his private papers he made some useful suggestions. The most notable was the idea that a rocket approaching the Earth should come in at a tangent to the Earth's surface, and use a horizontal reentry into the atmosphere to brake its fall. To dissipate the heat of reentry, he suggested an ablative heat shield. Both methods were later put into practice in the Mercury, Gemini, and Apollo manned spaceflight programs.

In 1959, *fourteen years after his death*, Congress honored Goddard with the Louis W. Hill Space Transportation Award. The Smithsonian Institution awarded him the coveted Langley Medal, and NASA named one of their main facilities **The Goddard Space Flight Center**. In 1960 the United States government awarded the Guggenheim Foundation and his widow, Mrs. Goddard, the sum of *one million dollars* in compensation for its use of *more than 200* of his patents.

● **Pedro Paulet**

Now we come to an interesting "might-have-been". In the 1969 book, *History of Rocketry and Space Travel*, the authors, Dr. Wernher von Braun and Frederick Ordway III, tell of a Peruvian chemical engineer named Pedro Paulet. From 1895 to 1897 Paulet was a student at the Institute of Applied Chemistry at the University of Paris where he *claimed* to have designed and tested the world's first liquid-fuel rocket motor. The problem is that he did it at home in his spare time, and for *no* logical reason, he didn't tell anyone about it until 1927.

In October of that year while living in Rome, and *presumably* spurred by the news of Goddard's success, he wrote a letter to *El Comercio* in Lima, Peru, claiming priority for his *own* design. *El Comercio* published the letter as an interesting article, which caught the attention of a young Russian engineer named Alexander Scherschevsky. Scherschevsky lived in Germany at the time, and paraphrased the article in his 1929 book titled *Die Rakete für Fahrt und Flug,* or *The Rocket for Travel and Flight.* According to Scherschevsky, Paulet's motor ran on gasoline and nitrogen peroxide, and worked like a pulse jet engine operating at 5 cycles per second. Constructed of steel, it weighed about 5 pounds, used a spark plug for ignition, and generated 200 pounds of thrust. After the publication of Scherschevsky's book, other writers picked up on the story with the result that, at least for a while, Paulet was acknowledged as the inventor of the liquid fuel rocket motor. Unfortunately, though both his story and his unique technology *seemed* credible, he was never able to produce the witnesses or the documentation needed to verify the truth of what he said, and he died in 1945 with his claims unproven.

As a postscript to this story, the Museo Aeronautico Del Peru (The Aeronautic Museum of Peru) has a website on the Internet. They refer to Paulet by his Peruvian name, *Pedro Paulet Mostajo*, and their homepage says that visitors to the museum in Lima will learn of his life and his work. At the time of this writing I could find no details about him on the website, but perhaps in the future something will appear.

● Hermann Oberth and The Germans

Hermann Oberth was born in Transylvania on June 25th, 1894. The son of a doctor, he too loved the works of Jules Verne. In his autobiography he says, *"At the age of 11, I received from my mother as a gift the famous books, **From the Earth to the Moon**, and **Journey Around the Moon** by Jules Verne, which I had read at least five or six times, and finally knew by heart."* Verne's imaginary moonship was fired from a giant gun, but by the age of 13 young Hermann had calculated that this would subject the astronauts to an acceleration of ***forty seven thousand times the force of gravity!*** *"They would be flattened into pancakes"*, he said. *"A cannon is **not** good for space flight. **It must be done with a rocket!**"*

Like Tsiolkovskii, Oberth was a dedicated reader. By the age of 15 he had taught himself Sir Isaac Newton's mathematics-of-change, called "Calculus", and worked out the basic mathematics of rocket propulsion. He graduated from high school in 1912. Then he studied medicine at the University of Munich where he continued to work on rocketry in his spare time. During World War I his medical education spared him from combat, and he was assigned to hospital duty for the duration. In 1917, with a well developed sense of things to come, he proposed the development of a long range, liquid-fuel missile to the German army. Politically unsophisticated, he wasn't nationalistic or hawkish. *He wasn't even a German citizen.* In his young and naive mind he simply thought that a rocket-bomb exploding in the middle of London would cause the British to surrender, and bring about a quick end to the war. To the good fortune of the rest of Europe at the time, his suggestion was ignored.

In 1918, immediately after the war, he married his high school sweetheart, Tilly, and continued his formal education in physics and math. By 1922, in pursuit of a doctorate, he'd organized all of his previous work into a paper which he titled *Die Rakete zu den Planetenraumen,* or *The Rocket to Planetary Space*. He submitted it to the University of Heidelberg as a Ph D thesis, but the University rejected it as being unrealistic. Among the reasons given, several professors who should have known better claimed, like the ill-informed *New York Times*, that a rocket wouldn't work in the vacuum of space *because it didn't have anything to push against.* The truth was that most of the work was beyond the faculty's understanding, and they weren't qualified to judge it. At about the same time, Europeans were beginning to talk about Robert Goddard. When Oberth read about him in a Heidelberg newspaper, he immediately wrote to Goddard in a broken-but-credible attempt at English, and said:

"Dear Sir:

Already many years I work at the problem to pass over the atmosphere of our earth by means of a rocket. When I was now publishing the results of my examination and calculations I learned by the newspaper, that I am not alone in my inquiries and that you, dear Sir, have already done much important works at this sphere. In spite of my efforts; I did not succeed in getting your books about this subject. Therefore I beg you, dear Sir, to let them have me. At once after coming out of my work I will be honored to send it to you, for I think that only by common work of the scholars of all nations can be solved this great problem.

Yours very truly,
Hermann Oberth
Stud. Math Heidelberg, Germany"

Goddard had never heard of Oberth. He was suspicious of Oberth's interest in his work, and he *reluctantly* sent him a copy of *A Method of Attaining Extreme Altitudes*. Possibly inspired by seeing Goddard's work in print, Oberth decided to turn *The Rocket to Planetary Space* into a book, but the publishers of that time thought it to be unmarketable, and they showed no interest. Not to be deterred, his loving wife, Tilly, presented him with a modest amount of money that she'd gradually *and secretly* saved since the beginning of their marriage, and in late 1923 *paid for the publication herself*. For the scientific community, the 92 page book covered all the mathematics of a rocket's operation, but it also contained enough *nontechnical* discussion to inspire the imagination of *the public*. The first printing sold out immediately, and the second and third printings were spoken for before they reached the book stores.

When Oberth sent a copy to Goddard as he had promised, Goddard was *very* upset. Goddard had always thought himself to be working alone. He considered the study of rocketry to be his own private domain, and he was alarmed to see Oberth publishing material that *he* considered privileged. For a short time thereafter he corresponded with Oberth. Then he cut off all communication as he gradually developed the *erroneous* belief that Oberth had stolen his ideas. In early 1925 Oberth took a teaching-job in Mediash, and learned for the first time about Konstantin Tsiolkovskii. Unlike Goddard, Tsiolkovskii *welcomed* correspondence with other scientists, and the two exchanged a series of encouraging and supportive letters.

By 1924 the Germans were suffering badly from the financial burden of the reparations demanded by the Versailles Treaty. Economically devastated, they were quick to embrace *anything* that promised a happier future. By skillfully riding this wave of German hope, a charismatic ex-aviator named Max Valier had become a popular and wealthy promoter of the glorious possibilities offered by science and technology. To attend a Max Valier lecture was like seeing Carl Sagan, Robert Redford, and P.T. Barnum all played by the same actor. Valier was impressed by Oberth's book, and asked his permission to rewrite the book in a nontechnical language that a layman could understand. Supportive of any effort to publicize the idea of space flight, Oberth agreed. Valier quickly finished the book, and published it under the title, *Der Vorstoss in den Weltenraum,* or *The Advance into Space.*

From 1925 to 1926 books by Oberth, Valier, and other authors generated a growing interest in space travel, and by 1927, societies dedicated to the study of rocketry were forming throughout the world. In Germany the most prominent was the "Verein für Raumschiffahrt", or "Society for Space Travel". The "VfR", as it came to be called, was officially founded on June 5th, 1927, and Hermann Oberth and Max Valier were charter members. By 1928 the membership had grown to 500, prompting the stuffy and *turf*-conscious "Society of German Engineers" to publish a series of attacks on Oberth's work.

Submitted to the Society Journal, *and signed by*, the society's "Privy Councilor Professor Dr. Lorenz of Danzig", these literary pogroms charged, among other things, that Oberth's rockets would never escape the Earth. When Oberth wrote a mathematically-perfect rebuttal, the Society refused to publish it on the grounds that it, "didn't have enough room in the Journal". Many years later Oberth's friend and colleague, Willy Ley, was told *off-the-record* that the *real* reason was that the Society could *not* allow the *eminent* Professor Dr. Lorenz to be contradicted by someone half his age. Not to worry though. On May 23rd, 1928, the Scientific Society for Aeronautics hosted a showdown debate, and the "Privy Councilor Professor Dr. Lorenz" was so thoroughly discredited that he never spoke of rocketry again.

On the very day of the Oberth-Lorenz debate, an event occurred that disgusted serious scientists, but helped the cause of rocketry in general. For several months, and without the VfR's knowledge, Max Valier had been working on a rocket-powered car with the German auto maker, Fritz von Opel. To avoid the time and expense of developing a liquid-fuel motor, they powered the car with standard, navy powder (i.e. *solid* fuel) rockets. Because the technology of these rockets was already well-established, the project was scientifically worthless, but it was a *great* publicity stunt. When Fritz von Opel drove the car at 125 miles per hour in front of 2,000 spectators at the Avus Speedway, the people of Germany went *wild*. As reporters extolled the triumph of German technology, an absolute *zoo* of rocket-powered daredevils took to the roads and the air in a heated effort to capitalize on what the newspapers were now calling "raketenrummel", or "rocket racket".

All this rocket-racket drew the attention of Fritz Lang, the famous director of the 1927 science-fiction film classic, *Metropolis*. Fritz Lang was the "Stephen Spielberg" of silent films, and the most popular director of his time. In 1929 Oberth took over the leadership of the VfR, and published an expanded version of *Die Rakete...*, titled *Wege zur Raumschiffahrt,* or *Road to Space Travel*. Lang read the new book. Then he commissioned his wife, Thea von Harbou, to write a fictional adventure script based on the technology described in the book. A popular and talented actress, Thea was also a *very* good writer, and the author of the book on which *Metropolis* was based. She titled the script *Frau im Mond,* or *Woman in the Moon*. Since Oberth was a world authority on rockets, Lang hired him as a scientific consultant on the film.

In the Fall of 1928 Lang brought Oberth to the giant UFA film studio in Berlin. All would have gone well had he limited Oberth to designing the sets. Unfortunately, someone in the publicity department got the bright idea that a real liquid-fuel rocket launch on the day of the film's premiere would help boost ticket sales. Unaware of the problems he would face, Oberth volunteered to build the rocket, and at least for Oberth, it was all downhill from there. He was a brilliant scientist, but his engineering skills were limited. He could calculate how a *theoretical* rocket would perform, but he couldn't predict what would happen to a *real* motor's metal when exposed to the heat of its operation. He knew nothing about welding, and nothing about metal alloys. Oberth was in trouble, but he lacked the "hands-on" experience to realize it.

To help him with the project he hired an unemployed engineer named Rudolph Nebel, and Alexander Scherschevsky, the writer-engineer who had popularized the claims of Pedro Paulet. Scherschevsky was, at least for the moment, "a man without a country". The Russians had sent him to Germany to study sailplanes, but he'd overstayed his assignment, and was afraid to go home. Oberth planned a rocket powered by liquid oxygen and gasoline, which he called the *Kegeldüse* because it had a cone-shaped combustion chamber. But several months of unexpected complications plus a rapidly approaching deadline convinced him that such a rocket would not be practical in the short run.

With less than a month to go he switched to what he *thought* would be a simpler design. His saving-plan was to place sticks of carbon on-end in a tube full of liquid oxygen. The sticks would burn down from the top, and by sizing them properly, the last of the carbon would be consumed just as the oxygen ran out. It sounded like a good idea, but further complications arose, and two weeks later the carbon rocket was going nowhere. As the movie premiere loomed, Oberth's anxiety grew. A week before the opening performance he threw up his hands and *quit*. Having no rocket to launch, the studio made excuses. Oberth returned to watch the premiere. Then thoroughly embarrassed, and plagued by a mounting sense of failure, he tinkered in the shop for a few more weeks, then abandoned the project for good.

Frau Im Mond opened on October 15th, 1929. Though Oberth failed to deliver the promised rocket, his movie set was, by all accounts, *wonderful*. The film was released in English-speaking countries under the titles, *The Girl in the Moon*, and *By Rocket to the Moon*. To increase the drama of the lift-off, Fritz Lang decided that the actors should count backward before the launch (i.e. 10, 9, 8, 7, 6, 5, 4, 3...etc.), and that the rocket should be fired when they reached zero. By introducing the world to this new cinematic device, Lang had assured his *own* place in the history of rocketry. As a strictly artistic "movie-maker", he had unwittingly suggested a thoroughly practical launch procedure that would be adopted and used by *generations* of rocket scientists to-come. Fritz Lang had invented **the countdown**.

In 1930 the VfR began to recruit *real engineers*. Rudolph Nebel joined at that time, and so did the son of the German Agriculture Minister, an 18-year-old *apprentice*-engineer named Wernher von Braun. With Oberth's help they perfected and finished the uncompleted *Kegeldüse*, and seeking the government's acknowledgment of its potential, successfully demonstrated it to the Reich Institute for Chemistry and Technology. Soon thereafter Oberth returned to Mediash, and resumed his job as a teacher. From late 1930 to 1937 Oberth stayed in Transylvania, where he continued to teach, write, lecture, and help other inventors. He built a few more rocket motors, but the lack of a nearby source for liquid oxygen (or even liquid air) prevented him from conducting any serious tests.

In September of 1930, with Oberth back in Transylvania, the VfR moved to an abandoned complex of military storage buildings on the north edge of Berlin, which they rented for the paltry equivalent of 4 U.S. dollars per month. The site included 2 square miles of undeveloped land that Nebel, with high hopes and *great* imagination, called the *Raketenflugplatz*, or *Rocket Aerodrome*. At this new location they made rapid progress, and by August of 1931 were flying a liquid fuel rocket, which they called a *Repulsor*, to altitudes as high as 3,500 feet.

But as often happens to people with good intentions, a major sequence of unexpected political events was eroding the ground on which they stood. For several years prior, a dangerous and charismatic politician named Adolph Hitler had been capturing the German imagination by claiming that *others* were to blame for all of Germany's troubles, and that *he alone* could restore Germany to its former glory. When a sudden and exponential rise in his popularity changed the political climate, more than half of the members quit. Then the VfR's funding vanished when a major contributor withdrew his help, and another was killed in an accident while vacationing in the mountains.

The Versailles Treaty, that *officially* ended World War I, prohibited Germany from making or possessing most of the weapons of war. But the framers of the treaty had, *ignorantly and foolishly*, neglected to include *rockets*. By 1932 the German Army had recognized this loophole, and decided to exploit it. Realizing that most of the rocket research at that time was being done by **amateurs**, the Army Weapons Department assigned Colonel Walter Becker to recruit Germany's most talented amateurs into a new Army rocket program. In 1933, following a visit to the VfR, Becker offered to fund the experiments of any member willing to work under Army supervision at the Kummersdorf proving ground. Greatly impressed by von Braun's work, he suggested that von Braun earn a Ph D, and he offered him the use of the Kummersdorf facilities to work on his thesis. Of all the potential candidates for the Kummersdorf program, only von Braun accepted. Shortly thereafter Hitler seized control of Germany, and declared himself the *Führer*, or *Leader*. In 1934 Hitler's "Gestapo" (short for "Geheimstaats Polizei", or "Secret State Police") closed down the VfR, and assigned most of the active members to jobs in the defense industry. In a phone conversation with the War Ministry, one of the officers in charge was heard to say *"Now I've all the rocket people safely on ice around here and can watch what they are doing."*

At Kummersdorf, Wernher von Braun and Becker's former assistant, Captain Walter Dornberger, headed up an excellent team of engineers. By December of 1934 they had flown a 4-1/2 foot rocket called an "A-2" to an altitude of 7,400 feet. Greatly impressed, Colonel Becker appropriated six million reichsmarks for further research. With money no longer a problem, they started work on an A-3. But the A-3 would be 25 feet tall, and they couldn't fly a rocket that big on the mainland. They were ten months into the search for a more remote site when something truly serendipitous occurred. While talking with his mother at Christmas, Werner Von Braun mentioned the need for a new place to work. Mrs. von Braun suggested a Baltic island called *Usedom* where her father used to hunt ducks. Dornberger found it perfect. The Army bought the land, and in 1937 the entire Kummersdorf team moved to a new facility on the north side of the island near a fishing village called Peenemünde.

After the move to Peenemünde the Army recruited additional engineers, including many of the men who had worked at the Raketenflugplatz. In December of 1937 they launched three A-3s from the nearby island of Greifswalder Oie, but all three went off course. They were just beginning to deal with the complexities of guidance, and von Braun later said that the trouble was due to a poor understanding of the principles involved. Though the A-3s' guidance systems failed, their motors were *very* successful. One of the rockets reached an altitude of 8 miles (40,000 feet!), and another flew 11 miles downrange before plunging into the sea. The goal of all this work was to ultimately produce what Von Braun called a *real* rocket, a 12 ton, 46-foot beast called an A-4. But *first* they needed a smaller rocket to perfect a workable guidance system. Since the A-4 designation was already taken, they called the guidance-test rocket an "A-5". They flew the first A-5 in the fall of 1938. By the fall of 1939 Hitler's Germany had attacked Poland, World War II had begun, and the A-5 was a complete success.

In 1938 Hermann Oberth naively accepted an offer to work on rockets at Germany's College of Engineering in Vienna. The pay was excellent, but the limited facilities and the way he was treated soon convinced him that the job's only purpose was to keep him from working for another country. When he tried to quit, he was told him that he'd seen too much of Third Reich technology to be *allowed* to quit. He was told to chose between an engineering job with German citizenship, *or a concentration camp*. Reluctantly he applied for German citizenship, but the bureaucracy worked *very* slowly. By the time he reached Peenemünde, the A-4 was nearly finished, and there wasn't much for him to do. To stay occupied, he worked up a proposal for a solid fuel antiaircraft rocket. But Peenemünde didn't have the facilities for solid fuel work, so the idea was turned over to WASAG, a famous manufacturer of commercial explosives. Oberth was then transferred to WASAG where he worked on solid fuel rockets until the end of the war.

Von Braun's engineers tested the first A-4 on June 13th, 1942. It exploded in a spectacular blossom of fire. They launched the second A-4 on August 16th, and it flew for 45 seconds. Then its metal fuselage ruptured, and it self-destructed in-flight. On October 3rd the *third* A-4 flew *perfectly*. It reached an altitude of 60 miles and a speed of 3,300 miles per hour. In less than 5 minutes it flew 120 miles downrange. At a celebration that evening Captain Dornberger said, "*...We have invaded space with our rocket and for the first time-mark this well-have used space as a bridge between two points on the Earth. We have proved rocket propulsion practicable for space travel. To land, sea, and air may now be added infinite space as a medium of future intercontinental traffic. This third day of October, 1942, is the first of a new era of transportation, that of space travel.*" Shortly thereafter, and *tragically*, Hitler renamed the A-4 his *Vengeance Weapon Two*, or "V2". He ordered it fitted it with a large explosive warhead; then used it *as the world's first ballistic missile* in a horrible and hateful reign of terror against England.

In 1945 after 3 years of *brutal* war, "The Allies" (which included England) defeated Germany, and shortly thereafter World War II ended. In May of 1945 Wernher von Braun, his brother Magnus, and dozens of other Peenemünde engineers surrendered to the Americans. In the Fall of 1945 they were brought to the U.S. through a program called *Operation Paperclip*, and by February of 1946 were at work in El Paso, Texas, laying the foundations of the new *American* space program.

Immediately after the war Hermann Oberth joined his family in the German town of Feucht, where he worked in a garden shop while writing a book titled *Man into Space*. In 1950 he finished his work at WASAG. From 1951 to 1953 he travelled through Italy and Switzerland, writing, consulting, and lecturing as he went. In 1955 Wernher von Braun, by then in charge of the U.S. Army missile program at Huntsville, Alabama, invited Oberth to join him there. Oberth worked at Huntsville until the launch of the U.S. Explorer satellite in 1958. Then he returned to Feucht to fulfil the requirements for a pension.

In 1961 Oberth came back to the U.S., and worked for Convair as a technical consultant on the Atlas rocket program. He retired in 1962 at the age of 68. In 1984 he wrote and published a book titled *Primer for Those Who Would Govern*, in which he expressed his thoughts on how a truly democratic society should be run. His student, friend, and colleague, Dr. Wernher von Braun, once said of Oberth: "*I myself owe him a debt of gratitude, not only for being the guiding light of my life, but also for my first contact with the theoretical and practical aspects of rocket technology and space travel.*" From 1960 through 1969 Von Braun played a major role in the development of America's giant "Saturn" rockets. Both men lived to see the completion of the Apollo Moon Project and the landing of the first men on the Moon. Herman Oberth lived to see the early flights of the Space Shuttle. Wernher von Braun died in 1977, and Herman Oberth died in 1989 at the age of 95.

The Attempted Destruction of Amateur Rocketry

Professionals in the fields of science and engineering often begin their careers without the benefit of formal training. They at first study on their own, *not* with the thought of earning a living, but from an natural interest in the chosen subject. Hermann Oberth was entirely self-taught when he worked out the mathematics of rocket propulsion, and Wernher von Braun had no formal training when he built his first rocket. From the Latin word, *amator*, or "lover" (*because they do it for the love of it*), the people who work in this way are called "amateurs". Of course Oberth and von Braun are extreme examples. Von Braun eventually obtained a formal college degree, and both men became the leaders in his field. Collectively their work, along with the work of many others, resulted in the development of the largest rocket ever built. Standing 363 feet tall, this incredible machine called the *Saturn Five* developed 7-1/2 million pounds of thrust (*75 times the power output of Hoover Dam*), and from 1969 through 1972, escaped the Earth's gravity, and took America's Apollo astronauts **to The Moon**. The simple fact that Oberth and Von Braun began their careers as **amateurs** illustrates how important it is to allow amateurs in all fields of science to pursue their activities with a *minimal* amount of interference.

In the United States, amateur rocketry, or the *nonprofessional* construction of rockets and rocket motors, dates back to the 1930s. People engaged in amateur rocketry make *their own* rocket propellant and build *their own* rocket motors. The making of the propellant, and the design, construction, and testing of the motors provide the valuable educational experience traditionally associated with the hobby. People engaged in amateur rocketry derive knowledge, enjoyment, **and their sense of accomplishment** from making everything **themselves**, and they **don't** want to buy commercially made rocket motors.

A distinctly *separate* hobby called *model rocketry* began in the 1950s. People involved in model rocketry use commercially made rocket motors. Both the motors and the rockets that use them are sold in toy stores and hobby shops, and before they are offered for sale they are tested and approved by an organization called the National Association of Rocketry (NAR for short). The NAR also conducts educational and recreational activities for its members, publishes a magazine, and helps promote model rocketry as a safe and worthwhile hobby.

In 1984 I wrote and published an instruction manual titled *Building Your Own Rocket Motors*, and shortly thereafter the NAR expressed *great* dismay over the book's existence. In the span of about two months they wrote a long complaint to *numerous* government agencies, and protested to all of the magazines in which I advertised. They published an editorial in their *own* magazine calling me "irresponsible". They *publicly* accused me of encouraging something called "basement bombing". They *publicly* called for all the magazines to refuse my advertising, and they said that my book should be thrown in the trash. I even caught two young men watching our house one day from behind a tree across the street. When I asked them if they were members of the NAR, one of them sheepishly admitted that he was. Then, with looks of considerable embarrassment, they walked back to their car, and drove away.

Thankfully, the magazine publishers ignored these people, and the few government agencies that paid me a visit went away smiling and shaking their heads. By encouraging people to build *homemade* rocket motors instead of buying them, I'd apparently upset a group of business interests who felt that my book was a threat to their sales and their profits. The NAR's magazine featured advertising for the model rocket industry's supposedly "safe" products, and some of the manufacturers included NAR literature and membership forms in their rocket kits. The NAR derived much of its membership and a good portion of its income from the model rocket industry, and when I finally understood the relationship, the motive behind the group's behavior seemed clear.

It is unusual for a writer to talk about things like this in a hobby book, but the hobby of *amateur* rocketry has been subjected to an ongoing campaign of destructive propaganda since the 1950s. This assault is unjustified, its effect on the hobby has been chilling, and I therefore think it appropriate to discuss some of the facts regarding its source. In the past 40 years a small industry has arisen that manufactures and sells *commercially* made rocket motors. The fact that it promotes its own well-being is appropriate. All industries do. What is *not* appropriate is that, though they don't represent the industry as a whole, a few special interests in and around the industry have been falsely alarming the authorities and the public with an ongoing disinformation campaign designed to frighten people *away* from amateur rocketry, put an *end* to amateur rocketry, and thereby *force* anyone who flies rockets to buy commercially made rocket motors.

For example, in 1987, as part of its "*Science Education Series*", Prentice Hall published a popular introduction to model rocketry titled *The Rocket Book*. They sold thousands of copies in bookstores and hobby shops throughout the country. On page 3 the authors (Cannon and Banks) warn that a "*study*" conducted by the American Institute of Aeronautics and Astronautics indicated that (quoted from the book), "*...rocketeers who made their own propellants had one chance in seven of killing themselves or causing an accident so serious that they would be maimed for life for each year that they practiced amateur rocketry*".

The American Institute of Aeronautics and Astronautics is a respected society of professional engineers. When the AIAA conducts a study they do it carefully and accurately. Of course anyone involved in amateur rocketry knows that the stated conclusion of *this* study is *completely* untrue. In an effort to resolve the conflict, and learn how the study was conducted, I tried to order a copy from the AIAA, and was told by their document search staff that *despite two intensive searches* they could find absolutely *no* record of any such study.

Since *The Rocket Book* was published in 1987, I then asked them to look up *anything* published prior to 1988 on the subject of safety in amateur rocketry, or just amateur rocketry or model rocketry in general. They found *nothing* published in the U.S., only three documents by the International Astronautical Federation, and *none* of them made any mention of safety issues or accidents. At the time of this writing, anyone who wants to verify this can contact the AIAA's document service at: **American Institute of Aeronautics and Astronautics, Aeroplus Access, 4th Floor, 85 John St., New York, NY 10038-2823**.

To understand why the authors of *The Rocket Book* would try to frighten their readers, and discredit amateur rocketry with a study *that apparently doesn't exist,* it might help to consider the following facts. All of the book's photographs are of *commercial* model rocket products, many with the company labels turned conspicuously toward the camera. Most of these photos were provided by a manufacturer whose commercial model rocket engines are extensively featured in the book. The book's cover states that one of its authors is a "consultant" to this company. The authors *dedicate* the book to this company's founder, and the book's forward is written by the *then*-president of the NAR.

The Rocket Book is thankfully out of print, but *thousands* of copies are still in circulation, and similar warnings are repeated in other books as well. Considering my own experience, before you believe them you might wish to find out if there is any real documentation for what the authors are saying. If any relevant studies have actually been done, try to find out whether they were *genuine* scientific studies, or merely the attempts of a few individuals to cloak a personal and self-serving opinion in the authority of an official organization. Try to obtain the *names* of the people who conducted these studies. Try to ascertain whether they had (or have) any affiliation with the model rocket industry, and try to find out whether or not they had (or have) a personal interest in the outcome. Insist on a detailed explanation of *exactly* how each study was conducted, and most important of all, *insist on seeing whatever factual data they have to back up their claims*.

To supplement these warnings, well meaning individuals, naively supportive of the views expressed in these books, have created and perpetuated *anti*-amateur rocketry propaganda of their own. A favorite tactic is to try to confuse homemade rocket motors with news reports about bombs or homemade explosives. If you insist on a *detailed* explanation of the incident in question, you will usually find that, though someone was indeed injured, the device that caused the injury was *not* a rocket. Therefore, when someone tells you that amateur rocketry is dangerous, politely but firmly insist that they provide *proof* of what they are saying.

If they respond by describing an accident involving a homemade rocket motor, ***ask them for the source of the story***, and insist that they provide ***specific details***. Ask them when and where the accident happened. ***Ask them for the victim's name and address***. Ask them ***exactly*** what he was doing, ***exactly*** what kind of a device he was making, and (this is very important) ***exactly what chemicals and other materials he was using*** (not just some of them, but ***all*** of them). A ***complete*** list of the chemicals and other components will often reveal that the device ***couldn't*** have been a rocket. ***Don't*** allow them to generalize. ***Don't*** accept less than a ***complete*** explanation, ***and don't accept excuses***. If they respond with solid, thoroughly detailed documentation that can be independently verified (and ***not*** with just an "estimate" or someone's "opinion"), ***you should take their warning seriously***. If they either cannot or will not provide these basic facts, then consider the possibility that what they are telling you might be fabricated, aka "**a lie**". *Of course* there are hazards associated with building homemade rocket motors, and to minimize or ignore them would be irresponsible. ***It is just as irresponsible to exaggerate***.

In recent years I've heard that the NAR is not as vehemently opposed to amateur rocketry as it used to be. Is this a simple case of battle fatigue? I hope not. With the passage of time both people and their organizations can change. If this apparent new attitude is genuine, then it's refreshing, and I hope that in years to come the NAR will take *affirmative* steps to distance itself as far as possible from the people and the mistaken policies of the past. Were such a course to be taken, it could help bring an end of an unfortunate situation, and the beginning of what *should* be a spirit of friendship and cooperation between the commercial and amateur forms of the hobby.

Accidents During the Fifties, and the Events that Followed

I was born in 1946, and readers of my generation might remember a series of 5 to 10 serious accidents that happened to people making homemade rockets in the late 1950s. These were the early years of the space program, and the news media was so attuned to *anything* involving rockets that, when these accidents occurred, they were reported by every newspaper in the country. Because so much time has passed, it would take an historian to reconstruct the details, but to the best of my memory, the essence of the story was always the same. Someone was either mixing chemicals, or packing them into a metal tube, when they exploded, and blew off a hand or an arm, or put out an eye. In one case I think that the person might have been killed.

The driving force behind this behavior was often a naive and youthful desire to become famous by inventing a powerful new rocket fuel, and this in itself was not a bad thing. What made it dangerous was the *erroneous* belief that, because of their perceived "power", *explosive* combinations of chemicals would work the best. At the beginning of this chapter ***in big bold letters*** I presented a list of chemicals that under most circumstances should be considered hazardous. Unless you really know what you are doing, these chemicals can get you into serious trouble. What people forget is that in the 1950s, without a prescription or any kind of a license, *you could buy them at any drug store*.

I know this for a fact because in 1958 my own father bought me a bottle of potassium chlorate at the old Bowser & Banks Pharmacy in Riverside, CA. When he told Jake Bowser that I was going to use it for rocket experiments, Jake warned him of the danger. A few days later my father had second thoughts, and out of concern for my safety, took it back and returned it. But many 12-year-olds didn't have such cautious fathers. The ease with which you could buy these chemicals resulted in thousands of kids, caught up in the romance and the excitement of the new space program, mixing up explosive combinations in their bedrooms and their parents' garages. *It was a dangerous situation. It was unique to its time. It is **not** the way we do things today, and nothing like it has happened since.*

In response to these accidents, the retail drug industry stopped selling these chemicals. By 1963 if you wanted to buy them, you had to go to a commercial chemical dealer. At the same time *responsible* hobbyists continued the *safe* pursuit of amateur rocketry as they had since the 1930s. Unfortunately, a young and aggressive model rocket industry, sensing an opportunity, played up these accidents. They capitalized on public ignorance of the distinction between *amateur* rocketry and what these kids were doing, and in the process, helped convince several states to declare the making of homemade rocket motors illegal.

For the next 20 years amateur rocketry survived through a few serious clubs, but their total membership remained small, and a lack of money kept them from accomplishing anything truly interesting. At the same time *model rocketry* and the market for *commercially* made rocket motors continued to grow. In 1984 when I first became involved in rocketry, a group of about 10 people were holding monthly rocket launches in the Mojave Desert at Lucerne Dry Lake here in Southern California. When I showed up one Saturday with my books and my *homemade* motors, what I was doing was considered both interesting and unique.

By 1986 the Lucerne launches were attracting two to three dozen people, and the rocket motors were getting larger. But so was their cost, and I remember being amazed one day when I learned that one participant had spent more than $100 on the motors for a single flight. When a product becomes too expensive to buy, people either do without it, or learn how to make it themselves, and by 1987 an increasing number of people were doing the latter.

The Tripoli Rocketry Association

By 1991 a well organized group called the "Tripoli Rocketry Association" had taken over activities at Lucerne. When I drove out to photograph one of their events in October of that year, I found more than 600 people in attendance. By the Summer of 1992 there were more than 1,000. Launches sponsored by Tripoli carry liability insurance, and therefore, as of this writing, only safety-certified commercially manufactured motors can be flown. Tripoli does *not*, however, censure its members for making homemade rocket motors. It simply requires them to do it on their own time, and avoid bringing them to officially sponsored events.

Chuck Rogers, a professional rocket engineer and one of Tripoli's past presidents, has told me that Tripoli's current membership stands at about 2,200. He says that, according to a recent poll, more than 500 members have expressed an interest in building and flying homemade rocket motors, and he guesses that somewhere between 100 and 200 are actively doing so. As a result, Tripoli's leaders are just now beginning to explore ways in which individual members might *someday* be able to safety-certify their homemade motors, and fly them at officially sponsored events.

The Tripoli Rocketry Association is organized into "prefectures", and there are currently more than 100 scattered throughout the U.S., Canada, Europe, and Australia. Tripoli was founded in 1964 by a group of high school students, but in the past 30 years it has undergone many changes. It is currently open to people of *any* age. The average member is 32 years old, and often has children of his own. Many members involve their entire families, and often combine a rocket launch with a weekend of camping. I won't provide an address or a phone number because the leadership is handled partly by volunteers and partly by elected officials whose addresses and phone numbers are constantly changing. I suggest instead that you look up their website on the Internet at **www.tripoli.org**.

The Reaction Research Society

An encouraging spin-off of the rapidly growing interest in *high power* rocketry has been a renewed interest in the truly advanced forms of the hobby, as embodied by the Reaction Research Society. Founded in 1942, the RRS has grown in recent years from a couple dozen members to about 250. It is based in Los Angeles, CA, and maintains a 40 acre test facility and launch site in the Mojave Desert just north of Edwards Air Force Base. At last count, seven members were building large, liquid fuel rockets.

As a group they are currently building a second block house on the property, and they recently poured the concrete foundation for the equipment that will be used to test a 10,000-pound-thrust liquid fuel motor currently under construction. If an amateur satellite is ever placed in orbit, these people may well be the ones who do it. If you'd like to get involved in some truly advanced and exciting projects, you should definitely contact The Reaction Research Society. They maintain a website on the Internet at www.rrs.org, and their mailing address is **Reaction Research Society, P.O. Box 90306, World Way Postal Center, Los Angeles, CA 90009**. Their current president, David Crisalli, tells me that they've maintained this box for the past 20 years, and have no intention of letting it go. Should that ever change, you can also locate them through the **Kern County Firemarshal's Office**, currently at **5642 Victor St., Bakersfield, CA 93308**.

About Black Powder

Because of its profound effect on the development of warfare, mining, and thereby heavy industry, **black powder**, also called "**gunpowder**", is considered by scholars in all fields to be one of most important inventions in the history of civilization. It is one of the oldest pyrotechnic mixtures known, and though its exact origins are uncertain, *recognizable* formulas for black powder first appear in the literature of the 13th Century. Black powder is a mixture of a chemical **oxidizer** (potassium nitrate or sodium nitrate), a simple **fuel** (a form of carbon such as coal or charcoal), and **sulfur**, which functions to *some* extent as a fuel, but mainly as a sort of melting-flux that helps the other ingredients burn faster. When combined in the proper proportions and ignited, its three components burn furiously, releasing **great** amounts of gas, smoke, and heat that are useful in dozens of military and industrial applications.

All the propellants in this book are either modified forms of black powder, or closely related mixtures. The popular perception is that black powder is an explosive used only in blasting, gunnery, and fireworks. But black powder makes an excellent rocket propellant as well. It is, in fact, the *exact* propellant used in the "safe" model rocket engines sold in hobby shops. The secret lies in packing it tightly into a solid stick, a form in which it burns at a fixed and predictable rate, and a form in which its burn rate can be carefully controlled.

As you can see in *Figure 1-1*, a loose, teaspoon-size pile of the KG3 propellant burns very quickly. Though it generates a great amount of heat, and a large puff of smoke and gas, *it does **not** explode*.

When properly triggered, high explosives like nitroglycerine undergo a sudden and violent decomposition of their entire molecular structure, even when loose and completely unconfined. But black powder and the propellants described in this book gain their explosive power *only* when burned in a tightly confined or totally sealed container, and are therefore *not* considered "high explosives".

Figure 1-1. A small pile of the loose **KG3** propellant burns very quickly. It generates a large amount of heat, gas, and smoke, but it does **not** explode.

● **Sensitivity to Impact**

In small quantities, black powder **made with potassium nitrate or sodium nitrate and a simple carbon fuel** is comparatively *insensitive* to impact, and I've verified this many times with the following experiment. You can try it yourself if you wish, but because it involves striking a piece of steel with a hammer, **you must wear safety goggles when performing this test**.

As shown in *Figure 1-2*, place a tiny pinch of commercial black powder or one of the propellants in this book on an anvil, or a large piece of iron or steel. Then begin striking it gently with a hammer. You will find *to you amazement* that you can increase the force of your blows to a point where you are in danger of chipping the hammer, and the powder will *not* explode. In fact, all you will succeed in doing is to pound it into a flat, gray-colored cake (*Figure 1-3*). Of course if you *miss* the powder and strike the anvil, the metal-to-metal impact could generate a spark that might ignite the powder, **so never use more than a small pinch of powder**, and be *very* careful when performing this test with other chemical mixtures, because unlike black powder, many of them are highly sensitive to impact.

Then why do the caps in a cap pistol make such a bang? Contrary to popular belief, the caps used in cap pistols are *not* made of black powder. The are made of *another* mixture similar in appearance but *completely* different in chemical composition. Antique muskets and flintlock rifles use black powder. What makes *them* work? The answer is that the hammer strikes a piece of flint that produces a spark. It is the *spark* that lights the powder, and *not* the impact of the hammer.

The fact is that *none* of the more common explosive fireworks, like firecrackers, cherry bombs, and M-80s, use black powder *for the very reason* that black powder is *not* sufficiently impact-sensitive or explosive enough to make them work. Black powder is used by the fireworks industry, but *not* for creating loud and violent explosions. It is used mainly as a skyrocket propellant, a lifting charge for mortars, and a bursting charge for scattering the stars and other special effects used in star shells. And, of course it is used in making fuse.

● Sensitivity to Sparks and Static Electricity

In its loose form, black powder is *extremely* sensitive to sparks. It can be ignited by even the *tiniest* spark. Common spark sources to avoid are *welding sparks, sparks from a grinding wheel, the sparks produced by striking together two pieces of iron or steel*, and *especially* **static electricity** sparks. If you get up in the morning, walk across the carpet, and see a spark jump from your finger to the light switch as you reach to turn it on, *this is not a good day to work with black powder*. A few years ago one of my readers toured a commercial rocket motor factory. He told me that they use an elaborate network of static electricity detectors. On days when this equipment tells them that the potential for static electricity is "high", *they shut the factory down*. I think that this is an *excellent* policy, and I'd like everyone who reads this book to adopt it.

When working with black powder, the greatest danger exists when you are sifting it or pouring it from one container into another. If you were standing directly over it, and if anything more than a small amount were accidentally ignited, it would not explode, but it *would* **burn** *so quickly and violently* **that you could probably** *not* **jump out of the way fast enough to avoid being seriously burned.** The most likely cause of such an ignition would be static electricity, and you can minimize the risk of a static electricity accident by following these rules.

1. **Always ground yourself** *before* **touching anything that contains black powder. You can do this by reaching down, and momentarily touching a water pipe or the ground.**

2. **When sifting black powder or pouring it from one container into another, place all the tools and containers on the ground, and** *let them to sit on the ground for several minutes before you begin.* This allows time for any static charge that these items might contain to drain away.

3. *Never* **operate a powder mill on top of a table, but** *always on the ground.* **If you are outside, set the mill on the dirt. If you have a patio, set it on the concrete.**

4. **Avoid working in dry and windy weather.**

5. **When handling black powder in its loose form,** *work slowly and deliberately.*

● Sensitivity to Friction

In 1992 I met another rocket maker who has worked extensively with black powder, and he told me the following story. He was loading *dry* black powder into a large motor one day when he became impatient with loading and tamping it in small doses. In an attempt to save time, he poured a *large* amount of powder into the motor casing, and tried to compact it all with one gigantic blow of the hammer.

As he rammed the tamp downward into the casing, the sudden compression of the air underneath forced a jet of air and loose powder upward between the sides of the tamp and the casing wall. The result was a momentary hissing sound followed by a loud "PHOOOM!" The tamp was blown high into the air. The powder that remained in the motor burned like a flare. The motor did *not* explode, and though startled, he was not injured. Of course the immediate question is, "What happened?"

We already know that black powder is relatively *insensitive* to impact. We also know that, when spread over the comparatively large area of the motor casing, the concentration of force and impact applied by his hammer was only a small fraction of that applied during the impact test described earlier. And we also know that the cushion of compressed air under the tamp would have prevented it from sharply impacting the powder already in the casing.

I can only make an educated guess at what actually happened, and my best guess is that the friction between the tamp moving downward and the thick film of dry powder moving upward generated enough heat to ignite the mixture. It is also possible that the jetting of the dry powder up the sides of the tamp acted like an electrostatic generator, creating a spark that ignited the powder. It would take a series of carefully designed experiments to find out exactly what happened, but an exact understanding of the mechanism isn't necessary. Just remember when loading your own motors to exercise common sense, and follow these rules.

1. *Don't* **use commercial black powder. Use** *only* **the** *slower-burning***, homemade propellants described in this book.** Though even commercial black powder is comparatively insensitive to impact, it burns *much faster* than these homemade propellants. If an accidental ignition should occur, the chances of an explosion and serious injury are correspondingly greater.

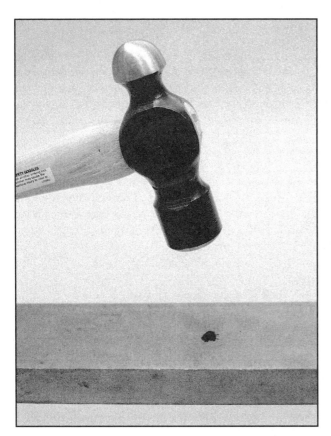 

Figure 1-2. A simple impact test for the **KG3** propellant. A tiny pinch is placed on a steel block and struck with a hammer. **Safety goggles must be worn during this test.**

Figure 1-3. The result of the impact test. The propellant does *not* explode. It is simply pounded into a flat, gray cake, and in this demonstration, stuck to the face of the hammer.

2. **Don't try to ram in a large amount of propellant with a single hammer blow.** When loading my own motors, I do it in small doses. I *first* give the tamp a few light taps to squeeze out any trapped air. *Then* I ram the propellant into a state of complete compaction.

3. ***Never*** **use a metal tamp.** ***Always*** **make tamps from a lightweight material like wood or plastic.** In the event of an accidental ignition, the tamp must be light enough to be blown out the front of the motor casing before the pressure in the casing can build to a dangerous level. Tamps made of metals like brass, steel, *and even aluminum*, are *heavy*. Their inherent resistance to motion, aka "inertia", can allow the pressure in the casing to build to the point of explosion.

4. **Never stand directly over the tamp or look directly down on it while you are working.**

● The Piezoelectric Effect

The **piezoelectric effect** is the tendency of some crystalline substances to generate an electric current when struck sharply. The most notable example is quartz, whose ability to respond in this way is used in electrical devices ranging from frequency-controlling crystals to piezoelectric cigarette lighters. Many years ago an explosives expert told me that there is *some* evidence that potassium nitrate crystals are slightly piezoelectric. If this is true, it would be theoretically possible for a large barrel of black powder to contain enough crystals to generate an *internal* spark if dropped or shocked in just the right way.

As far as he knew at the time no one had performed the experiments needed to test this theory. He seemed to think it more likely that the known cases of dropped barrels exploding were caused by the sparks generated when the iron barrel hoops hit the ground. In any case it's an interesting theory, and one that might cause a person to use extra care when handling black powder in large amounts.

● Storage of Black Powder

When you buy black powder at a gun shop, it usually comes in a metal can. The theory says that in the event of a lightning strike or a fallen power line, the can will act like a Faraday cage, conduct the electricity *around* the powder, and reduce the chance of ignition. Whether or not this really works is open to question, but it never hurts to be safe. So when storing your own black powder, *keep it in a lightweight metal container.* If the powder ever ignites, you want the container to blow its lid and *not* explode, *so look for a light weight can with a loose fitting top, like a thin, sheet metal, tea or cookie box.*

● Disposal of Black Powder

To safely dispose of black powder, dissolve it in water. Fill a bucket with warm water and a dash of laundry detergent. Stir the powder in with a spoon. Let it soak for a few hours, and stir it again to insure that it's completely dissolved. Then filter the solution through a sieve lined with paper towels. The sulfur and the charcoal will remain in the towels, which you can place in a plastic bag, and put in the trash. Then use the nitrate solution as a fertilizer in your garden, or dispose of it in the same way that you'd normally dispose of *any* unused nitrate fertilizer.

● Handling and Storage of Finished Motors

Once black powder has been compressed into a solid stick, its sensitivity to ignition is reduced, and finished rocket motors with the solid propellant sealed inside are even less hazardous. When handling and storing them, you should follow the same safety procedures recommended for storing and handling commercially-made model rocket engines, which are:

1. **Keep finished rocket motors away from all fire and flame, and do not smoke in their presence.**

2. **In case of a fire near or among the rocket motors, use water or a foam fire extinguisher to prevent their ignition.**

3. **Store finished rocket motors in a cool, dry place, and never expose them to temperatures above 150° Fahrenheit.**

4. **Dispose of damaged, defective, or unwanted motors by soaking them in water.**

● Safety During Propellant Preparation and Motor Construction

1. **Work out of doors whenever possible.** If you confine your mixing and loading activities to a workbench in the back yard, you reduce the risk of a fire in the event of an accidental ignition.

2. **Never keep large amounts of rocket propellant on your workbench. Bring *only* what you need for the immediate task at hand.**

3. **Never operate a powder mill indoors. *Always* set it on the ground *outside* at a safe distance from your house and other buildings, and surround it with a sandbag wall (*Figure 1-4*).** In the event of an explosion, the sand will absorb most of the heavy debris, and everything else will be directed upward out the top of the enclosure. At the time of this writing, empty sand bags cost 25 cents each, and you can buy them at hardware-lumber yards like Home Depot and Lowe's. Fill each bag 1/2 full, fold over the open end, lay it flat, and butt the folded end against the next bag in the row. As you finish each row, you can flatten and level it by walking on its upper surface. When you add a *new* row, stagger the joints as you would if you were building a brick wall. The powder mill's electric cord passes through a short length of 2" PVC water pipe placed in the bottom row.

    When you make this enclosure, leave plenty of air space around the powder mill. The enclosure in the photo is *square*, and it's sufficient for a 1/2 gallon mill. It is 8 rows high with 8 bags per row. It took 64 bags, and a little more than "a yard" of sand to make. For a 1 gallon mill make a taller, *rectangular* enclosure by adding 3 more rows, and an extra bag along two sides of each row. With *10* bags per row it will take 110 bags to make. The extra bags will more than double the enclosure's internal volume. For a *permanent* installation, dig a pit in the ground, and line it with cement blocks.

Figure 1-4. A sandbag enclosure **greatly** reduces the chance of injury or property damage if the powder mill ever explodes. Sandbags are inexpensive. In the event of an explosion, they absorb most of the heavy debris, and everything else is directed upward out the top of the enclosure.

4. **Mix *only* enough propellant for the motors that you plan to build in the next day or two, and avoid storing large quantities of unused propellant for any length of time.**

5. **Wear a respirator or a dust mask when handling or working with propellants.** Though the chemicals in these propellants are nontoxic or mildly toxic, *all of them* are irritating to the lungs and nasal passages. **Long term exposure to charcoal, coal dust, and carbon black can cause black lung and other serious respiratory diseases.**

6. **Handle finished rocket motors gently. Black powder propellant grains are hard and brittle. They can easily be broken or cracked. A cracked propellant grain will cause a motor to explode shortly after ignition.** The long and narrow motors like the NV6-83-G44 should be stored and transported in plastic bubble wrap (See the **PACKAGING SUPPLIES** headings in *The Yellow Pages*). **If you accidentally drop a motor, you should *assume* that the propellant grain has cracked, and *you should destroy the motor by soaking it in water.***

7. **Keep all rocket propellants and finished rocket motors away from children.** Children are fascinated by rockets, and that is good. But if you intend to involve anyone who doesn't own a valid driver's license, be sure that you lay down a strict set of safety rules, and ***never*** let them mix or handle rocket propellants without close supervision. I've always had a problem with the idea that people magically become mature at 18 to 21 years of age. I think it is more sensible to tie the making of homemade rocket motors to the possession of a valid driver's license.

If the State you live in thinks you are mature enough to operate a 4,000 pound killing machine (aka a car), you are *probably* responsible enough to build homemade rocket motors. If, by the same token, you are 40 years old, and your license has been revoked for drunk driving, you are clearly *not* a responsible person. You are a danger to yourself and the people around you, and you have no business engaging in amateur rocketry.

● Aren't You Afraid That a Terrorist Will Use a Homemade Rocket?

My answer to this question is *only in the movies*. In real life homemade rockets are *worthless* as weapons. First of all they are very inaccurate. When fired horizontally, small variations in performance cause most of them to either overshoot their target or fall unacceptably short. Also consider that the fact that the process of making and testing rockets is exactly the kind of high-profile activity that a terrorist avoids. It takes unacceptable amounts of time. It involves noisy, smoky experiments, and it leaves a trail of material purchases that is easy for the authorities to follow. In the few cases where terrorists have *successfully* used rockets, they've been *professionally* made, and either stolen, or purchased on the black market. From a terrorist's point of view, it is simpler and cheaper to park a truck filled with explosives next to a public building, then quietly walk away.

Additionally, the rocket motors in *this* book have *progressive* thrust-time curves. Their low starting thrust results in a beautiful and *very* realistic lift-off in vertical flight. But if you try to launch them at a low angle from a tube (like a "Bazooka"), as they exit the tube they will drop unpredictably, or fall to the ground. I've designed these rocket motors to carry cameras and scientific equipment in *vertical* flight. They are the exact *opposite* of the kind used in weapons.

● Won't Someone Use Homemade Black Powder to Make a Bomb?

The people who fear *this* eventuality have a limited knowledge of chemistry. What they don't understand is that you *don't* need a well known explosive like black powder to make a violent explosion. Even if black powder were outlawed, the supermarkets and hardware stores of America contain *dozens* of products that are *just as powerful*, the most obvious example being *matches*. Since the bombing at the 1996 Summer Olympics, there have been so many descriptions of pipe bombs in the news that I'm sure I'm not revealing anything new when I explain that, for less than $100, anyone anywhere can buy the pipe fittings and enough matches to make a similar device. Likewise, anyone who saw the 1995 medical thriller, *Outbreak*, starring Dustin Hoffman, will remember the horrendous explosion at the end of the film caused by a device called a "fuel-air" bomb. Fuel-air weapons actually exist, and are powered by a fine mist of air and *gasoline*. Frustrating though it is, the kind of weak and dysfunctional personalities that make and use bombs know all of this, and at least for now, aside from "good intelligence" or the threat of punishment, there is little you can do to stop them.

Therefore, outlawing hobbies like amateur rocketry, preventing the public from making homemade rocket propellant, and limiting the public's access to anything that *might* be misused would be *completely* ineffective, and it would also be *wrong*. For every person who abuses these things there are *thousands* who put them to good and creative use. For example, we all know that in 1996 ammonium nitrate was used to bomb the Murrah federal building in Oklahoma City. But ammonium nitrate is a cheap and excellent fertilizer, and it's also a *precursor* chemical used in the production of many *other* chemicals. A *healthy and vibrant* America *needs* ammonium nitrate, and later in this book I will teach you how, with home-lab techniques, to convert ammonium nitrate into *sodium* nitrate, the main ingredient in the *slow-burning* NG6 and NV6 propellants.

Since I started this work in 1983 I've met at least a dozen professional rocket engineers. *Every single one of them* built his first rocket motor by the time he was seventeen, and several have told me that it was their early experience with *homemade* rocket motors that lead them to pursue rocket engineering as a career. Most of the people who tinker with these things do so in a *responsible* manner with *benevolent* intentions. As a group they are unusually intelligent, self motivated, and highly creative. When allowed to pursue their interests freely and without interference, their creativity blossoms in ways that benefit America as a whole. These people represent the *best* among us. To deny them access to the knowledge and materials needed to pursue their hobby would do our country far more harm than good. Instead of trying to *restrict* our citizens' involvement with technology, **we should establish more severe punishments for those who misuse it!**

Can I Sell the Rocket Motors that I Make?

You can sell the rocket motors that you make **only with the proper licenses**. **To do otherwise violates both Federal and State laws, and the penalties for breaking these laws are *severe*.** If you want to *sell* your rocket motors, you *must* contact the Bureau of Alcohol, Tobacco, and Firearms in Washington D.C., and your own State Firemarshal's office. They can explain to you the procedures that you need to follow to become a legitimate, licensed manufacturer of either model rocket engines or high power rocket motors. The business of making and selling rocket motors is interesting, and you *don't* need expensive equipment or a degree in rocket engineering to be successful. You can start with simple hand tools, and learn as you go. The fact is that, as of this writing, at least **three** currently successful rocket motor companies were started as small part time businesses by people with **no** formal education in chemistry or engineering.

From time to time there have been suggestions from within the industry that licensing and motor certification be made more difficult. *Predictably*, the people promoting these ideas say that they are concerned about "safety". But it's worth noting here that if licensing and certification were made costly to new competitors, the market position of the *already-established* businesses would be *strengthened*. Competition may be hard on the individual manufacturers, but it's good for the hobby as a whole. Since 1991 the natural competition within the industry has resulted in at least one completely new motor technology, and a flood of interesting new motor designs.

What is the Safety Record so Far?

As of the year, 2000, since the 1984 publication of my first book, my company, the Teleflite Corporation, has sold 10,000 copies of that book, and enough cardboard tubes to make 250,000 homemade rocket motors. Of course not everyone who buys the book actually *makes* a rocket motor. But if just a *tenth* of them do, if just a *tenth* of the cardboard tubes have actually been used, that amounts to 1,000 people making a total of 25,000 rocket motors over a span of 17 years. In 1985 and 1986 several of my readers attended hobby shows where people opposed to the making of homemade rocket motors had exhibits. In each case, when my readers mentioned my book, they were warned that making homemade rocket motors was *very* dangerous, and that people who shared this view (hoping to discredit the book) were actively searching for someone who'd been injured by following the book's instructions. Until 1991 I received sporadic reports that the search continued. Since 1992 I have heard nothing. It is now the year, 2000, and despite these efforts to find one, there has *never* been a report of an accident or injury resulting from any book ever written by me, or published by my company, the Teleflite Corporation.

What if a Serious Accident Happens?

Though people concerned about the future of amateur rocketry are understandably reluctant to think about it, somewhere, someday, there's bound to be an accident in which someone is injured or killed, either by a rocket or while making one. In a society of 250 million people, *all* activities (*including golf!*) experience *some* level of injury and death. When such an accident occurs, the people in positions of responsibility should do their best to find out what happened, and keep it from happening again. Their efforts should be limited, however, *to the specific problem that caused the accident*, and should *not* include anything of a global nature that cripples the hobby or results in its demise.

Many years ago, here in Southern California, a 2-year-old boy was struck in the head and *killed* by a model airplane flown by his father. This did *not* result in a clamor to outlaw model airplanes. People are regularly injured and killed while riding skate boards, in-line skates, motorcycles, and jet skis, yet these activities continue to flourish as they should. Once in a rare-while a car careens out of control at a race track, and kills or injures a group of spectators, but we don't ban auto racing. Similarly, private airplanes sometimes crash, killing everyone on-board and people on the ground as well, *but we don't outlaw private aviation.*

In each case we recognize that these activities play an important role in our society. We value them because they are educational and recreational, and are therefore **good for the mind and good for the soul**. We make them as safe as is practical, and we let them continue. We *could* pander to the fears of our most timid citizens, and limit the activities of everyone accordingly. But we *don't*, because to do so would create a world so limited in spirit and dimension that it wouldn't be worth living in. At the turn of the millennium the average American is *not* well educated, has a short attention span, and little or no curiosity about the world around him. He wants to be well fed, stylishly clothed, and perpetually entertained, and it's increasingly difficult to get him interested in *anything* related to science or engineering. This is *not* good for America's future, and anyone who understands this must also realize that Americans *need* to have hobbies like amateur rocketry. If an accident happens tomorrow, *if there are several serious accidents every year*, the safety record of this excellent and valuable hobby will still be *ten* times as good as the record of the "acceptable" hobbies mentioned above.

In light of the long and aggressive campaign *against* amateur rocketry, it is important to understand that small solid fuel rocket motors are easy to make. *Some* of the forces opposed to amateur rocketry see this as an *economic* issue. They think that if they could talk the government into making homemade rocket motors illegal, then anyone who wants to fly rockets would have to *buy* the motors. The market for commercially made motors would be secured, and both sales and profits would increase. You should also understand that, in today's competitive business climate, organizations which *claim* to act in the public interest can be little more than public relations fronts for the industries they serve. That the words, *safe* and *unsafe*, can be little more than code words for *what's good for business* and *what isn't*. Finally realize that even a respected professional organization may have within its ranks a few members who inappropriately use its name, power, and prestige to promote their own selfish agenda. Their motives can range from childish concerns about "turf" to greed and economic self interest, and *anyone* who engages in this behavior should be exposed and discredited.

If and when a serious accident happens, the forces opposed to amateur rocketry will *probably* demand the hobby's demise. The authorities will respond, and the future of amateur rocketry will be subject to the intelligence and good will of whoever has to deal with the problem. It is my hope that these people will *not* take action based on an isolated event, particularly if that event resulted from a failure to observe basic safety procedures. It is my hope that they realize the value of amateur rocketry, and consider its long and excellent safety record. It is also my hope that they ignore the clamor of the selfish interests, and *avoid* the foolish temptation to simply outlaw the hobby, or restrict it in a destructive or unreasonable way.

I finished the preceding pages in the last months of the year 2000, but as I write these final words, it is Sunday, Sept. 16th, 2001. I awakened five days ago to the news of the most monstrous terrorist attack in the history of the world. The subhuman filth that committed this heinous crime did *not* use rockets. They hijacked four large passenger jets, then crashed them into the Pentagon and the World Trade Center. These cheaply-made, production-line murderers are *not* great warriors. They are gullible, brooding, and *infantile* boys-in-the-bodies-of-men. And regardless of what they babble, they are *not* sacrificing their lives for Islam. They are *selfishly* dying to reach an imagined Paradise-for-the-simple-minded, filled, they *foolishly* believe, with "black-eyed virgins" who, in payment for their hideous deeds, will immediately marry them, then fawn over them and eternally satisfy their most private adolescent urges. Though they say that their motive is "martyrdom", it's actually an obnoxious and malignant blend of narcissism, stupidity, and testosterone run amok.

They are indoctrinated in this bizarre and pathetic fantasy by a self-absorbed cult of evil old men, whose massive and inappropriate sense of self-importance would *never* allow them to realize that their radical beliefs and the hatred they feel are born of their own *personal* failings and their own defective thinking. They rage instead against our freedom of thought, the modern world, and everything that America stands for.

These ranting, demonic lunatics long ago abandoned the idea of forcing us to think like they do. They want to kill us all. If they can't destroy us with bombs, chemicals, or biological weapons, they will try to frighten us into limiting our freedoms, our enthusiasm, our creativity, our intelligence, **our success**, and thereby our influence on the cultures around us. Our superior technology will of course help us defeat them, so the discouragement of public interest in science and technology is high on their agenda. On a cultural level, hobbies like amateur rocketry accomplish the exact *opposite* of what these evil, sworn, and dedicated enemies want. So the restriction or elimination of amateur rocketry or the materials needed to pursue the hobby would, I am sure, fit nicely into their plans.

2. BASIC CONCEPTS

Basic Rocket Function

Most rocket motors are powered by the chemical reaction between a **fuel** which is burned and an **oxidizer** that supplies the oxygen needed for burning. In a solid fuel rocket motor the fuel and the oxidizer are premixed. This blend of chemicals is called a **propellant**, and when molded into a solid shape inside of a rocket motor, it is called a **propellant grain**. From room temperature to a safe temperature above, the propellant's chemicals peacefully coexist, allowing the rocket motor to be handled, transported, and safely stored. When raised to the proper ignition temperature, those same chemicals begin a violent but *predictable* combustion reaction that spreads through the propellant at a controlled and predetermined rate.

As the atoms of fuel and oxygen combine, some of their electrons drop into lower energy states, and the *lost* energy is released as light (the bright and impressive-looking flame) and a *great* amount of *heat*. The heat is, in fact, so intense that it raises the temperature of the reaction products by several thousand degrees, causing most of them to expand into a gas. Because these hot gases occupy a much larger volume than the original propellant, a tremendous pressure develops that forces them out the motor's exhaust nozzle at supersonic speed. The motor is placed in the rocket so that its nozzle points *backward*, and since *Newton's* **Third Law of Motion** states that *for every action, there is an equal and opposite reaction*, a pushing force called **thrust** occurs that drives the rocket *forward*.

Cardboard Motor Casings

A solid fuel rocket motor's **casing** is the cylindrical tube that contains all of the motor's components, and confines the hot combustion gases generated during its operation. A motor casing can be made of steel, aluminum, fiberglass, phenolic, or any material capable of withstanding the conditions under which the motor operates. The casings for *these* motors are made of cardboard (multiple layers of paper and glue). Cardboard is low in cost, and it has the resilience required for this particular building method. It also has the burst strength needed to withstand the pressures at which these motors operate. Because some of the paper is burned away during motor operation, a cardboard casing is *not* reloadable. It should be used once, then discarded.

Clay Nozzles

A rocket motor's **nozzle** is the part of the motor through which the exhaust gases escape. A nozzle can be made of steel, graphite, ceramic, or any material with the necessary hardness and resistance to heat. The nozzles in *these* motors are made from a dry-rammed mixture of clay, grog, and wax. It is inexpensive and easy to form, and it has the hardness and the refractory properties needed to retain its shape during motor operation. Because a rammed, clay nozzle becomes an integral part of the casing during the motor building process, and because it experiences substantial erosion during motor operation, a clay nozzle *cannot* be reused. It must be discarded along with the casing after the motor has been fired.

The De Laval Rocket Nozzle

In the late 19th century a well known engineer named Carl Gustav De Laval invented the first, practical steam turbine engine, *and along with it*, the special nozzle used to maximize the speed at which the steam hit the turbine blades. The nozzle's design was soon adapted by Robert Goddard and other engineers for use in *rocket* motors, where it became known as **the De Laval nozzle**.

A De Laval nozzle is made with a precisely engineered venturi shape that maximizes the speed of the motor's exhaust. Assuming that the other design parameters are fixed, this maximizes the motor's performance and efficiency. The mathematics of De Laval nozzles are a standard part of all rocket engineering texts. But the narrow neck of the nozzle is *very* short, and when made of the ceramic mixture described in *this* book, it erodes out so quickly during motor operation that all the design benefits are lost. Knowing that some of the people who read this book will want to experiment with De Laval nozzles, I spent some extra time during the project, and developed a method of mounting real, stainless steel De Laval nozzles in cardboard tubes. They can be quickly machined by anyone with a metal lathe, and the process of installing one takes less than 5 minutes.

Once the nozzle is in place, the propellant and the other components are loaded in the normal manner. The result is a rocket motor that performs according to classic nozzle engineering principles, and a little more efficiently than a motor with a clay nozzle. Because a De Laval nozzle erodes out only one or two thousandths of an inch with each firing, it can be reused dozens of times. After many firings, when accumulated erosion causes a noticeable drop in motor performance, a small decrease in the propellant's baking soda content restores performance to its previous level. **Chapter 10** contains complete instructions on how to make and install De Laval nozzles.

Basic Terminology

Thrust, abbreviated **F**, is the pushing force generated by a rocket motor, and thrust may or may not vary during a motor's operation. Whether or not it varies, and how much it varies depends on the motor's design.

Burn time, abbreviated **T**, is a motor's total time of operation from the beginning of thrust to the end of thrust.

Total impulse, abbreviated I_t, is the total amount of energy generated by a rocket motor from the beginning of thrust to the end of thrust. You can figure it out by breaking up a motor's thrust-time curve (**page 22**) into a series of tiny thrust-time segments, and adding them all up (a process called "integrating" in Calculus). To work it out arithmetically you multiply the motor's average thrust by its burn time. In the English system of measurement the total impulse is expressed in **pound-seconds**. In the metric system it is expressed in **Newton-seconds**. The metric-to-English conversion ratio is 4.45 Newtons per pound, so one Newton-second equals 0.225 pound-seconds.

Propellant mass, abbreviated m_p, is the total amount of propellant burned during a rocket motor's operation.

Average thrust, abbreviated F_a, is the value obtained by mathematically averaging a motor's thrust over its entire burn time. To calculate a motor's average thrust, you divide the motor's total impulse by its burn time.

EXAMPLE 2-1:

A rocket motor has a total impulse (I_t) of 80 lb.-secs. (356 Nt.-secs.) and a burn time of 4 seconds. Its average thrust, (F_a) is calculated as follows:

$$F_a = I_t / T$$

$$F_a = 80 \text{ lb.-secs.} / 4 \text{ secs.} = 20 \text{ lbs.} \quad or \quad F_a = 356 \text{ Nt.-secs.} / 4 \text{ secs.} = 89 \text{ Nts.}$$

Specific impulse, abbreviated I_{sp}, is the standard measure of how efficient a propellant is at generating thrust in a well designed rocket motor. You can calculate the I_{sp} of a given propellant-motor combination by dividing the motor's total impulse by its propellant mass. The answer first appears in "pound-seconds per pound" (lb.-secs./lb.), but the "lbs." above and below the dividing line cancel out, leaving just the "secs." Therefore, though it may sound strange, a propellant's specific impulse is expressed in *seconds*. Because a propellant with a high specific impulse takes a rocket farther than one with a low specific impulse, you can think of specific impulse as the approximate equivalent of gas mileage in a car.

EXAMPLE 2-2:

The rocket motor in EXAMPLE 2-1 carries 1.25 lbs. of propellant, and we already know that its total impulse is 80 lb-secs., or 356 Nt.-secs. The specific impulse is calculated as follows:

$$I_{sp} = I_t / m_p$$

$I_{sp} = 80$ lb.-secs. / 1.25 lbs. = 64 lb.-secs. / lb. = 64 secs., *or* $I_{sp} = 356$ Nt.-secs. / 1.25 lbs. = 284.8 Nt.-secs. / lb.

But 284.8 Nts. = 64 lbs., so the answer works out again to 64 lb.-secs./lb., or 64 secs.

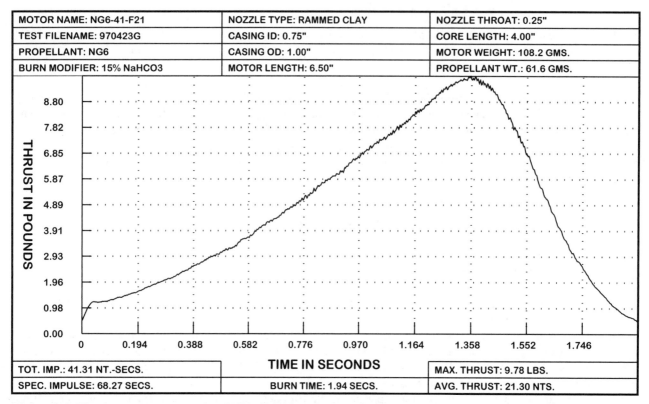

Figure 2-1. The thrust-time curve of an *NG6-41-F21* rocket motor. A thrust-time curve provides the vital information needed to evaluate a rocket motor's performance and predict what it will do during an actual flight.

Casing i.d. is the *inside* diameter of a rocket motor casing. In the motors in *this* book, it equals the o.d. (outside diameter) of the propellant grain. In the time delay calculation on **page 29**, I refer to the casing i.d. as "D_{cas}".

Casing o.d. is the *outside* diameter of a rocket motor casing. The difference between a casing's o.d. and its i.d., when divided by 2, equals its wall thickness. The wall thickness is important because it directly effects the casing's strength and how much operating pressure the casing can withstand. If the wall is too thin, it will rupture at normal operating pressure. If the wall is too thick, it unnecessarily increases the motor's size and weight.

Motor length in *this* book refers to the total length of a rocket motor's *casing*, including the hollow volume in front of the forward bulkhead.

The **nozzle throat** is the narrowest part of the exhaust hole in a rocket motor's nozzle. In the *clay* nozzles it amounts to a small, cylindrical passage, terminating at the beginning of the nozzle's **divergent taper**. In the De Laval nozzles, to reduce friction and improve performance, this passage is shortened to a narrow, ring-shaped zone just wide enough to separate the **convergent** and **divergent tapers**.

Motor weight in *this* book is the total weight of a rocket motor *before* it is fired. It amounts to the combined weights of all the motor's parts, including the cardboard casing, the clay nozzle and forward bulkhead, and the propellant and time delay charge. Knowing a motor's total weight is important when trying to predict how a rocket will perform during an actual flight.

A **thrust-time curve** is a graphic plot of how a rocket motor's thrust varies with time. Thrust-time curves are useful because they provide the information needed to calculate important things like a rocket's speed and altitude. For my own work I developed the standard form in *Figure 2-1*. This example illustrates the thrust-time curve of an **NG6-41-F21** rocket motor.

Units along the horizontal axis represent the time in seconds from the beginning of thrust, and units along the vertical axis represent the thrust in *pounds* that the motor is generating at any given moment. The printed information above and below the curve provide all the motor's vital statistics. To a trained eye, the size and shape of the curve combined with the other information provide an instant understanding of the motor's capabilities. This data can then be used to compare one motor's performance to another's, or predict the flight performance of a rocket that is powered by the motor.

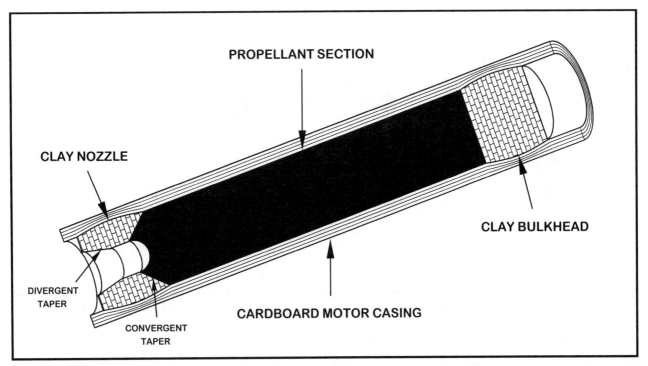

Figure 2-2. *A cutaway drawing of an end burner. The propellant grain is a solid, cylindrical stick with its sides sealed tightly to the casing wall. A dimple is formed in the rear end just large enough to accept an igniter.*

Grain geometry is the term used to describe the shape of a rocket motor's propellant grain. Because a propellant grain's *initial* shape determines the path that the flame follows after ignition, a motor's grain geometry determines what its chamber pressure and thrust will be during each moment of its operation.

End Burners

The simplest grain geometry is a solid, cylindrical stick of propellant with its sides sealed tightly to the casing wall. To facilitate ignition, a small dimple is formed in the grain's rear end, just big enough to accept an igniter. *Figure 2-2* is a cutaway drawing of one of these motors. It's called an **end burner**, and *Figure 2-3* shows what happens when it burns.

Immediately after ignition the flame spreads outward in the form of a hemisphere with a surface area *greater* than the propellant grain's cross section. This creates an initial surge of high pressure and high thrust that quickly diminish as the shape of the burning surface changes into a shallow cup. At that point the pressure and thrust level off, and should theoretically remain constant for the rest of the burn.

When properly made, end burners produce a constant level of thrust for a comparatively long time. But they have the *disadvantage* that the inside wall of the casing is exposed to open flame for the duration of the burn. Because the casings for the motors in *this* book are made of *paper*, the implications are obvious. If you make the burn time too long, you'll burn through the casing wall. Also, for the chamber pressure and thrust to remain constant, it is *absolutely imperative* that an end burner's propellant burn at a *perfectly* uniform rate.

Small end burners made with *homogeneous* propellants work fairly well. But as end burning propellant grains are made larger, their cup-shaped burning surfaces grow increasingly uneven. This phenomenon is called "cavitation". It happens regardless of how evenly the propellant is mixed, and it causes unacceptable variations in chamber pressure and thrust. When building a *large* rocket motor, the solution is to adopt a *core* burning configuration. Because rocket engineers understand this well, all of the larger commercial and military solid fuel motors, including the giant boosters on the Space Shuttle, use some variation of the *core* burning design.

Time constraints involved with the publication of this book, a possibly unsolvable engineering problem, and difficulty in obtaining the proper cardboard tubes prevented me from developing a *workable* end burner. But I *was* able to develop the tooling and the basic method of construction for those who want to experiment on their own. I discuss end burners at length in **Chapter 10**.

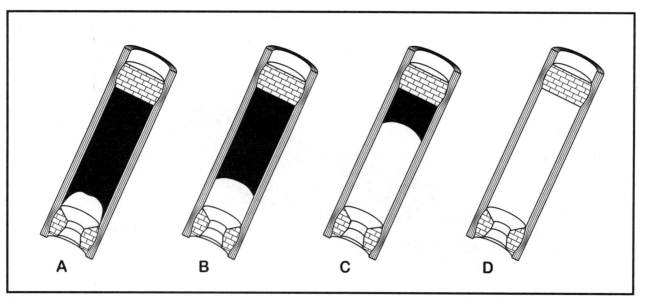

Figure 2-3. *The flame in an end burner travels in a straight line from the point of ignition to the forward bulkhead. The dome-shaped flame front, formed immediately after ignition, generates a surge of high pressure and high thrust. As the dome changes into a shallow cup, both chamber pressure and thrust are reduced, and thereafter remain constant.*

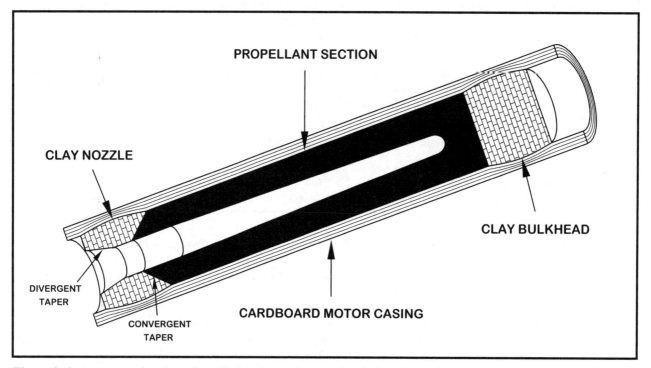

Figure 2-4. *A cutaway drawing of a cylindrical core burner. A cylindrical core burner has a cylindrical bore hole, or "core", running right up the center of the propellant grain.*

Cylindrical Core Burners

The most common grain geometry has a hollow bore hole, or **core**, running right up the center of the propellant grain. In its simplest form, the core is shaped like a slightly tapered cylinder. *Figure 2-4* is a cutaway drawing of one of these motors. It's called a **cylindrical core burner**, and *Figure 2-5* shows what happens when it burns. Ignition starts at the core's forward end, and *instantly* spreads to the rear, lighting the length of the core as it goes. From that point on the flame spreads outward in the shape of an expanding cylinder, consuming the last of the propellant as it reaches the inside of the casing wall. In motors with small nozzles and short cores, the chamber pressure and thrust start out low and reach their maximums at burnout. In motors with larger nozzles and longer cores the initial values are higher. In motors with *very* large nozzles and *very* long cores the starting pressure and starting thrust can exceed 50% of their maximums.

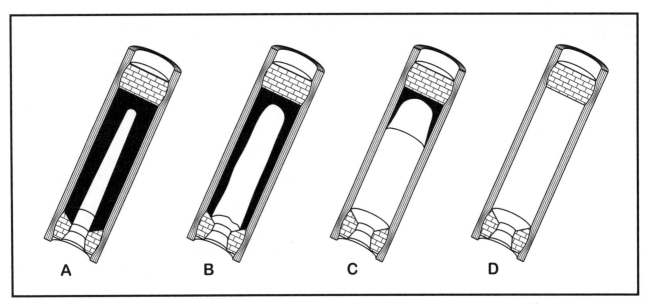

Figure 2-5. The flame in a cylindrical core burner spreads outward in the form of an expanding cylinder. The ongoing increase in the amount of propellant burning at any moment causes a steady increase in chamber pressure and thrust as the burn proceeds.

A motor's **core length** is the length of the propellant grain's bore hole, as measured from the boundary between the nozzle clay and where the propellant begins, to where the hole terminates at the front of the motor. A motor's core length is critical because it directly controls the amount of propellant burning at any given moment. If the core is too short, the motor performs poorly. If the core is too long, the maximum chamber pressure can exceed the motor's design limits, and cause the motor to explode.

Because the burning surface in a core burner is *larger* than the burning surface in an end burner, a core burner requires a slower burning propellant. Because the distance that the flame travels is *short* (from the edge of the core to the inside of the casing wall), *small* core burners have comparatively short burn times. *All* core burners have the *advantage* that the layer of unburned propellant between the flame and the casing wall insulates the casing from the heat of the flame until the final moments of the burn.

Figures 2-6, *2-7*, and *2-8* are examples of the thrust-time curves of an end burner, a short cylindrical core burner, and a long cylindrical cure burner. In each illustration the points **A** through **D** correspond to examples **A** through **D** in *Figures 2-3* and *2-5*. The rocket industry uses dozens of other grain geometries, but the building techniques described in *this* book will limit you to these three.

Time Delays

After its motor stops firing, a rocket coasts upward, slowing as it goes. Assuming that the flight has been vertical, it eventually comes to a stop, turns over, and falls back down. The turnover point is called the **apogee**, and in most *low* altitude flights, the apogee is the best place to eject the parachute. The simplest way to do it is to pop off the nose cone, and force the chute out with a small puff of gas. The exact timing is important, and though an electronic timer is the most reliable, a pyrotechnic delay works *reasonably* well if you make it properly. The placement of the delay depends on the motor's grain geometry, and *Figure 2-9* shows its location in an end burner.

As you can see, a pyrotechnic delay is a carefully measured length of *slow*-burning propellant packed in front of the main propellant grain. It ends at a small hole through the motor's forward bulkhead. When the main propellant grain is used up, the motor's chamber pressure drops back to atmospheric (or close to it). The flame starts consuming the time delay, and *immediately* slows down to a fixed and predetermined rate. After a calculated number of seconds, it reaches the hole. Then it flashes *through* the hole to the front of the motor, where it ignites a charge of loose powder that generates enough gas to push out the parachute. To calculate the length of the delay, you multiply the delay's *open-air* burn rate (its burn rate at atmospheric pressure) by the desired number of seconds.

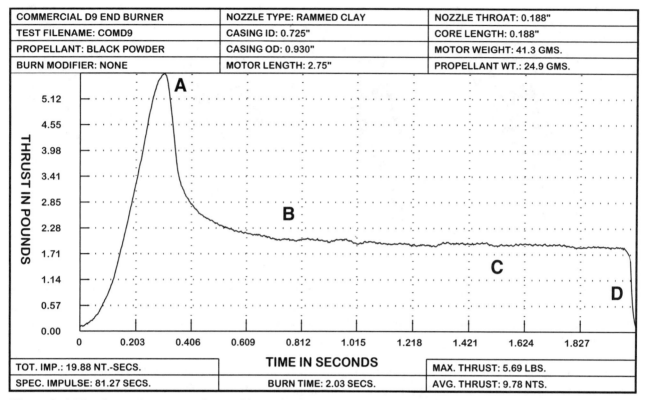

Figure 2-6. The thrust-time curve of an end burner. After an initial surge of high thrust, motor thrust drops back to a level where it remains constant for the duration of the burn. End burners are good for applications requiring low levels of thrust, and comparatively long burn times. Points **A** through **D** correspond to drawings **A** through **D** in **Figure 2-3**.

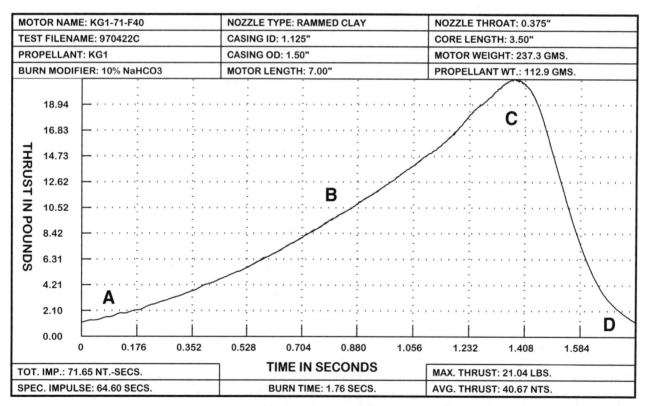

Figure 2-7. The thrust-time curve of a short cylindrical core burner. Immediately after ignition, motor thrust increases steadily, and reaches its maximum when the flame reaches the inside of the casing wall. At any given moment the remaining, unburned propellant acts as an insulator, protecting the casing from the hot flame until the final moments of the burn. The core burning configuration is therefore particularly well suited for the paper motor casings described in this book. Points **A** through **D** correspond to drawings **A** through **D** in **Figure 2-5**.

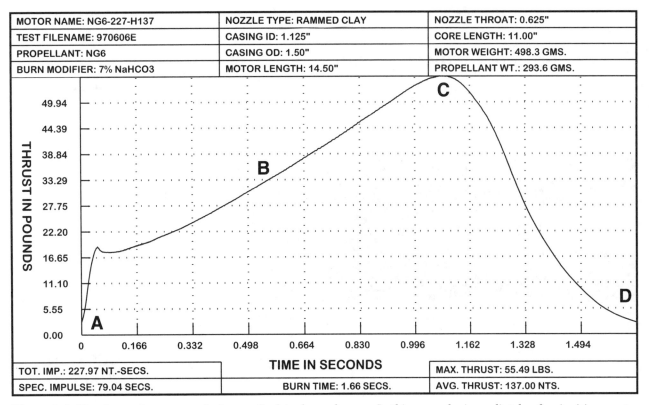

Figure 2-8. The thrust-time curve of a long cylindrical core burner. In this example, immediately after ignition, motor thrust increases to about a third of its maximum, and thereafter rises gradually. The high starting thrust makes a long cylindrical core burner a good choice for lifting a heavy payload. Points **A** through **D** correspond to drawings **A** through **D** in *Figure 2-5*.

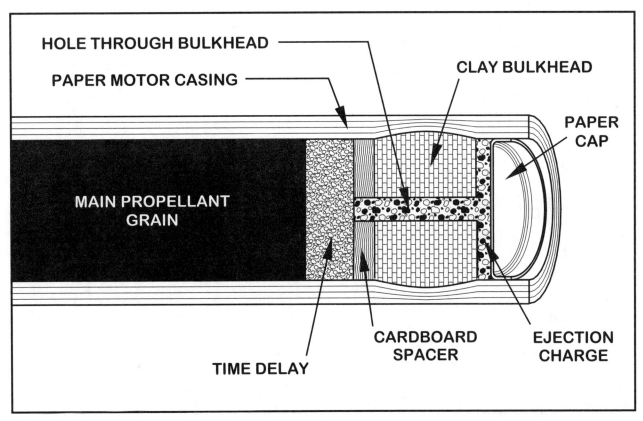

Figure 2-9. In an end burner, the beginning of the time delay is located at the forward end of the propellant grain.

27

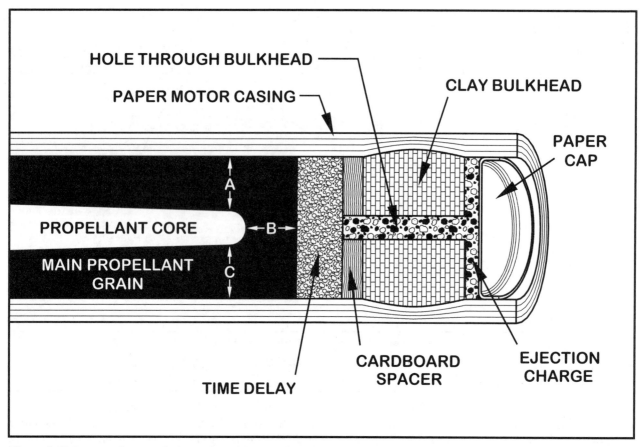

Figure 2-10. In a core burner, the beginning of the time delay is located at a distance ahead of the core equal to the propellant's thickness between the front edge of the core and the casing wall. Distances **A**, **B**, and **C** should all be the same.

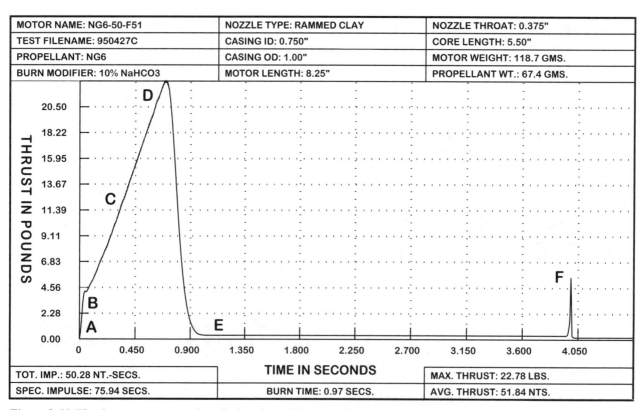

Figure 2-11. The thrust-time curve of a cylindrical core burner with a 3 second time delay and a parachute ejection charge. Points **A** through **F** correspond to the 6 drawings labeled **A** through **F** in *Figure 2-12*.

EXAMPLE 2-3:

You've chosen a slow burning propellant to use as a time delay. Its open air burn rate is 0.12"/sec., and you want the total time delay to be 6 seconds. You calculate the length as follows:

Length of time delay = open air burn rate x desired delay time. Therefore:

Length of time delay = (0.12" / sec.) x 6 secs. = 0.72"

The time delay should therefore be 0.72" long.

Figure 2-10 shows a time delay's placement in a cylindrical core burner. The length of the delay is the same, but as you can see, it is *not* placed directly at the end of the propellant core. It starts at a distance *ahead* of the core equal to propellant's thickness between the front of the core and the casing wall.

Cylindrical core burners are made this way because, by the time the flame has traveled the distance from the ignition point to the casing wall, it has burned an *equal* distance *forward*. This distance equals 1/2 the inside diameter of the casing minus 1/2 the diameter of the front end of the core hole, and I express it mathematically as follows:

$$L = 1/2 D_{cas} - 1/2 D_{core}$$

where D_{cas} is the inside diameter of the motor casing, D_{core} is the diameter of the forward end of the core, and L is the distance between the forward end of the core and the beginning of the time delay.

EXAMPLE 2-4:

You're building a core burner with a casing whose inside diameter is 0.75". The forward end of the propellant core is 0.30" diameter. The distance from the forward end of the core to the beginning of the time delay should be:

$$L = 1/2 D_{cas} - 1/2 D_{core}$$

or

$$L = 1/2 (0.75") - 1/2 (0.30") = 0.375 - 0.150 = 0.225"$$

The time delay should therefore begin 0.225" (about 7/32") ahead of the forward end of the propellant core.

You can use any of the propellants in this book as a time delay, but because a time delay adds weight to a motor, it is wise to chose one of the slower burning propellants. A delay made from a slow burning propellant will be physically shorter, and therefore smaller and *lighter* than a delay made from a fast burning propellant. The *best* in this regard are the **NG6** and **NV6** propellants made with sodium nitrate and carbon black.

The amount of loose powder needed to pop out a parachute can vary greatly. It depends on the total volume of the rocket's parachute compartment, the volume of any hollow space in the nose cone, and how tightly the nosecone fits into the body tube. It is hard to calculate, because it depends on how much free air space the compartment contains, and *that* can vary according to things like the size of the parachute, and how tightly or loosely the "chute" is rolled. My advice is, for a given rocket design, to set up a "dummy" body tube on the ground complete with nose cone and parachute. Then, starting with a *small* amount of powder, experiment by electrically igniting various amounts of powder until the desired effect is achieved.

Single Stage Rockets

A **single stage rocket** uses a single rocket motor or a cluster of motors to lift its payload to the desired altitude, and most low altitude sounding rockets are of the single stage type. *Figure 2-12* shows what happens during the flight of a single stage rocket. Below each rocket drawing is a drawing of its motor, and you'll notice that I've labeled the drawings with the letters, **A** through **F**.

If you look at *Figure 2-11*, you'll see the thrust-time curve of a long, cylindrical core burner. I've included a time delay at the end of the motor's burn, and the small pulse generated by the parachute ejection charge. In this example the delay is 3 seconds long. I've marked 6 points along the curve with the letters, **A** through **F**. Each of these points corresponds to one of the drawings in *Figure 2-12*, and each drawing shows what the rocket and its motor are doing at that point on the curve.

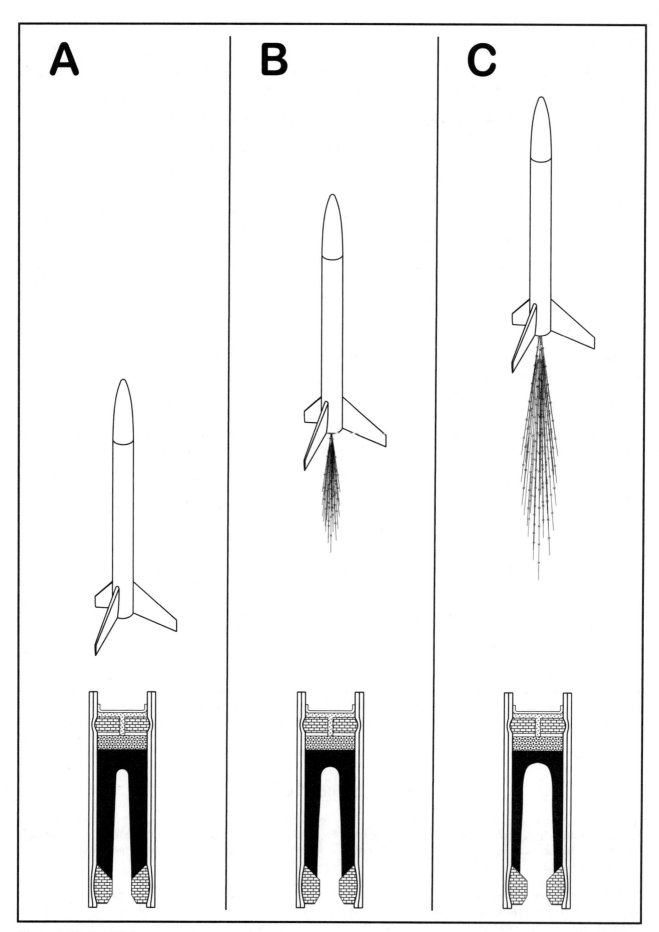

Figure 2-12*. The flight of a single stage rocket.*

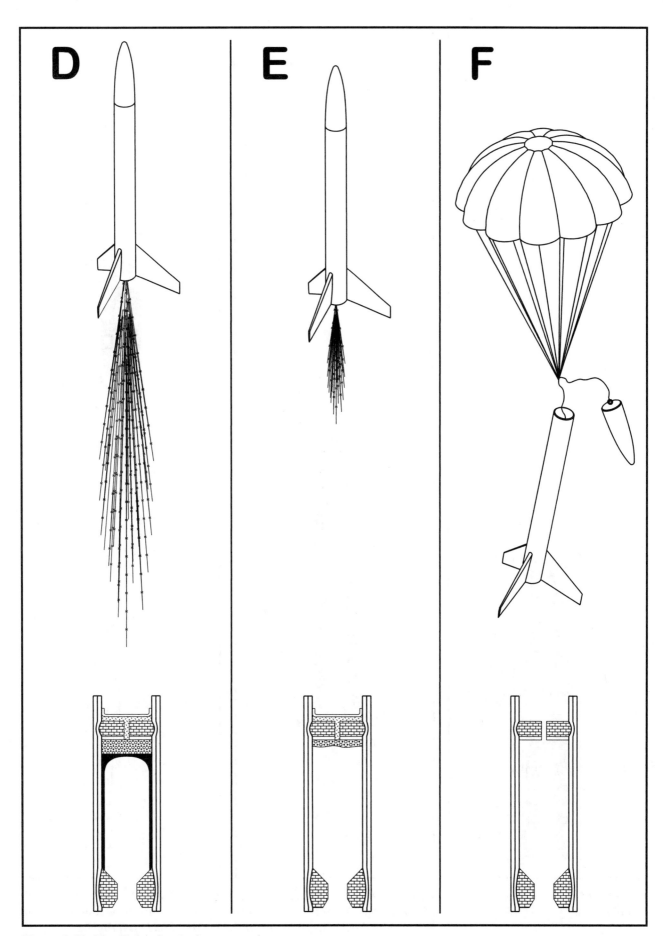

***Figure 2-12**. The flight of a single stage rocket.*

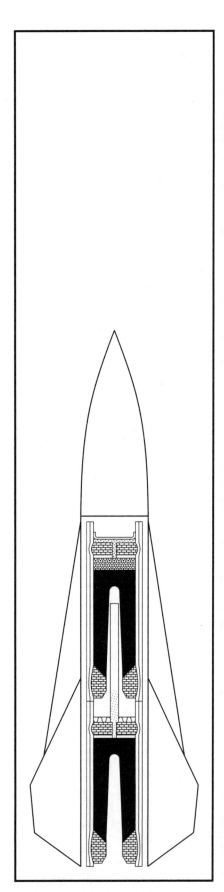

Figure 2-13. A two stage rocket just before lift-off.

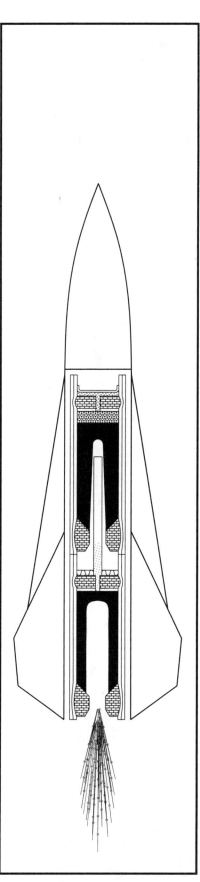

Figure 2-14. The propellant is beginning to burn. The motor's thrust has exceeded the rocket's weight, and the rocket has lifted off the ground.

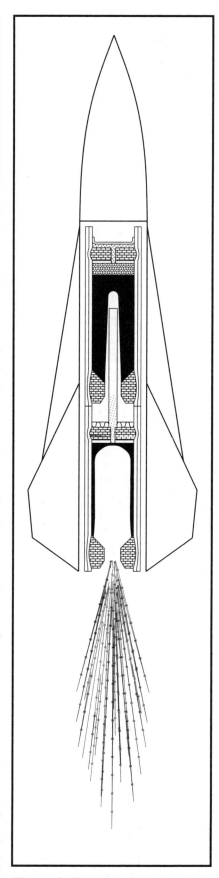

Figure 2-15. Most of the booster's propellant has been consumed, and the rocket is in full flight.

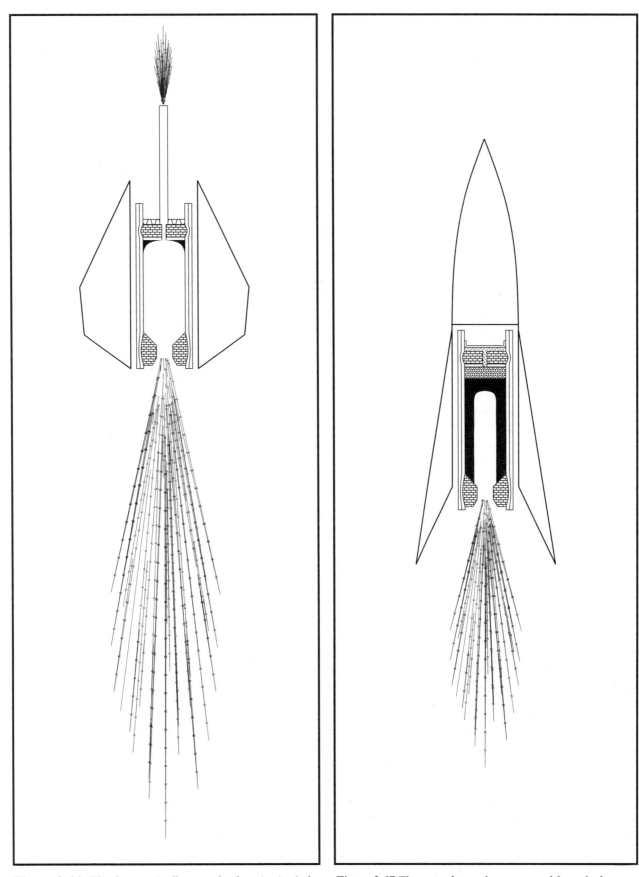

Figure 2-16. *The booster's flame tube has ignited the second stage motor, and the booster has dropped away.*

Figure 2-17. *The second stage has separated from the booster, and it's beginning to generate thrust. It will eventually reach a higher altitude than either itself or the booster would have reached, had either been launched alone.*

At point **A** the rocket is sitting on the launch pad, fully loaded and ready to fly. At point **B** the motor has ignited, and the propellant is beginning to burn. The motor has developed enough thrust to overcome the rocket's weight, and the rocket has lifted off the ground. At point **C** the motor's chamber pressure is rapidly increasing, and so is the amount of propellant burning at each moment. The motor's thrust is also increasing, and the rocket is accelerating upward at an ever-increasing rate.

At point **D** the motor's thrust and the rocket's acceleration have reached their maximums, and the last of the propellant is being consumed. At point **E** the flame has moved into the time delay, and it's burning slowly toward the hole in the forward bulkhead. At point **F** the flame has reached the hole, flashed *through* the hole, and ignited the ejection charge. The ejection charge has popped off the nose cone and pushed out the parachute. Motors with different grain geometries produce different thrust-time curves, but the basic sequence of events is always the same.

Two Stage Rockets

A **two stage rocket** is a vehicle made of *two* rockets stacked on top of one another. The lower rocket, called the **first stage**, or **booster**, fires first, lifting and accelerating the entire assembly until it burns out and falls away. Then the upper rocket, called the **second stage**, fires, and continues on to the intended altitude. Two stage rockets are a challenge to build, and interesting to fly. To an inexperienced observer the rocket goes through a normal lift-off-accelerate-coast sequence. Then to the observer's surprise, the booster drops off and parachutes back to the ground, while the second stage fires, repeats the accelerate-coast sequence, and releases its *own* parachute at a higher altitude.

The advantage of a two stage rocket is that, by the time the second stage fires, it is already high in the air and moving rapidly upward. The result is that the combination reaches a greater altitude than either rocket would have reached on its own. *Figure 2-13* is a cutaway drawing of a two stage rocket as it appears before ignition. If you look closely, you'll see a small pipe extending from the hole in the booster's forward bulkhead into the core of the second stage motor. It's made out of paper, but a soda straw will work. I call it a **flame tube**, and it transfers the fire from the front of the booster into the core of the second stage motor. I'll discuss it later in detail.

In *Figure 2-14* the rocket is lifting off the ground. In *Figure 2-15* the rocket is in full flight, and most of the booster's propellant has been consumed. In *Figures 2-16* and *2-17* the booster and the second stage have separated. The flame tube has transferred the fire from the booster's motor into the second stage motor, The second stage motor has ignited, and the second stage is on its way.

Motor Classification

When working with rocket motors, you need a way of ranking them according to their performance. This book uses the system developed by the model rocket industry in the 1960s. It was originally intended for motors up through **D**, but the makers of high power rocket motors now use it too. It arranges motors according to their total impulse, their average thrust, and the time delay before parachute ejection. In the English system, rocket thrust is measured in pounds, but *this* system is metric, so the thrust is measured in "Newtons". The metric-to-English conversion ratio is 4.45 Newtons per pound, so one Newton equals 0.225 pound.

Figure 2-18 shows a commercially manufactured model rocket motor of the type sold in hobby shops. This one is made by Estes Industries of Penrose, Colorado, and it's classified according to the system described above. Notice the letter-number code stamped directly above the company logo. It says "**D12-3**".

A few pages ago I defined "total impulse". The letter, **D**, refers to this value, and indicates that it lies somewhere between 10.01 and 20.00 Newton-seconds (abbreviated "Nt.-secs."). The number, **12**, *after* the letter, **D**, is the average thrust of the motor in Newtons. 12 Newtons in the metric system equals about 2.7 lbs. in the English system, so the motor's average thrust is 2.7 lbs. Most commercial model rocket motors use a small, pyrotechnic charge to push out the rocket's parachute. The final number in the triad (a **3** in this case) is the time delay in seconds between the end of the motor's burn and the moment of parachute ejection. A **0** (that's a *zero*) means that there is *no* delay, and that the motor is a booster. A **P** means that there is *no* delay, and that the front of the motor is completely *plugged*. A plugged motor does *not* function as a booster, and will *not* eject a parachute. Plugged motors are used in rocket powered gliders, or rockets with clusters of motors, where one of the *other* motors contains the charge needed to fire the second stage or eject the parachute.

*Figure 2-18. A commercially manufactured **Estes D12** model rocket motor (shown larger than actual size). The letter-number code printed on the side indicates the total impulse, the average thrust, and the time delay (in seconds) before parachute ejection.*

I've illustrated the entire classification system in *Figure 2-19* on **page 36**, and I've calculated the total impulse ranges in the English *and* metric systems. As you can see, each time you go up a letter in the alphabet, the motor's total impulse range doubles. For rocket motors using the same propellant, the total propellant weight doubles as well. I've taken it as far as the **J** classification, but as of this writing, motors as large as **P** have been successfully built and flown.

Of course the letter classification for total impulse provides only a *range*, and doesn't give an exact figure. In the example of the Estes motor the question arises as to whether the total impulse is 10.01 Nt.-secs., 20.00 Nt.-secs., or something in between. In 1991 Fred Brennion of Rogers' Aeroscience adopted a logical solution to the problem by placing the number for the motor's total impulse (in Nt.-secs.) in *front* of the letter designation.

Since the Estes motor has a total impulse of 19.60 Nt.-secs., Fred would round that number off to 19, and relabel the motor a **19-D12-3**. Rogers' Aeroscience now uses this labeling method in their rocket performance prediction programs, and the practice is gradually being adopted by other people in the industry.

For my own work, because the type of propellant is important, I give each propellant a letter-number classification, and place *that* figure in front of all the others. If the motor in this example happened to run on the **KG3** propellant, it would be a **KG3-19-D12-6**. But because the motors in *this* book are all homemade, and you can make the time delays any length you wish, I'd drop the number for the time delay, and call the motor a **KG3-19-D12**.

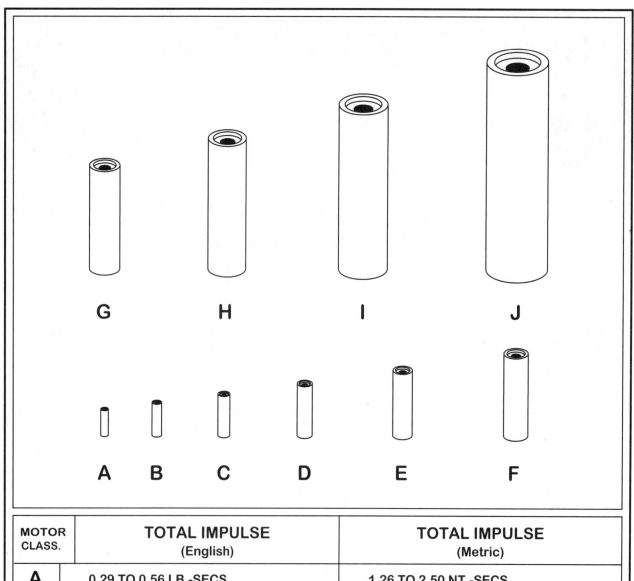

MOTOR CLASS.	TOTAL IMPULSE (English)	TOTAL IMPULSE (Metric)
A	0.29 TO 0.56 LB.-SECS.	1.26 TO 2.50 NT.-SECS.
B	0.57 TO 1.12 LB.-SECS.	2.51 TO 5.00 NT.-SECS.
C	1.13 TO 2.24 LB.-SECS.	5.01 TO 10.00 NT.-SECS.
D	2.25 TO 4.48 LB.-SECS.	10.01 TO 20.00 NT.-SECS.
E	4.49 TO 8.96 LB.-SECS.	20.01 TO 40.00 NT.-SECS.
F	8.97 TO 17.92 LB.-SECS.	40.01 TO 80.00 NT.-SECS.
G	17.93 TO 35.96 LB.-SECS.	80.01 TO 160.00 NT.-SECS.
H	35.97 TO 71.92 LB.-SECS.	160.01 TO 320.00 NT.-SECS.
I	71.93 TO 143.83 LBS.-SECS.	320.01 TO 640.00 NT.-SECS.
J	143.84 TO 287.65 LB.-SECS.	640.01 TO 1280.00 NT.-SECS.

Figure 2-19. The rocket motor classification system used by the commercial model rocket and high power rocket industries. Each new letter in the alphabet doubles the size of the previous motor. This chart shows total impulse ranges up through J, but the system can be extended indefinitely.

3. TOOLS YOU BUY

Weighing Equipment

When weighing the chemicals for a rocket propellant, you must *always* use an accurate scale because any deviation from the intended formula will make the rocket motor perform poorly. When I started this work I experimented with common spring scales, but the results were inconsistent and disappointing. Experience has taught me that, for propellant batches smaller than 2 kilograms, the scale should be accurate to 1/10 of a gram.

● A Triple Beam Balance

The most practical and cost-effective scale is a **triple beam balance**, and the most popular ones are made by a company called "Ohaus". You can buy a *new* balance from any scale or lab equipment dealer, and you'll find used balances at auction sites on the Internet. At the time of this writing the most popular site is **www.ebay.com**. Type the name, "Ohaus", into the eBay title-search engine. If you've taken a class in chemistry, you already know how to use a triple beam balance. For those of you who don't, here is a brief explanation.

● Adjusting the Balance

A triple beam balance gets its name from the three sliding weights that make it work. The weights can be moved back and forth along three separate scales, or "beams", that read from 0 to 10 gms., 0 to 100 gms., and 0 to 500 gms. Before you can use the balance you have to adjust it to zero. To do so, place it on a solid, level surface, and remove everything from the sample pan. Then move the three weights back to their zero points (*Figure 3-1*). With a gentle, downward tap, start the pointer at the right end of the balance rocking, and watch where it comes to rest. If the scale is properly adjusted, the pointer will stop on the alignment mark to the right. If the scale is *out* of adjustment, it will come to rest a little above or below the mark (*Figure 3-2*).

To zero the balance, look for the small, chrome knob under the sample pan at the left end of the balance (*Figure 3-3*). This knob is actually a small weight, and screwing it in *toward* the center of the balance makes the pointer drop. Screwing it out *away* from the center makes the pointer rise. Using a process of trial and error, screw the knob in and out, tapping the pointer each time until it comes to rest on the alignment mark. When the pointer and the mark line up properly (*Figure 3-4*), the balance is ready to use.

● Operating the Balance

When weighing powdered chemicals, you need something to put them in. An empty, plastic, soft margarine tub works fine, but you must *always* remember to subtract the weight of the tub from the scale reading to obtain the actual weight of the chemicals *in* the tub. For this reason, for future reference, you need to weigh the tub ahead of time, and write that weight *on* the tub with an indelible marking pen. The procedure for doing this goes as follows.

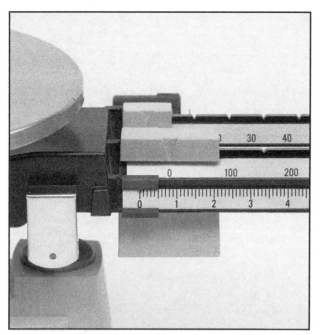

Figure 3-1. To zero a triple beam balance, first move the three sliding weights to their zero points.

Figure 3-2. If the balance is out of adjustment, the pointer at the end of the beam will come to rest either above or below the alignment mark.

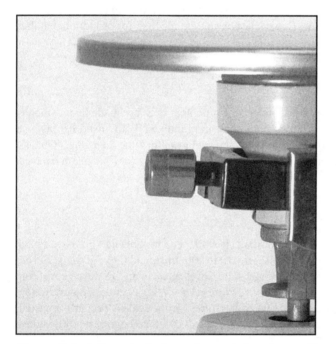

Figure 3-3. Adjust the balance by screwing the small chrome knob under the sample pan in and out as required.

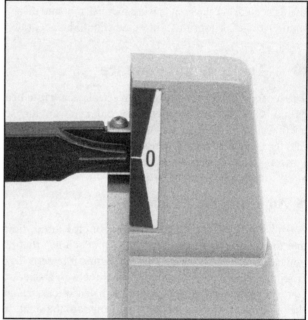

Figure 3-4. When the pointer comes to rest on the alignment mark, the balance is properly adjusted and ready to use.

Place the tub on the sample pan, and move the largest weight to the 100 gm. notch. This makes the pointer drop, so you know that the tub weighs *less* than 100 gms. Move the large weight back to 0, and try the same thing with the medium-size weight. Slide it a notch at a time to the right until the pointer drops again. In *Figure 3-5* you can see that placing it at the 50 gm. notch first causes this to happen. This means that the tub weighs somewhere between 40 and 50 gms.

To find out *exactly* what the tub weighs, move the medium-size weight back to the 40 gm. notch. Then slide the smallest weight back and forth until the pointer comes to rest on the alignment mark. To calculate the weight of the tub, add up the readings from all three scales. In *Figure 3-6* you can see that the reading on the 0 - 500 gm. scale is 0 gms. The reading on the 0 - 100 gm. scale is 40 gms., and the reading on the 0 - 10 gm. scale is 4.7 gms. The tub in this example therefore weighs (0 + 40 + 4.7) = 44.7 gms.

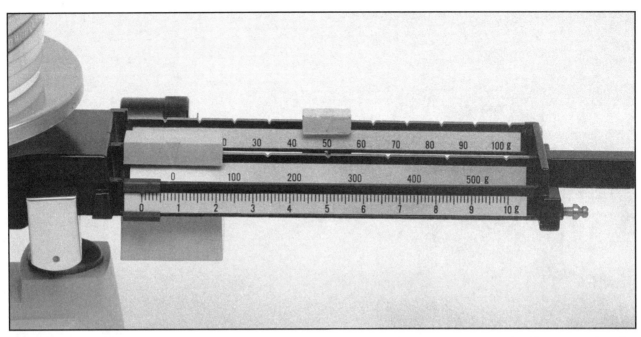

Figure 3-5. Moving the medium size weight to the 40 gram notch has no effect, but moving it to the 50 gram notch makes the pointer drop, so you know that the tub weighs between 40 and 50 grams.

To weigh out a specific amount of chemical, place the tub on the sample pan, and set the sliding weights for the desired weight *plus* 44.7 gms. Then fill the tub with the chemical until the pointer comes to rest on the alignment mark.

EXAMPLE 3-1:

You wish to weigh out 375.0 gms. of potassium nitrate. The total weight of the nitrate plus the tub will be:

375.0 gms. + 44.7 gms. = 419.7 gms.

Set the sliding weights so that they total 419.7 gms. Then place the tub on the scale, and fill it with potassium nitrate until the pointer comes to rest on the alignment mark. The tub now contains exactly 375.0 gms. of potassium nitrate.

● **Helpful Accessories**

Ohaus sells an optional accessory called a **tare beam**. It amounts to a fourth beam with its own sliding weight that you can move to the right or the left to compensate for the weight of the tub. When properly adjusted, you can dispense with the arithmetic, and make direct readings of your weight measurements. Another optional accessory is a sample scoop with its own carefully calibrated counterweight. You place the scoop on the sample pan, hang the counterweight on a peg at the *right* end of the balance, and the net result is the desired reading of zero. Forever thereafter, you use the *scoop* as the chemical container, and as with the tare beam, you can make direct measurements without having to account for the weight of the container.

The basic Ohaus balance weighs up to 610 gms. (about 1-1/3 lbs.). If you need to weigh more than 610 gms., you can buy a set of attachment weights that hang on a *second* peg at the right end of the balance. With these added weights the capacity increases to 2610 gms. (about 5-3/4 lbs). My personal experience has been that, though I own a set of these weights, unless I am mixing a 1000 gm. batch of propellant, I don't use them.

At the time of this writing a basic Ohaus triple beam balance (*without* accessories) costs about $125. If you go to the classified ads and the auction sites on the Internet, you can buy them second hand for much less.

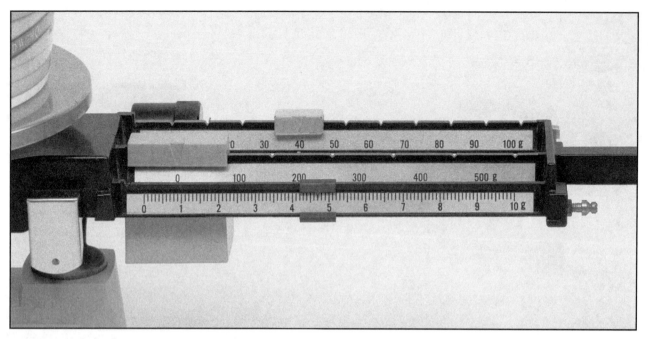

Figure 3-6. To find the weight of the tub, add up the readings from each of the three sliding weight scales. In this case the answer is 44.7 grams.

● Reloading Scales

Reloading scales (*Figure 3-7*) are used by sportsmen to weigh out small amounts of rifle and pistol powder, and you can buy one at a gun shop or a sporting goods store. Reloading scales are inexpensive, but their capacity is limited to 100 gms. or less. This is alright if you are mixing 50 to 100 gram batches of propellant, but for anything more, reloading scales will *not* work. To weigh 500 gms., you can weigh out five 100 gm. doses, but you'll *probably* build up a cumulative error, and the error *could* be beyond the limits acceptable for your work. If you already own a reloading scale, go ahead and try it. If it *doesn't* work, then buy something larger.

● Other Scales

Other suitable scales are **Harvard-Trip balances**, and **electronic digital scales** (if they are accurate to 1/10 of a gram). The Harvard-Trip devices are mechanical like their triple beam cousins, but less convenient to use. They are primarily designed for comparing the weight of one thing to another. Without an accurate set of comparison weights, their capacity is limited. For people on limited budgets, I've designed a **homemade centigram balance** that weighs up to 200 gms. with an accuracy of 1/100 of a gram. It is *ten times* as accurate as a triple beam balance, and it costs $15 to make. For a nominal fee you can download a complete set of plans from the **Teleflite Corporation** website at **www.teleflite.com** or **www.amateur-rocketry.com**. The electronic scales are the most convenient. The new ones are expensive, but like the triple beam balances there is a large market in used equipment. Electronic scales can be zeroed with the push of a button, and because you don't have to wait for a beam to come to rest, you can work *much* faster than you can with a mechanical scale.

● Scale Maintenance

All *mechanical* scales have to be serviced periodically to maintain their accuracy. If you don't know how to do it, a scale dealer can do it for you. It usually involves disassembly, a cleaning of the scale's critical parts, and a recalibration of the weights that make it work. Keep your scale indoors. Keep it clean, and keep it covered when you're not using it. Treating it gently will minimize the need for service.

Powder Mills

● The Manufacture of Commercial Black Powder

Anyone interested in rockets is usually curious to know how commercial black powder is made, so I'm starting this discussion with a brief description of the process that I've condensed from a chapter in *The Chemistry of Powder and Explosives* by T.L. Davis. It goes approximately as follows.

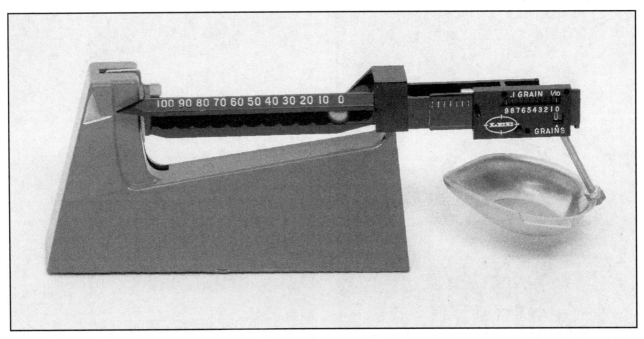

Figure 3-7. *A reloading scale is very accurate, but its capacity is limited to 50 or 100 grams. It is suitable **only** for weighing very small amounts of chemical.*

225 pounds of potassium nitrate, 45 pounds of charcoal, and 30 pounds of sulfur are crudely mixed in a large, rotating barrel, then dampened with water, or stirred into a thick hot water solution, and spread on the floor to cool and dry. The resultant, damp mass is placed in a large, donut-shaped channel, and crushed for 3 hours under two, revolving, *ten ton*, iron wheels. This thick, damp paste, called "clinker", or "wheel cake", is hydraulically pressed at 1,200 pounds per square inch between heavy aluminum plates, resulting in a series of "press cakes" about 2 feet square and 3/4 of an inch thick.

The press cakes are cracked and granulated between crushing rollers in a device called a "corning mill". The results are sieved, dried, and rounded by tumbling again in a wooden barrel with a small amount of graphite to blacken and polish the grains. The finished product is then separated into different grain sizes, and packaged for sale. As the corning mill operates, it generates a by-product of fine, black powder dust, which is collected, packaged, and sold separately as the "meal powder" used in fireworks.

● Can I use a Mortar and Pestle?

The equipment needed to do this is beyond the means of an amateur, and when people read about it, they often ask if they can accomplish the same thing with a **mortar and pestle**. The answer is ***no***. A mortar and pestle (*Figure 3-8*) is a small, hand operated device. To mix even an ounce of material thoroughly enough would take at least an hour of meticulous, hand grinding. Also, it is *absolutely* vital that the consistency and burn rate of your propellants be identical from batch to batch. Because a mortar and pestle is worked *by hand*, it is *impossible* to achieve consistent results.

● Can I use a Blender?

People also ask if they can powder the chemicals in a **blender** (*Figure 3-9*), and *again* the answer is ***no***. In the process of developing the propellants for my first book, I experimented extensively with blenders. Blenders are designed to work with liquids, and *not* dry powders. The blades that do the work are mounted on a shaft that fits through a bearing in the bottom of the blender jar. Liquids lubricate the bearing, but dry powders are abrasive, and quickly *destroy* the bearing. Also, a blender pulls the liquid down through the center of the blade assembly. Then it forces it up the sides of the jar in a continual donut-shaped cycle. Long before a chemical is sufficiently powdered to be useful it begins sticking to the sides of the jar. After a few minutes *everything* sticks to the jar while the blades spin uselessly in open air. You can turn the blender off, reach down with a stick, poke the material *carefully* back down into the blades, and turn the blender back on. But within seconds the powder is sticking again. If you repeat this maneuver several dozen times, you'll end up with a few ounces of something that *looks* usable. But despite its appearance, its quality and grain size will be inconsistent from batch to batch, and it will therefore be *useless* for making a rocket propellant.

*Figure 3-8. A mortar and pestal is **not suitable** for mixing rocket propellant.*

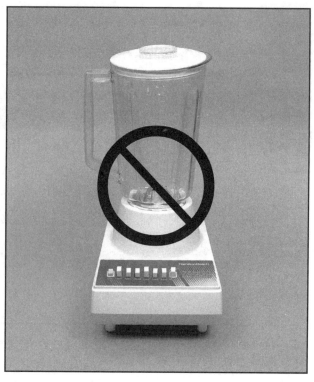

*Figure 3-9. A kitchen blender is **not suitable** for mixing rocket propellant.*

If you *do* experiment with a blender, *never* use it for mixing chemicals together. Use it *only* for powdering the individual ingredients, and keep a supply of spare bearings and blade assemblies on hand to replace the ones that wear out.

WARNING! *NEVER* **USE A BLENDER TO MIX PROPELLANT CHEMICALS TOGETHER! THE FRICTION BETWEEN THE DRY CHEMICALS AND THE BLENDER'S MOVING PARTS WILL GENERATE ENOUGH HEAT TO IGNITE THE PROPELLANT,** *AND THE BLENDER WILL EXPLODE***!**

Ball Mills and Rock Tumblers

A **ball mill** is a jar with a tight fitting lid, partially filled with metal or ceramic pellets, and laid on its side on a motorized base. As the jar revolves, the pellets inside the jar tumble over one another. If you add a mixture of dry chemicals to the jar, the tumbling pellets will grind them to a powder, and mix them together. Ball mills come in many sizes ranging from the tiny mills that powder pharmaceuticals to the giant mills that crush metal ores in the mining industry.

Before World War II ball mills were strictly *industrial* tools and were far too expensive for the hobby market. *After* The War a dramatic increase in leisure time resulted in the birth of many new hobbies including the hobby of rock collecting. Since many of the new-found rocks were of near-gem quality, the collectors (aka "rock hounds") started making jewelry out of them, and by 1955 a new industry had evolved to supply the necessary tools. When the tool makers realized that a ball mill could be used to polish these stones, they developed an inexpensive mill called a **rock tumbler** that sells on the *hobby* market for 1/4 to 1/10 the price of an industrial mill.

The burn rate of a powdered propellant is partially determined by how thoroughly it is mixed, and I've already emphasized that the burn rate has to remain constant from batch to batch. A rock tumbler's electric motor turns at a fixed and constant

Tumbler Size	Approx. Jar Volume	Number of Pellets	Weight of Pellets	Pellet Dimensions	Propellant Batch Size
"Rolling Stones" Toy Tumbler	1-1/2 cups	26 pellets	12 oz.	1/2" x 1/2"	93.8 grams
1-1/2 Lb.	1 pint	34 pellets	1 lb.	1/2" x 1/2"	125 grams
3 Lb.	1 quart	68 pellets	2 lbs.	1/2" x 1/2"	250 grams
4-1/2 Lb.	1-1/2 quarts	31 pellets	3 lbs.	3/4" x 3/4"	375 grams
6 Lb.	1/2 gallon	41 pellets	4 lbs.	3/4" x 3/4"	500 grams
12 Lb.	1 gallon	82 pellets	8 lbs.	3/4" x 3/4"	1 kilogram

Figure 3-10. Comparison chart for rock tumblers. This chart lists the approximate jar volume, the recommended size and number of milling pellets, the weight of the pellets, and the size of the propellant batch that each machine will make.

rate, so the jar containing the chemicals makes an exact number of revolutions every hour. Assuming that all *other* parameters are fixed, if you mix two batches of propellant in a rock tumbler, and run each batch for the same length of time, the burn rates of both will be the same. A rock tumbler is therefore *ideal* for making powdered rocket propellant, and with the exception of the **KS** propellant, *all* the propellants in this book should be made in a rock tumbler.

There are many brands on the market, but because of their low price, their good engineering, and their solid rubber jars, I like the ones made by **Lortone Inc.**, **2856 N.W. Market St.**, **Seattle**, **WA 98107**. As of this writing I've used three of their tumblers for 12 years, and I've run each one for *at least* a thousand hours without a breakdown. I like them so well that, at least for now, my company, **The Teleflite Corporation**, features them exclusively in its catalogue of rocket making supplies. Of course as time passes, circumstances might change, but as of the year, 2004, you can order a Lortone tumbler from Teleflite's website at **www.teleflite.com**, or **www.amateur-rocketry.com**.

To find a Lortone tumbler where you live, look in *The Yellow Pages* under **LAPIDARIES**, **LAPIDARY EQUIPMENT**, or **ROCK SHOPS**. Or contact Lortone, and ask them for the name and phone number of the nearest dealer. Should Lortone ever go out of business, there are many other brands. If you can't find a "rock shop", check with department stores and toy stores. For used machines, try the Internet auction sites like **www.ebay.com**.

Rock tumblers come in many sizes, and the rock-hobby industry ranks them according to how many pounds of gemstones they can process. The dealers who sell them call them "3 pound tumblers", "12 pound tumblers", etc. But because *you* want to know how much *rocket propellant* a tumbler will make, I've drawn up the chart in *Figure 3-10* that compares tumbler "poundage" with jar volume and propellant production for the most popular sizes. *Figure 3-11* shows two of the tumblers made by Lortone, and *Figure 3-12* is a photo of a small but usable machine that I bought at Toys 'R' Us.

● How Big a Tumbler Should I Buy?

A half pound of propellant will make eight **D** motors, so a 3 pound tumbler with a 1 quart jar is sufficient for making anything *up* to a **D**. If you want to make bigger motors, you'll need a bigger machine. A **G** motor (including the time delay) takes 5 to 6 ounces of propellant. With two pounds of propellant, you can make 5 or 6 **G**s, so a 12 pound tumbler with a one gallon jar would then be a good choice.

Of course the **KS** propellant is mixed by hand, and you can powder the potassium nitrate by precipitation with alcohol (**Chapter 7**). My best advice is to read this book first, decide which motors you're going to make, and *then* buy a tumbler that meets your needs.

Figure 3-11. Two rock tumblers made by Lortone Inc. of Seattle, WA. The Model 3A (on the left) makes 250 grams. of propellant per batch. The Model QT-12 (on the right) makes 1 kilogram.

● Setting up a Powder Mill

Important note. Beginning with this page, I'll talk about rock tumblers *strictly* in terms of their ability to process powdered rocket propellant. So from this point on I'll refer to all rock tumblers as "powder mills." Experience has taught me that to *thoroughly* powder and *thoroughly* mix the chemicals, the milling pellets have to be *heavy*. And they also have to be *hard*, so that they don't abrade away and contaminate the finished product.

> **And especially important in this case, the pellets must be *non-sparking*. A spark generated inside the mill will ignite the propellant, and cause the mill to explode!**

For making the pellets, *metal* is an obvious choice, **but you can't use iron or steel, because things made of iron and steel generate *sparks* when struck together.** Copper works, but it's very soft, and it's hard to find in barstock form. Aluminum is too light, and lead is *so* soft that it quickly grinds away and contaminates the finished propellant. In the presence of the propellant's oxidizer, the powdered lead turns to *lead oxide*. Lead oxide is **toxic.** Your tools, your shop, your clothes, *and your rocket motors' exhaust* will be contaminated with lead oxide, a known cause of **lead poisoning**.

The best choice is **brass** (made of copper and zinc). Machine shops make things out of brass, and the places where they buy their metal usually sell brass. To find brass where you live, go to *The Yellow Pages*, and look under **METALS**. You'll need either 1/2" dia. or 3/4" dia. round, brass barstock. If you can't find a dealer in the phone book, call a machine shop, and ask them where *they* buy it.

The length and the diameter of each pellet should be equal, so make the 1/2" dia. pellets 1/2" long, and make the 3/4" dia. pellets 3/4" long. If you have a metal cutting band saw, you can cut the pellets yourself. If you are *very* ambitious with a hand hack saw, you can make them yourself. Otherwise, the place where you buy the brass will *usually* cut them for you for an additional fee; typically about half-again what you paid for the metal.

Before you can use the pellets, you have to deburr them. You can do this with a hand file, or by running their edges against a belt sander, and you also have to clean them thoroughly. Industrial metals are shipped with a light coating of oil to prevent corrosion. Hand scrubbing with hot water and detergent works, or you can rinse them in acetone or lacquer thinner.

Figure 3-12. A "Rolling Stones" rock tumbler. As of 2002, it is made by Natural Sciences Industries Ltd. of Far Rockaway, New York. Small but effective, it makes 94 grams of propellant per batch. I bought this one new at Toys "R" Us, but look for them at yard sales for $10 or less. The top edge of the tumbler jar has some plastic mold marks that should be sanded off to make a better seal. Also, apply a light coating of Vaseline or grease to the lid's rubber O-ring.

Referring back to *Figure 3-10*, you'll find a complete list of the *number* of pellets, the total *weight* of the pellets, and the *size* of the pellets needed for each powder mill, and it's probably not advisable to second guess these figures. To save a little money, you can try brass balls, or brass nuts or bolts from a scrap yard, and these things will work. But they will *probably* mix the propellant either slower or faster than the cylindrical pellets, and you'll have to experiment with your milling times to match my results.

To specify an *exact* milling time and to turn the mill on and off on schedule, you'll need a **small appliance timer**. Small appliance timers cost $5 to $10. They are used to turn the lights on and off when you're not at home, and you'll find them at hardware-lumber yards like Home Depot and Lowe's, and big department stores like Kmart and Walmart. Look in the electrical or housewares department, or ask one of the clerks. *Figure 3-13* shows a complete powder mill with the pellets, the open jar ready for the chemicals, and the motorized base plugged into the timer and ready to go.

● Powder Mill Maintenance

To keep a mill running on time you have to keep it clean and lubricated. The motors that drive these mills use sleeve-type bearings. When the bearings dry out or get gummed up, the motors slow down, and *that* degrades the milling process. You must also keep a mill's rollers clean and lubricated. These parts *also* use sleeve bearings that will drag on the motor and slow it down if they get dirty or dry.

Rubber and plastic drive belts stretch after a while, and it's important to replace them when they wear out. At the time of this writing, the Lortone model QT-6, QT-66, and QT-12 machines use cogged, v-shaped timing belts that eliminate this problem, but many other brands use standard, rubber O-rings. If your mill *doesn't* use a timing belt, you *might* be able to find one that fits. You can buy cogged timing belts from the places that sell pulleys, gears and bearings. Check *The Yellow Pages* under the **GEARS** and **BEARINGS** headings.

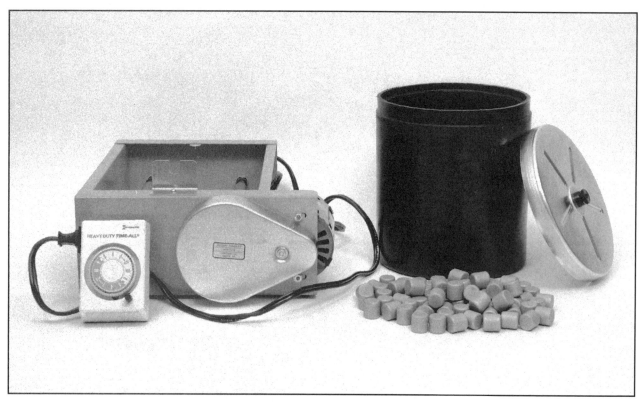

Figure 3-13. A complete powder mill setup with a rock tumbler, the tumbling pellets, and the timer, all plugged in and ready to go.

After years of operation, a *Lortone* mill's rubber jar becomes hard and polished, and it starts to slip. The fix for the problem is **fan belt dressing**, and you can buy it at an auto parts store. The dressing that I use comes in a spray can, and I get it at **Pep Boys**. Unlike most brands, it's a clear liquid (*not* black and sticky). I wipe the jar with a light coating, and let it dry for an hour. Special chemicals in the dressing soften the rubber and restore its natural gripping qualities.

If, after several treatments, you see a black, rubber residue sticking to the mill's rollers, you can wipe it off with a rag soaked in acetone. But be careful not to wipe too vigorously or too long. Acetone softens the surface of the plastic from which the rollers are made. If you do this carelessly, you can damage them.

The brass pellets will last for many years, and they'll clean themselves as they mix and grind the chemicals. But if you stop using the mill for any length of time, wash and dry the jar and the pellets before placing them in storage. If you don't do this, in the presence of residual oxidizer, the pellets will develop a crusty, black oxide coating. To remove this coating, put the pellets back in the mill, fill the jar half full of water, add a handful of sand, and run the mill for an hour. The sand will grind away the coating, and the pellets will come out clean.

Vacuum Pumps

Important note. You do *not* need a vacuum pump to make the rocket motors in this book. All but 6 of these motors run on propellants that contain a glue called a "binder". During the loading process you soften the binder with a small amount of solvent. Before a motor can be fired the solvent has to evaporate, and these motors will *rapidly* dry in open air. When acetone is used, the smallest take 12 hours, and the largest take 72. *If you're in a hurry*, and you want to use a motor on the same day that you made it, you can reduce the drying time to as little as 20 minutes with a vacuum pump.

Important note. The pump you buy *must* be the high vacuum or refrigeration type. The diaphragm pumps and the combination compressor/vacuum pumps *will not work*.

Figure 3-14 is a photo of a small, belt-driven pump. It's a Welch model 1400, and it's typical of the small pumps used in chemistry and science labs. *Figure 3-15* is a photo of a Lammert direct-drive pump. Direct-drive pumps are used in laboratories too, but because they are more compact, and therefore portable, they are also popular with the people who service air conditioners and refrigeration equipment.

Figure 3-14. *A Welch model 1400 belt driven vacuum pump. The model 1400 costs $1,200 new, but I found this one, and an even bigger pump in a "leak detector" that I bought at an auction for $100.*

New vacuum pumps are expensive. The year, 2000, W.W. Grainger Industrial Equipment catalogue lists a Welch 1400 for $1268.00, and Grainger's equivalent of the Lammert costs $742. Grainger's cheapest is a 3 cfm direct drive pump made by Dayton for $316.45. Grainger's has more than 350 locations throughout the country, and as of this writing, they have a website on the Internet at **www.grainger.com**. If you live near a big city, you'll *probably* find a Grainger outlet in the business section of the city's phone book. Grainger's sells only to businesses, and *not* to the public. If you don't own a business, find a friend who does, or see if the place where you work will deal with them for you. If you're shopping for a *new* vacuum pump, get your first quote from Grainger's. Then look in ***The Yellow Pages*** under **VACUUM EQUIPMENT**, and see if anyone can beat their price.

Fortunately there's a large market in *used* vacuum pumps, and if you know how to shop, you can save a *tremendous* amount of money. Start by calling the people who service and repair vacuum pumps. Look in ***The Yellow Pages*** under headings like **VACUUM EQUIPMENT, SERVICE & REPAIR.** These places can often sell you a rebuilt pump for half the price of a new one. A friend of mine who repairs and installs vacuum pumps for a living tells me that the Welch and Kinney belt drive pumps are good, and among the direct drive pumps, Leybold-Hereaus and Trivac are the best.

For *used* pumps, check the industrial surplus stores and the Internet auction sites like **www.ebay.com**. Shops that are going out of business often sell pumps for 10 to 20 cents on the dollar. If you're savvy about used scientific equipment, you can buy vacuum pumps for even less. Start attending aerospace and scientific equipment auctions, and look for a "vacuum station" or a "leak detector". Both devices are about the size and shape of a washing machine, and unless they've been cannibalized, they'll have at least one good pump inside. People who don't know this often pass them up. The Welch pump in the photo came from a leak detector that had *two* pumps, and I paid $100 for the whole thing.

Important note. *Many vacuum pumps have seals made of a rubber called "buna" (aka "nitrile")*. Alcohol has no effect on buna, but acetone *dissolves* buna, so pumps with buna seals *cannot* be used to evaporate acetone. It is therefore important that, before you buy a pump, you read the rest of this book, and decide which solvent you're going to use. Motors made with acetone dry the fastest, but if you plan to use a vacuum pump, the pump's seals *must* be made of a special rubber called "**Viton**". Before you buy a pump, call the manufacturer, and *verify* that the seals are made of Viton. If they are *not* made of Viton, you'll have to replace them. Anyone who works on vacuum pumps can do it for you. If you tell them that the pump will be used to evaporate acetone, they will *immediately* know what to do.

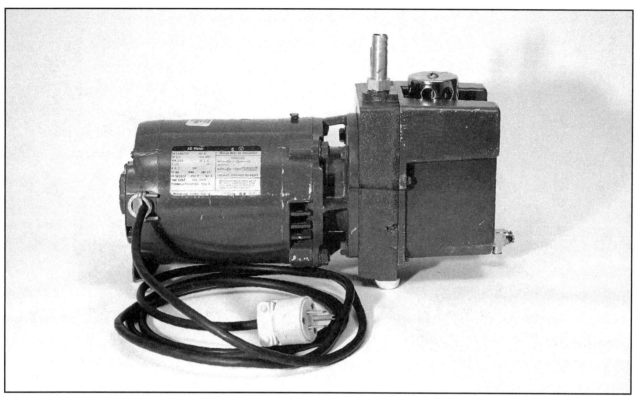

Figure 3-15. *An example of a small, direct-drive vacuum pump. I bought it from a machine shop that was going out of business. This model is no longer made, and some of the parts are not available, but I paid just $35 for it, so it was still a good buy.*

Disposable Syringes

Before the propellants are loaded, all but the **KS** propellant must be dampened with a small but *exact* amount of solvent. The best tool for measuring and dispensing the solvent is a **disposable veterinary syringe**. (*Figure 3-16*). To find these syringes, look in *The Yellow Pages* under **VETERINARY SUPPLIES**.

Vet-supply dealers sell everything from animal medicines to special feeds and tack. They normally sell disposable syringes in sizes from 3cc. to 60cc., and you can buy them singly, or in large boxes. They are made of materials that are impervious to acetone, so they're ideal for this application. But before you can use a syringe, *you must take the following precaution.*

WARNING! DISPOSABLE SYRINGES HAVE *VERY* SHARP NEEDLES THAT *MUST* BE DULLED BEFORE THEY CAN SAFELY BE USED IN A WORKSHOP ENVIRONMENT.

You can do this with a file or a piece of sandpaper. *Figure 3-17* illustrates the situation. The needle on the left appears as originally purchased, and you can see the sharp and dangerous point. The one on the right has been flattened by rubbing it on a piece of sandpaper. It is now *reasonably* safe, and ready to use. The syringes that I buy are made by **Monoject**, a division of **Sherwood Medical** in St. Louis, MO. Some come with needles, and others require that you buy the needles separately.

● How to use a Syringe

To get an accurate measurement from a syringe you have to fill it properly, and here's how to do it. Dip the tip of the needle into the liquid (*Figure 3-18*), and draw in *more* liquid than you need by pulling back on the plunger.

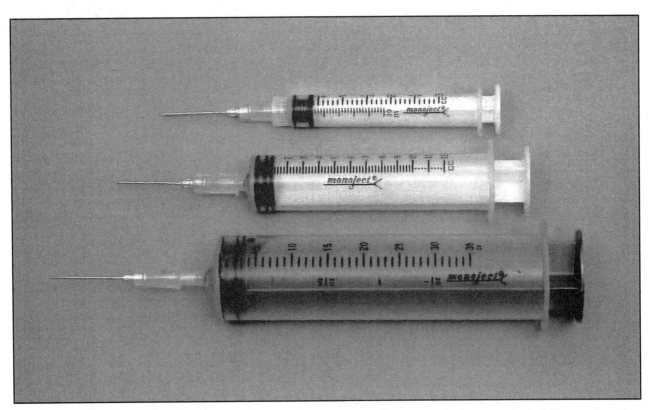

Figure 3-16. Disposable syringes are used to accurately measure small amounts of acetone or alcohol. You can buy them from a veterinary supply dealer. The sizes you need will depend upon the size of the motors you are building.

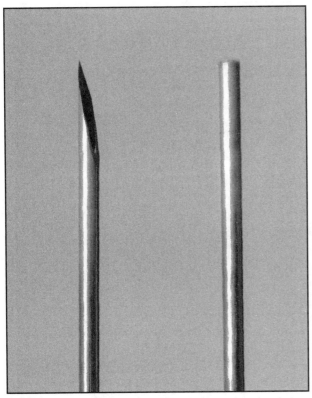

*Figure 3-17. To make a syringe's sharp pointed needle safe, it **must** be dulled with sandpaper or a file. The one on the left is brand new, and you can see the potential hazard. The one on the right has been sanded as described, and is now safe to use.*

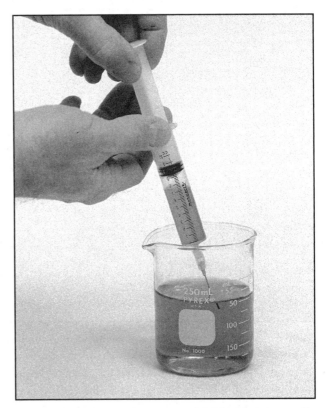

Figure 3-18. To fill a syringe, hold the tip of the needle under the surface of the liquid, and draw in more than you need by pulling back on the plunger.

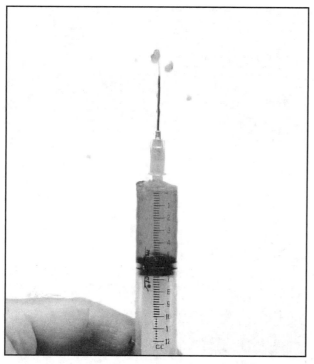

Figure 3-19. After filling the syringe, hold it upright, and rap on the side with your fingernail to loosen any trapped air. Then force the air out by pushing up on the plunger until nothing but liquid comes out.

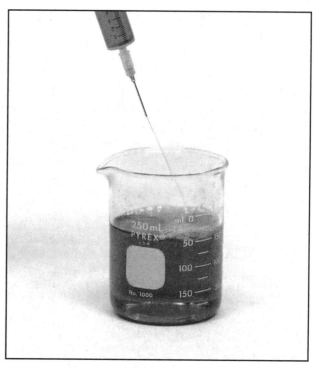

Figure 3-20. Then hold the syringe over the solution container, and force out any excess liquid over and above the amount that you need.

Figure 3-21. Polyethylene cups are useful, because they are flexible and unaffected by acetone or alcohol.

Then hold the syringe in front of you with the needle pointing *up*, and rap on the body with your finger nail to loosen any trapped air. With the syringe still in the vertical position, force the air out by pushing *upward* on the plunger until the liquid begins to flow (**Figure 3-19**). Then point the syringe *down* over the solution container, and force out any *excess* liquid over and above the amount that you need (**Figure 3-20**).

As of this writing, even when purchased singly, most of the Monoject syringes cost less than a dollar, and you can use each one 50 to 100 times before it wears out. The size you need will depend on the amount of solvent you'll be measuring, and *that* will depend on the size of the motors you are building. Some trial and error will be necessary.

Figure 3-22. A collection of other useful things; all found at a supermarket and a hardware store.

Figure 3-23. A plastic kitchen colander and a bucket are used to separate the finished propellant from the milling pellets.

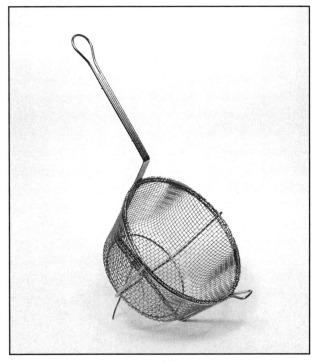

*Figure 3-24. A stainless steel french fry basket separates the chemicals **much** faster than a colander. I bought the one in this photo at Smart & Final Iris.*

Mixing Cups

Mixing cups are used for blending the solvent into the rocket propellant prior to loading. The ones I use are made of polyethylene plastic. I like polyethylene because it is flexible and unaffected by alcohol or acetone. *Figure 3-21* is a photo of *two* polyethylene cups. I bought the one on the left from an industrial plastic dealer, and I bought the one on the right at a variety store. For *large* amounts of propellant, I use a large, soft margarine or whipped cream container. I'm not sure about other brands, but the ones that come with Imperial Margarine, Blue Bonnet Margarine, and Cool Whip work fine.

Other Handy Things

Other handy things (*Figure 3-22*) are oven or refrigerator racks, spoons, forks, stainless steel pots, measuring cups, sieves, a kitchen colander, plastic buckets, and those little bamboo Bar-B-Que skewers that you buy at the supermarket. A common table fork (the kind you eat with) is used to mix the solvent into the rocket propellant. During my first day of motor building, I developed a blister on one of the fingers of the hand that I used to hold the fork. I solved the problem by wrapping the handle with vinyl electrical tape as shown in the photo. If you do this before you start, you'll save yourself some discomfort.

To separate the finished propellant from the milling pellets, you sift it through a plastic colander into a bucket. If you buy the right size colander and bucket (*Figure 3-23*), the two will fit together nicely.

Plastic colanders are cheap, and the work alright, but they sift *slowly*, and they don't last very long. Restaurants equipment dealers sell a stainless steel tool called a **french fry basket** (*Figure 3-24*). I bought the one in the photo at **Smart & Final Iris**. These baskets are expensive ($17 for the one in the picture), but they are *very* sturdy, and because the mesh is so big, they sift very quickly.

4. TOOLS YOU MAKE

Figure 4-1. *A nozzle mold centers the core spindle, and forms the nozzle's divergent exit taper. The one in this photo was made on a metal lathe. If you don't have a lathe, you can improvise with things like steel washers, or pieces of round aluminum barstock.*

General Comments

The rocket motors in this book are built with hand tools, and this chapter explains how to make them. The best tools are made of steel or nylon on a metal lathe. To make a top quality casing retainer, you'll need a milling machine and a welder. Properly made tools last the longest, and a properly made nozzle mold produces motors that actually work better.

For people who *don't* have metal working equipment, I'll explain how to improvise with things like plastic pipe, hose clamps, and wooden dowels. These improvised tools are primitive, but they work just fine, and they'll provide you with enough experience to know if you want to make a further investment in the hobby.

If you want a *better* set of tools, you'll at least need a metal lathe, or you'll need someone with a metal lathe to make the tools *for* you. If you choose the latter, you can pay a machine shop. But first look for someone who *owns* a lathe. Most neighborhoods have at least one hobbyist, or maybe a retired machinist with a lathe in his garage, who will make things for people at greatly reduced prices. If you *can't* find someone with a lathe, check with high schools and community colleges. Many schools offer adult night classes in machine shop, and some of my readers have taken advantage of these classes to make a set of tools, and learn a new skill in the bargain.

Figure 4-2. When the nozzle mold is removed, it leaves behind a divergent exit taper that aids the expansion of the motor's exhaust into the atmosphere.

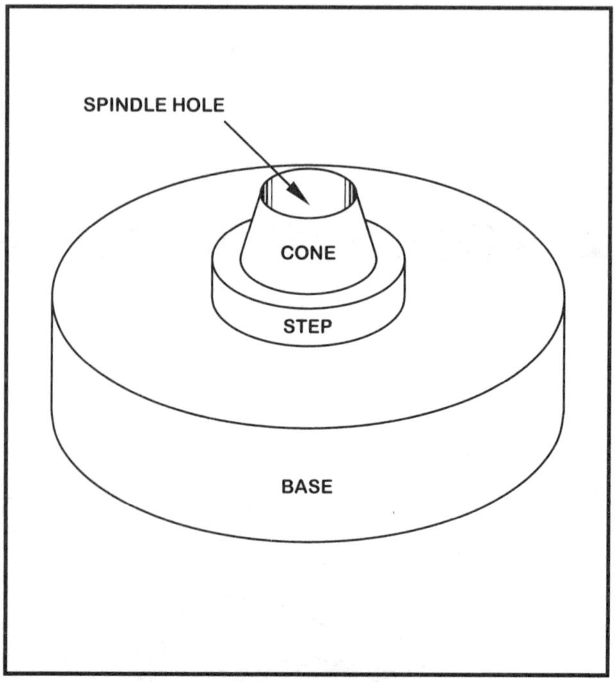

Figure 4-3. The parts diagram for a nozzle mold.

Nozzle Molds

A *clay* nozzle is made by ramming dry nozzle clay against a **nozzle mold** (*Figure 4-1*). When the rocket motor is finished, the nozzle mold and another tool, called a core spindle, are removed. The result is a hard, ceramic bulkhead at the rear of the motor, with a central hole for the rocket exhaust. A female, cone-shaped divergent exit taper (*Figure 4-2*) aids the expansion of the motor's exhaust into the atmosphere, and this facilitated expansion increases the motor's exhaust velocity and thrust.

The best material for a nozzle mold is *stainless steel*. Regular steel is cheaper, but in the presence of the propellant's oxidizer, a tool made of regular steel will quickly rust if you don't keep it clean. You can get by with aluminum or brass, but because these metals are soft, a tool made of aluminum or brass will quickly wear out. Wood and plastic will *not* work, because they are *not* strong enough to withstand the rigors of the ramming and loading process.

Important note. *When shopping for steel, **ask for an alloy that is easy to machine**. When buying regular steel, ask for "cold rolled" steel. When buying stainless steel, ask for "type 304". Type 304 stainless is more expensive than cold rolled, but you only need a pound or two. For a job this small, the added cost will amount to a few dollars at most.*

Important note. *Most metal dealers have bins of what they call "remnants", which they sell at reduced prices. These are the cutoff scraps left over from large, industrial orders. If you can find what you need in a remnant bin, you'll save some money.*

The following instructions explain the setups and procedures that I use to make a nozzle mold, and **Chapter 5** contains *all* the drawings of *all* the molds needed to make *all* the motors in this book. In the pages ahead I refer to a nozzle mold's parts, and I've illustrated these parts in *Figure 4-3*. To make a nozzle mold, refer to *Figure 4-3*, and proceed as follows.

Begin by cutting a piece of round, stainless steel barstock to the appropriate size. The strategy in machine work is to always start with a piece of metal that's a little bigger than the finished product. Then trim away the excess metal during the machining process. In machine shop jargon, the thing you are machining is called "the workpiece".

Figure 4-4. The total height from the bottom of the body to the top of the cone is summed up. The workpiece is trimmed to this length with a series of face cuts.

Figure 4-5. A series of side cuts trims the outside diameter of the step back to the face of the body.

Figure 4-6. Using the cross feed screw, the lathe tool is slowly backed out across the face of the body. This cleans up the body, and machines it to its finished dimension.

Figure 4-7. A series of side cuts generates the outside diameter of the cone's base, and machines it back to the face of the step.

1. Mount the workpiece in a 3-Jaw lathe chuck, and true up the end with a light facing cut. Then turn it around, and true up the opposite end in the same manner. When mounting it in the chuck this second time, use the following procedure. From the drawing, read the thickness of the nozzle mold's body, and mount the workpiece so that a little *less* than that length of material is gripped by the jaws of the chuck.

Figure 4-8. Using the cross feed screw, the lathe tool is slowly backed out across the face of the step. This cleans up the step, and machines it to its finished dimension.

Figure 4-9. With a center drill mounted in the lathe's tail stock drill chuck, the top of the cone is center-drilled.

Figure 4-10. The spindle hole is started with a slightly undersize drill; then finished with the exact size drill indicated in the drawing.

Figure 4-11. The lathe's compound is set at a 15° angle, and locked in place. An eyeball setting is good enough.

When you do this, be sure to mount it *squarely*. Turn the lathe on, and see if it wobbles as it spins. If it *does* wobble, turn the lathe *off*, loosen the chuck, and wiggle the workpiece until it is properly aligned. Some trial and error will be necessary. Then from the drawing, read the total height of the nozzle mold from the bottom of the body to the top of the cone. With a series of face cuts, trim the workpiece back to that length (*Figure 4-4*).

Figure 4-12. The sides of the cone are cut by hand feeding the lathe tool with the lathe's compound screw.

Figure 4-13. When the cone's base meets the face of the step, the cone is finished.

2. With a series of side cuts, machine the outside diameter of the step back to the front face of the body (*Figure 4-5*). If the lathe has an adjustable carriage stop, it will be helpful to use it when doing this.

3. With the lathe's cross feed screw, slowly back the tool out across the face of the body. This cleans up the body, and at the same time, machines the body to its finished dimension (*Figure 4-6*).

4. With a second series of side cuts, generate the diameter of the base of the cone, and machine it back to the face of the step (*Figure 4-7*).

5. With the lathe's cross feed screw, slowly back the tool out across the face of the step. This cleans up the step, and at the same time, machines the step to its finished dimension (*Figure 4-8*).

Figure 4-14. The top of the cone, and all sharp edges are deburred with a file.

6. Mount a center drill in the tailstock drill chuck, and center-drill the top of the cone (*Figure 4-9*). Switch to a slightly undersize drill, and rough-drill the spindle hole (*Figure 4-10*). Then finish-drill the spindle hole with the *exact* size drill indicated in the drawing, and use tapping fluid or coolant to produce as smooth a finish as possible.

7. As shown in *Figure 4-11*, set the lathe's compound at a **15 degree angle** (an eyeball setting is good enough). Cut the sides of the cone by hand feeding with the compound's lead screw, and use the lathe's cross feed screw to advance the tool into the work with each succeeding cut. *Figure 4-12* shows the work half done. Keep cutting until the base of the cone meets the step (*Figure 4-13*).

8. With the lathe turned on and the workpiece spinning, deburr the sharp edge at the top of the cone with a fine file (*Figure 4-14*). Then use the file to smooth the other sharp edges. With the completion of this final step, the nozzle mold is finished.

Figure 4-15. Two metal disks are cut with a hacksaw and a miter box.

Figure 4-16. A 3/8" dia. hole is drilled through the center of each disk.

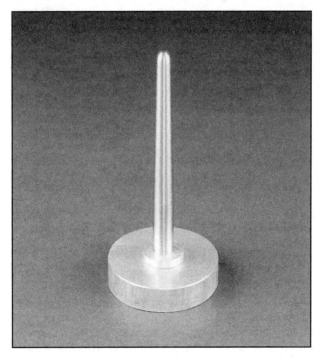

Figure 4-17. The two disks are stacked on the work surface, and the core spindle is inserted.

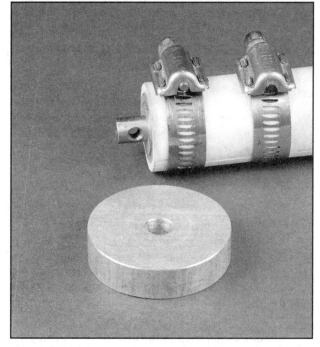

Figure 4-18. The finished motor (in the casing retainer) is lifted off the work surface, leaving the large disk behind.

● Tolerances

Check your measurements with a micrometer or a vernier caliper as you go. A drilled hole is good enough for the spindle hole. An eyeball setting on the compound angle for the cone is fine, and a tolerance of plus or minus .005" is more than adequate for all the other dimensions. The outside diameter of the body is *not* critical; it can remain the original o.d. of the metal you started with. Please note that this is not the only way to make a nozzle mold. This is just the way that *I* do it. If you're already experienced in machine work, you might prefer another approach.

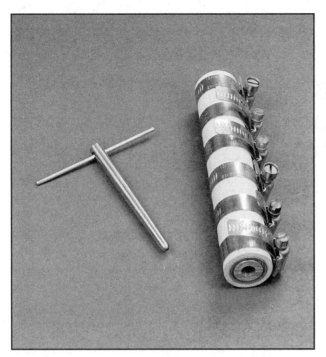 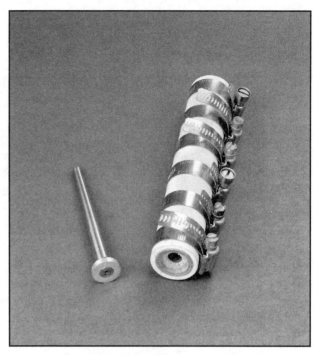

Figure 4-19. With a twisting motion, a T-handle cut from a piece of piano wire pulls out the core spindle.

Figure 4-20. The large end of the spindle is used to wiggle out and remove the small disk.

Improvised Centering Rings

To make a nozzle mold *without* a metal lathe, you can improvise. It is not practical to make an divergent taper cone, but you *can* make a set of **centering rings** for the core spindle. You'll need two round pieces of brass, aluminum, or steel of suitable diameter, a drill press, and a bandsaw or a hand operated hacksaw with a miter box. Here's how to do it.

1. With a hacksaw or a metal cutting bandsaw, cut a 1/8" thick disk from the end of a piece of 3/4" dia., round, metal barstock. Aluminum is the easiest to cut, and though many woodworkers don't know it, you can cut aluminum on a *wood* cutting bandsaw with a *sharp*, standard, wood cutting blade. You can make the cut square by clamping the barstock to the saw's miter gauge, set at a 90 degree angle. Then, using the same tools and technique, cut a 1/2" thick disk from a piece of 2" diameter barstock. If you do it with a hacksaw, clamping the bar into a cheap, carpenter's miter box (*Figure 4-15*) will steady the saw, and insure that the cut is reasonably straight.

2. Using the best means available, locate the center of each disk. Clamp each disk into a drill press vise, and drill a 3/8" diameter hole through its center (*Figure 4-16*).

You can now substitute these two rings for the 3/4" x 3/8" nozzle mold illustrated in drawing number **98-010** on **page 99**. Here's how to use them.

1. Place the large ring on the work surface, and lay the small ring on top of it. Insert the core spindle (*Figure 4-17*). Press the motor casing (inside the casing retainer) down over the assembly, just as you would if you were using a nozzle mold. Then form the nozzle, and load the motor in the usual manner. Due to the lack of a divergent taper, motors made in this way will take a little more clay, so check your depths as you work, and verify that all the motor's internal dimensions are correct.

2. When the loading process is finished, lift the motor (in the casing retainer) off the work surface, leaving the large ring behind (*Figure 4-18*). Using a thick piece of piano wire (aka "model airplane landing gear wire") as a T-handle, pull out the core spindle (*Figure 4-19*). With the large end of the core spindle, or a rod of the same diameter, wiggle and pull out the small ring (*Figure 4-20*). Remove the motor from the casing retainer, and set it on a rack to dry.

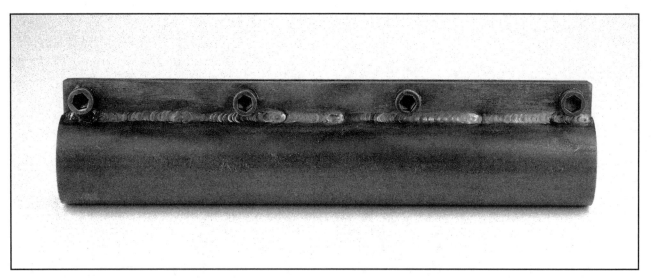

Figure 4-21. *A casing retainer keeps the cardboard motor casing from splitting during the loading process. This retainer is made of steel, but you can improvise with plastic pipe and hose clamps.*

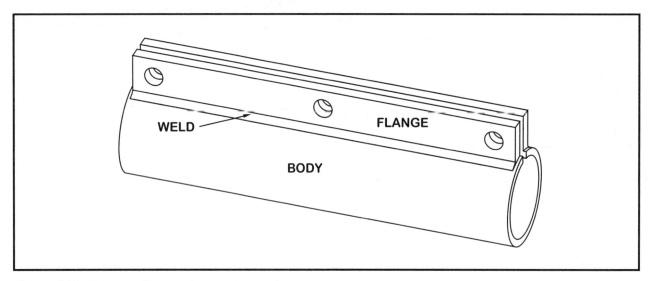

Figure 4-22. *The parts diagram for a casing retainer.*

Casing Retainers

To make a dense and solid propellant grain and a tight fitting nozzle that won't blow out, you have to ram in the propellant and the nozzle clay with considerable force. The expansive pressure generated when doing this is enough to split a cardboard tube, and you have to wrap something around the tube to hold it together. I call the tool that protects the tube a **casing retainer** (*Figure 4-21*). A casing retainer is a piece of pipe with a slot cut lengthwise, so you can slip it over a motor casing and pinch it closed. All the motors in this book are designed to fit loosely into standard sizes of schedule 40 pipe. Why schedule 40? Because pipe is made in five standard wall thicknesses ranging from schedule 5 to schedule 160. Schedule 5 is thin like the exhaust pipe on a car, while schedule 160 has walls so thick that it doesn't even look like pipe. Schedule 40 is thick enough to be strong enough, but thin enough to be pinched shut with hand-operated clamps, and that makes it ideal for this purpose.

I've been asked many times why I didn't design these rocket motors to fit standard, rocket industry motor mounts. And the answer is that, had I done so, their casing retainers would have to be bored from solid metal barstock; an expensive and time consuming process. Designing the motors to fit into standard sizes of pipe resulted in a major savings in the cost of tooling up to make them. Schedule 40 pipe comes in galvanized steel, black-lacquered steel, stainless steel, brass, aluminum, and plastic. Metal and pipe dealers sell most of these types. You can buy the plastic pipe and the black-lacquered steel pipe at a hardware store. The following instructions explain the setups and procedures that I use to make a casing retainer, and **Chapter 5** contains all the drawings of all the retainers needed to make all the motors in this book.

Important note. *Black steel pipe is coated with a thick, black lacquer to keep it from rusting. Before you begin, wipe off this coating with a rag soaked in lacquer thinner, acetone, or M.E.K.*

The photos that accompany the following instructions show a casing retainer being made for a **G** motor, but the procedure is the same regardless of the tool's size. In *Figure 4-22* I've illustrated a casing retainer's parts. To make a casing retainer, refer to *Figure 4-22*, and proceed as follows.

1. Cut a piece of black-lacquered (*not* galvanized), Schedule 40 steel pipe to length. Then square up, finish, and deburr the ends on a metal lathe (*Figure 4-23*). Ideally, you should make the casing retainer slightly longer than the motor you are building. If you're making motors of more than one length, economics might make this impractical. If this is the case, make a retainer that fits the *longest* motor. Though a little inconvenient, you can then make *all* of the motors the full length of the retainer, and trim the shorter ones back to the proper length when they're finished.

Figure 4-23. A piece of Schedule 40 steel pipe is cut longer than required; then machined to the proper length, and deburred on a metal lathe.

Figure 4-24. On a milling machine, the pipe is laid on a pair of V-blocks parallel with the long axis of the table; then clamped in place with a piece of round, steel barstock, and a pair of hold-downs.

2. On a milling machine, lay the pipe on a pair of V-blocks, parallel with the long axis of the milling table. Insert a piece of round, steel barstock through the pipe, and clamp it in place with a pair of hold-downs (*Figure 4-24*). On a suitable arbor in the vertical milling head, mount a side milling cutter or a stack of slitting saws, and cut a slot completely through the wall of the pipe along its entire length (*Figure 4-25*). **Important note.** *Steel pipe is made with a welded seam down the inside. When you cut the slot, orient the pipe so that the seam is cut away in the process.* Deburr the edges of the slot, inside and out, and proceed to **step 3**. From this point on I'll call this finished piece "*the body*".

Figure 4-25. A slot is cut along the length of the pipe with a side milling cutter or a stack of slitting saws.

Figure 4-26. The holes for the clamping bolt threads are drilled. Starting each hole with a center drill insures its accurate location.

Figure 4-27. The threads for the clamping bolts are cut with a hand tap held straight by a dead center mounted in the milling machine's drill chuck.

3. Cut the blanks for the flanges about 1/8" shorter than the body, and clamp one of them in the milling vise. Mount a drill chuck with the proper size drill in the vertical milling head. Refer to the hole locations indicated in the drawing, and lay out and drill the holes for the clamping bolt threads (*Figure 4-26*). To assure the accurate position of each hole, start it with a center drill, and finish it with a standard drill of the indicated size. Then replace the drill with a dead center, and using the center to keep the tap straight with respect to the hole, hand-tap the threads for the bolts (*Figure 4-27*).

4. Lay the finished flange against the blank flange. Carefully align the two, and clamp them together. With the proper size transfer punch, mark the positions of the holes for the clamping bolts *on the blank flange*. Take the punch-marked *blank* flange to a drill press, and drill the holes for the clamping bolts.

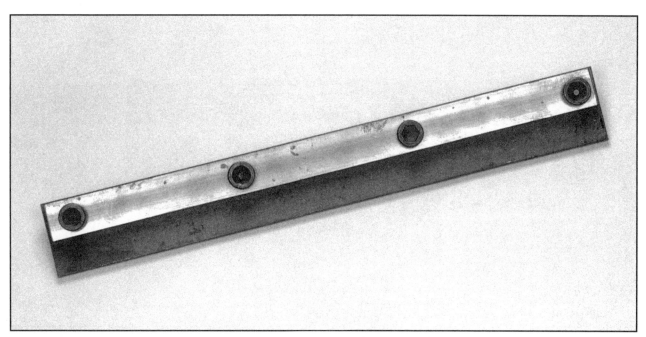

Figure 4-28. The two flanges are bolted together with the welding spacer sandwiched in between.

Figure 4-29. The body-and-flange assembly, clamped in a vise, and ready for welding.

5. Cut the blank for the welding spacer. Align it *with*, and clamp it *to*, one of the finished flanges. With the proper size transfer punch, mark the positions of the clamping bolt holes. Then, just as you did with the flanges, drill the holes in the welding spacer, and make them the diameter indicated in the drawing. **Important note.** *The thickness of the welding spacer should be equal to the width of the slot that you cut in the body.*

6. Assemble the flanges and the welding spacer with the clamping bolts (***Figure 4-28***). Then slide the welding spacer *into* the slot in the body, and clamp the whole assembly in a vise, pinching the slot down *tight* on the welding spacer. If the casing retainer is *longer* than the jaws of the vise, add C-clamps as needed (***Figure 4-29***).

7. TIG-weld the flanges to the body, but *don't* let the weld penetrate through to the welding spacer.

Figure 4-30. The casing retainer is finished by welding the flanges to the body. TIG-welding produces the neatest job, and a combination of pulsed and backstep techniques keeps the retainer from warping as it cools.

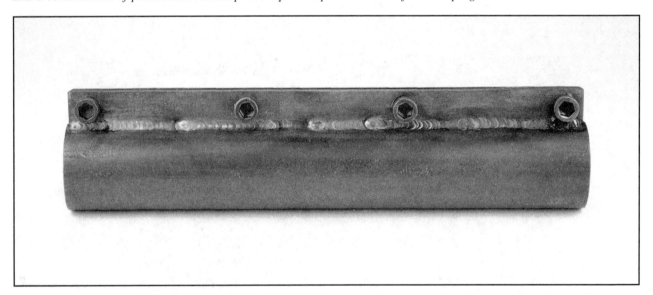

Figure 4-31. The clamping bolts and the welding spacer are removed. When the clamping bolts are replaced, the casing retainer is finished.

    If you're not an experienced welder, welding the flanges can be tricky. If you do it wrong, the casing retainer will warp when it cools. Since I ruined my first retainer, I now take them all to a *professional* welder, who TIG-welds them (tungsten-inert gas) with a combination of pulse and backstep techniques that *prevent* warping. As you can see by the one in *Figures 4-30* and *4-31*, he does a beautiful job.

8. Remove the clamping bolts, pull out the welding spacer (*Figure 4-31*), and the casing retainer is finished.

Though I'd like to claim the credit, I'm not the originator of this design. All the retainers in my first book were the plastic-and-hose clamp type described ahead. A Seattle rocket builder named Brad Gray started making them like this. *His* flanges are made of angle iron. What I've done here is to take his improvement even farther, and make them out of thick, rectangular barstock, giving more thickness and meat to the threads for the clamping bolts.

● Tolerances

The tolerances on a casing retainer are not critical, but you *do* have to be careful about the positions of the holes. If they don't line up properly, the bolts won't fit. It is also important to machine the ends of the body true and square.

Figure 4-32. The table saw's blade is set just high enough to cut through the wall of the plastic pipe.

Figure 4-33. When you make the cut, use a wooden "pusher" to protect your fingers.

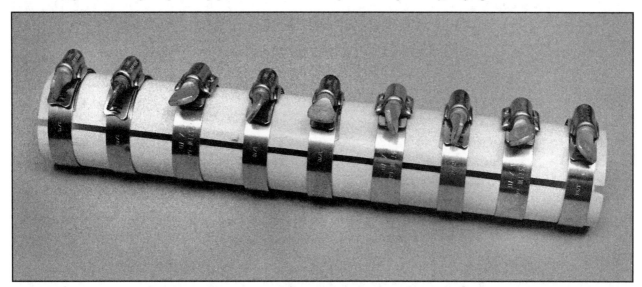

Figure 4-34. A series of hose clamps pinches the plastic casing retainer shut. Because of the number of screws you have to tighten, this improvised retainer is time consuming to use. But it's easy to make, and it works just fine.

An Improvised Plastic Casing Retainer

To make a casing retainer *without* metal working tools you can improvise. The improvised retainer is not as convenient to use as the welded steel kind, but it works just fine, as the 10,000 readers of my first book can verify. To make an *improvised* retainer, proceed as follows.

Cut a piece of Schedule 40, PVC plastic pipe to the proper length, and square up the ends. Then cut a slot of the proper width along the entire length of the pipe. Plastic can be cut with woodworking tools, so if you're careful, you can make the cut on a table saw. Set the blade to cut just a little deeper than the wall thickness of the pipe (*Figure 4-32*), and set the distance of the saw's fence from the blade at half the outside diameter of the pipe. Because you're cutting only one side of the pipe, you'll have to *temporarily* remove the saw's blade guard, *and replace it when you are finished.* **Warning! It is dangerous to push the pipe through the blade by hand, so make a wooden "pusher" to protect your fingers.** I made the pusher in *Figure 4-33* by gluing a small block of plywood to the end of a piece of wooden "two-by-four". A second board, clamped to the saw's table, keeps the pipe from wandering sideways during the cut.

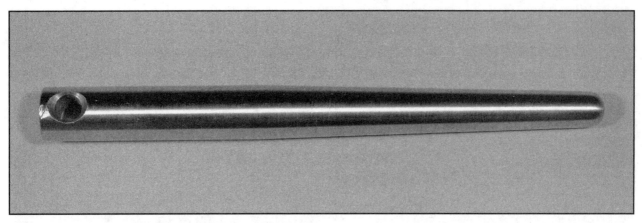

Figure 4-35. A core spindle forms the rocket motor's nozzle throat, and the hollow bore hole through its propellant grain. Usable spindles can be made from brass or aluminum, but stainless steel spindles are the best.

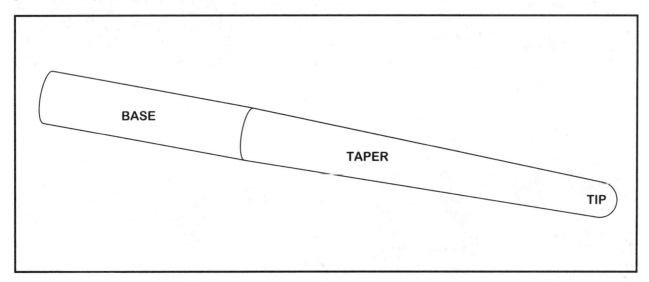

Figure 4-36. The parts diagram for a core spindle.

If the slot is too narrow, wiggle the pipe around, and widen it with a second cut. If you make a mistake, try again. Plastic pipe is cheap, and if you ruin a piece or two, it's no great loss. To pinch the retainer shut, use stainless steel hose clamps. You can buy them at a hardware store. Line them up side-by-side as shown in *Figure 4-34*.

Core Spindles

A **core spindle** (*Figure 4-35*) forms a rocket motor's nozzle throat, and the hollow bore hole through the motor's propellant grain. During the loading process the propellant and the nozzle clay are packed tightly *around* the spindle. When the motor is finished, the spindle is removed, leaving a long, narrow hole of the proper dimensions.

The best material for a core spindle is *stainless steel*. You can get by with aluminum or brass, but because these metals are soft, a spindle made of aluminum or brass will quickly wear out. Core spindles are slightly tapered. It would simplify things if they didn't have to be, but without this taper, they get jammed in so tight during the loading process that you can't get them out. The taper isn't critical, but it *does* have to be steep enough to allow for easy removal of the spindle when you are done. On the other hand, all the motors in this book are designed to have their igniters placed at the *front* of their propellant cores. If the front of the core is too narrow, the igniter won't fit. So the *small* end of the spindle (the tip) has to be *wider* than the igniters that you plan to use.

The following instructions explain the setups and procedures that I use to make a core spindle, and **Chapter 5** contains all the drawings of all the spindles needed to make all the motors in this book.

Figure 4-37. A piece of stainless steel rod is cut 1/4" longer than indicated, and the ends are trued up on a metal lathe.

Figure 4-38. The ends of the workpiece are center-drilled.

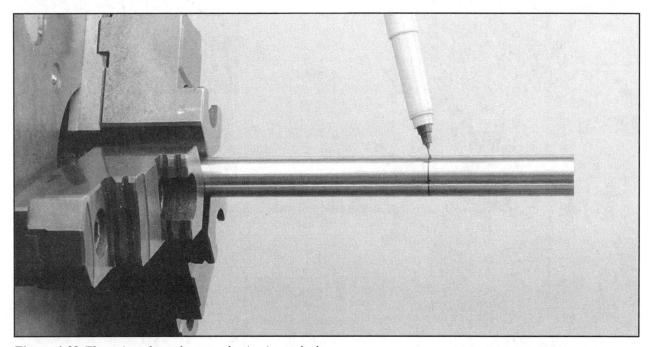
Figure 4-39. The point where the taper begins is marked.

In the pages ahead I refer to a core spindle's parts, and I've illustrated these parts in *Figure 4-36*. To make a core spindle, refer to *Figure 4-36*, and proceed as follows.

1. Cut a round, stainless steel rod of the proper diameter 1/4" longer than the length indicated in the drawing. Then mount it in a metal lathe, and true up the ends (*Figure 4-37*). With a *very small* center drill, center-drill each end (*Figure 4-38*). From this point on, I'll call this rod "the workpiece".

2. Loosen the chuck, and slide the workpiece out. Retighten the chuck, and mark the place where the taper begins. If you disengage the headstock spindle, you can spin the workpiece against the pen, and make a ring-shaped mark like the one in *Figure 4-39*. Then loosen the chuck, and remove the workpiece from the lathe.

Figure 4-40. *A precision reference bar is mounted firmly between the chuck and the tailstock center.*

Figure 4-41. *The combination reference plate-and-sine bar is clamped firmly to the side of the lathe's compound.*

3. *Big* lathes have tailstocks that can be adjusted either *toward* or *away from* the operator, thereby centering or decentering the tailstock. If your lathe has this capability, take the steps needed to adjust the tailstock to *dead center* in perfect alignment with the headstock. Then obtain a piece of round, precision diameter steel barstock at least 8" long. It can be a length of drill rod or precision ground shafting, or you can make it yourself. The only requirements

*Figure 4-42. With the proper combination of feeler gauges and gauge blocks held at the **right** end of the sine bar, the sine bar is pushed gently against the reference bar, forcing the lathe's compound to rotate to the desired angle.*

are that it be smoothly machined, and that its diameter not vary by more than .005" along its entire length. Mount it in the lathe chuck, true up the ends, and center-drill *one* end. I call this finished device a **reference bar**. It will become a permanent accessory in your tool kit.

4. Replace the tailstock drill chuck with a live center, and remount the reference bar *firmly* between the chuck and the live center (*Figure 4-40*). Use the tailstock screw to take up any slack, and lock it in place.

5. Remove the lathe's tool post, and loosen the compound so that it swivels freely. Obtain or make a piece of flat, metal plate of suitable size, with surfaces as close to parallel as possible, and clamp it to the side of the compound. As with the reference bar, I keep one of these in my tool kit, and I call it a **reference plate**.

   Clamp a **sine bar** to the reference plate, and clamp the sine bar-reference plate assembly to the side of the compound (*Figure 4-41*). A sine bar is a hardened, flat, precision-ground steel bar with two perfectly round steel dowels spaced an exact distance apart. The most popular sizes are 2.5000" and 5.0000". You can buy a new import for $15 to $25, but sine bars are so common in machine work that you'll find them at yard sales and swap meets for $2 to $5. I'll assume in this demonstration that the length of the sine bar is 5.0000" (rounded to 5.000").

6. Now refer to the spindle drawing, and calculate the sine bar offset as follows:

$$\text{Offset} = ((D_b - D_t) / 2) \times (L_s / L_t)$$

where D_b is the diameter of the spindle's body, D_t is the diameter of its tip, L_s is the length of the sine bar, and L_t is the length of the taper as indicated in the drawing.

> *EXAMPLE 4-1:*
>
> *Refer to drawing No. 98-071-C on page 139, and assume that you're using a 5.000" sine bar. Note that the spindle's body is 0.562" dia., that the tip is 0.375" dia., and that the tapered portion of the spindle is 4.75" long.*
>
> $$\text{Offset} = ((D_b - D_t) / 2) \times (L_s / L_t)$$
>
> or
>
> $$\text{Offset} = ((0.562" - 0.375") / 2) \times (5.000" / 4.750") = (0.187" / 2) \times 1.053 = 0.094" \times 1.053 = 0.098"$$
>
> *The offset for a 5.000" sine bar should therefore be 0.098"*

Figure 4-43. The first cut is made by hand feeding the lathe tool **toward** the headstock with the lathe's compound screw.

Figure 4-44. The second cut begins where the first cut left off.

7. Select a combination of gauge blocks and/or feeler gauges that total 0.098". As you face the lathe, hold them together between the reference bar and the *right* dowel of the sine bar. With the lathe's crossfeed screw (*Figure 4-42*), gently push the sine bar up against the reference bar. As you do so, the compound will automatically rotate to the proper taper angle. If it sticks and *doesn't* rotate, lubricate the contact points with a few drops of oil. Being careful not to disturb the compound's angle, lock it down. Then remove the reference bar, the sine bar, and the reference plate. Remount the tool post, and proceed to **step 8**.

8. Remount the workpiece in the lathe chuck with the *base* end against the tailstock's live center, and not more than 4 inches of its length exposed between the chuck and the tailstock. Take up any slack with the tailstock screw, lock the screw in place, and *disengage* the lathe's carriage feed mechanism. Mount a *left-hand* cutting tool in the lathe's tool post, and turn the lathe on. Starting at the mark where the taper begins, lock the carriage in place, *hand*-feed the tool toward the headstock with the lathe's compound screw, and cut the taper *toward* the headstock as far as the compound will move (*Figure 4-43*).

Important note. *The total feed distance on most lathe compounds is quite short. On my own lathe it's only 4 inches. When cutting tapers longer than 4 inches (as in this example), do it in steps. Move the workpiece a little farther out each time, and pick up where you left off with each new cut.*

9. Turn the lathe off, and loosen the chuck. Slide the tailstock back, and slide the workpiece out so that the live center supports it again. Retighten the chuck, and take up any slack with the tailstock screw. Crank the compound back to the starting point, and move the lathe carriage so that the cutting tool is about 1/4" to the right of where the first cut ended. Turn the lathe on, and with the crossfeed screw, move the cutting tool toward the workpiece until it just touches the surface. Then with the compound screw, hand-feed the tool toward the headstock *again*, making the *second* cut (*Figure 4-44*).

10. Repeat **step 9** until you've cut the taper to within 1/2" of the spindle's tip. Each new cut will be deeper than the one before. When the depth becomes too great to cut in one pass, do it in two or three.

11. When you're finished, turn the lathe off, loosen the chuck, and reverse the workpiece, so that the base end is mounted in the chuck, and the tip is supported by the live center. Retighten the chuck, take up any slack with the tailstock screw, and use the carriage feed to trim down the last bit of metal on the tip (*Figure 4-45*).

Figure 4-45. The workpiece is reversed, and the carriage feed is used to trim down the tip.

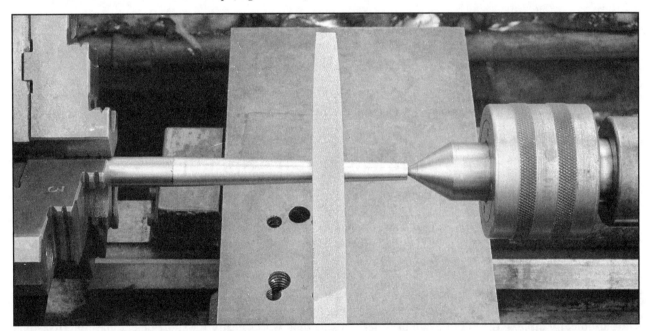

Figure 4-46. With the lathe running at high speed, a mill bastard file or a mill fine file smooths out most of the marks left by the lathe tool. The results of this step are not critical. It is sufficient to remove the worst of the marks.

13. With the lathe running at high speed, use a mill bastard file or a mill fine file (*Figure 4-46*) to smooth out the worst of the marks left by the lathe tool. A sharp lathe tool will reduce the need for this step. When machining stainless steel, I use a carbide tool with a sharp point, and I grind a fairly steep cutting edge angle into its upper surface. Then I round the point ever so slightly with a green stone or a diamond wheel.

14. Remember that you originally cut the workpiece 1/4" longer than necessary. That was to accommodate the hole that you drilled for the tailstock center. Now you have to grind off that excess metal. To do so, hand-rotate the tip against a grinding wheel, grinding it back, and crudely rounding it at the same time (*Figure 4-47*). Check it as you go with a ruler or a vernier caliper, and grind it back to the total spindle length indicated in the drawing.

15. To finish the tip, mount the workpiece back in the lathe. Running the lathe at high speed, work the tip smooth, round, and even with a mill bastard file or a mill fine file.

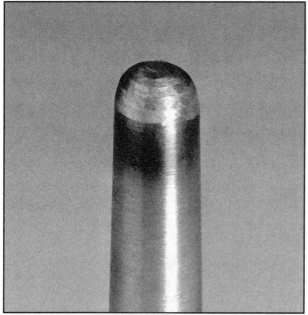

Figure 4-47. The tip of the spindle is ground to length, and crudely rounded on a grinding wheel.

Figure 4-48. The spindle is polished with #220 sandpaper.

16. With the spindle still spinning in the lathe, polish it with #220 sandpaper (*Figure 4-48*). Then drill a hole of suitable diameter near the end of the base, and countersink the edges of the hole (*Figure 4-49*). When removing the spindle from a finished motor, you can insert a steel rod or a Phillips screwdriver through this hole to help you twist and pull it out.

Important note. *The preceding method is used only to machine tapers that are longer than the total travel of the lathe's compound. You can make shorter tapers in one pass by reversing the workpiece (so that the base is held in the jaws of the chuck), and feeding the tool in the opposite direction, **toward** the tailstock.*

An Improvised Core Spindle

To make a core spindle *without* a metal lathe, *again* you can improvise.

1. Cut a piece of stainless steel rod slightly longer than the length indicated in the drawing. Square up the ends on a belt sander, and with a scribe or a sharp, indelible marking pen, mark the point where the taper begins.

2. Mount the base end of the rod in an electric hand drill. Turn the drill on, and with the rod spinning, grind the taper to the proper length and shape by working it back and forth against a running belt sander (*Figure 4-50*). The grinding action might heat the metal enough to burn away the mark, and you can minimize the heat by periodically dipping the rod in a cup of water. This method works easily for short tapers, but requires some practice for the long ones. As with the slot in the plastic casing retainer, you might waste a piece or two before you get it right.

Tamps

Tamps (*Figure 4-51*) are used to ram the nozzle clay and the rocket propellant into the motor casings. Tamps have to be durable so that they last, but they also have to be light, so that they transmit the force of your hammer blows to the material being loaded. *These two requirements completely rule out the use of metals*, a fact that surprises people unfamiliar with the motor building process.

Tamps made of metal fail in two ways. First of all, the striking end quickly becomes cracked and mushroomed from metal fatigue. Then the *mass* of the metal absorbs the energy of the hammer, and *that* prevents the propellant from being packed as densely as it should. Inadequate density makes the propellant burn too fast, and *that* causes massive nozzle erosion, or a complete blowout of the motor casing wall.

Figure 4-49. A hole is drilled near the base of the spindle, and the edges of the hole are countersunk.

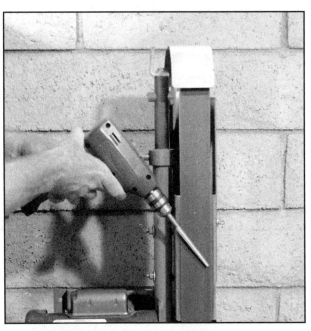

Figure 4-50. An improvised core spindle is ground to shape by spinning it in a hand drill, while running it against a belt sander.

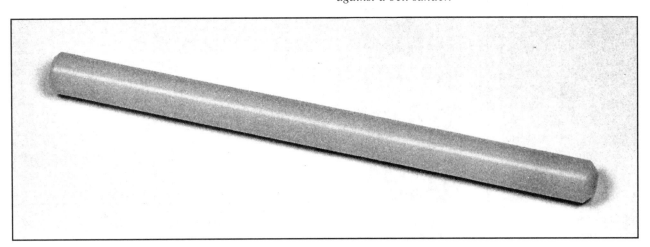

Figure 4-51. Nylon tamps are used to ram the nozzle clay and the powdered rocket propellant into the motor casings.

Warning! Tamps made of metal are dangerous for the following reason. In the event of an accidental ignition, the tamp must be light enough to be blown out the front of the motor casing before the pressure beneath it can build to a dangerous level. Tamps made of metals like brass, steel, and, even aluminum, are *heavy*! Their inherent resistance to motion (ask "inertia") can allow the pressure inside the casing to build to the point of *explosion*!

The fireworks industry has traditionally used wooden tamps. When I began this work, I started with wood, but I quickly learned that wood doesn't last, and I embarked on a search for something stronger. On a hunch, I experimented with plastics, and quickly found that **Nylon** works *beautifully*. Those of you familiar with plastics might suggest Delrin (generic name "acetyl"), but my experiments with Delrin were disappointing. Though Delrin machines well, it shatters too easily when struck with a hammer. Nylon is *absolutely amazing*. It machines easily. It doesn't mushroom. When struck with a hammer, it *doesn't* shatter. It is impervious to solvents, and a tamp made of Nylon lasts for a *long, long* time. Most people think of Nylon as the synthetic cloth used in parachutes and backpacks. But you can also buy it in sheet, block, and barstock form from industrial plastic dealers. To find Nylon barstock, look in *The Yellow Pages* under headings like **PLASTICS: ROD, TUBE & SHEET**. The place where I buy it sells it by the foot, and they add on a small cutting charge for anything less than a full length. The large diameters are expensive, but well worth the investment.

Important note. *When machining Nylon, use a standard, high speed steel cutter, but grind it with a steep clearance, and a steep cutting edge angle (**Figure 4-53**). Then round the point very slightly on an oil stone.* For boring Nylon, use a spade-shaped wood boring bit called a **Speedbor** (*Figure 4-54*). Speedbors come in regular 6", and extra long 16"

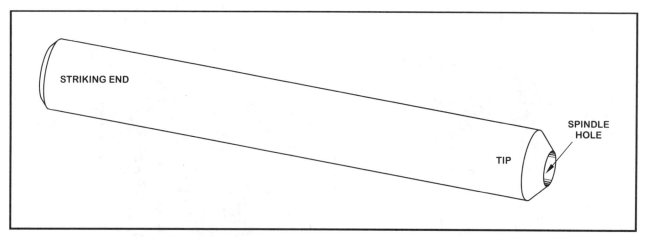

Figure 4-52. The parts diagram for a tamp.

Figure 4-53. The lathe tool for cutting Nylon is ground with a steep clearance and cutting angle, and a slightly rounded point.

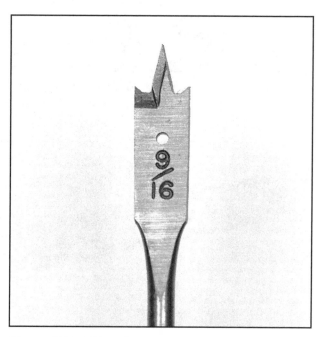

Figure 4-54. A "Speedbor" wood boring bit is the best tool for boring deep holes in Nylon.

lengths, and most hardware stores sell them. The larger Speedbors have two pointed ears on the cutting end. Before you use one, grind these ears off, and, making sure to maintain the same cutting edge angle, grind a slight bevel in their place (*Figure 4-55*). This prevents the creation of a deep groove where propellant might collect inside the tamp, and it also helps keep the bit centered when boring deep holes.

Some tamps have solid ends. Others are bored either partially or completely through. The shorter tamps are made in one piece, and the longer tamps are made in 2 or 3 pieces and plugged together. The one in the following example is representative, except for the fact that I've shortened it to fit the photographs. The following instructions explain the setups and procedures that I use to make a tamp, and **Chapter 5** contains all the drawings of all the tamps needed to make all the motors in this book. In the pages ahead I refer to a tamp's parts, and I've illustrated these parts in *Figure 4-52*. To make a Nylon tamp, refer to *Figure 4-52*, and proceed as follows.

1. Cut 2 pieces of 1-1/8" dia., round Nylon barstock slightly longer than necessary. Mount them in a lathe, machine their ends flat, and center-drill their ends (*Figure 4-56*).

2. Replace the lathe's drill chuck with a live center. Then loosen the lathe chuck, slide one of the bars out about 8 inches, and support the center-drilled end with the live center. Take up any slack with the tailstock screw, and lock the screw in place.

Figure 4-55. Before I use a Speedbor, I grind off the two points, and replace them with a slight bevel. This helps keep the bit centered when boring deep holes.

Figure 4-56. The end of a round Nylon bar is machined flat, and center-drilled on a lathe.

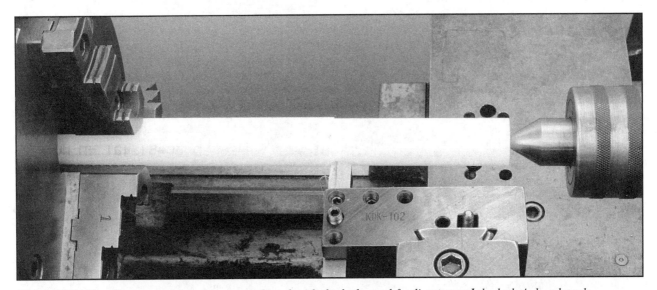
*Figure 4-57. The diameter of the tamp is machined with the lathe tool feeding **toward** the lathe's headstock.*

3. Set the tool bit for a depth of cut that trims the bar down to the diameter indicated in the drawing. Turn the lathe on a medium to high speed, and machine the diameter with the carriage feeding *toward the headstock* (*Figure 4-57*). As the cut proceeds, a long, nylon thread will form. It will probably wrap itself around the bar, but you can cut it with a razor blade and pull it off when you're finished. Turn the lathe off, move the tailstock back 8 inches, and slide the workpiece out until its end is *again* supported by the live center.

4. Repeat **step 3** until all but the short portion held in the lathe chuck has been machined to the proper diameter. Then reverse the bar, and finish the end. This multistep method of machining the diameter is necessary because Nylon is very flexible. If you try to machine more than 8 inches at a time, the bar will flex and vibrate, and the tool will chatter and ruin the finish. Repeat **steps 2** through **4** with the second bar.

5. Now it's time to bore the spindle holes. Replace the live center with a drill chuck. Mount a Speedbor in the drill chuck. Running the lathe at a very *slow* speed, use the tailstock's feed screw to bore the spindle holes to the depths indicated in the drawing (*Figure 4-58*). As you work, back the tool out every 1/4" or so, and clear the chips from the hole. If they build up inside, they will jam and melt from the heat of friction. Note here that you'll bore one piece *completely* through, and the other only *partially* through.

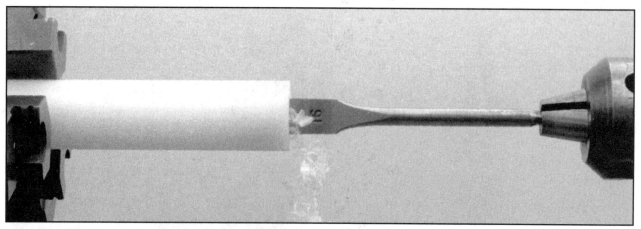

Figure 4-58. The tamp's spindle hole is bored with a Speedbor wood boring bit.

Figure 4-59. A single face cut cleans up any scarring left by the boring bit.

Figure 4-60. The socket is bored to the dimensions indicated in the drawing.

6. When boring the spindle hole, if you've scarred up the end of the bar with the boring bit, clean it up with a face cut as shown in **Figure 4-59**.

7. The *longer* tamps are plugged together from two short sections. To make the *socket*, mount the *fully* bored piece back in the lathe chuck. Slide the end out about two inches. Replace the lathe's cutting tool with a lathe *boring* tool, and bore the socket to the inside diameter and depth indicated in the drawing (**Figure 4-60**).

8. To make the *plug*, mount the *partially* bored piece in the lathe chuck, *bored end out*. Slide the end out about two inches. Replace the lathe's boring tool with a *left-hand cutting* tool, and machine the plug to the outside diameter and length indicated in the drawing (**Figure 4-61**).

Important note. *Before you machine the plug, measure the inside diameter of the socket, and make the outside diameter of the plug 8 to 10 thousandths of an inch larger than the inside diameter of the socket. When you hammer the two together, this will insure a tight, permanent fit.*

9. Reverse the workpiece, and cut a 45 degree bevel on the solid, striking end (**Figure 4-62**).

Figure 4-61. The plug is machined 8 to 10 thousandths of an inch **larger** in diameter than the inside diameter of the socket.

Figure 4-62. A bevel is machined on the tamp's striking end.

10. Fit the plug into the socket, and hammer the two together (*Figure 4-63*). Place the entire assembly back in the lathe *with the bored end out*. Machine a 45 degree angle on the tamp's nose, and the tamp is finished.

Important note. *When machined on a lathe, Nylon produces tough, stringy chips and flash that you have to clean off when you are finished. The inside edge of a bored hole can be shaved clean with a deburring tool or a countersink mounted in the tailstock drill chuck. Outside edges can be trimmed with a file while the lathe is running.*

The tamp in the previous example is just one of the several types used to make the motors in this book. The "starting tamp" used to ram in the first dose of nozzle clay has a flat tip, but the basic machining procedure is the same. As you use a tamp, its nose will become scarred and damaged. When the damage becomes bothersome, put it back in the lathe and resurface it.

Wooden Tamps - A Homemade Boring Jig

To make a tamp *without* a metal lathe, you can improvise with a hardwood dowel. But the dowel you choose must be *absolutely straight*. The dowels sold by hobby shops, lumber yards, and hardware stores are often warped. When shopping for a dowel, sort through the whole pile, and pick the straightest one you can find.

Like the nylon tamps, most of the wooden tamps have holes either bored or drilled through their centers. You can drill a short hole by eye, but a deep hole is tricky, and you'll need a **boring jig** to keep the Speedbor running straight. To make a boring jig, proceed as follows.

1. Buy a length of Schedule 40, PVC plastic water pipe with an inside diameter a little *larger* than the diameter of the dowel, and buy a slip-fit plastic pipe cap to match. By the best means available, locate the center of the cap, and drill a hole through the center just a tiny bit *larger* than the shank of the Speedbor (*Figure 4-65*).

2. Cut a piece of plastic pipe about one foot long, and press the cap onto one end. Insert the Speedbor through the hole in the cap, and mount the shank of the Speedbor in an electric hand drill. *Figure 4-66* illustrates these parts as they appear separately, and fully assembled.

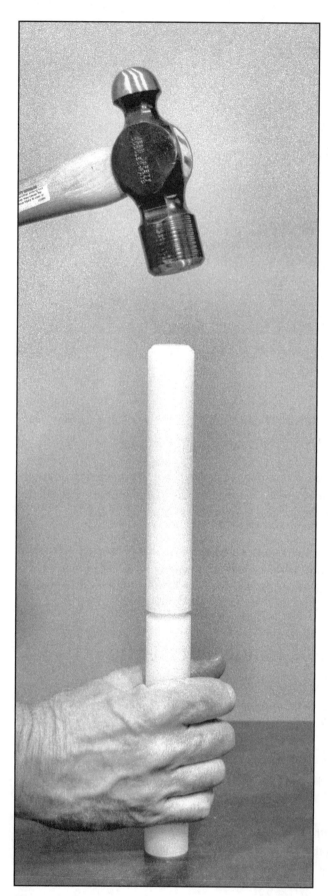

Figure 4-63. The two halves of the tamp are permanently hammered together.

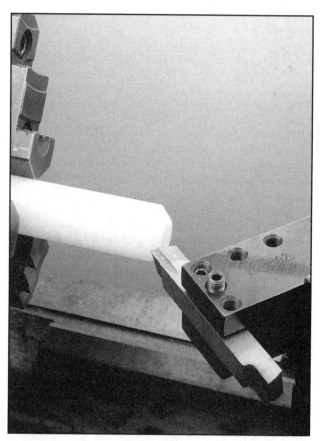

Figure 4-64. When the tip of the tamp is machined to a 45 degree angle, the tamp is finished.

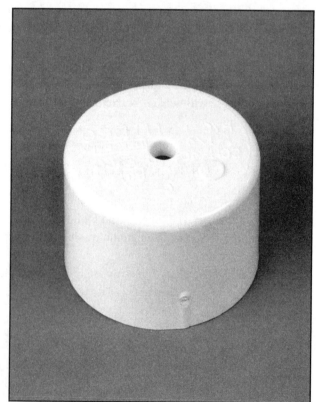

Figure 4-65. A hole is drilled through the pipe cap, a little larger than the diameter of the boring bit shank.

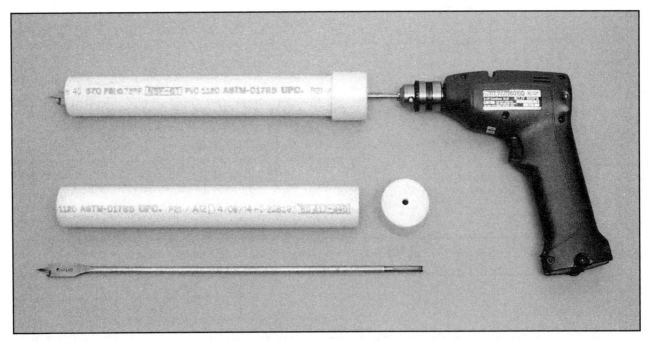

Figure 4-66. Below: the separate parts of an improvised boring jig are a plastic pipe, a matching pipe cap, and a long-shank Speedbor wood boring bit. Above: the boring jig assembled, and mounted in an electric hand drill.

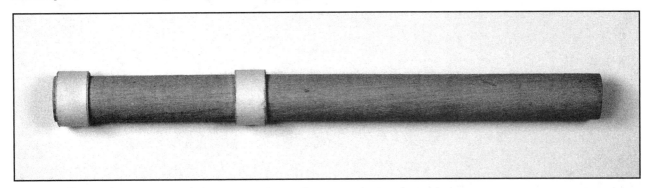

Figure 4-67. Masking tape rings around one end of the dowel build it up to the inside diameter of the boring jig.

An Improvised Wooden Tamp

To use the jig, and make a wooden tamp, proceed as follows.

1. Saw the dowel to the proper length, and sand the ends square and smooth. Mount the dowel in a wood lathe, and sand its diameter down to a point where it slides in and out of the motor casing *loosely* without binding.

2. Near one end of the dowel, build up two rings of masking tape, several inches apart, until they slide smoothly into the jig without wobbling (*Figure 4-67*).

3. Using a folded rag to protect the dowel from damage, grip the dowel *gently* in a vise, and drill a small starter hole in the center of one end. Press the tip of the Speedbor into the starter hole (*Figure 4-68*), and *keep it there* while you slide the jig *over* the two tape rings. Then start up the drill, and bore the hole (*Figure 4-69*).

4. You can, if you wish, sand a 45 degree cone on the tamp's tip by hand rotating it against a running belt sander. But wood is fragile, and the sharp edge around the spindle hole will quickly break away. A *flat* nose tamp reduces motor performance, but a flat nose tamp will last longer.

5. Spray the tamp with a coat of varnish or paint, *but choose a finish that is hard, and impervious to the alcohol or the acetone in the propellant*. A paint salesman can advise you. The paint seals the wood grain, and makes the tamp last longer. *Figure 4-70* shows a wooden tamp, finished and ready to use.

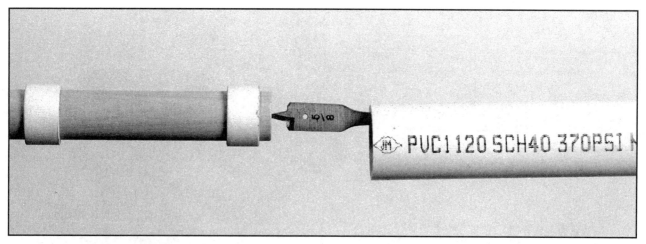

Figure 4-68. The hole is started by pressing the point of the boring bit into the center of the end of the dowel.

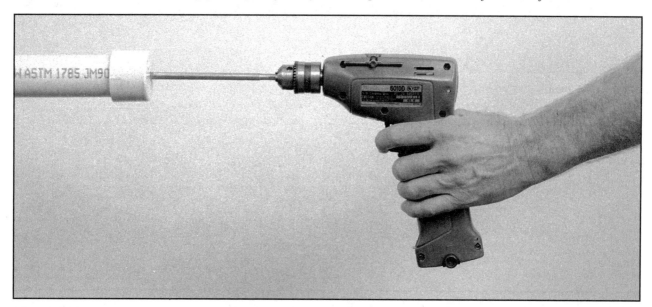

Figure 4-69. While the hole is bored with an electric hand drill, the boring jig keeps the bit running straight.

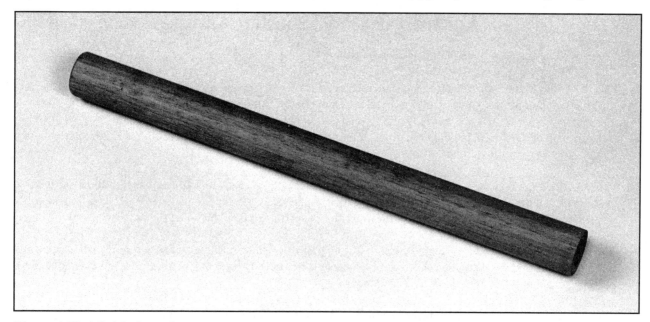

Figure 4-70. A finished wooden tamp.

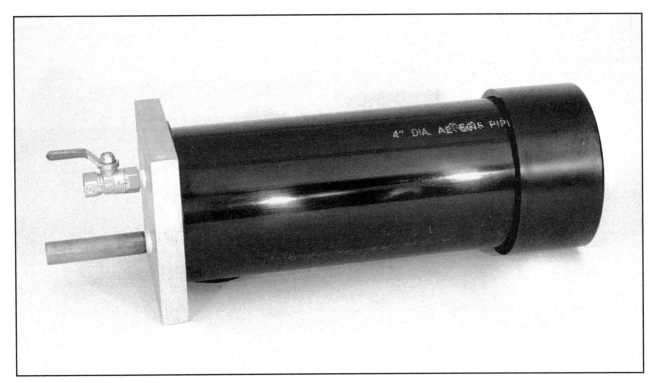

Figure 4-71. A homemade vacuum chamber.

Boring the end grain of a hardwood dowel is time consuming. The hole shown in the photos took 15 minutes. Withdraw the Speedbor periodically, and dump out the sawdust. The tape rings keep the dowel aligned with the bit, and sliding the pipe cap as far back from the end of the dowel as you can keeps the bit running straight. Of course the tool is crude, but with a little practice, you can make a wooden tamp good enough to build most of the motors in this book.

A Homemade Vacuum Chamber

Important note. You do *not* need a vacuum chamber to make these rocket motors. Most of them run on propellants that contain a glue called a "binder". During the loading process you soften the binder with a small amount of solvent. Before a motor can be fired the solvent has to evaporate, and these motors will *rapidly* dry in open air. When acetone is used, the smallest dry in 12 hours, and the largest take 72. If, however, you want to *speed the process even further*, you can do it with a vacuum pump and a homemade vacuum chamber.

MATERIALS LIST

1. One piece 4" dia. schedule 40 PVC or ABS plastic pipe
2. One matching 4" dia. plastic pipe cap
3. One piece machinable aluminum, 5" square and at least 1/2" thick
4. One 3/8" x 4" brass pipe nipple
5. One 1/4" x 1" brass pipe nipple (also called a "close" nipple)
6. One 1/4" ball valve
7. One piece rubber gasket material 5" square
8. The appropriate glue for either PVC or ABS plastic pipe

Figure 4-71 shows a vacuum chamber that I built with regular woodworking tools. It is made from a piece of 4" dia., schedule 40 plastic pipe and a matching pipe cap. You can use ABS or PVC, but ABS is cheaper. The lid is made of aluminum. The lid gasket is cut from a piece of standard, rubber gasket material, and the hose connection is made from a 3/8" brass pipe nipple. A 1/4" ball valve lets air back into the chamber. *Figure 4-72* is an exploded view of the chamber's components.

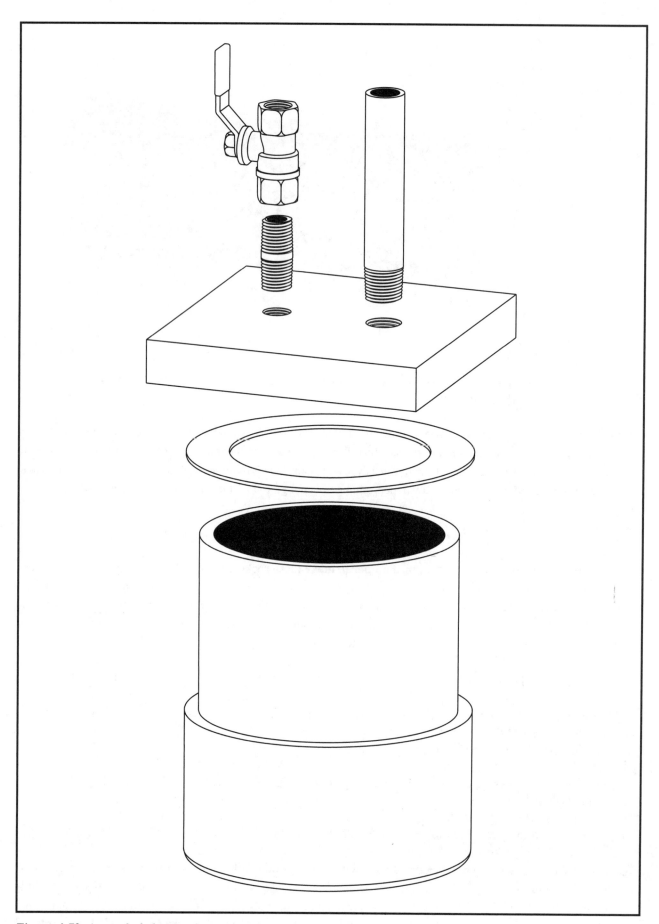

Figure 4-72. An exploded view of a homemade vacuum chamber.

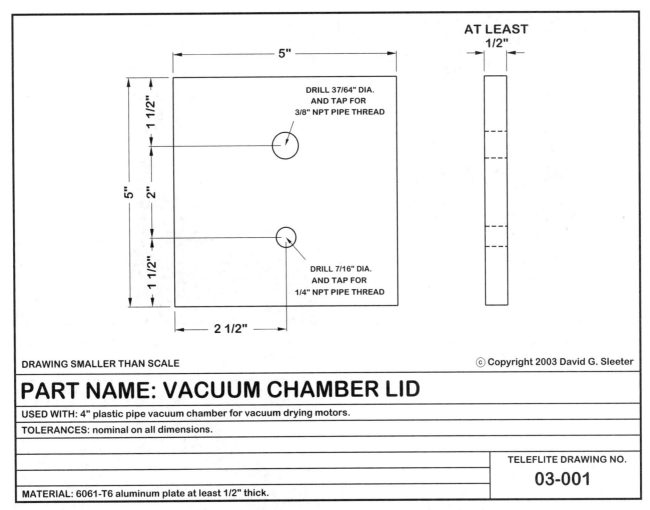

Figure 4-73. The shop drawing for the vacuum chamber lid.

You can buy the aluminum from a metal dealer, but be sure to ask for an alloy that's *machinable*. 6061-T6 is ideal. Like you did with the steel, check the dealer's remnant bin before you pay full price for a new piece. Hardware stores sell the 4" ABS, and irrigation stores sell the 4" PVC. Whichever you choose, be sure to use the right glue. Some glues are made for ABS, some are made for PVC, and some will work with both. If you feel unsure, ask a salesman. You'll find everything else at a hardware store. To make a homemade vacuum chamber, proceed as follows.

1. Make the aluminum lid per the drawing in *Figure 4-73*. Screw one end of the 1/4" x 1" brass nipple into the ball valve, and screw the other end into the matching hole in the lid. Saw the threads off one end of the 3/8" x 4" nipple, and deburr the cut end with a file. With a pair of "vise-grips" (aka "locking pliers"), screw the threaded end into the matching hole in the lid, and use Teflon plumber's tape on *all* threads to insure an airtight fit. *Figure 4-74* shows the finished assembly.

2. Cut the rubber gasket 3-1/2" i.d. x 5" o.d.

3. Cut the plastic pipe to length, and true up and smooth the ends. You can do this on a metal lathe *if the lathe is big enough*. Otherwise, you can do it with a flat block of metal or wood, and some #120 sandpaper. Hold the pipe firmly in an upright position, wrap the sandpaper around the block, and sand both ends of the pipe until they are smooth and flat, and all the saw marks have been removed (*Figure 4-75*).

4. Glue the pipe cap onto one end of the pipe, and *use plenty of glue*. Lather it thickly onto the inside wall of the cap. Then *immediately* press the cap onto the pipe, and hammer it on *tight* with a rubber mallet (*Figure 4-76*). **Important note.** *Let the glue to dry for a full 24 hours before placing the chamber under a vacuum.*

Important note. *Some ABS pipe is made from a three-layer laminate with solid inner and outer walls, and a foamed "cellular" core. If this is what you are using, seal the exposed foam with epoxy glue. Rub the glue firmly into the ends of the pipe, and let it harden. Then smooth the sealed ends with fine sandpaper.*

Figure 4-74. The brass nipple and ball valve are screwed into the lid. Teflon plumbers tape insures a tight seal.

Figure 4-75. Sandpaper and a block of wood are used to smooth the ends of the plastic pipe.

Figure 4-76. The cap is hammered into place with a rubber mallet.

Figure 4-77. The rubber gasket is pasted to the chamber's lid with Vaseline or silicone vacuum grease.

To use the chamber, coat both sides of the rubber gasket and the brass nipple on the chamber's lid with Vaseline. Vaseline works *temporarily*, but it quickly dries in a vacuum. As soon as you can, buy a tube of **silicone vacuum grease**. You'll find it at laboratory-chemical dealers, and the places that sell and service vacuum equipment. It's a little expensive, but it won't evaporate in a vacuum, and you'll use it in such small amounts that a single tube will last for many years.

Connect the vacuum pump's hose to the nipple, and close the ball valve. Place the motors-to-be-dried inside the chamber. Stick the rubber gasket onto the lid (*Figure 4-77*), and hold the lid against the end of the chamber while turning on the vacuum pump. Vacuum forces alone will keep the lid in place. No clamps are needed. When the motors are dry, turn off the pump, and open the valve. When the chamber reaches atmospheric pressure, the lid will automatically release. The vacuum chamber shown here is 12" long, but you can make it any length you wish. I have another one about 3 feet long that I use for drying **H** and **I** motors. Buy the pump and the hose *before* you make the lid. Depending on the size of the hose that fits the pump, you might need a larger or smaller pipe nipple than the one shown in this example.

Figure 4-78. Homemade sieves.

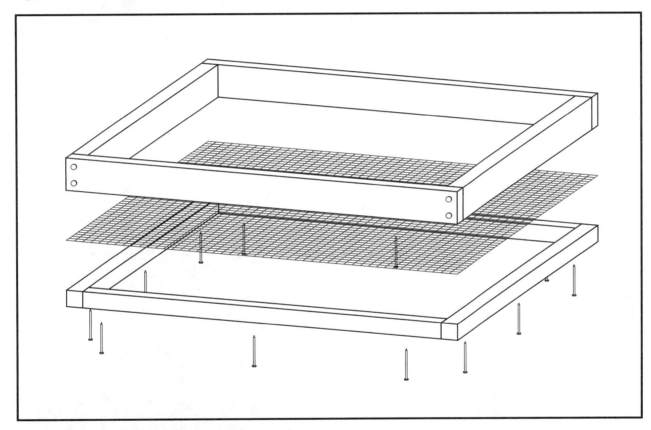

Figure 4-79. An exploded view of a homemade sieve.

Homemade Sieves

Homemade **sieves** are used to granulate oxidizers, and separate crushed coal and charcoal into different particle sizes. You'll need them to process the Mesquite charcoal for the **KG3** propellant, and possibly to granulate homemade sodium nitrate (**page 183**).

Figure 4-78 is a photo of the 3 sieves you'll need. You make them by stapling 18" squares of aluminum window screen or hardware cloth onto 18" square frames made of "one-by-two" pine. If you've never heard of "hardware cloth", try to imagine what giant window screen would look like. You can buy it at a hardware store, and it comes in several mesh sizes. You'll need the 1/8" mesh and the 1/4" mesh. Do a neat job, but don't invest a lot of time or money in making these sieves look nice. They are strictly utilitarian. When you use *anything* made of wood around coal or charcoal, it gets dirty and awful-looking pretty fast. *Figure 4-79* is an exploded view of a sieve that show's how it's nailed and glued together. In the chapters on the chemicals and chemical preparation, I'll tell you which sieves to use, and how to use them.

A Homemade Coal Crusher

The gadget in *Figure 4-80* is a homemade **coal crusher**. It amounts to a piece of 4" galvanized water pipe, 2 feet long with a flat cap on one end, and a piece of 2" pipe, 4 feet long with a cap on one end. The small pipe fits inside the large pipe. To make it work, you drop chunks of coal or charcoal into the large pipe, hammer down on them with the small pipe, and dump the results into the sieves described on the previous page. It's primitive, *but it works*. *New* pipe is *expensive*, but if you're willing to settle for something that's rusty and ugly, you can find what you need at a scrap yard.

Figure 4-80. *A homemade coal crusher.*

5. TOOL DRAWINGS

Important Note. The following pages contain *all* the drawings of *all* the tools needed to make *all* the motors in this book, *so please don't be intimidated by their great number.* Only 8 tools are needed for any particular design, and each motor drawing has an **INDEX OF REQUIRED TOOLING** that refers back to these drawings, and tells you exactly which ones you need. People not experienced in machine work will find complete toolmaking instructions in **Chapter 4**.

Casing Retainer Notes

Make casing retainers from plain, black, *ungalvanized* steel pipe, and TIG weld the flanges with pulse and backstep techniques to prevent warping. Keep the length of each retainer within + or - 0.020" of the length shown in the drawing, and make the flanges about 1/8" shorter than the body. A casing retainer must sit squarely on the nozzle mold during the loading process. An overhanging flange makes the retainer tilt, *and a tilted retainer produces a motor with a nozzle **that is out of alignment***. If you accidentally make a retainer with a flange that overhangs, grind the flange back until it *doesn't* overhang.

A casing retainer's length should equal the length of the *longest* motor you plan to build. You can make a short motor with a *long* retainer, but you *can't* make a long motor with a *short* retainer. Look at the motor designs carefully, and consider this fact before you begin. In each motor drawing you'll notice that I recommend a casing retainer that is *longer* than the motor in the drawing. *I do this to accommodate a time delay, which increases the motor's length.* Instructions for making casing retainers begin on **page 60**. Casing retainer drawings begin on **page 90**. *If you don't have access to metal working tools, the instructions on **pages 65-66** will show you how to improvise.*

Nozzle Mold Notes

Make nozzle molds from stainless steel, and keep all dimensions but the base diameter and the height of the cone within + or - 0.005" of the dimensions shown in the drawing. The height of the cone will evolve naturally during the machining process. To put it simply, the height of the cone will be *whatever it turns out to be*. The base diameter is nominal. It can be the original diameter of the material from which the mold is made.

Nozzle molds 98-006 through 98-013 are made in one piece, but the molds for the I-65, the 1-1/8" DeLaval motors, and the end burners are made in *two* pieces with a base and an insert. To machine the cone on a *large*, one-piece mold, you would have to remove a large amount of metal *around* the cone. The base-insert configuration allows you to make the cone-shaped insert from a smaller piece of metal. This saves time and material, and you might want to consider the base-insert configuration for the other molds as well. The choice is a judgement call based on experience and personal preference.

Set screws are optional. One or two set screws lock a nozzle mold and a core spindle together. When the mold and the spindle are locked together, you can use the mold as a knob to grip when removing the spindle from a finished motor. But the set screws permanently scar the base of the spindle, so set the screws should *not* be installed until you are *sure* that the spindle-mold combination you've chosen is permanent. Instructions for making nozzle molds begin on **page 54**. *If you don't have access to a metal lathe, the instructions on **page 59** will show you how to improvise.*

Core Spindle Notes

Make core spindles from stainless steel. The base diameter of the spindle determines the diameter of the motor's nozzle throat, *so the base diameter is always important*. Fortunately, stainless steel rod is manufactured to close tolerances. If you choose a rod who's fractional diameter (i.e. 5/32", 3/8", 9/16", etc.) equals the fractional diameter of the nozzle throat shown in the motor drawing, that diameter should be sufficient. If the rod is slightly oversize, and jams in the nozzle mold's hole, place the rod in a metal lathe, and with the lathe running on high speed, cut, file, or sand the base down as needed.

When a motor is finished, the spindle has to be removed. Spindles from 5/32" through 5/16" dia. can be pulled out with pliers or "vise-grips", or permanently locked to their nozzle molds with set screws. In the latter case, the nozzle mold doubles as a knob to grip during removal. The holes in the bases of the larger spindles are optional. If you drill a hole through a spindle's base, you can insert a metal rod to use as a T-handle. If you later decide to permanently lock the spindle to the mold with set screws, the position of the hole might conflict with the set screw locations. The decision about what to do is a judgement call based on experience and personal preference.

Instructions for making core spindles begin on **page 66**. Core spindle drawings begin on **page 121**. Each drawing contains 2 to 6 spindles, labeled **A**, **B**, **C**... etc., whose lengths increase in 1/4" or 1/2" increments. For your convenience I've included the 5 inch sine bar offsets needed for machining the tapers, and I've labeled each spindle accordingly. *If you don't have access to a metal lathe, the instructions on page 72 will show you how to improvise.*

Tamp Notes

Make tamps from solid, round, Nylon barstock. Ideally, a tamp and its spindle hole should be made just long enough for the motor you are building, *and no longer*. Because Nylon is resilient, a *long* tamp is "springy", and it doesn't transmit your hammer blows as effectively as a *short* tamp. If you have a choice, a *short* tamp is *always* better than a long tamp. Drill 11/64", 21/64", and "F" size holes with an **aircraft extension bit**. An aircraft extension bit is a standard drill with an extra long shank. Aircraft extension bits come in 6" and 12" lengths, and you can buy them from an industrial tool dealer. Look in *The Yellow Pages* under headings like **MACHINE TOOLS**, or **TOOLS & MACHINERY**. The flutes on a standard drill carry away the chips as the drill turns, so a hole made with a standard drill can be completed in one pass. When drilling a *deep* hole with an aircraft extension bit, the extended shank gets in the way, and you have to *manually* clear the chips by repeatedly backing the drill out of the hole as you go. Drills 11/64" through "F" should be backed out and cleared every 1/8" to 1/4".

We all know that drills come in standard fractional diameters, like 1/8", 5/32", 1/4", etc., but drills are made in number and letter sizes too. Number drills and letter drills provide all the important diameters not covered by the fractional drills. Number drills are numbered **1** through **80**, with **80** being the smallest, and letter drills are designated **A** through **Z**, with **A** being the smallest. An **F**-size drill is 0.257" dia.; just a little bigger than 1/4". You can buy an **F**-size drill from an industrial tool dealer.

Drill holes 3/8" dia. and larger with a **Speedbor** wood boring bit. Because of its rather crude form and method of operation, a Speedbor *usually* cuts a hole slightly larger than its designated size. If it *doesn't*, and the tamp jams on the base of the spindle when you try to use it, enlarge the first inch of the spindle hole by 1/64" to 1/32". Speedbors come in standard 6" lengths, and extra long 16" lengths. The 6" Speedbors are used for holes up to 5" long, and most hardware stores sell them. The 16" Speedbors are used for holes up to 15" long, and they might be harder to find. Here in Moreno Valley, a nearby Home Depot sells them, but the "Mom & Pop" hardware store *does not*. If you *can't* find one at a hardware store, check with industrial tool dealers. The 16" Speedbors are used by building and electrical contractors, and anyone who sells tools to the building trades should have them.

If your lathe is big enough, you can make even the largest and longest tamps from a single piece of Nylon. *People with small lathes* can make them in two or more sections, and plug them together. You'll find plug and socket details in drawings **98-107** and **98-108** on **pages 170 - 171**. When made of *Nylon*, the plug should be about 0.010" larger in diameter than the socket to insure a tight fit. *Loose* plug and socket combinations can be tightened by shimming with aluminum foil or small pieces of paper. Instructions for making Nylon tamps begin on **page 72**. Nylon tamp drawings begin on **page 145**. *If you don't have access to a metal lathe, the instructions beginning on page 77 will show you how to improvise.*

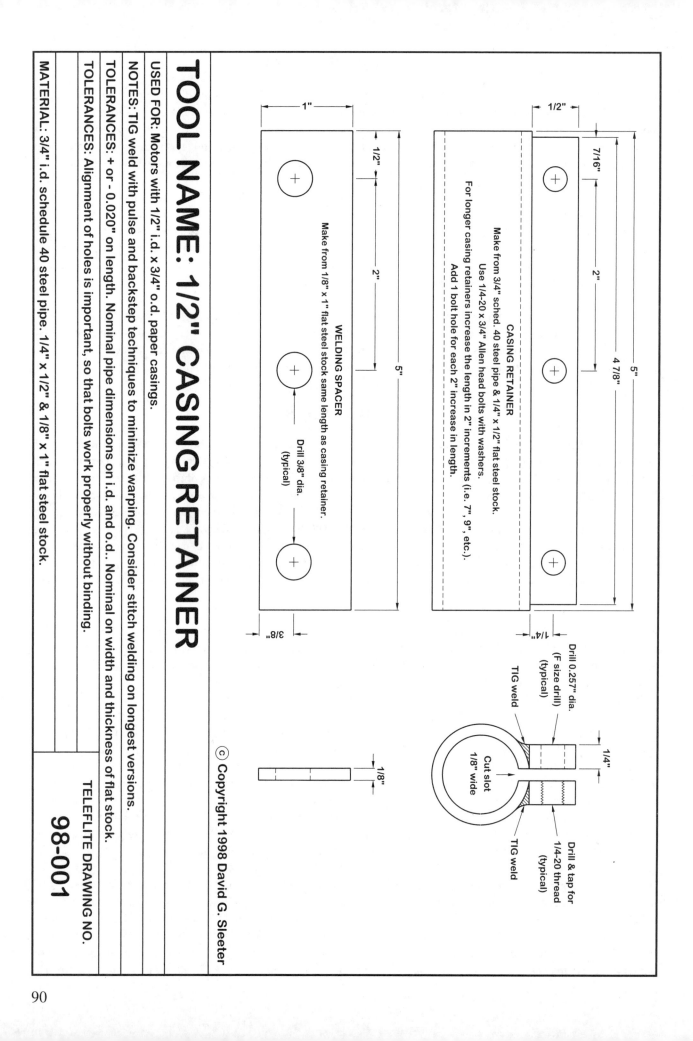

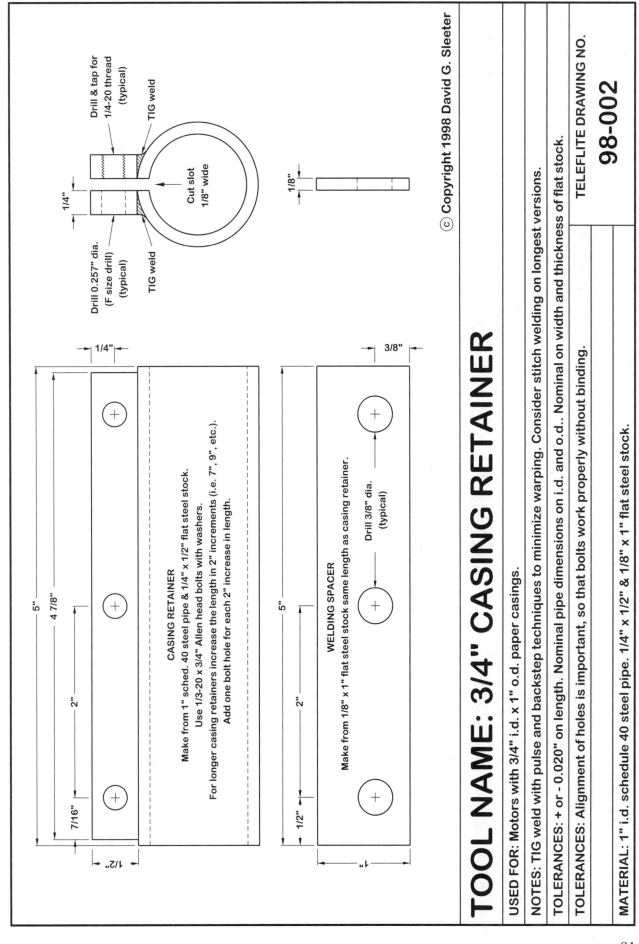

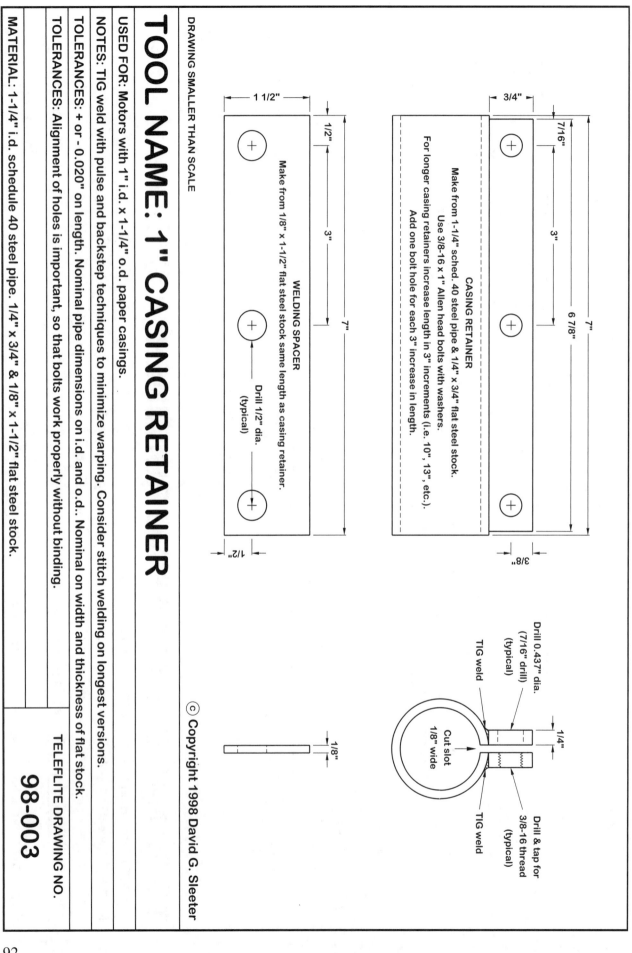

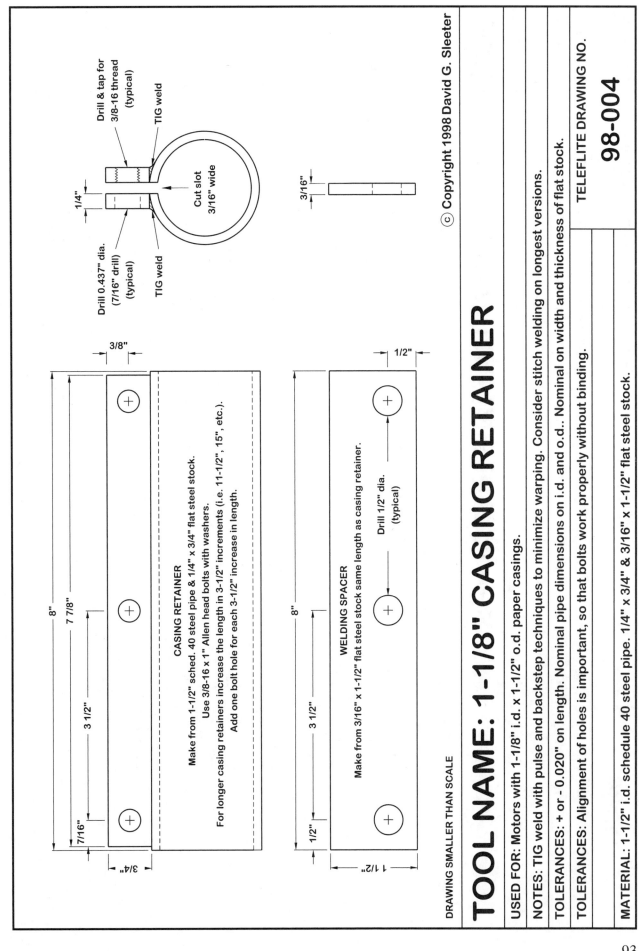

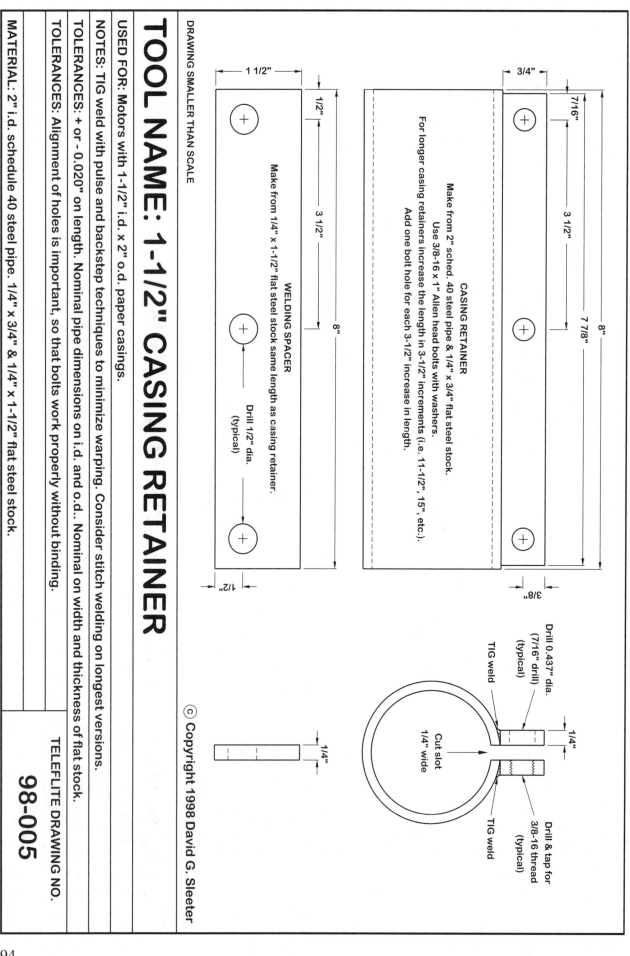

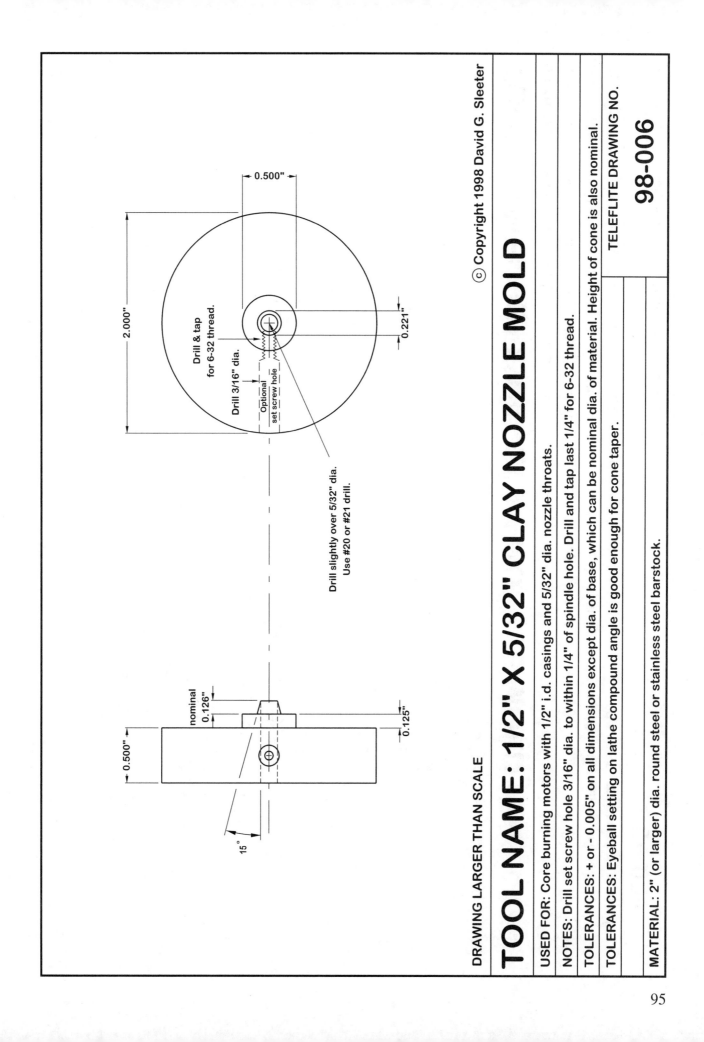

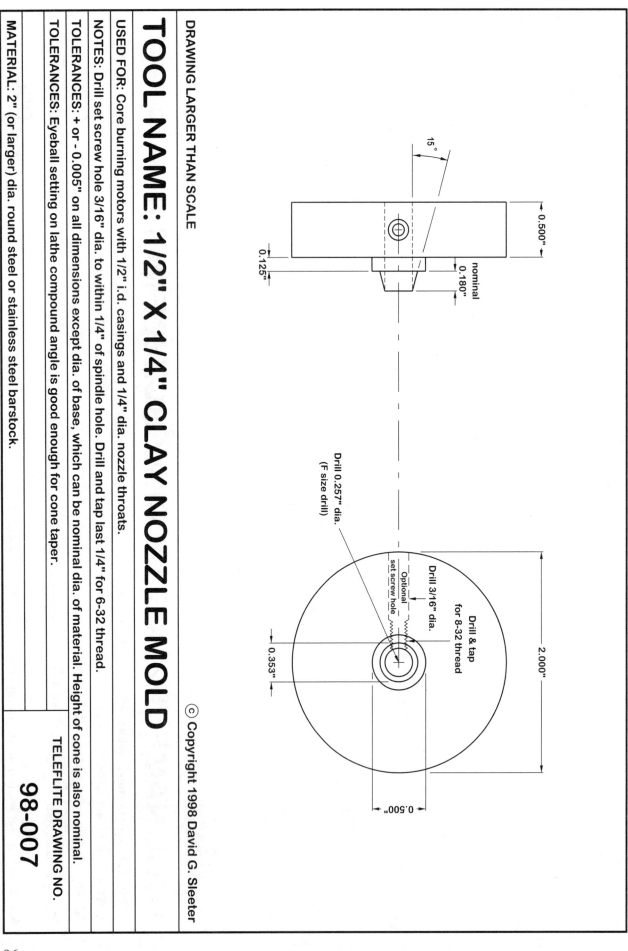

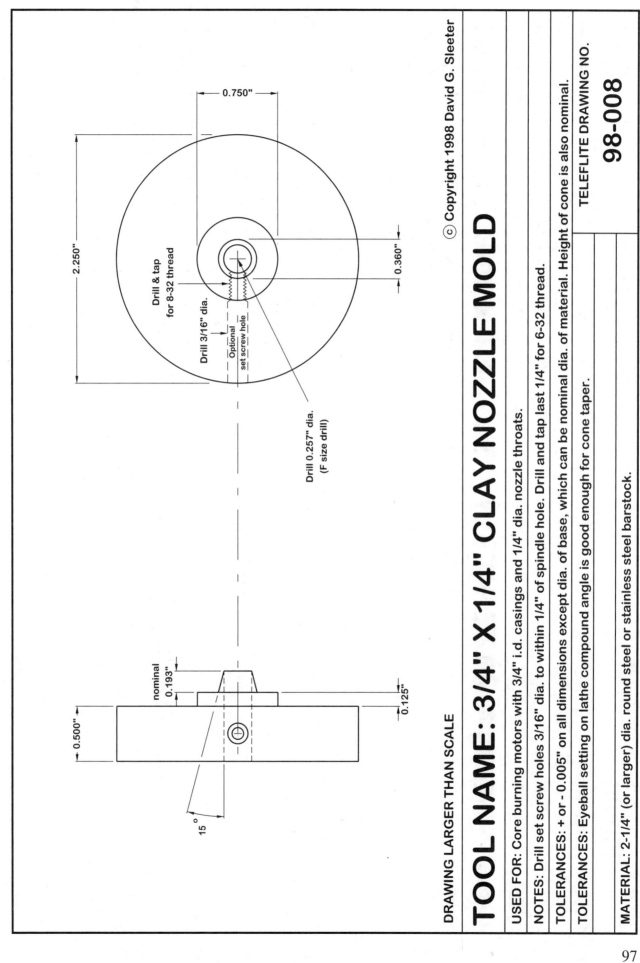

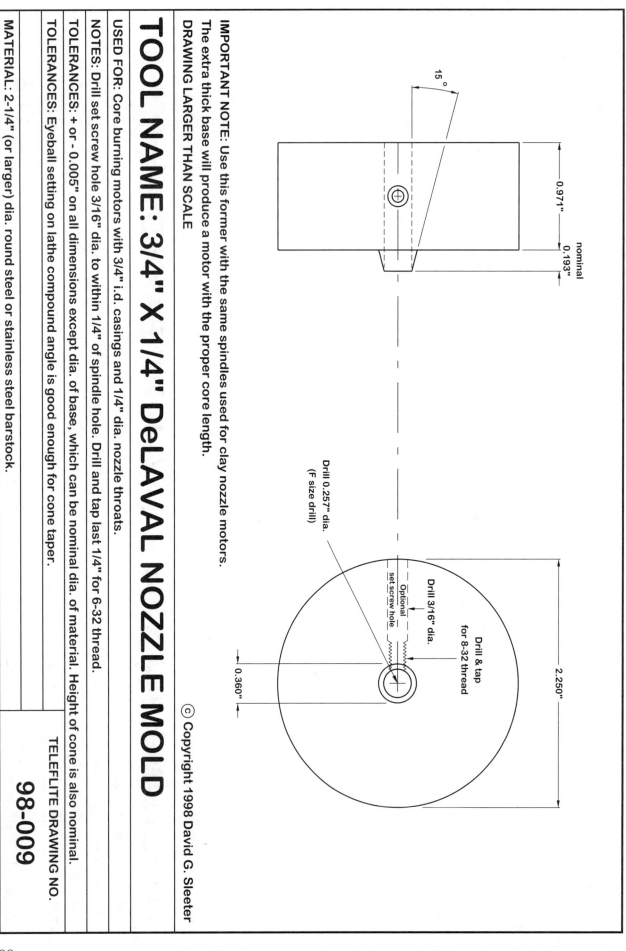

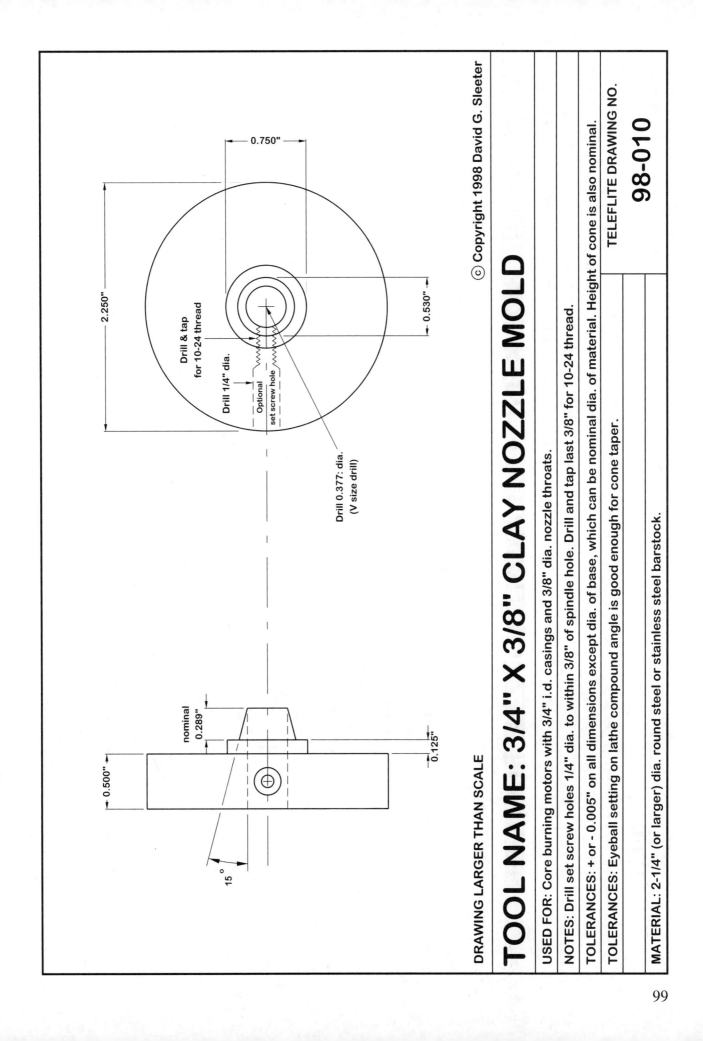

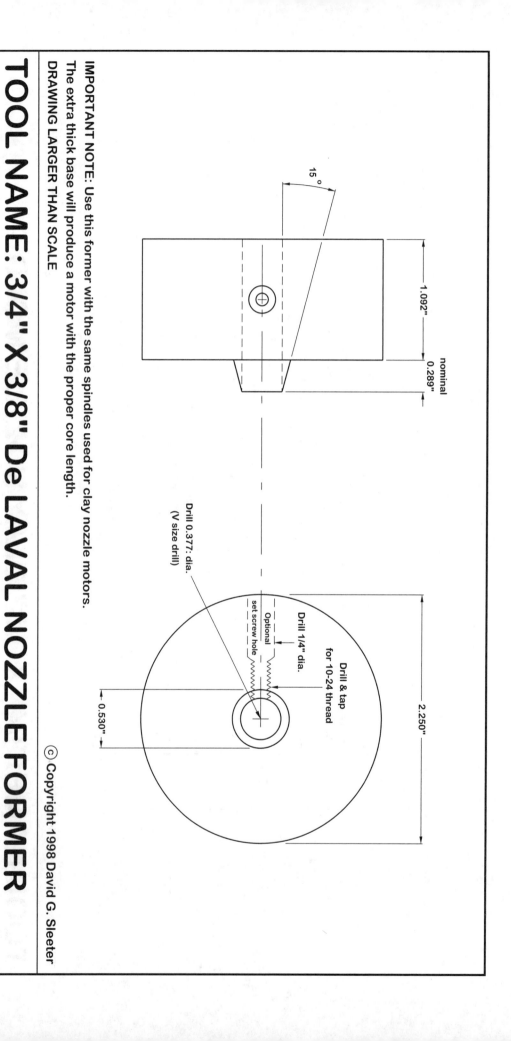

TOOL NAME: 3/4" X 3/8" De LAVAL NOZZLE FORMER

IMPORTANT NOTE: Use this former with the same spindles used for clay nozzle motors. The extra thick base will produce a motor with the proper core length.

DRAWING LARGER THAN SCALE

USED FOR: Core burning motors with 3/4" i.d. casings and 3/8" dia. nozzle throats.

NOTES: Drill set screw hole 1/4" dia. to within 3/8" of spindle hole. Drill and tap last 3/8" for 10-24 thread.

TOLERANCES: + or - 0.005" on all dimensions except dia. of base, which can be nominal dia. of material. Height of cone is also nominal.

TOLERANCES: Eyeball setting on lathe compound angle is good enough for cone taper.

MATERIAL: 2-1/4" (or larger) dia. round steel or stainless steel barstock.

© Copyright 1998 David G. Sleeter

TELEFLITE DRAWING NO.

98-011

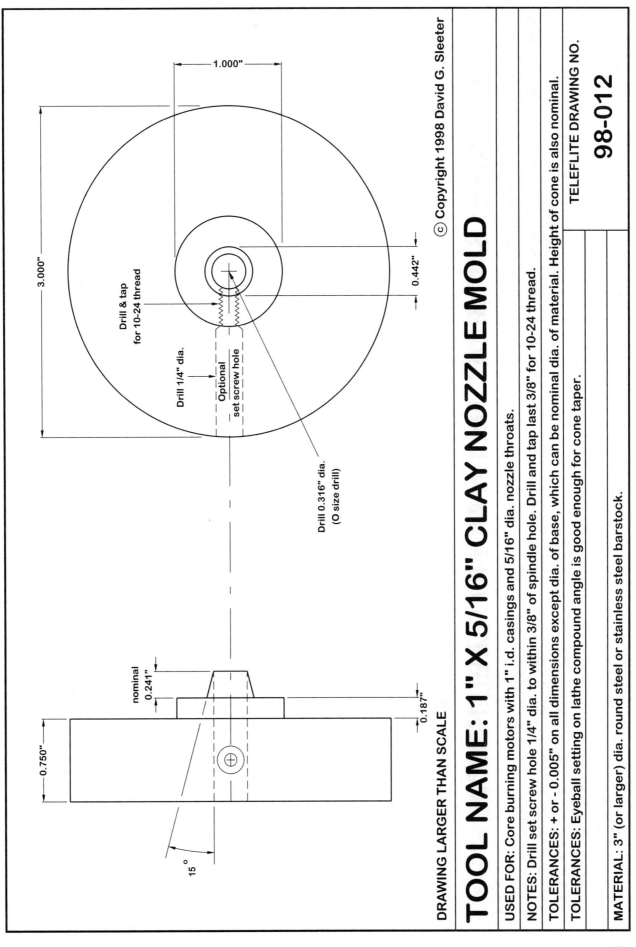

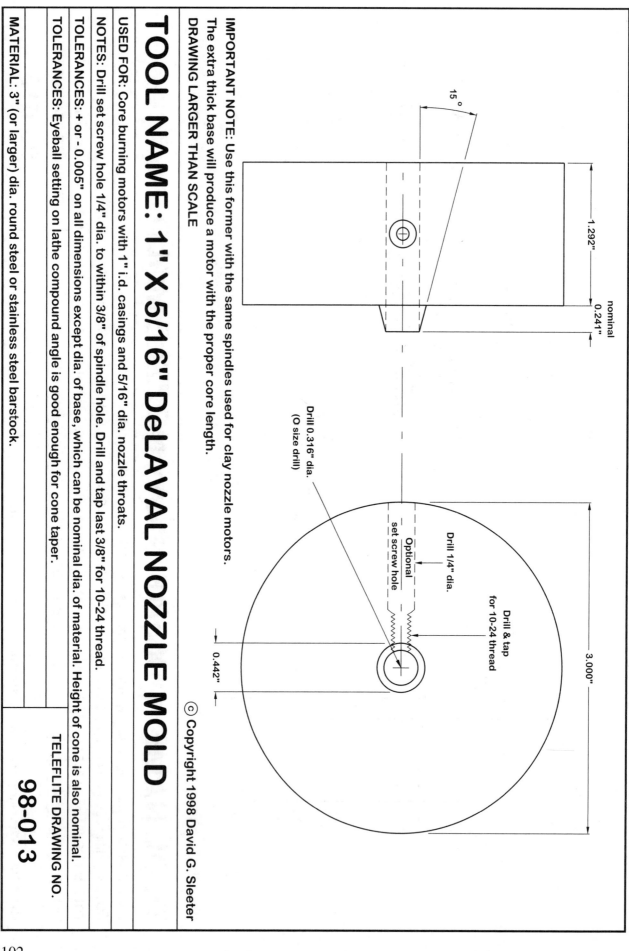

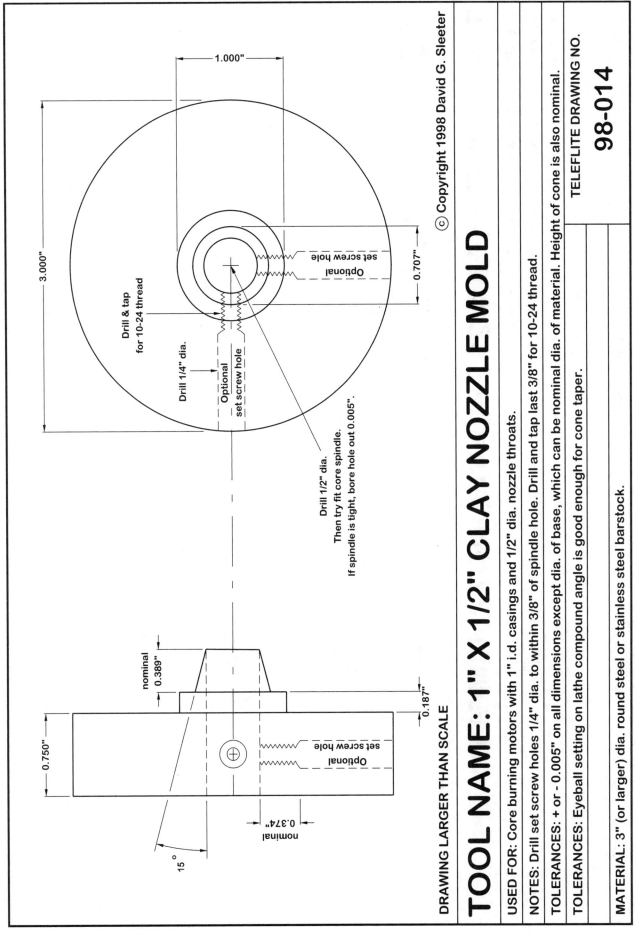

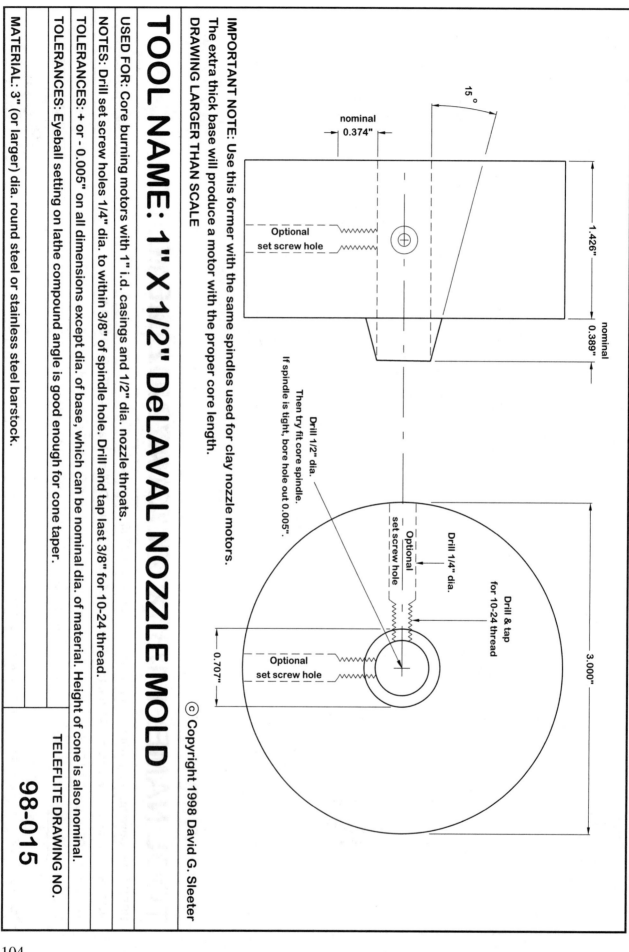

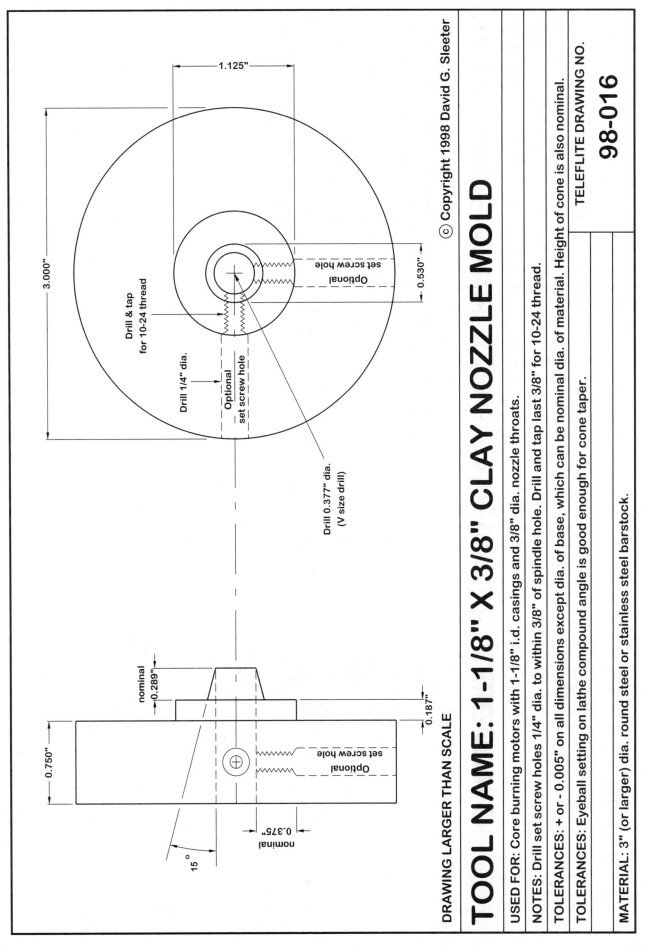

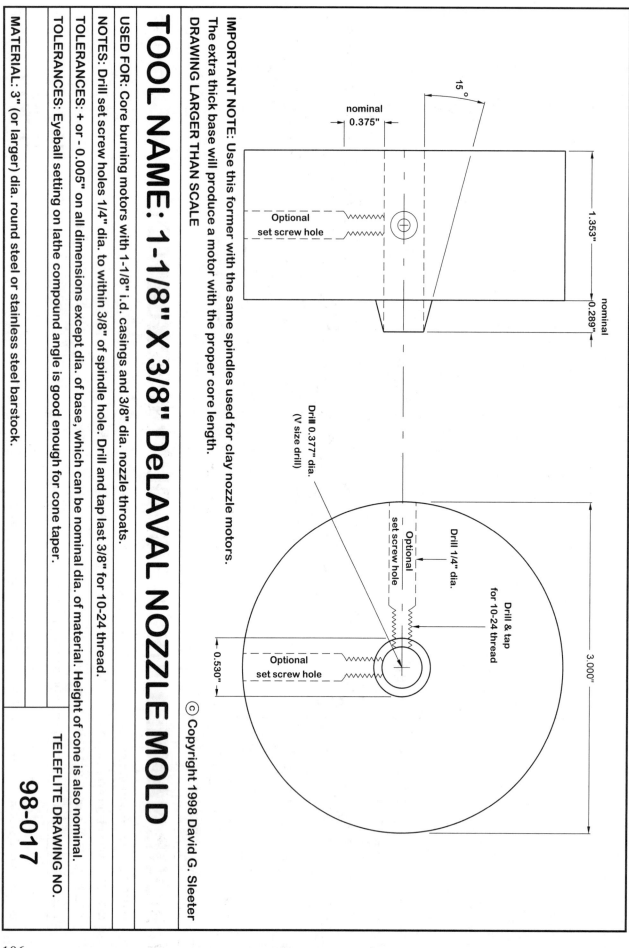

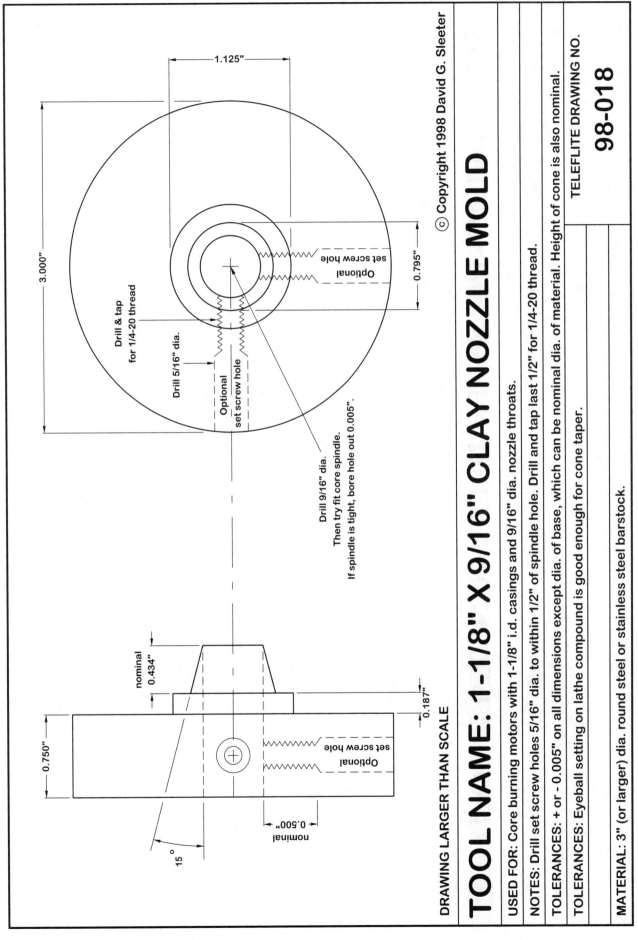

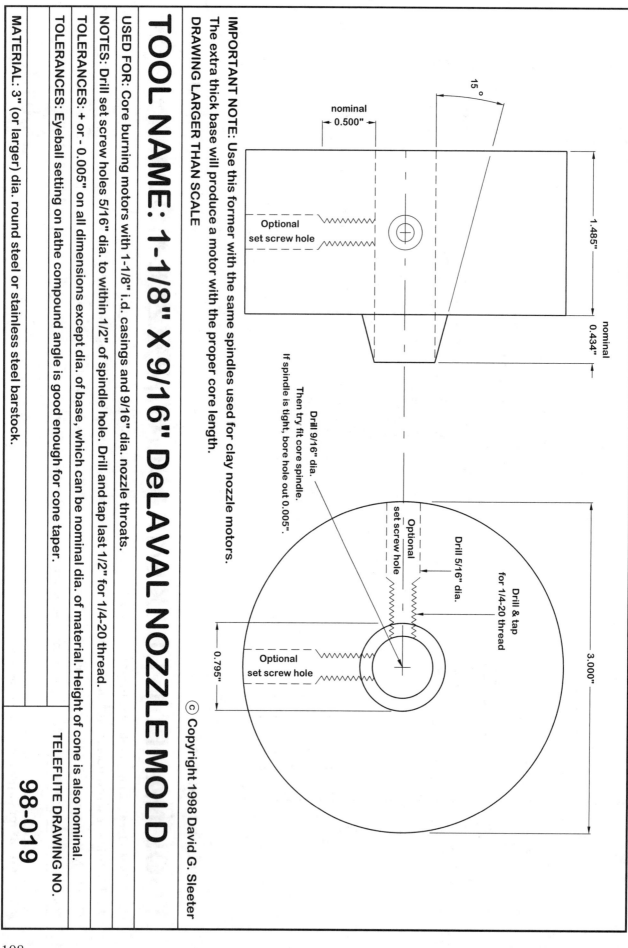

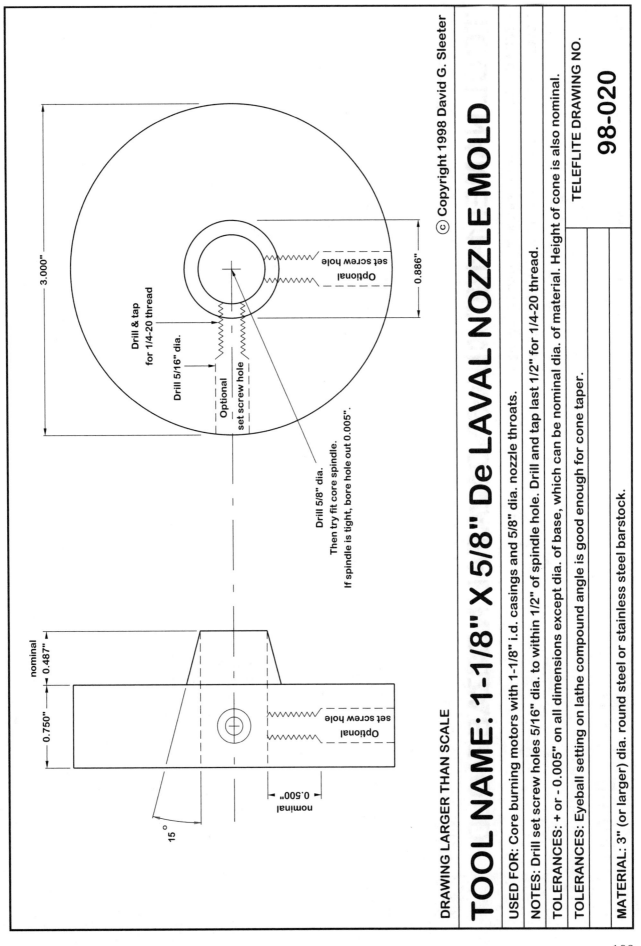

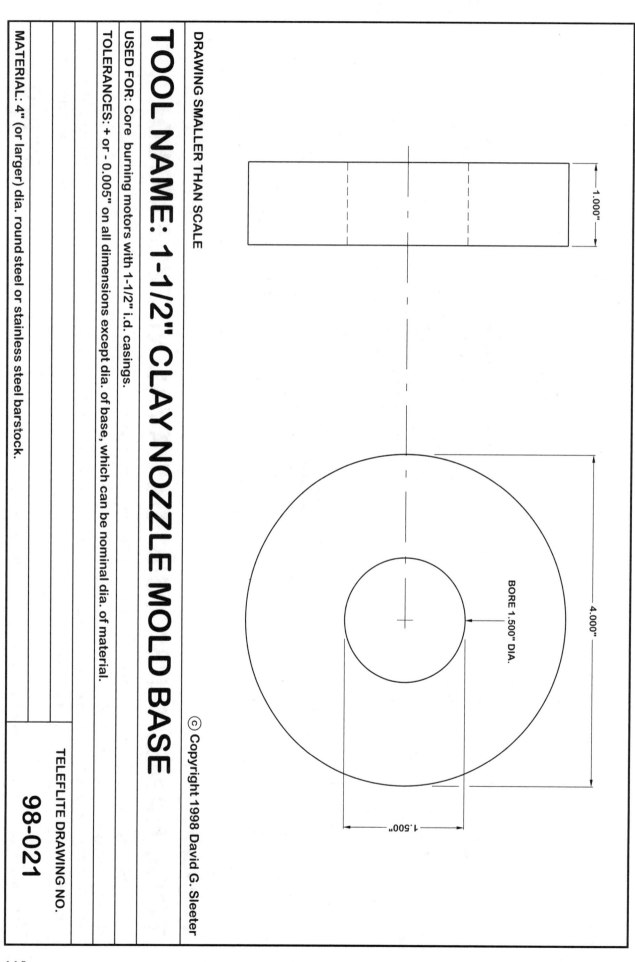

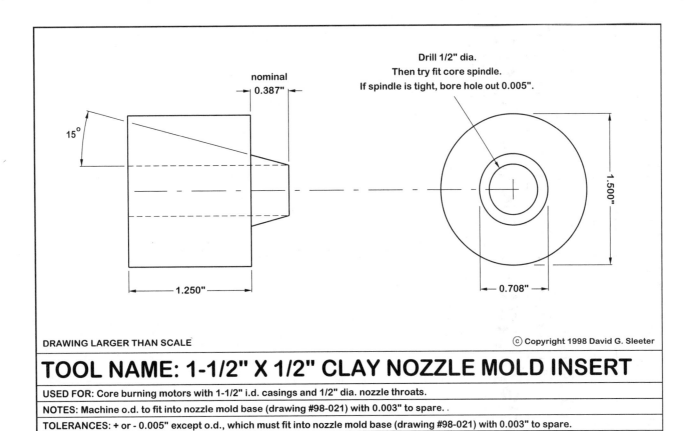

TOOL NAME: 1-1/2" X 1/2" CLAY NOZZLE MOLD INSERT

USED FOR: Core burning motors with 1-1/2" i.d. casings and 1/2" dia. nozzle throats.	
NOTES: Machine o.d. to fit into nozzle mold base (drawing #98-021) with 0.003" to spare.	
TOLERANCES: + or - 0.005" except o.d., which must fit into nozzle mold base (drawing #98-021) with 0.003" to spare.	
TOLERANCES: Eyeball setting on lathe compound angle is good enough for cone taper.	TELEFLITE DRAWING NO. 98-022
MATERIAL: 1-9/16" (or larger) dia. round steel or stainless steel barstock.	

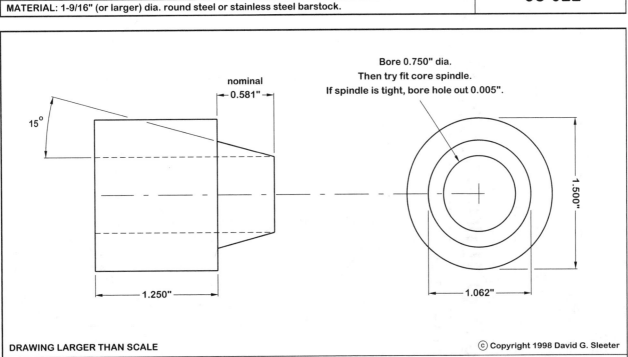

TOOL NAME: 1-1/2" X 3/4" CLAY NOZZLE MOLD INSERT

USED FOR: Core burning motors with 1-1/2" i.d. casings and 3/4" dia. nozzle throats.	
NOTES: Machine o.d. to fit into nozzle mold base (drawing #98-021) with 0.003" to spare.	
TOLERANCES: + or - 0.005" except o.d., which must fit into nozzle mold base (drawing #98-021) with 0.003" to spare.	
TOLERANCES: Eyeball setting on lathe compound angle is good enough for cone taper.	TELEFLITE DRAWING NO. 98-023
MATERIAL: 1-9/16" (or larger) dia. round steel or stainless steel barstock.	

TOOL NAME: 1-1/2" DeLAVAL NOZZLE MOLD BASE

USED FOR: Core burning motors with 1-1/2" i.d. casings.

TOLERANCES: + or - 0.005" on all dimensions except dia. of base, which can be nominal dia. of material.

MATERIAL: 4" (or larger) dia. round steel or stainless steel barstock.

DRAWING SMALLER THAN SCALE

© Copyright 1998 David G. Sleeter

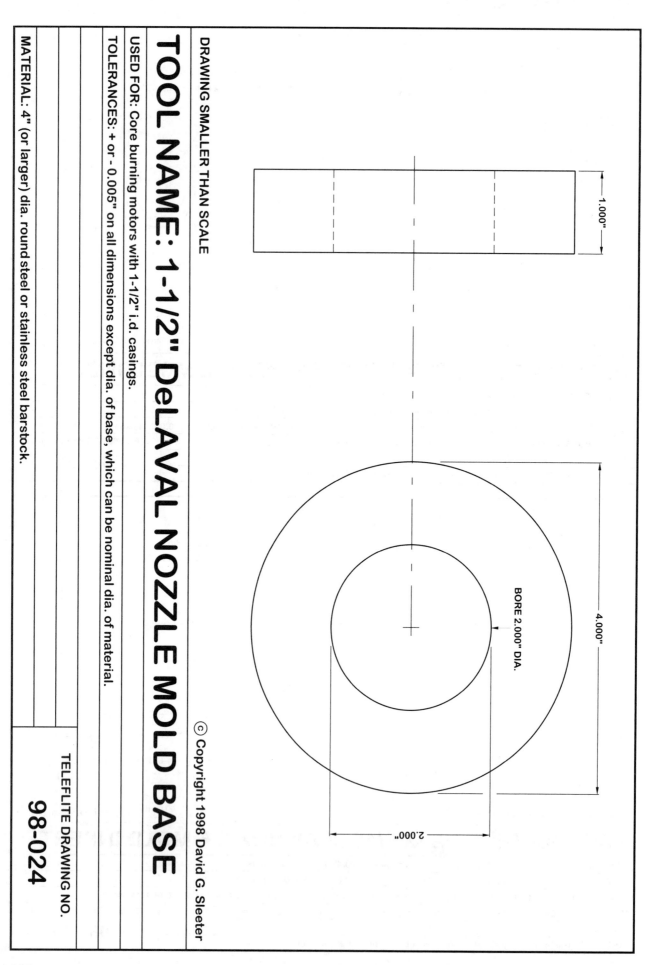

- 1.000"
- 4.000"
- BORE 2.000" DIA.
- 2.000"

TELEFLITE DRAWING NO. 98-024

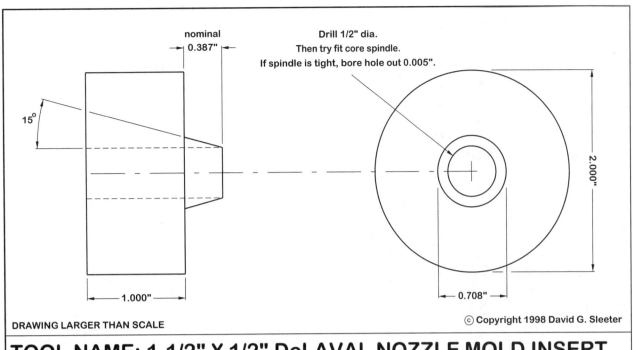

TOOL NAME: 1-1/2" X 1/2" DeLAVAL NOZZLE MOLD INSERT

USED FOR: Core burning motors with 1-1/2" i.d. casings & 1/2" dia. nozzle throats.	
NOTES: Machine o.d. to fit into nozzle mold base (drawing #98-024) with 0.003" to spare.	
TOLERANCES: + or - 0.005" except o.d., which must fit into nozzle mold base (drawing #98-024) with 0.003" to spare.	
TOLERANCES: Eyeball setting on lathe compound angle is good enough for cone taper.	TELEFLITE DRAWING NO. **98-025**
MATERIAL: 2-1/8" (or larger) dia. round steel or stainless steel barstock.	

TOOL NAME: 1-1/2" X 3/4" De LAVAL NOZZLE MOLD INSERT

USED FOR: Core burning motors with 1-1/2" i.d. casings and 3/4" dia. nozzle throats.	
NOTES: Machine o.d. to fit into nozzle mold base (drawing #98-024) with 0.003" to spare.	
TOLERANCES: + or - 0.005" except o.d., which must fit into nozzle mold base (drawing #98-024) with 0.003" to spare.	
TOLERANCES: Eyeball setting on lathe compound angle is good enough for cone taper.	TELEFLITE DRAWING NO. **98-026**
MATERIAL: 2-1/8" (or larger) dia. round steel or stainless steel barstock.	

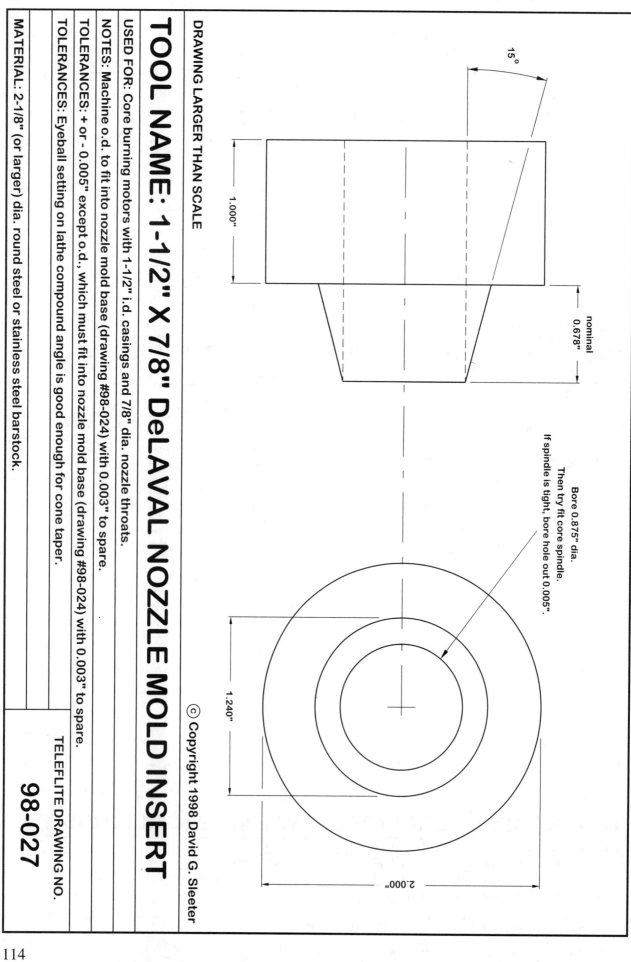

TOOL NAME: 1-1/2" X 7/8" DeLAVAL NOZZLE MOLD INSERT

DRAWING LARGER THAN SCALE

© Copyright 1998 David G. Sleeter

15°
1.000"
nominal 0.678"

Bore 0.875" dia.
Then try fit core spindle.
If spindle is tight, bore hole out 0.005".

1.240"
2.000"

USED FOR: Core burning motors with 1-1/2" i.d. casings and 7/8" dia. nozzle throats.

NOTES: Machine o.d. to fit into nozzle mold base (drawing #98-024) with 0.003" to spare.

TOLERANCES: + or - 0.005" except o.d., which must fit into nozzle mold base (drawing #98-024) with 0.003" to spare.

TOLERANCES: Eyeball setting on lathe compound angle is good enough for cone taper.

MATERIAL: 2-1/8" (or larger) dia. round steel or stainless steel barstock.

TELEFLITE DRAWING NO.
98-027

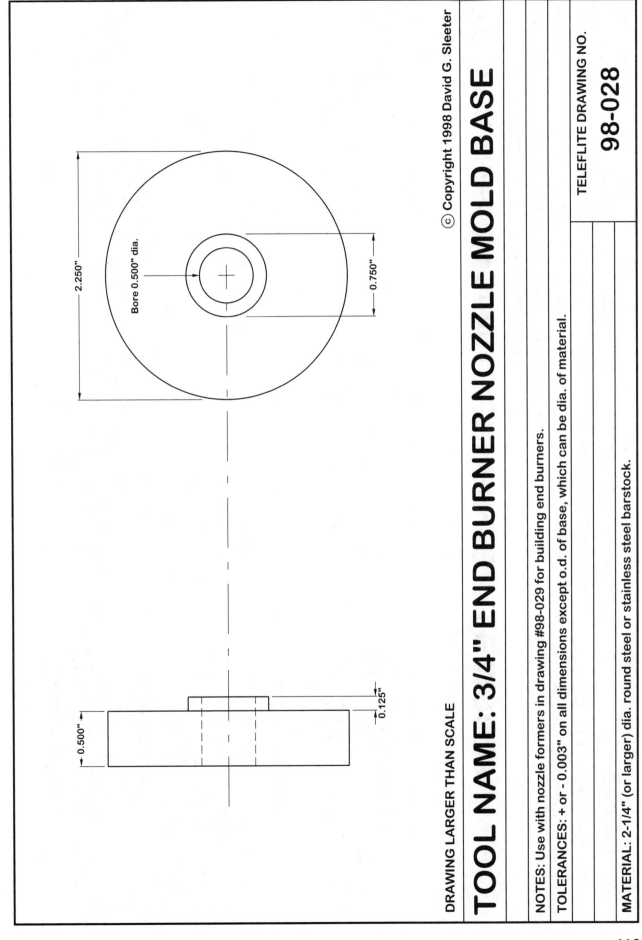

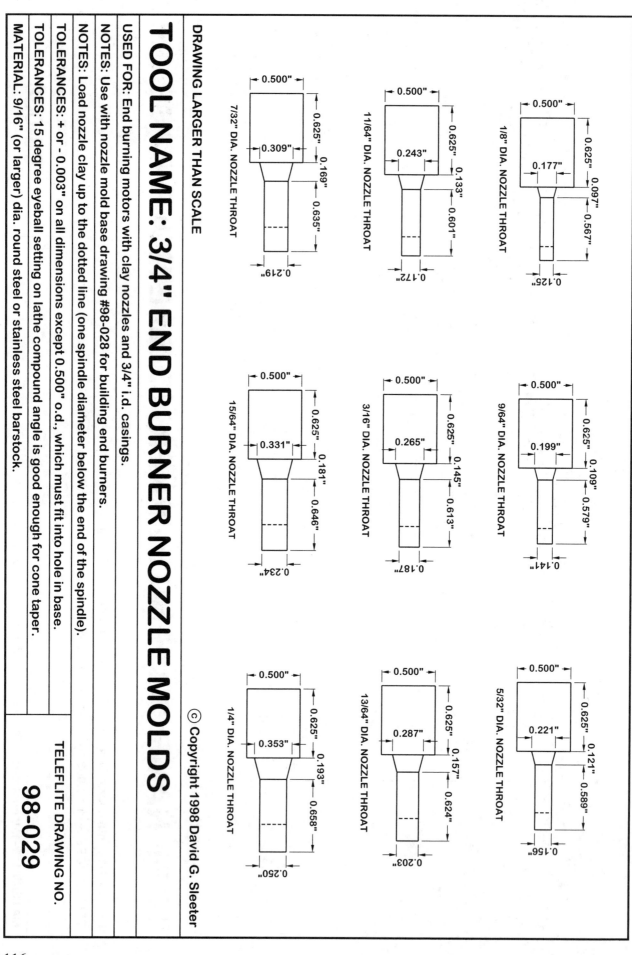

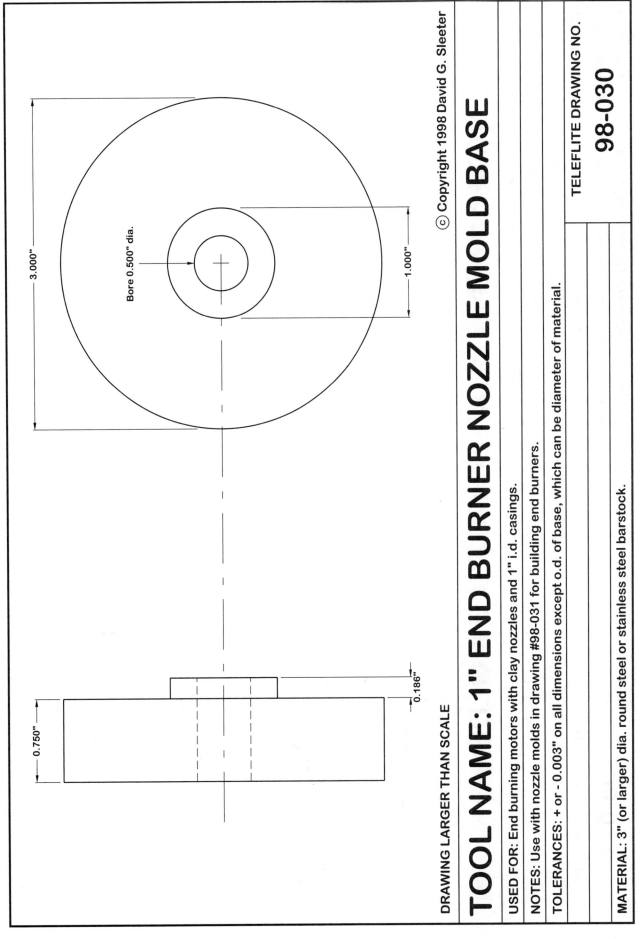

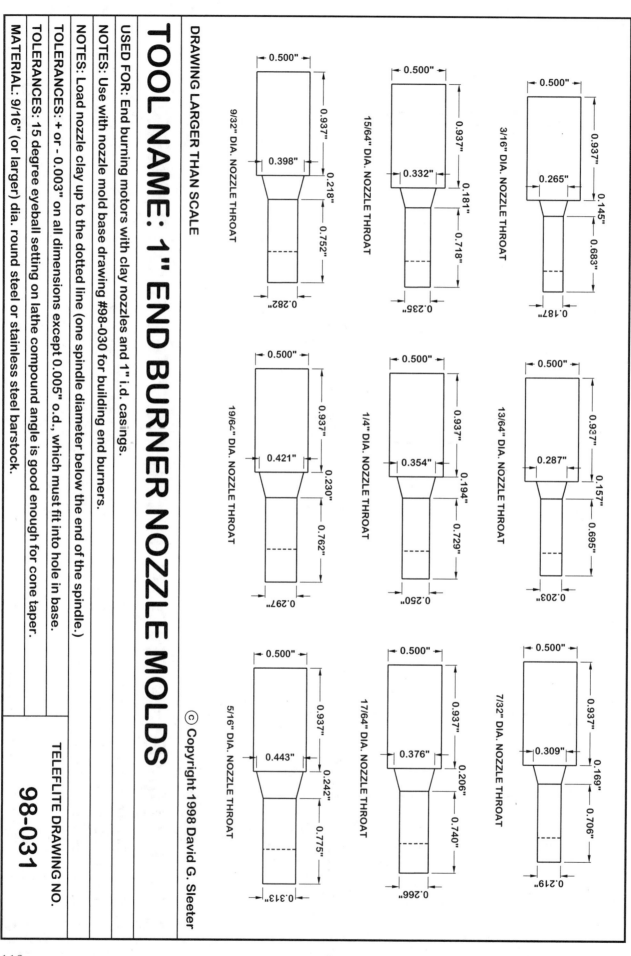

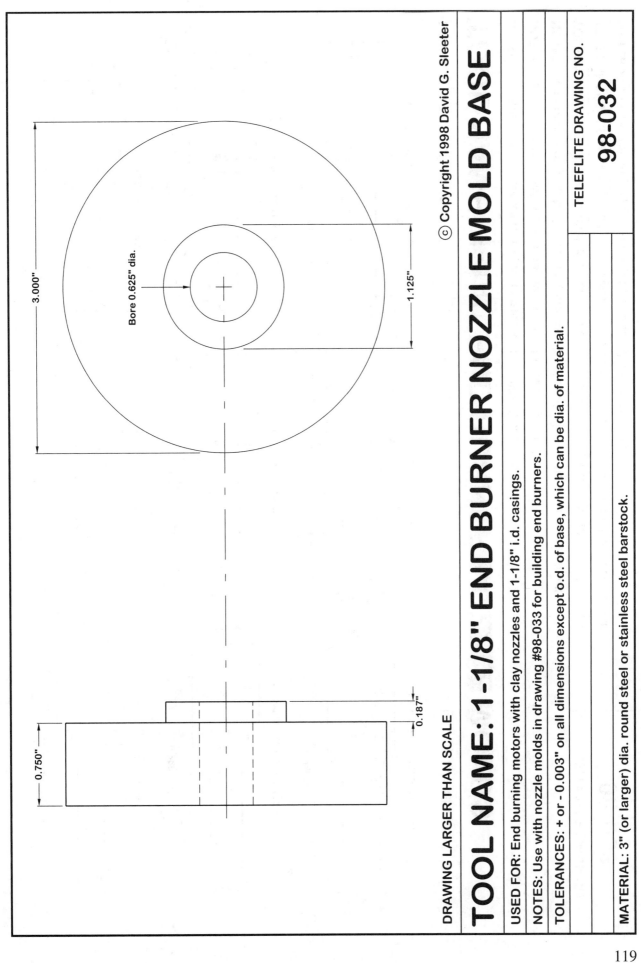

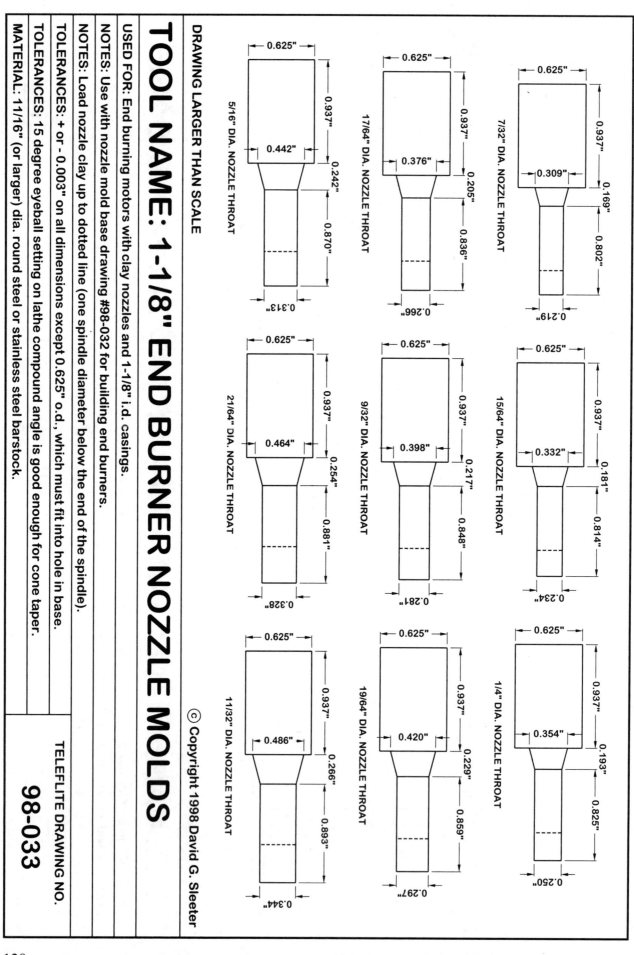

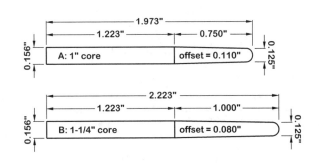

DRAWING LARGER THAN SCALE © Copyright 1998 David G. Sleeter

TOOL NAME: 1/2" X 5/32" CORE SPINDLES

USED FOR: Core burning motors with 1/2" i.d. casings and 5/32" dia. nozzle throats.	
TOLERANCES: + or - 0.050" on overall length. + or - 0.125" on point where taper begins.	
TOLERANCES: + or - 0.010" on dia. of tip. Nominal 5/32" dia. of raw material for base dia.	
	TELEFLITE DRAWING NO. **98-034**
MATERIAL: 5/32" dia. round stainless steel rod.	

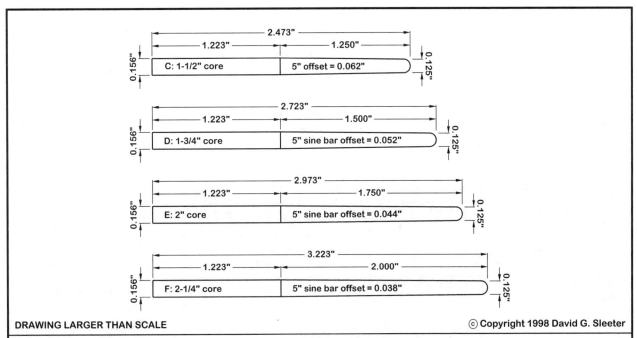

DRAWING LARGER THAN SCALE © Copyright 1998 David G. Sleeter

TOOL NAME: 1/2" X 5/32" CORE SPINDLES

USED FOR: Core burning motors with 1/2" i.d. casings and 5/32" dia. nozzle throats.	
TOLERANCES: + or - 0.050" on overall length. + or - 0.125" on point where taper begins.	
TOLERANCES: + or - 0.010" on dia. of tip. Nominal 5/32" dia. of raw material for base dia.	
	TELEFLITE DRAWING NO. **98-035**
MATERIAL: 5/32" dia. round stainless steel rod.	

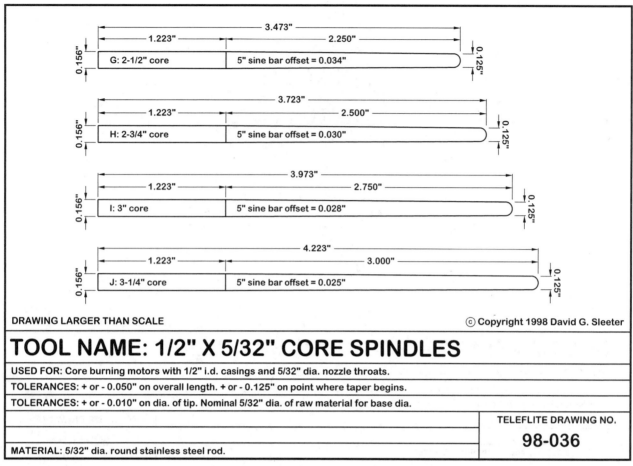

TOOL NAME: 1/2" X 5/32" CORE SPINDLES

USED FOR: Core burning motors with 1/2" i.d. casings and 5/32" dia. nozzle throats.
TOLERANCES: + or - 0.050" on overall length. + or - 0.125" on point where taper begins.
TOLERANCES: + or - 0.010" on dia. of tip. Nominal 5/32" dia. of raw material for base dia.
MATERIAL: 5/32" dia. round stainless steel rod.

TELEFLITE DRAWING NO. 98-036

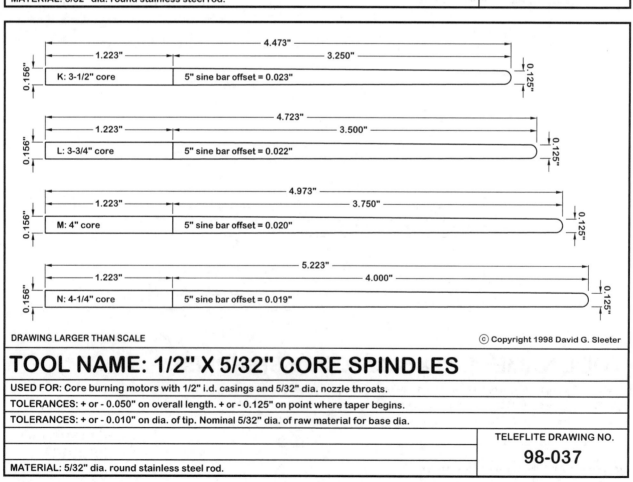

TOOL NAME: 1/2" X 5/32" CORE SPINDLES

USED FOR: Core burning motors with 1/2" i.d. casings and 5/32" dia. nozzle throats.
TOLERANCES: + or - 0.050" on overall length. + or - 0.125" on point where taper begins.
TOLERANCES: + or - 0.010" on dia. of tip. Nominal 5/32" dia. of raw material for base dia.
MATERIAL: 5/32" dia. round stainless steel rod.

TELEFLITE DRAWING NO. 98-037

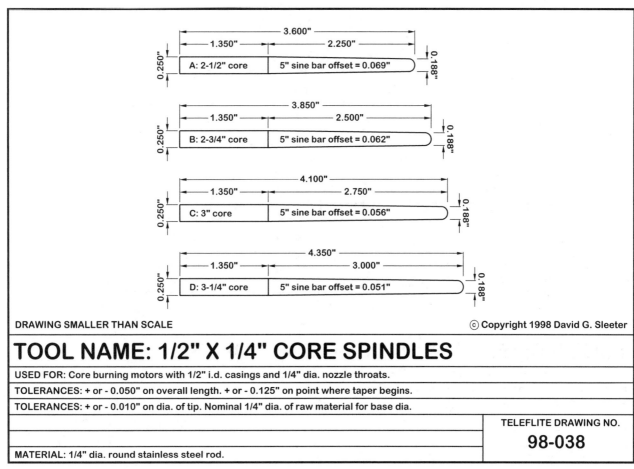

DRAWING SMALLER THAN SCALE © Copyright 1998 David G. Sleeter

TOOL NAME: 1/2" X 1/4" CORE SPINDLES

USED FOR: Core burning motors with 1/2" i.d. casings and 1/4" dia. nozzle throats.
TOLERANCES: + or - 0.050" on overall length. + or - 0.125" on point where taper begins.
TOLERANCES: + or - 0.010" on dia. of tip. Nominal 1/4" dia. of raw material for base dia.

TELEFLITE DRAWING NO. **98-038**

MATERIAL: 1/4" dia. round stainless steel rod.

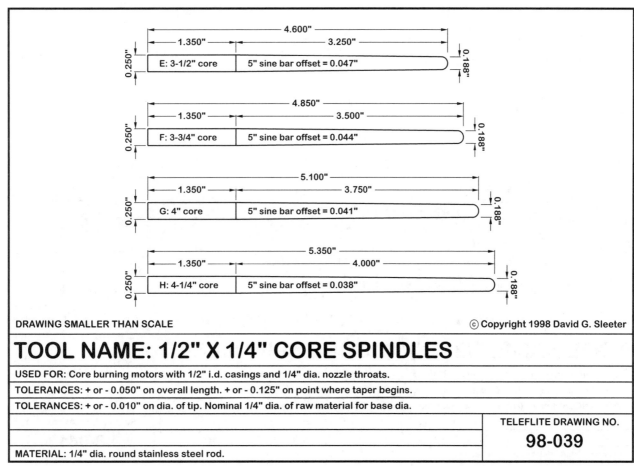

DRAWING SMALLER THAN SCALE © Copyright 1998 David G. Sleeter

TOOL NAME: 1/2" X 1/4" CORE SPINDLES

USED FOR: Core burning motors with 1/2" i.d. casings and 1/4" dia. nozzle throats.
TOLERANCES: + or - 0.050" on overall length. + or - 0.125" on point where taper begins.
TOLERANCES: + or - 0.010" on dia. of tip. Nominal 1/4" dia. of raw material for base dia.

TELEFLITE DRAWING NO. **98-039**

MATERIAL: 1/4" dia. round stainless steel rod.

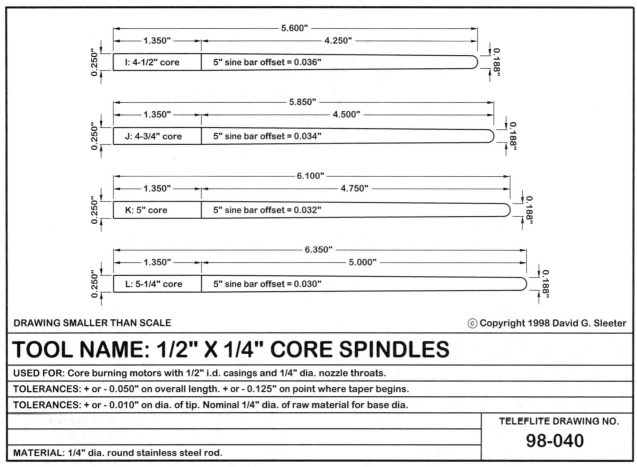

TOOL NAME: 1/2" X 1/4" CORE SPINDLES

USED FOR: Core burning motors with 1/2" i.d. casings and 1/4" dia. nozzle throats.
TOLERANCES: + or - 0.050" on overall length. + or - 0.125" on point where taper begins.
TOLERANCES: + or - 0.010" on dia. of tip. Nominal 1/4" dia. of raw material for base dia.

MATERIAL: 1/4" dia. round stainless steel rod.

TELEFLITE DRAWING NO. **98-040**

© Copyright 1998 David G. Sleeter

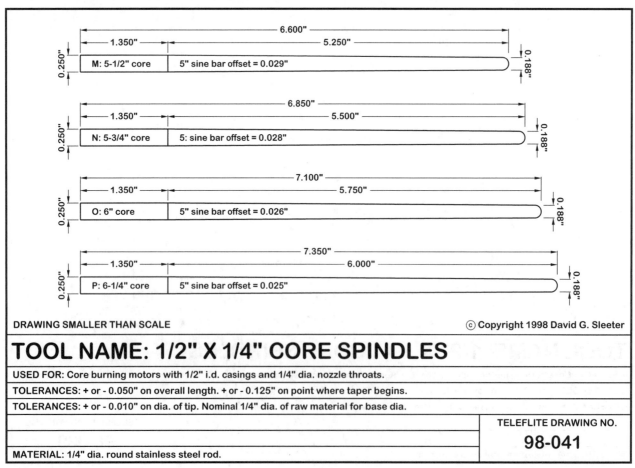

TOOL NAME: 1/2" X 1/4" CORE SPINDLES

USED FOR: Core burning motors with 1/2" i.d. casings and 1/4" dia. nozzle throats.
TOLERANCES: + or - 0.050" on overall length. + or - 0.125" on point where taper begins.
TOLERANCES: + or - 0.010" on dia. of tip. Nominal 1/4" dia. of raw material for base dia.

MATERIAL: 1/4" dia. round stainless steel rod.

TELEFLITE DRAWING NO. **98-041**

© Copyright 1998 David G. Sleeter

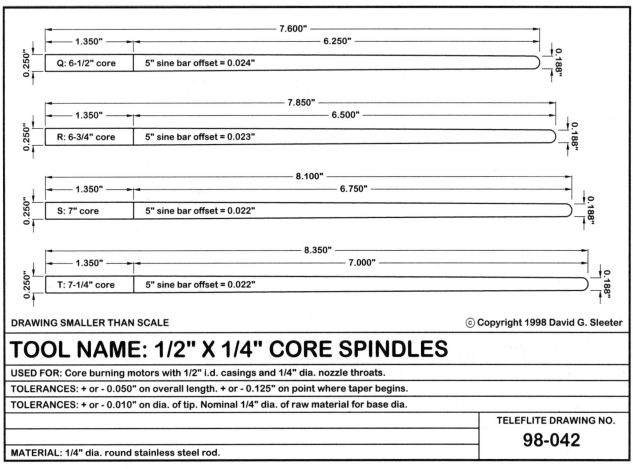

TOOL NAME: 1/2" X 1/4" CORE SPINDLES

USED FOR: Core burning motors with 1/2" i.d. casings and 1/4" dia. nozzle throats.
TOLERANCES: + or - 0.050" on overall length. + or - 0.125" on point where taper begins.
TOLERANCES: + or - 0.010" on dia. of tip. Nominal 1/4" dia. of raw material for base dia.
MATERIAL: 1/4" dia. round stainless steel rod.

TELEFLITE DRAWING NO. **98-042**

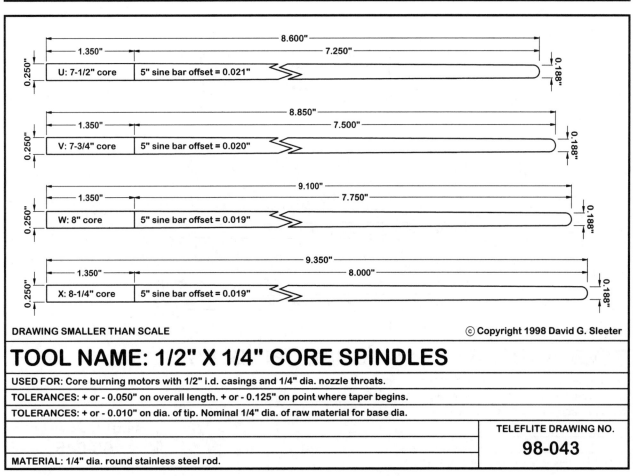

TOOL NAME: 1/2" X 1/4" CORE SPINDLES

USED FOR: Core burning motors with 1/2" i.d. casings and 1/4" dia. nozzle throats.
TOLERANCES: + or - 0.050" on overall length. + or - 0.125" on point where taper begins.
TOLERANCES: + or - 0.010" on dia. of tip. Nominal 1/4" dia. of raw material for base dia.
MATERIAL: 1/4" dia. round stainless steel rod.

TELEFLITE DRAWING NO. **98-043**

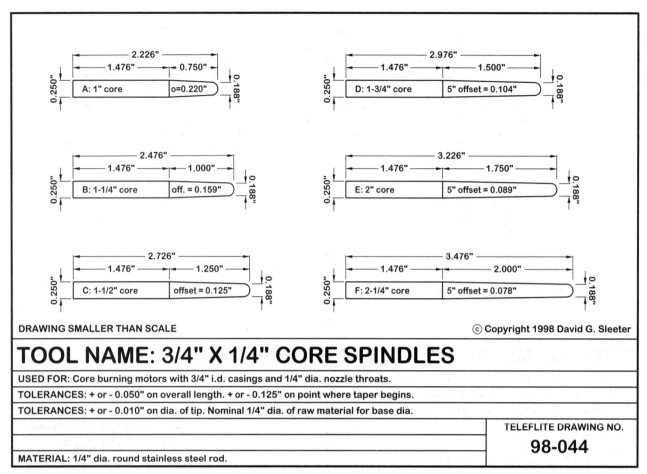

DRAWING SMALLER THAN SCALE © Copyright 1998 David G. Sleeter

TOOL NAME: 3/4" X 1/4" CORE SPINDLES

USED FOR: Core burning motors with 3/4" i.d. casings and 1/4" dia. nozzle throats.

TOLERANCES: + or - 0.050" on overall length. + or - 0.125" on point where taper begins.

TOLERANCES: + or - 0.010" on dia. of tip. Nominal 1/4" dia. of raw material for base dia.

MATERIAL: 1/4" dia. round stainless steel rod.

TELEFLITE DRAWING NO. **98-044**

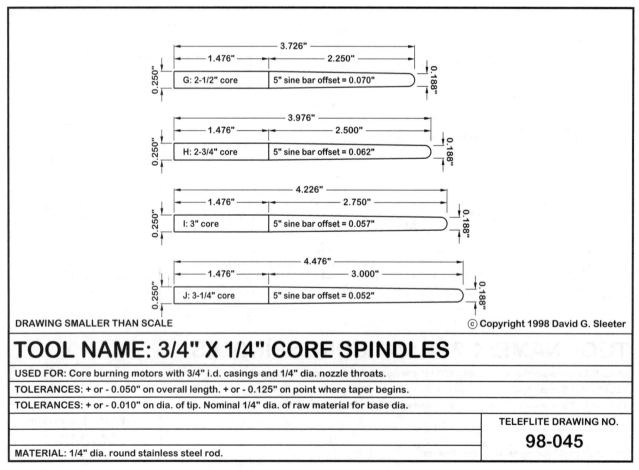

DRAWING SMALLER THAN SCALE © Copyright 1998 David G. Sleeter

TOOL NAME: 3/4" X 1/4" CORE SPINDLES

USED FOR: Core burning motors with 3/4" i.d. casings and 1/4" dia. nozzle throats.

TOLERANCES: + or - 0.050" on overall length. + or - 0.125" on point where taper begins.

TOLERANCES: + or - 0.010" on dia. of tip. Nominal 1/4" dia. of raw material for base dia.

MATERIAL: 1/4" dia. round stainless steel rod.

TELEFLITE DRAWING NO. **98-045**

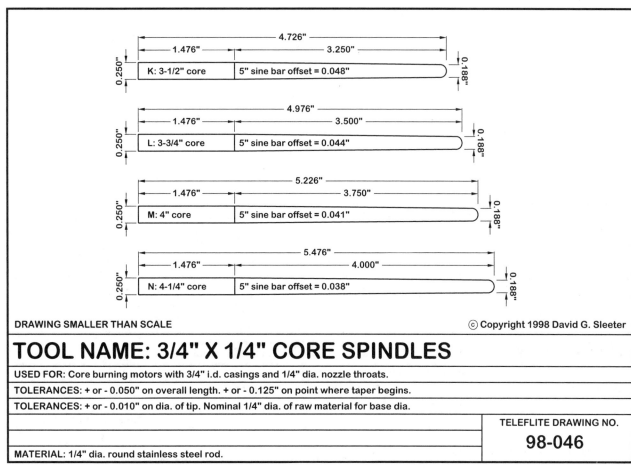

TOOL NAME: 3/4" X 1/4" CORE SPINDLES

USED FOR: Core burning motors with 3/4" i.d. casings and 1/4" dia. nozzle throats.
TOLERANCES: + or - 0.050" on overall length. + or - 0.125" on point where taper begins.
TOLERANCES: + or - 0.010" on dia. of tip. Nominal 1/4" dia. of raw material for base dia.
MATERIAL: 1/4" dia. round stainless steel rod.

TELEFLITE DRAWING NO. **98-046**

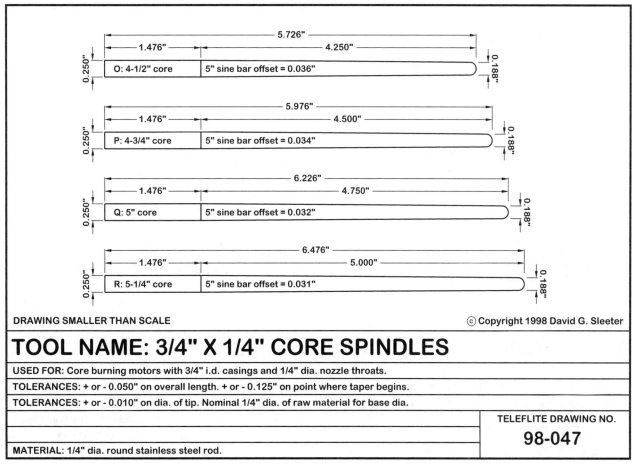

TOOL NAME: 3/4" X 1/4" CORE SPINDLES

USED FOR: Core burning motors with 3/4" i.d. casings and 1/4" dia. nozzle throats.
TOLERANCES: + or - 0.050" on overall length. + or - 0.125" on point where taper begins.
TOLERANCES: + or - 0.010" on dia. of tip. Nominal 1/4" dia. of raw material for base dia.
MATERIAL: 1/4" dia. round stainless steel rod.

TELEFLITE DRAWING NO. **98-047**

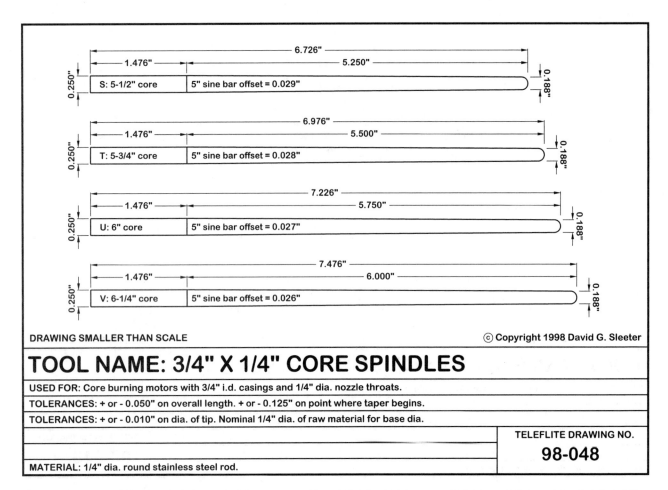

TOOL NAME: 3/4" X 1/4" CORE SPINDLES

USED FOR: Core burning motors with 3/4" i.d. casings and 1/4" dia. nozzle throats.

TOLERANCES: + or - 0.050" on overall length. + or - 0.125" on point where taper begins.

TOLERANCES: + or - 0.010" on dia. of tip. Nominal 1/4" dia. of raw material for base dia.

MATERIAL: 1/4" dia. round stainless steel rod.

TELEFLITE DRAWING NO. **98-048**

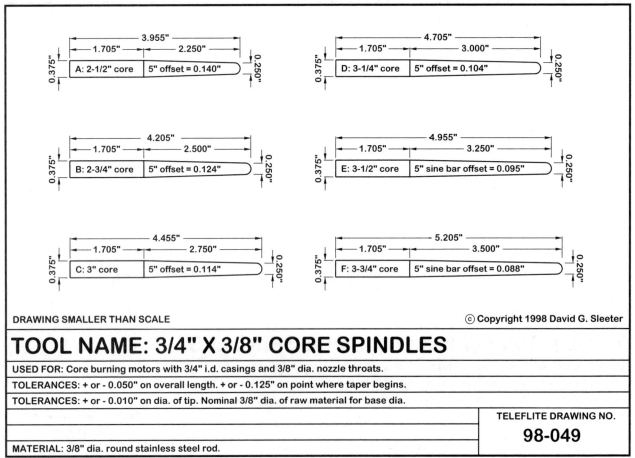

TOOL NAME: 3/4" X 3/8" CORE SPINDLES

USED FOR: Core burning motors with 3/4" i.d. casings and 3/8" dia. nozzle throats.

TOLERANCES: + or - 0.050" on overall length. + or - 0.125" on point where taper begins.

TOLERANCES: + or - 0.010" on dia. of tip. Nominal 3/8" dia. of raw material for base dia.

MATERIAL: 3/8" dia. round stainless steel rod.

TELEFLITE DRAWING NO. **98-049**

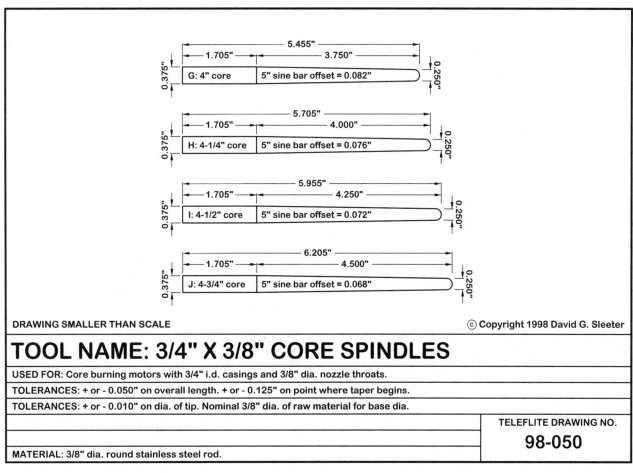

TOOL NAME: 3/4" X 3/8" CORE SPINDLES

USED FOR: Core burning motors with 3/4" i.d. casings and 3/8" dia. nozzle throats.

TOLERANCES: + or - 0.050" on overall length. + or - 0.125" on point where taper begins.

TOLERANCES: + or - 0.010" on dia. of tip. Nominal 3/8" dia. of raw material for base dia.

TELEFLITE DRAWING NO. **98-050**

MATERIAL: 3/8" dia. round stainless steel rod.

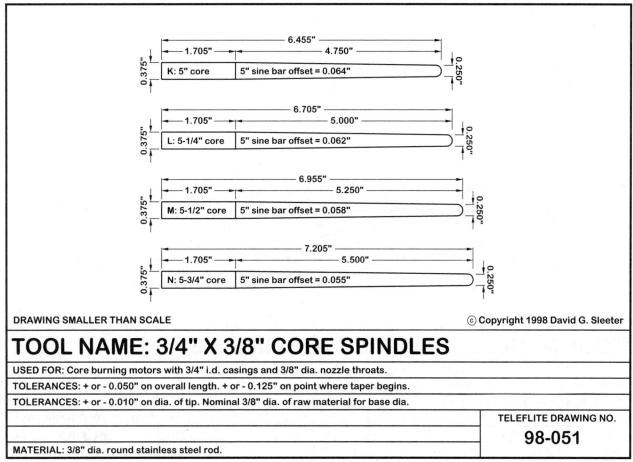

TOOL NAME: 3/4" X 3/8" CORE SPINDLES

USED FOR: Core burning motors with 3/4" i.d. casings and 3/8" dia. nozzle throats.

TOLERANCES: + or - 0.050" on overall length. + or - 0.125" on point where taper begins.

TOLERANCES: + or - 0.010" on dia. of tip. Nominal 3/8" dia. of raw material for base dia.

TELEFLITE DRAWING NO. **98-051**

MATERIAL: 3/8" dia. round stainless steel rod.

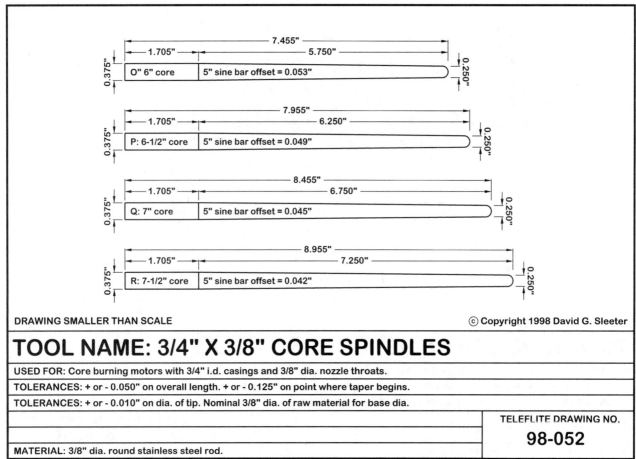

TOOL NAME: 3/4" X 3/8" CORE SPINDLES

USED FOR: Core burning motors with 3/4" i.d. casings and 3/8" dia. nozzle throats.
TOLERANCES: + or - 0.050" on overall length. + or - 0.125" on point where taper begins.
TOLERANCES: + or - 0.010" on dia. of tip. Nominal 3/8" dia. of raw material for base dia.

TELEFLITE DRAWING NO. 98-052

MATERIAL: 3/8" dia. round stainless steel rod.

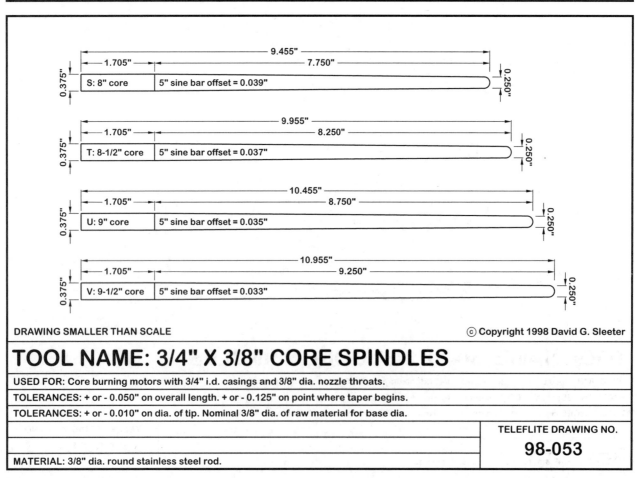

TOOL NAME: 3/4" X 3/8" CORE SPINDLES

USED FOR: Core burning motors with 3/4" i.d. casings and 3/8" dia. nozzle throats.
TOLERANCES: + or - 0.050" on overall length. + or - 0.125" on point where taper begins.
TOLERANCES: + or - 0.010" on dia. of tip. Nominal 3/8" dia. of raw material for base dia.

TELEFLITE DRAWING NO. 98-053

MATERIAL: 3/8" dia. round stainless steel rod.

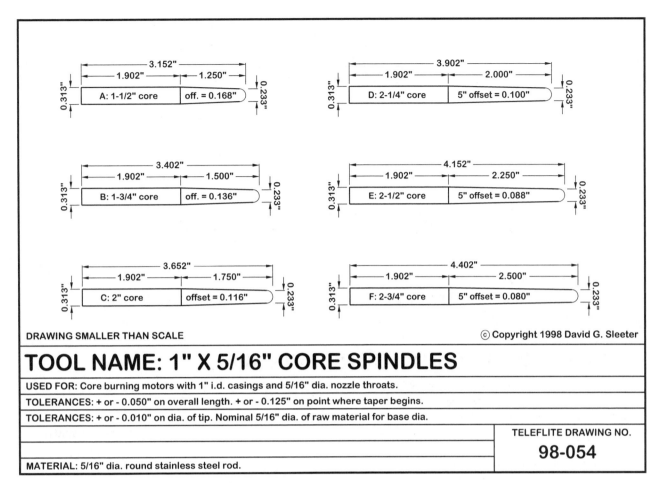

DRAWING SMALLER THAN SCALE © Copyright 1998 David G. Sleeter

TOOL NAME: 1" X 5/16" CORE SPINDLES

USED FOR: Core burning motors with 1" i.d. casings and 5/16" dia. nozzle throats.

TOLERANCES: + or - 0.050" on overall length. + or - 0.125" on point where taper begins.

TOLERANCES: + or - 0.010" on dia. of tip. Nominal 5/16" dia. of raw material for base dia.

TELEFLITE DRAWING NO. **98-054**

MATERIAL: 5/16" dia. round stainless steel rod.

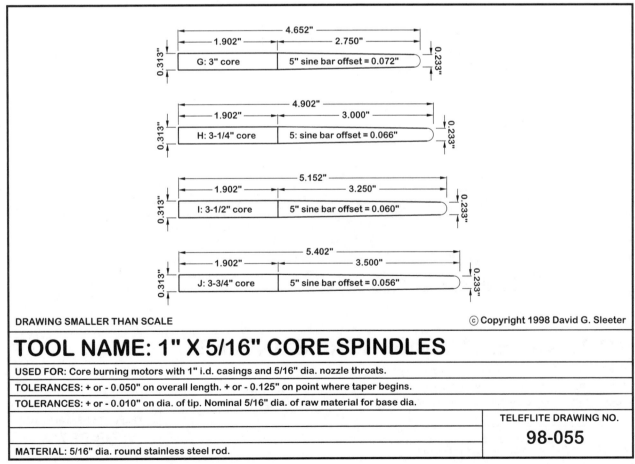

DRAWING SMALLER THAN SCALE © Copyright 1998 David G. Sleeter

TOOL NAME: 1" X 5/16" CORE SPINDLES

USED FOR: Core burning motors with 1" i.d. casings and 5/16" dia. nozzle throats.

TOLERANCES: + or - 0.050" on overall length. + or - 0.125" on point where taper begins.

TOLERANCES: + or - 0.010" on dia. of tip. Nominal 5/16" dia. of raw material for base dia.

TELEFLITE DRAWING NO. **98-055**

MATERIAL: 5/16" dia. round stainless steel rod.

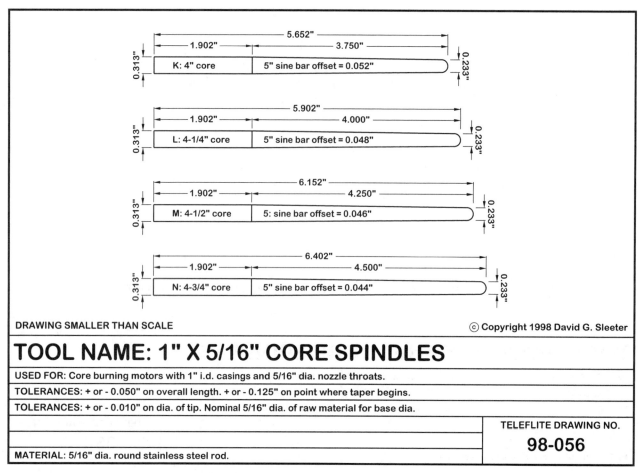

TOOL NAME: 1" X 5/16" CORE SPINDLES

USED FOR: Core burning motors with 1" i.d. casings and 5/16" dia. nozzle throats.

TOLERANCES: + or - 0.050" on overall length. + or - 0.125" on point where taper begins.

TOLERANCES: + or - 0.010" on dia. of tip. Nominal 5/16" dia. of raw material for base dia.

TELEFLITE DRAWING NO.
98-056

MATERIAL: 5/16" dia. round stainless steel rod.

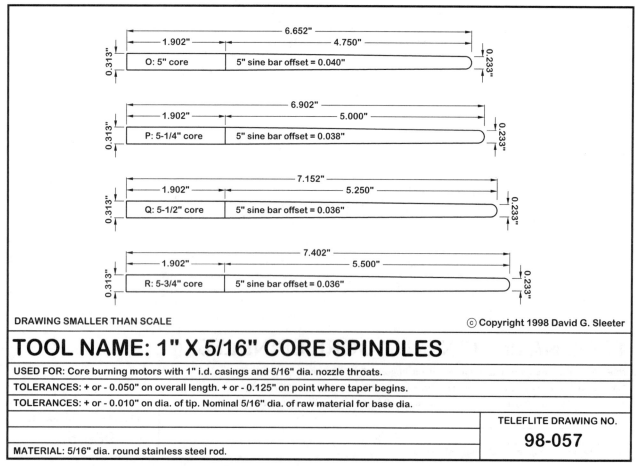

TOOL NAME: 1" X 5/16" CORE SPINDLES

USED FOR: Core burning motors with 1" i.d. casings and 5/16" dia. nozzle throats.

TOLERANCES: + or - 0.050" on overall length. + or - 0.125" on point where taper begins.

TOLERANCES: + or - 0.010" on dia. of tip. Nominal 5/16" dia. of raw material for base dia.

TELEFLITE DRAWING NO.
98-057

MATERIAL: 5/16" dia. round stainless steel rod.

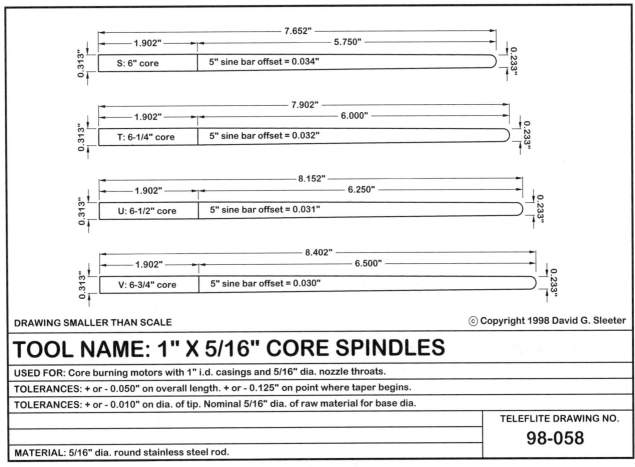

TOOL NAME: 1" X 5/16" CORE SPINDLES

USED FOR: Core burning motors with 1" i.d. casings and 5/16" dia. nozzle throats.
TOLERANCES: + or - 0.050" on overall length. + or - 0.125" on point where taper begins.
TOLERANCES: + or - 0.010" on dia. of tip. Nominal 5/16" dia. of raw material for base dia.

TELEFLITE DRAWING NO. **98-058**

MATERIAL: 5/16" dia. round stainless steel rod.

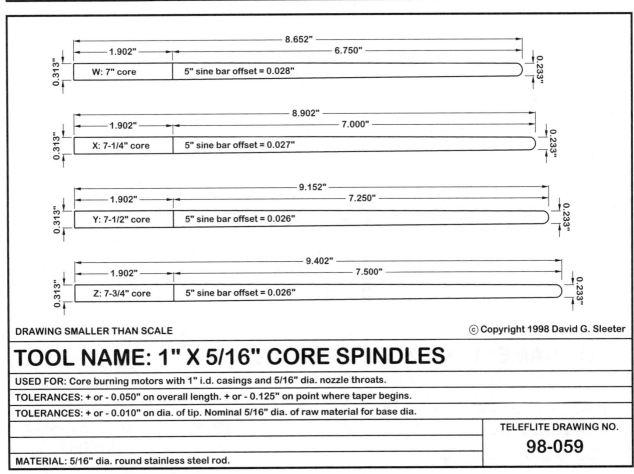

TOOL NAME: 1" X 5/16" CORE SPINDLES

USED FOR: Core burning motors with 1" i.d. casings and 5/16" dia. nozzle throats.
TOLERANCES: + or - 0.050" on overall length. + or - 0.125" on point where taper begins.
TOLERANCES: + or - 0.010" on dia. of tip. Nominal 5/16" dia. of raw material for base dia.

TELEFLITE DRAWING NO. **98-059**

MATERIAL: 5/16" dia. round stainless steel rod.

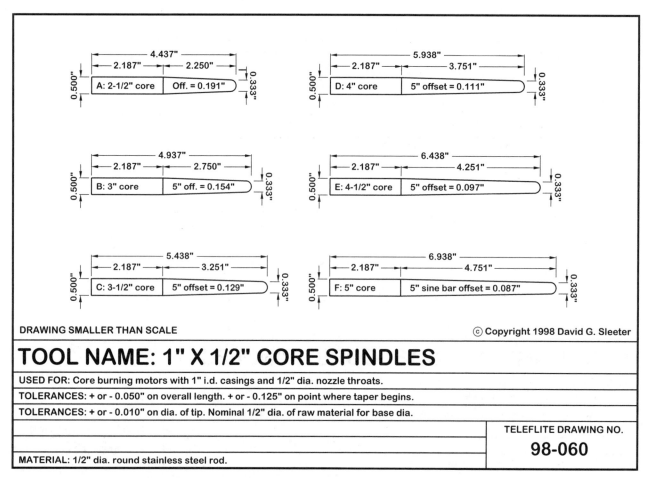

TOOL NAME: 1" X 1/2" CORE SPINDLES

USED FOR: Core burning motors with 1" i.d. casings and 1/2" dia. nozzle throats.
TOLERANCES: + or - 0.050" on overall length. + or - 0.125" on point where taper begins.
TOLERANCES: + or - 0.010" on dia. of tip. Nominal 1/2" dia. of raw material for base dia.

TELEFLITE DRAWING NO. **98-060**

MATERIAL: 1/2" dia. round stainless steel rod.

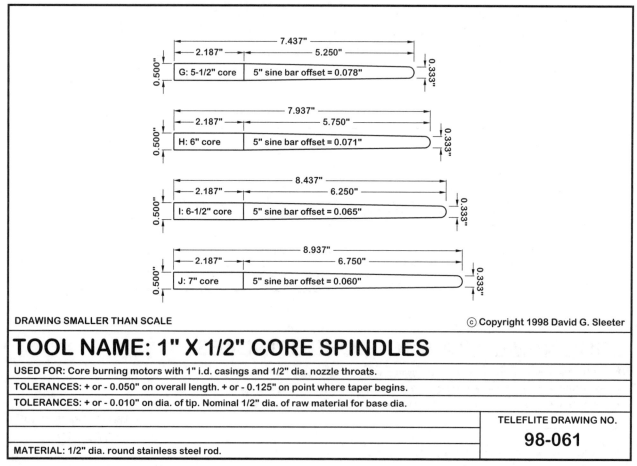

TOOL NAME: 1" X 1/2" CORE SPINDLES

USED FOR: Core burning motors with 1" i.d. casings and 1/2" dia. nozzle throats.
TOLERANCES: + or - 0.050" on overall length. + or - 0.125" on point where taper begins.
TOLERANCES: + or - 0.010" on dia. of tip. Nominal 1/2" dia. of raw material for base dia.

TELEFLITE DRAWING NO. **98-061**

MATERIAL: 1/2" dia. round stainless steel rod.

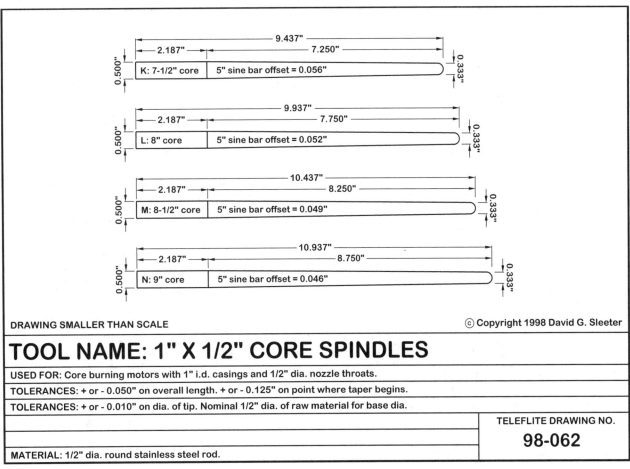

TOOL NAME: 1" X 1/2" CORE SPINDLES

USED FOR: Core burning motors with 1" i.d. casings and 1/2" dia. nozzle throats.
TOLERANCES: + or - 0.050" on overall length. + or - 0.125" on point where taper begins.
TOLERANCES: + or - 0.010" on dia. of tip. Nominal 1/2" dia. of raw material for base dia.

TELEFLITE DRAWING NO. **98-062**

MATERIAL: 1/2" dia. round stainless steel rod.

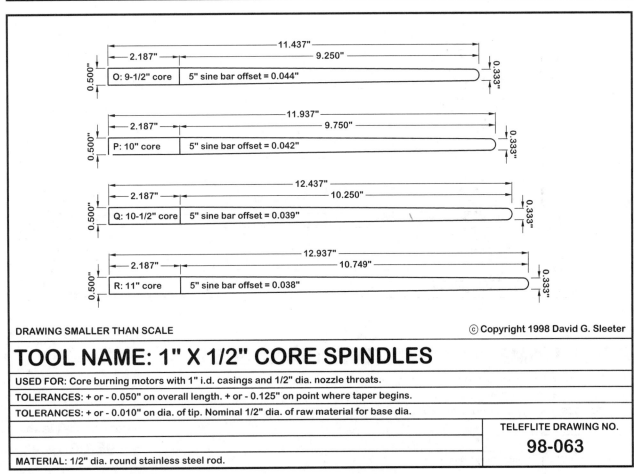

TOOL NAME: 1" X 1/2" CORE SPINDLES

USED FOR: Core burning motors with 1" i.d. casings and 1/2" dia. nozzle throats.
TOLERANCES: + or - 0.050" on overall length. + or - 0.125" on point where taper begins.
TOLERANCES: + or - 0.010" on dia. of tip. Nominal 1/2" dia. of raw material for base dia.

TELEFLITE DRAWING NO. **98-063**

MATERIAL: 1/2" dia. round stainless steel rod.

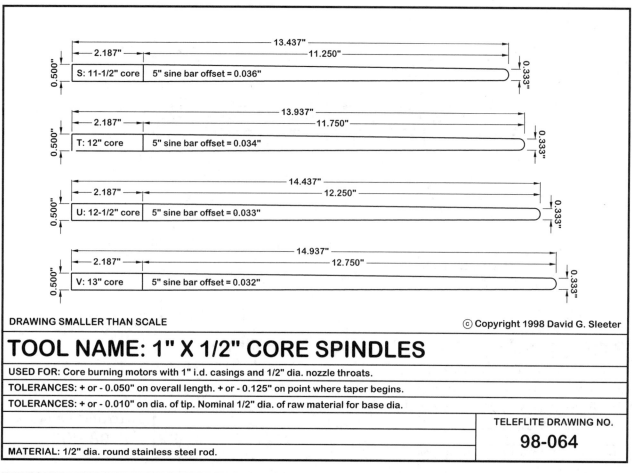

DRAWING SMALLER THAN SCALE © Copyright 1998 David G. Sleeter

TOOL NAME: 1" X 1/2" CORE SPINDLES

USED FOR: Core burning motors with 1" i.d. casings and 1/2" dia. nozzle throats.
TOLERANCES: + or - 0.050" on overall length. + or - 0.125" on point where taper begins.
TOLERANCES: + or - 0.010" on dia. of tip. Nominal 1/2" dia. of raw material for base dia.

TELEFLITE DRAWING NO. **98-064**

MATERIAL: 1/2" dia. round stainless steel rod.

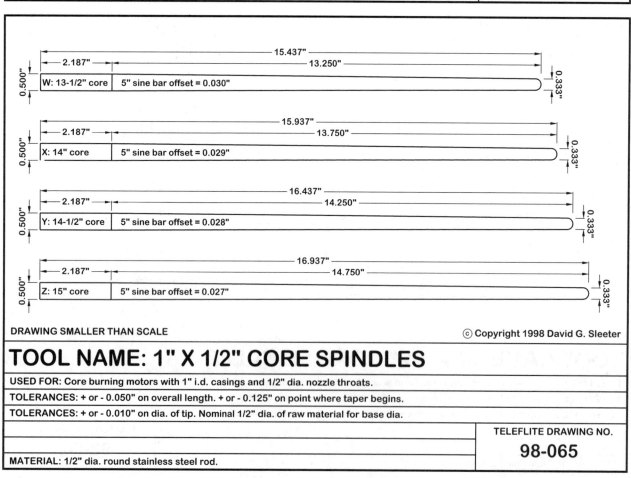

DRAWING SMALLER THAN SCALE © Copyright 1998 David G. Sleeter

TOOL NAME: 1" X 1/2" CORE SPINDLES

USED FOR: Core burning motors with 1" i.d. casings and 1/2" dia. nozzle throats.
TOLERANCES: + or - 0.050" on overall length. + or - 0.125" on point where taper begins.
TOLERANCES: + or - 0.010" on dia. of tip. Nominal 1/2" dia. of raw material for base dia.

TELEFLITE DRAWING NO. **98-065**

MATERIAL: 1/2" dia. round stainless steel rod.

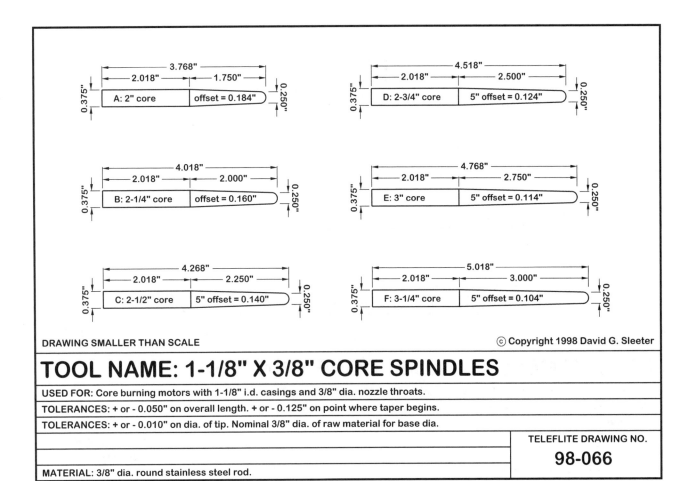

TOOL NAME: 1-1/8" X 3/8" CORE SPINDLES

USED FOR: Core burning motors with 1-1/8" i.d. casings and 3/8" dia. nozzle throats.

TOLERANCES: + or - 0.050" on overall length. + or - 0.125" on point where taper begins.

TOLERANCES: + or - 0.010" on dia. of tip. Nominal 3/8" dia. of raw material for base dia.

TELEFLITE DRAWING NO. **98-066**

MATERIAL: 3/8" dia. round stainless steel rod.

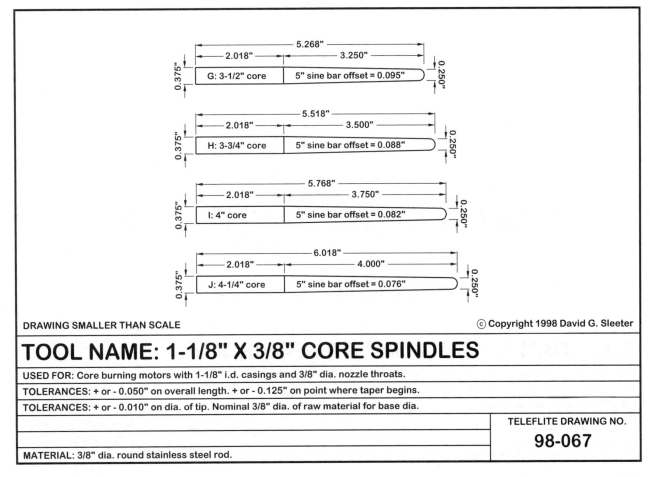

TOOL NAME: 1-1/8" X 3/8" CORE SPINDLES

USED FOR: Core burning motors with 1-1/8" i.d. casings and 3/8" dia. nozzle throats.

TOLERANCES: + or - 0.050" on overall length. + or - 0.125" on point where taper begins.

TOLERANCES: + or - 0.010" on dia. of tip. Nominal 3/8" dia. of raw material for base dia.

TELEFLITE DRAWING NO. **98-067**

MATERIAL: 3/8" dia. round stainless steel rod.

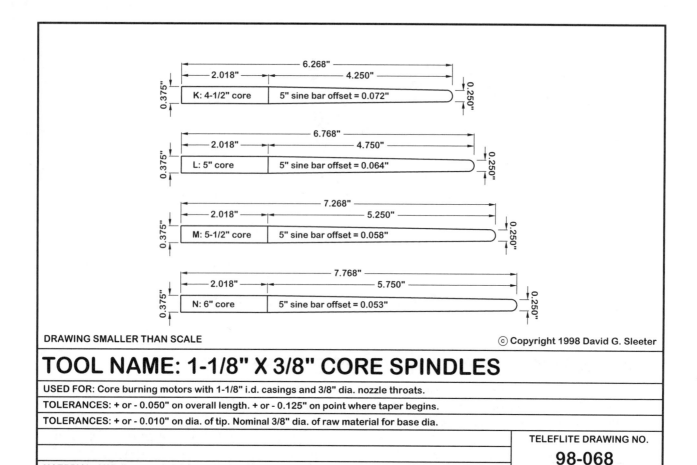

DRAWING SMALLER THAN SCALE © Copyright 1998 David G. Sleeter

TOOL NAME: 1-1/8" X 3/8" CORE SPINDLES

| USED FOR: Core burning motors with 1-1/8" i.d. casings and 3/8" dia. nozzle throats. |
| TOLERANCES: + or - 0.050" on overall length. + or - 0.125" on point where taper begins. |
| TOLERANCES: + or - 0.010" on dia. of tip. Nominal 3/8" dia. of raw material for base dia. |

TELEFLITE DRAWING NO.
98-068

MATERIAL: 3/8" dia. round stainless steel rod.

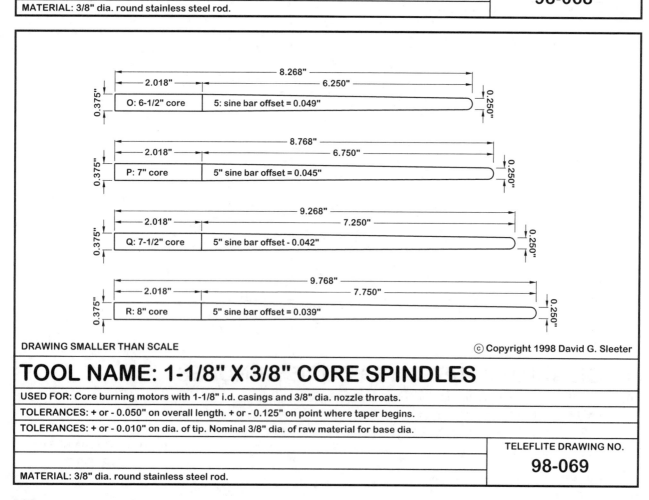

DRAWING SMALLER THAN SCALE © Copyright 1998 David G. Sleeter

TOOL NAME: 1-1/8" X 3/8" CORE SPINDLES

| USED FOR: Core burning motors with 1-1/8" i.d. casings and 3/8" dia. nozzle throats. |
| TOLERANCES: + or - 0.050" on overall length. + or - 0.125" on point where taper begins. |
| TOLERANCES: + or - 0.010" on dia. of tip. Nominal 3/8" dia. of raw material for base dia. |

TELEFLITE DRAWING NO.
98-069

MATERIAL: 3/8" dia. round stainless steel rod.

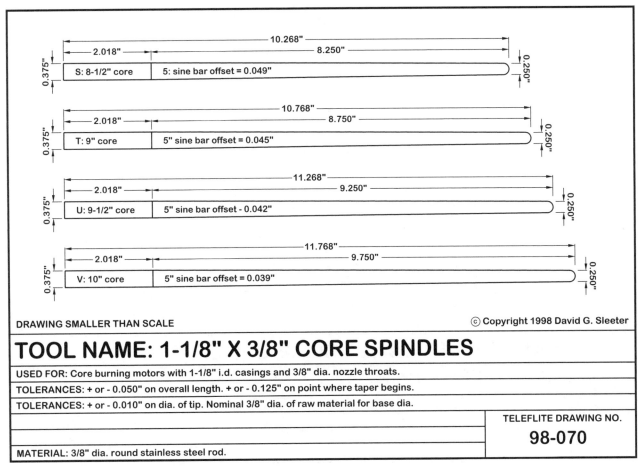

TOOL NAME: 1-1/8" X 3/8" CORE SPINDLES

USED FOR: Core burning motors with 1-1/8" i.d. casings and 3/8" dia. nozzle throats.
TOLERANCES: + or - 0.050" on overall length. + or - 0.125" on point where taper begins.
TOLERANCES: + or - 0.010" on dia. of tip. Nominal 3/8" dia. of raw material for base dia.

TELEFLITE DRAWING NO. **98-070**

MATERIAL: 3/8" dia. round stainless steel rod.

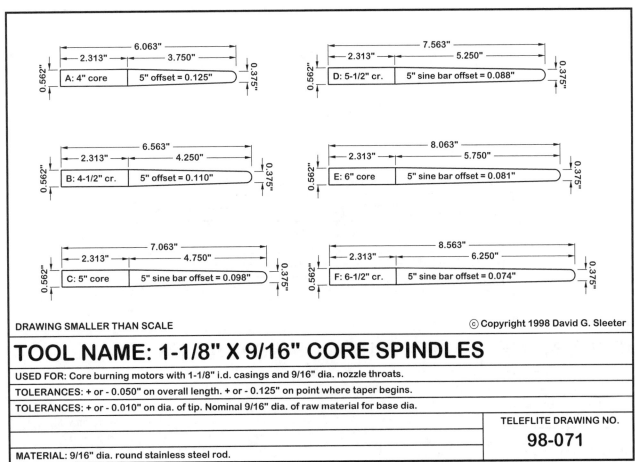

TOOL NAME: 1-1/8" X 9/16" CORE SPINDLES

USED FOR: Core burning motors with 1-1/8" i.d. casings and 9/16" dia. nozzle throats.
TOLERANCES: + or - 0.050" on overall length. + or - 0.125" on point where taper begins.
TOLERANCES: + or - 0.010" on dia. of tip. Nominal 9/16" dia. of raw material for base dia.

TELEFLITE DRAWING NO. **98-071**

MATERIAL: 9/16" dia. round stainless steel rod.

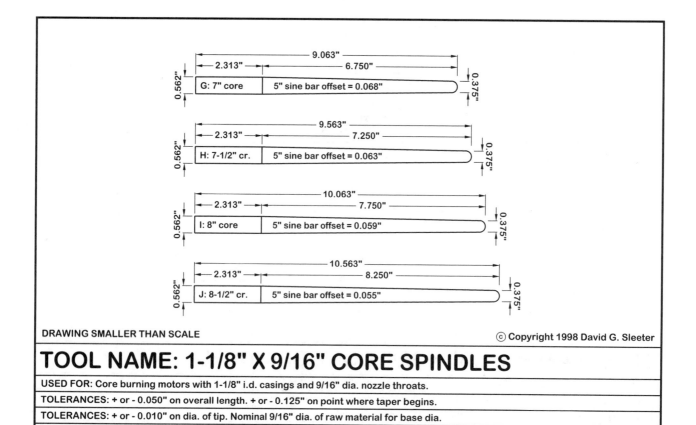

DRAWING SMALLER THAN SCALE © Copyright 1998 David G. Sleeter

TOOL NAME: 1-1/8" X 9/16" CORE SPINDLES

USED FOR: Core burning motors with 1-1/8" i.d. casings and 9/16" dia. nozzle throats.
TOLERANCES: + or - 0.050" on overall length. + or - 0.125" on point where taper begins.
TOLERANCES: + or - 0.010" on dia. of tip. Nominal 9/16" dia. of raw material for base dia.

MATERIAL: 9/16" dia. round stainless steel rod.

TELEFLITE DRAWING NO. **98-072**

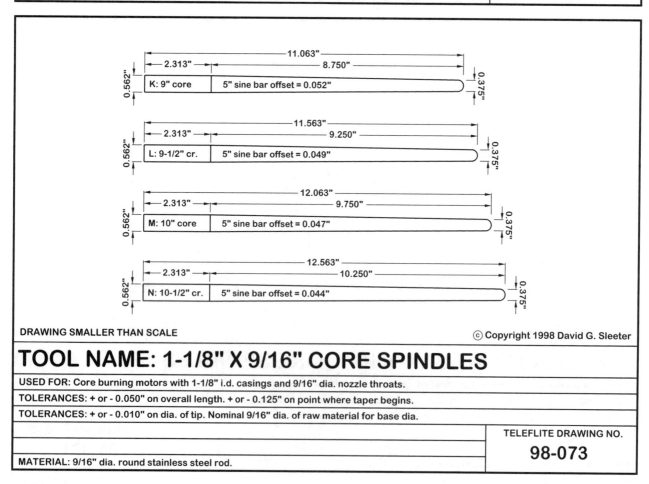

DRAWING SMALLER THAN SCALE © Copyright 1998 David G. Sleeter

TOOL NAME: 1-1/8" X 9/16" CORE SPINDLES

USED FOR: Core burning motors with 1-1/8" i.d. casings and 9/16" dia. nozzle throats.
TOLERANCES: + or - 0.050" on overall length. + or - 0.125" on point where taper begins.
TOLERANCES: + or - 0.010" on dia. of tip. Nominal 9/16" dia. of raw material for base dia.

MATERIAL: 9/16" dia. round stainless steel rod.

TELEFLITE DRAWING NO. **98-073**

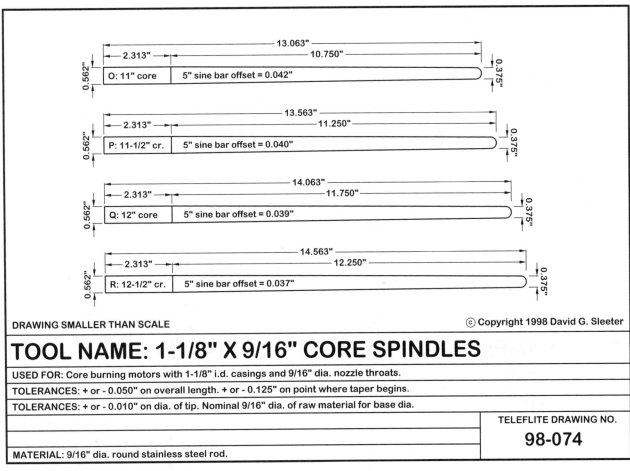

TOOL NAME: 1-1/8" X 9/16" CORE SPINDLES

USED FOR: Core burning motors with 1-1/8" i.d. casings and 9/16" dia. nozzle throats.
TOLERANCES: + or - 0.050" on overall length. + or - 0.125" on point where taper begins.
TOLERANCES: + or - 0.010" on dia. of tip. Nominal 9/16" dia. of raw material for base dia.

TELEFLITE DRAWING NO. **98-074**

MATERIAL: 9/16" dia. round stainless steel rod.

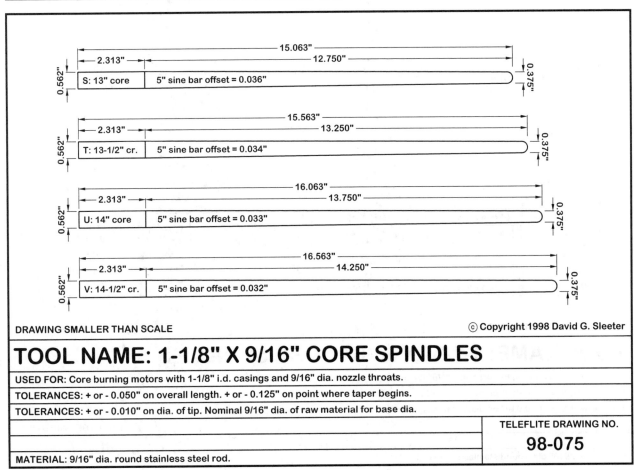

TOOL NAME: 1-1/8" X 9/16" CORE SPINDLES

USED FOR: Core burning motors with 1-1/8" i.d. casings and 9/16" dia. nozzle throats.
TOLERANCES: + or - 0.050" on overall length. + or - 0.125" on point where taper begins.
TOLERANCES: + or - 0.010" on dia. of tip. Nominal 9/16" dia. of raw material for base dia.

TELEFLITE DRAWING NO. **98-075**

MATERIAL: 9/16" dia. round stainless steel rod.

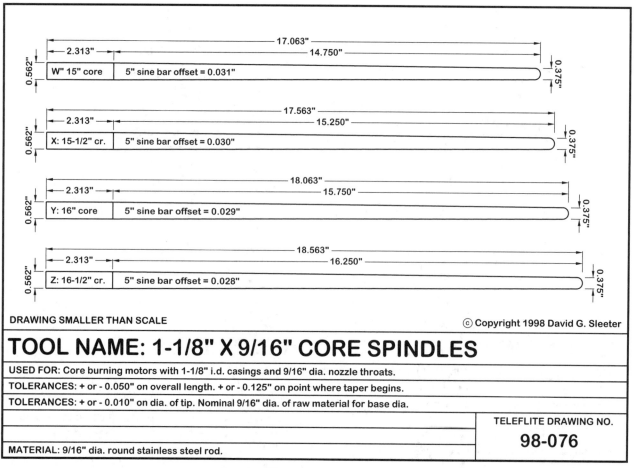

TOOL NAME: 1-1/8" X 9/16" CORE SPINDLES

USED FOR: Core burning motors with 1-1/8" i.d. casings and 9/16" dia. nozzle throats.
TOLERANCES: + or - 0.050" on overall length. + or - 0.125" on point where taper begins.
TOLERANCES: + or - 0.010" on dia. of tip. Nominal 9/16" dia. of raw material for base dia.
MATERIAL: 9/16" dia. round stainless steel rod.

TELEFLITE DRAWING NO. 98-076

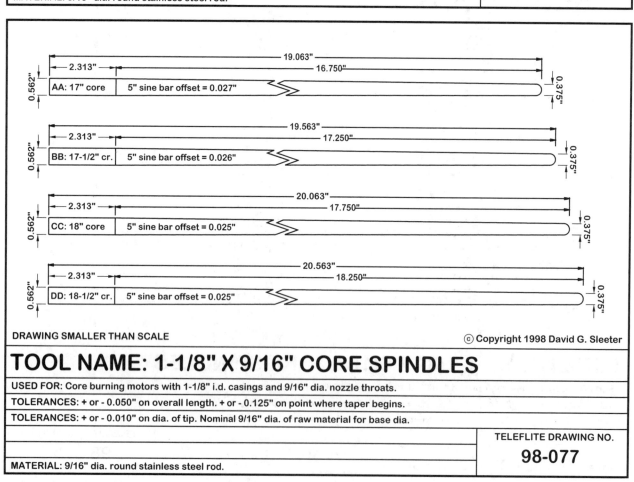

TOOL NAME: 1-1/8" X 9/16" CORE SPINDLES

USED FOR: Core burning motors with 1-1/8" i.d. casings and 9/16" dia. nozzle throats.
TOLERANCES: + or - 0.050" on overall length. + or - 0.125" on point where taper begins.
TOLERANCES: + or - 0.010" on dia. of tip. Nominal 9/16" dia. of raw material for base dia.
MATERIAL: 9/16" dia. round stainless steel rod.

TELEFLITE DRAWING NO. 98-077

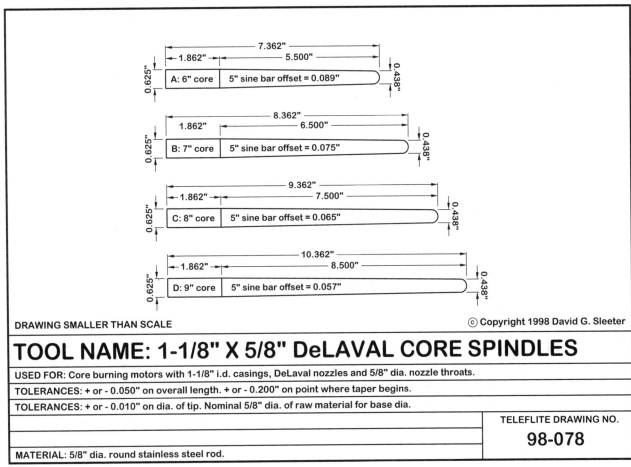

DRAWING SMALLER THAN SCALE © Copyright 1998 David G. Sleeter

TOOL NAME: 1-1/8" X 5/8" DeLAVAL CORE SPINDLES

USED FOR: Core burning motors with 1-1/8" i.d. casings, DeLaval nozzles and 5/8" dia. nozzle throats.
TOLERANCES: + or - 0.050" on overall length. + or - 0.200" on point where taper begins.
TOLERANCES: + or - 0.010" on dia. of tip. Nominal 5/8" dia. of raw material for base dia.

TELEFLITE DRAWING NO.
98-078

MATERIAL: 5/8" dia. round stainless steel rod.

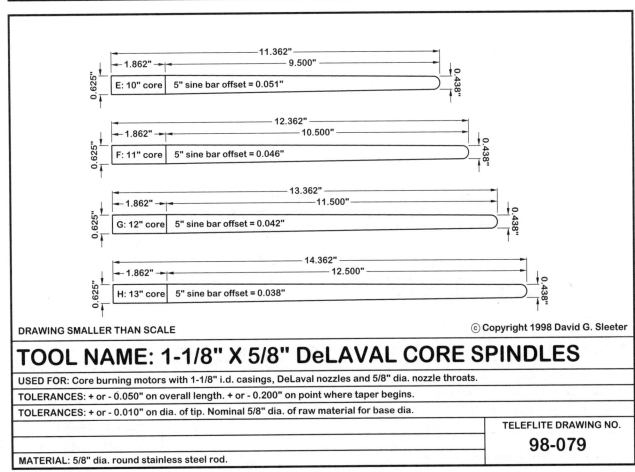

DRAWING SMALLER THAN SCALE © Copyright 1998 David G. Sleeter

TOOL NAME: 1-1/8" X 5/8" DeLAVAL CORE SPINDLES

USED FOR: Core burning motors with 1-1/8" i.d. casings, DeLaval nozzles and 5/8" dia. nozzle throats.
TOLERANCES: + or - 0.050" on overall length. + or - 0.200" on point where taper begins.
TOLERANCES: + or - 0.010" on dia. of tip. Nominal 5/8" dia. of raw material for base dia.

TELEFLITE DRAWING NO.
98-079

MATERIAL: 5/8" dia. round stainless steel rod.

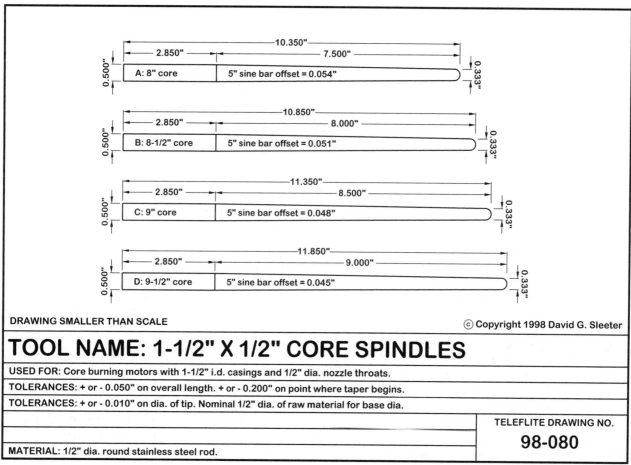

TOOL NAME: 1-1/2" X 1/2" CORE SPINDLES

USED FOR: Core burning motors with 1-1/2" i.d. casings and 1/2" dia. nozzle throats.

TOLERANCES: + or - 0.050" on overall length. + or - 0.200" on point where taper begins.

TOLERANCES: + or - 0.010" on dia. of tip. Nominal 1/2" dia. of raw material for base dia.

MATERIAL: 1/2" dia. round stainless steel rod.

TELEFLITE DRAWING NO. **98-080**

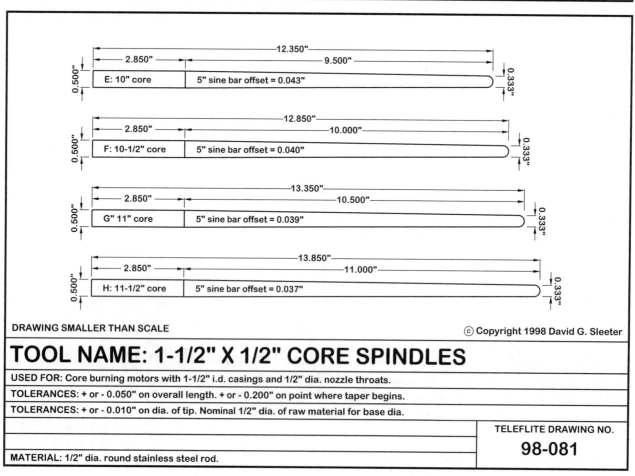

TOOL NAME: 1-1/2" X 1/2" CORE SPINDLES

USED FOR: Core burning motors with 1-1/2" i.d. casings and 1/2" dia. nozzle throats.

TOLERANCES: + or - 0.050" on overall length. + or - 0.200" on point where taper begins.

TOLERANCES: + or - 0.010" on dia. of tip. Nominal 1/2" dia. of raw material for base dia.

MATERIAL: 1/2" dia. round stainless steel rod.

TELEFLITE DRAWING NO. **98-081**

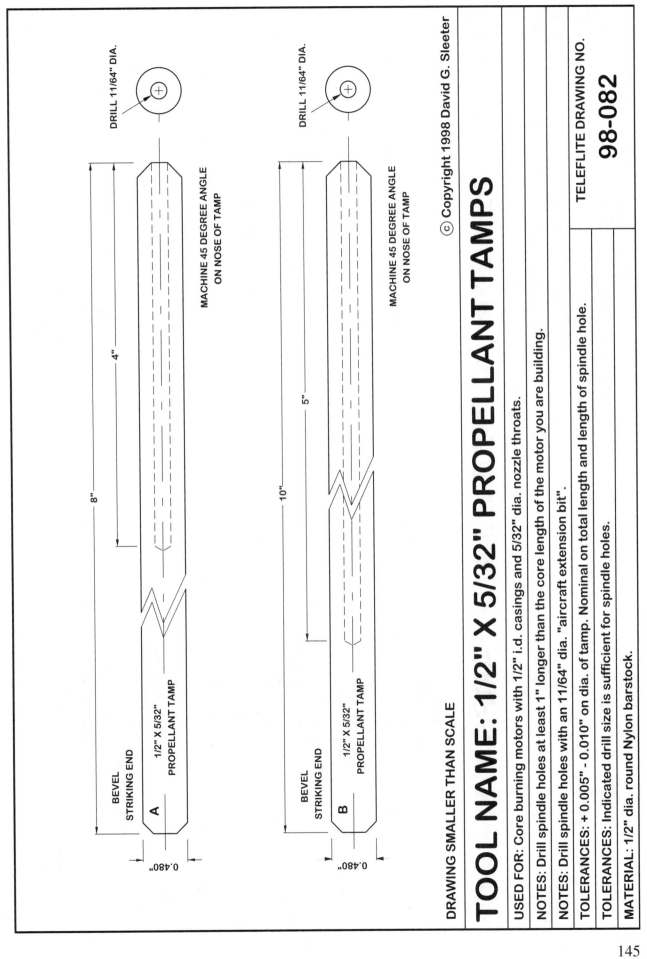

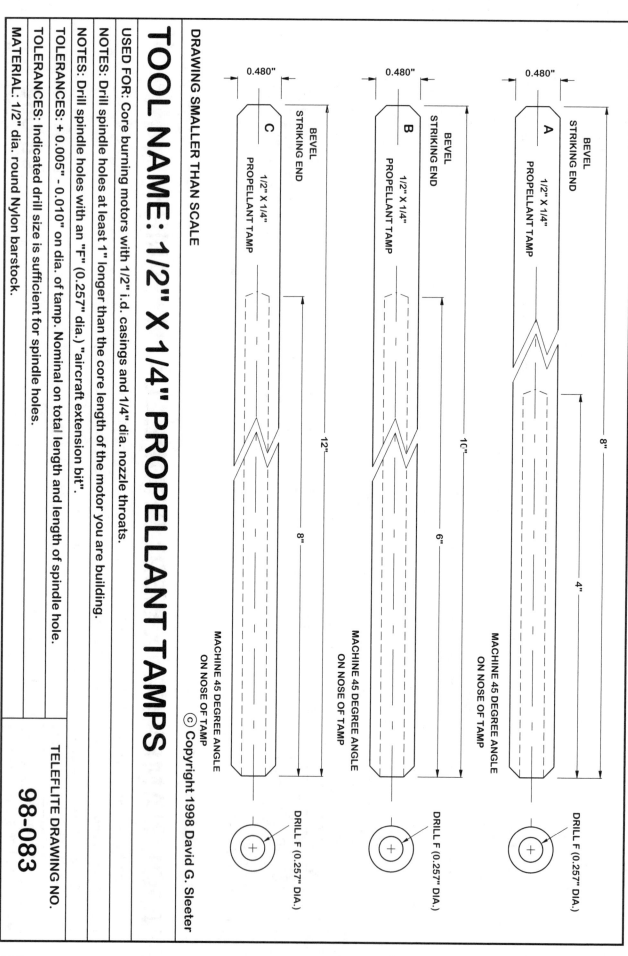

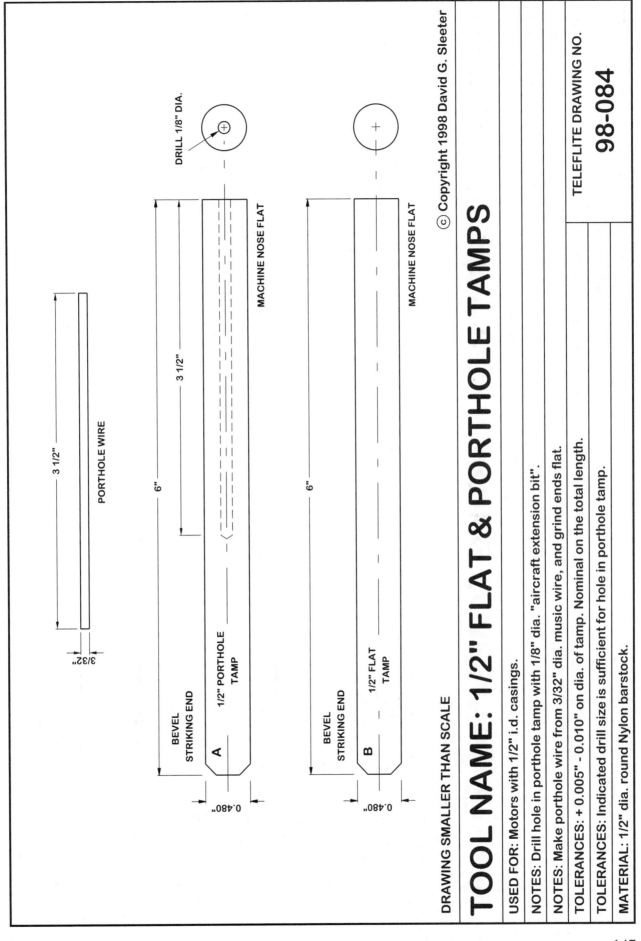

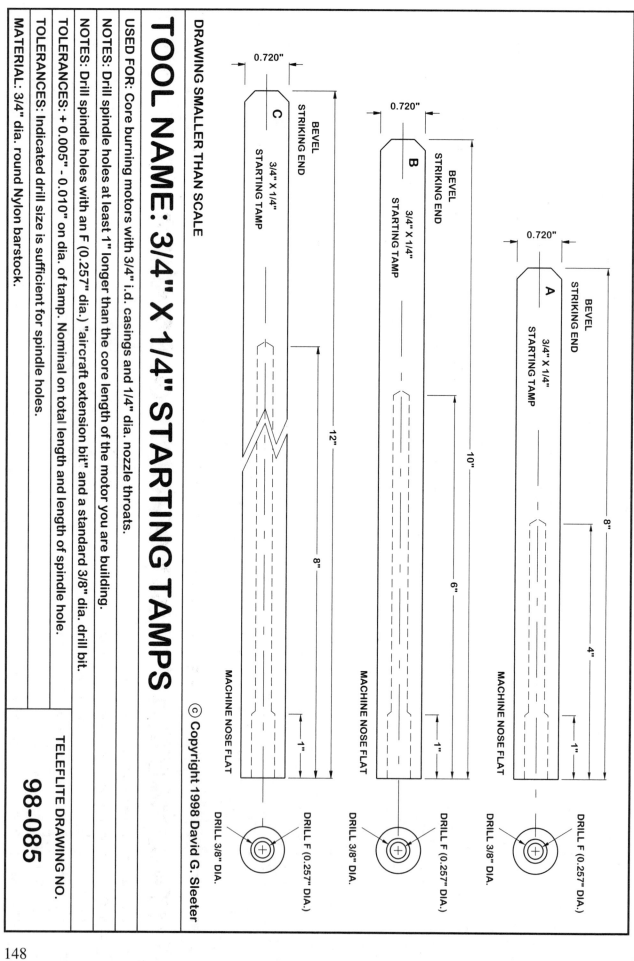

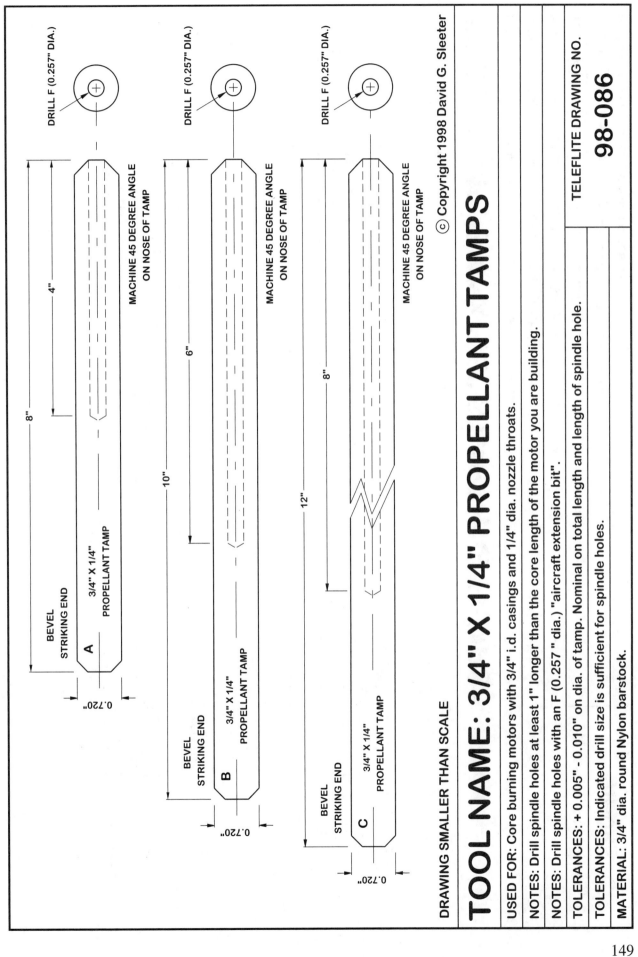

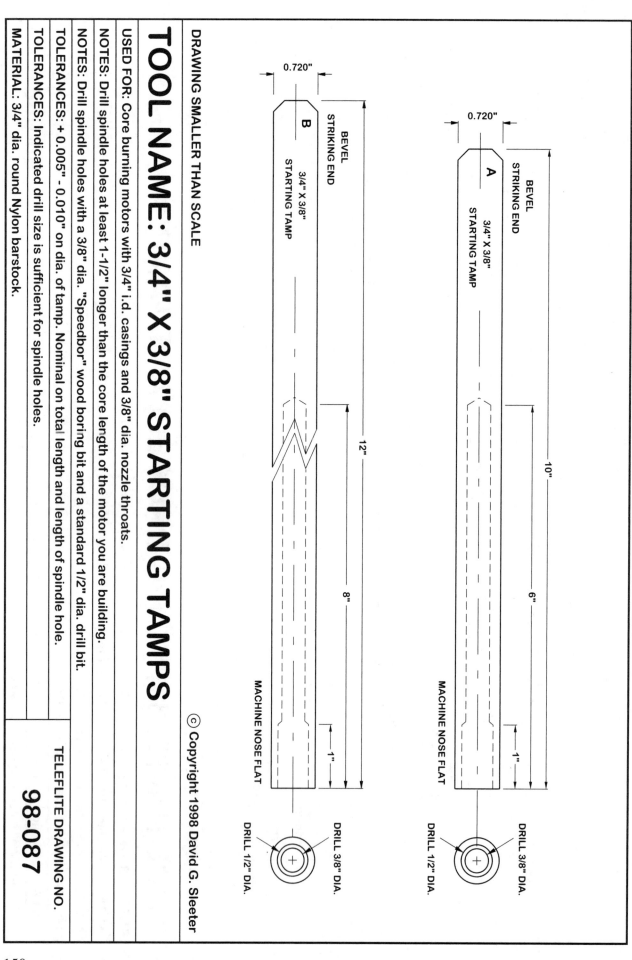

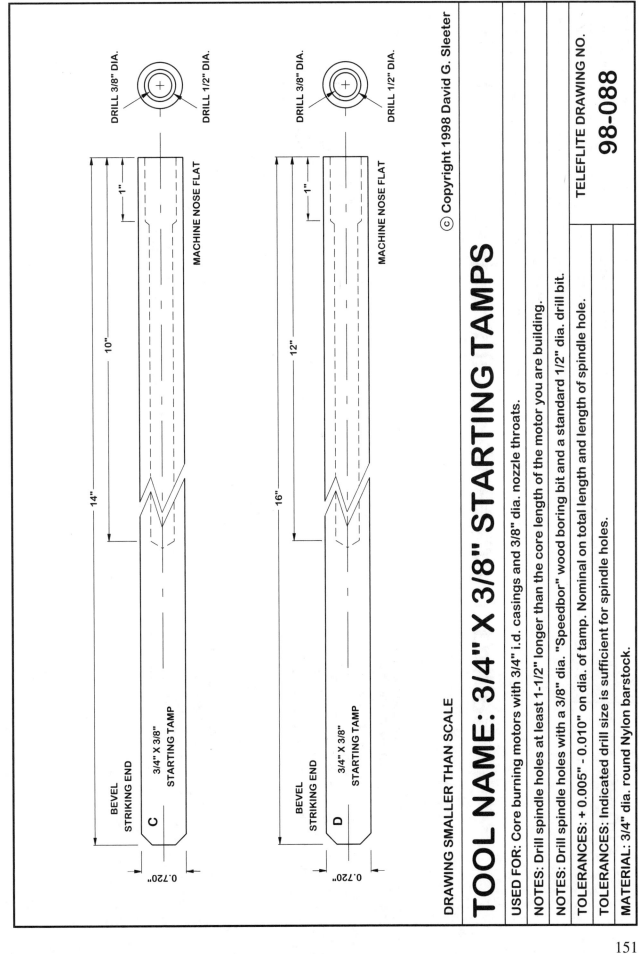

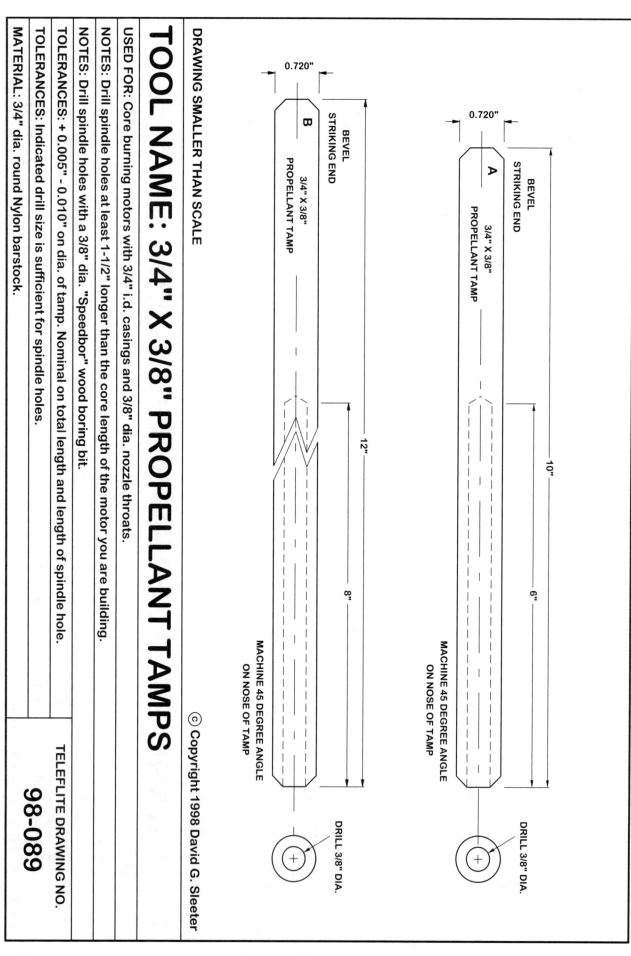

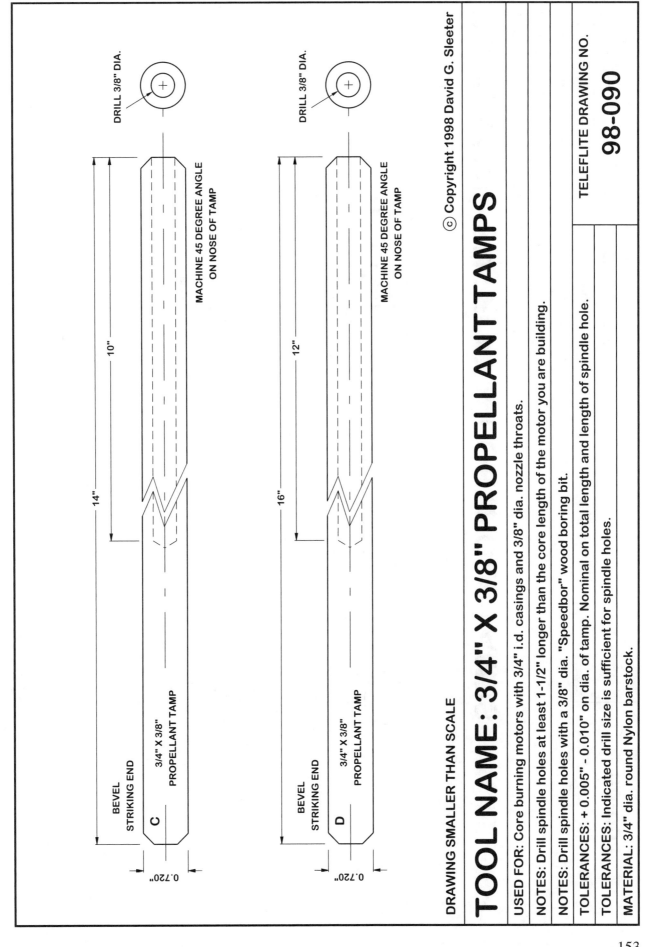

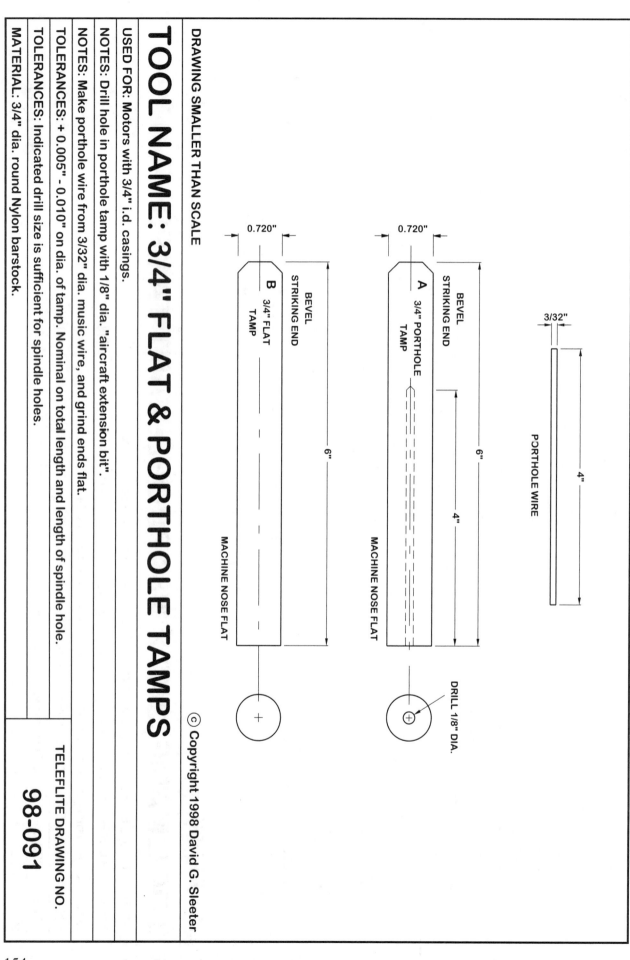

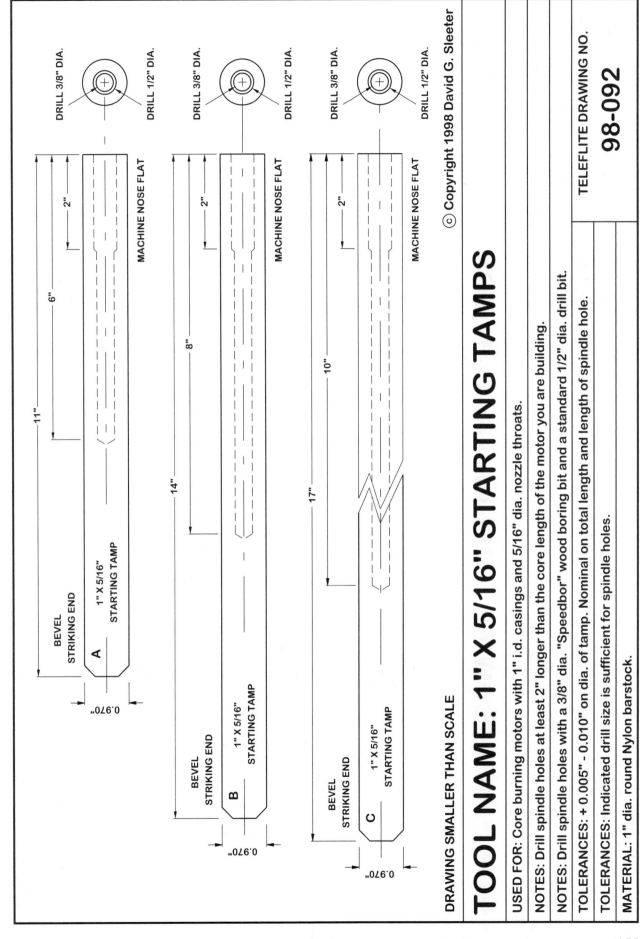

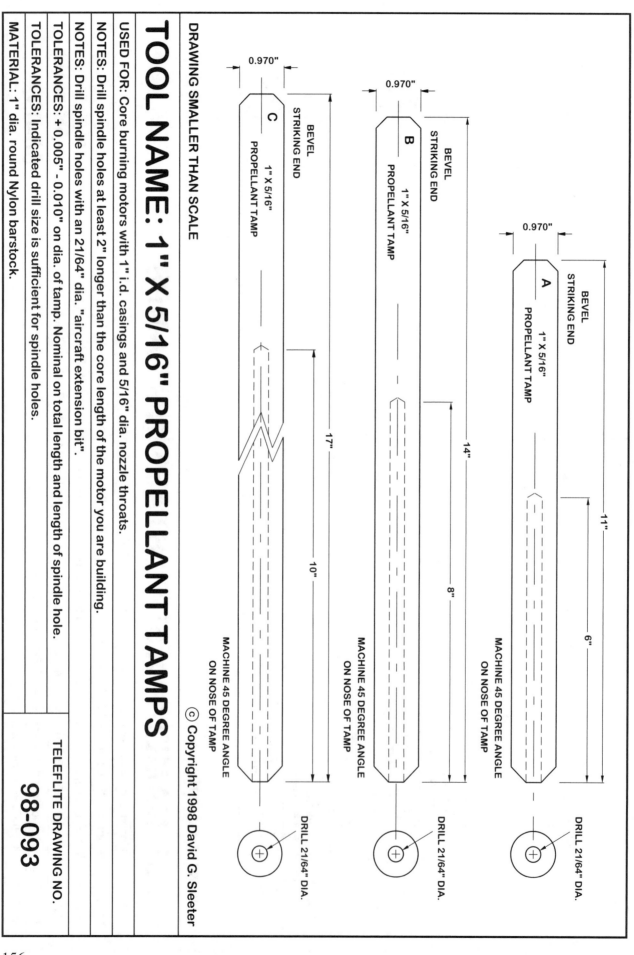

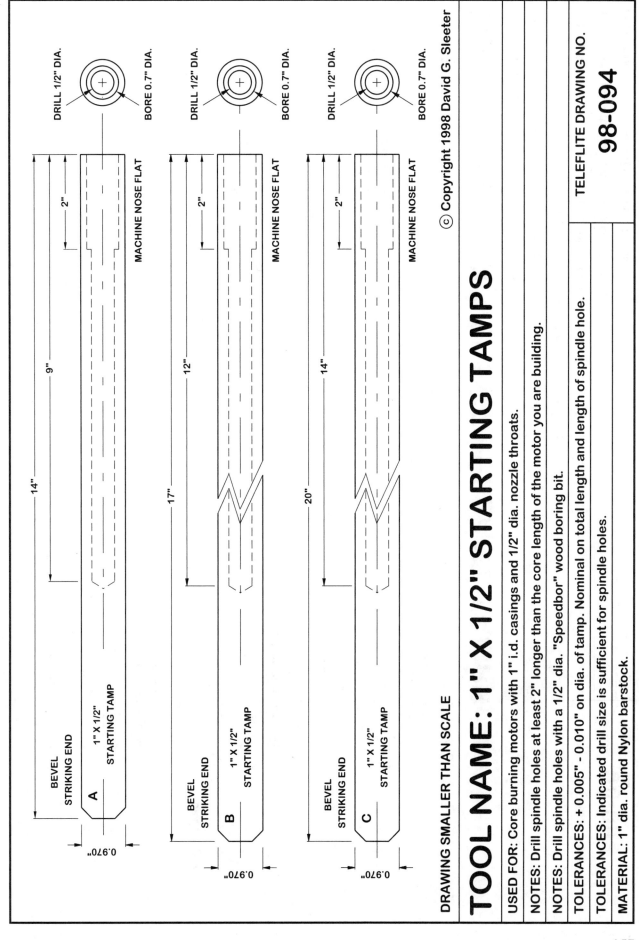

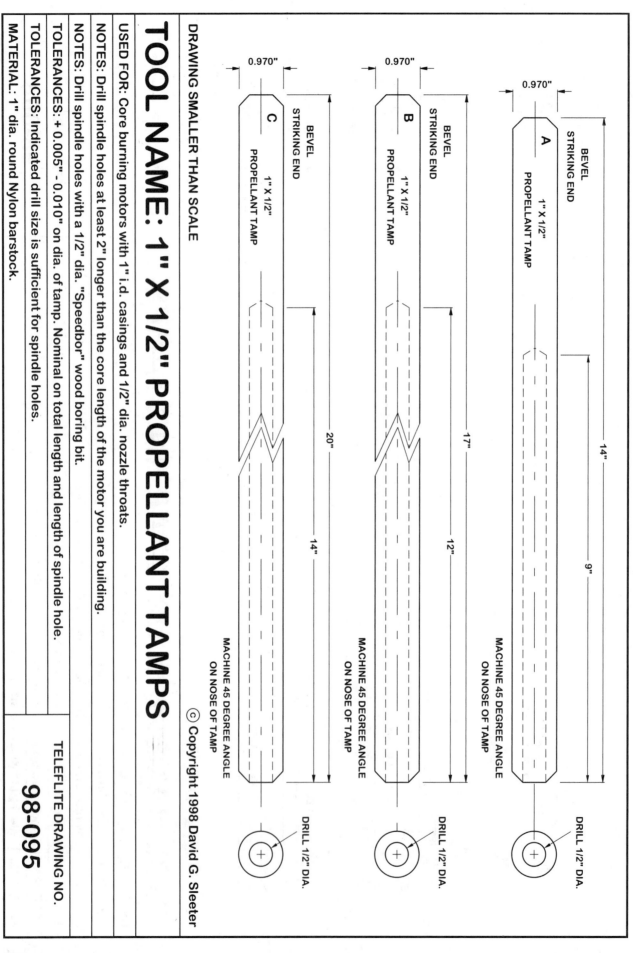

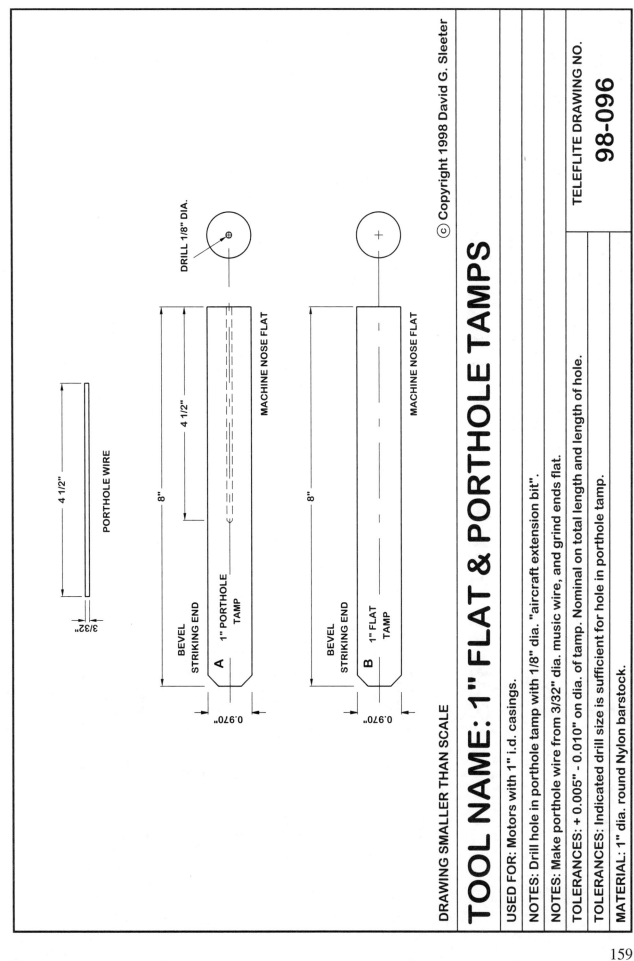

DRAWING SMALLER THAN SCALE © Copyright 1998 David G. Sleeter

TOOL NAME: 1" FLAT & PORTHOLE TAMPS

USED FOR: Motors with 1" i.d. casings.

NOTES: Drill hole in porthole tamp with 1/8" dia. "aircraft extension bit".

NOTES: Make porthole wire from 3/32" dia. music wire, and grind ends flat.

TOLERANCES: + 0.005" - 0.010" on dia. of tamp. Nominal on total length and length of hole.

TOLERANCES: Indicated drill size is sufficient for hole in porthole tamp.

MATERIAL: 1" dia. round Nylon barstock.

TELEFLITE DRAWING NO.

98-096

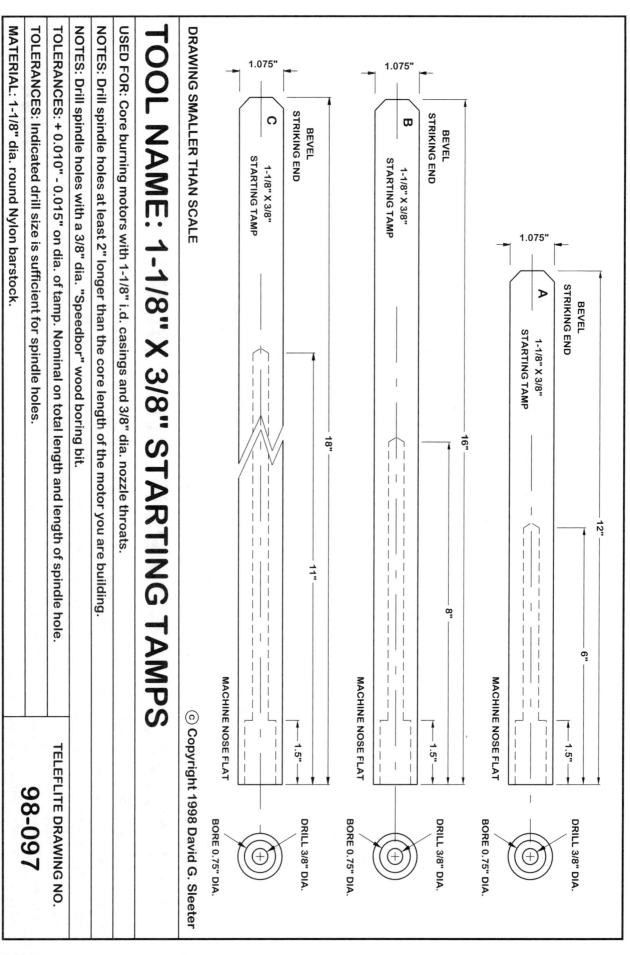

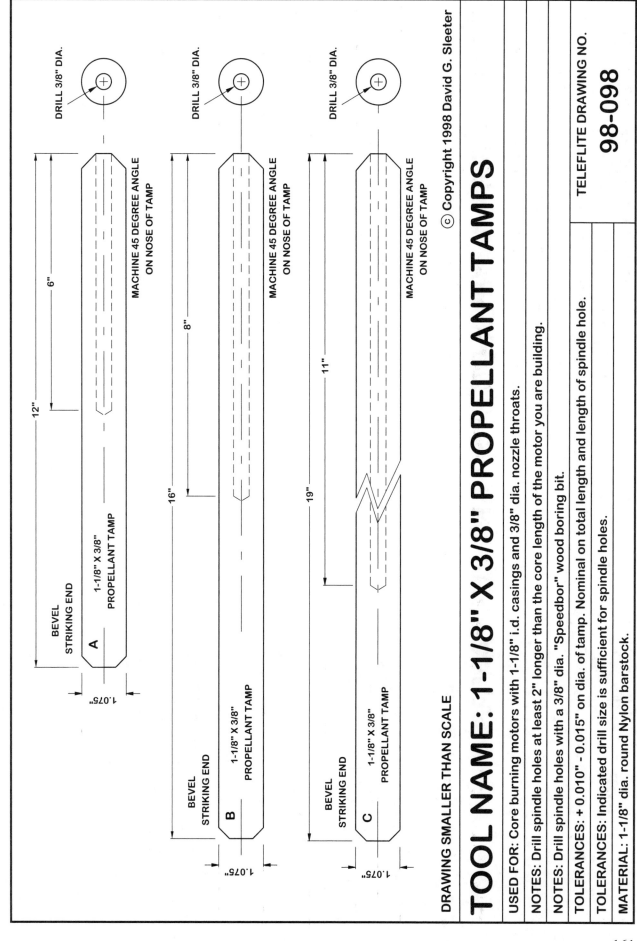

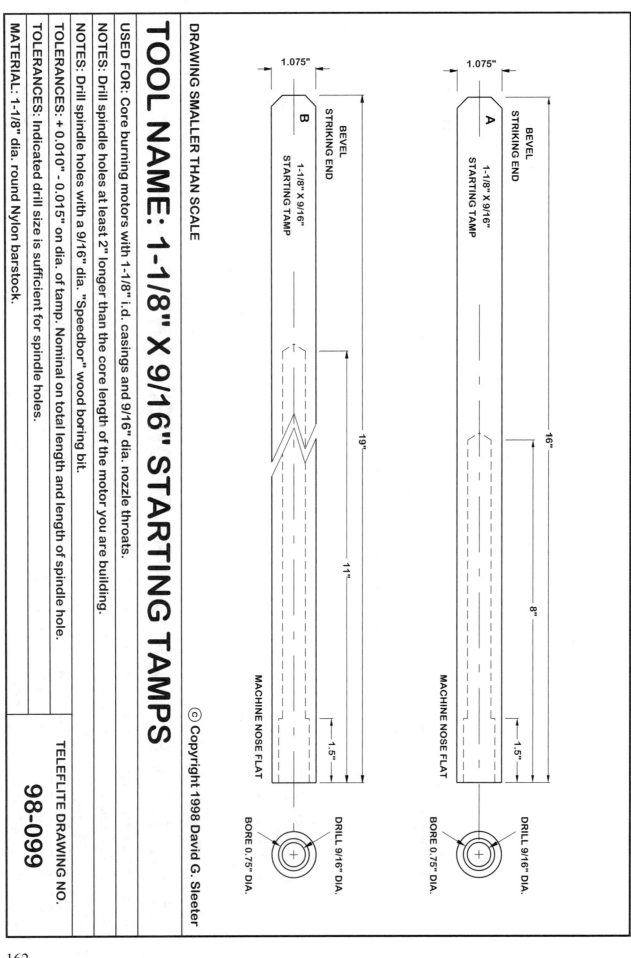

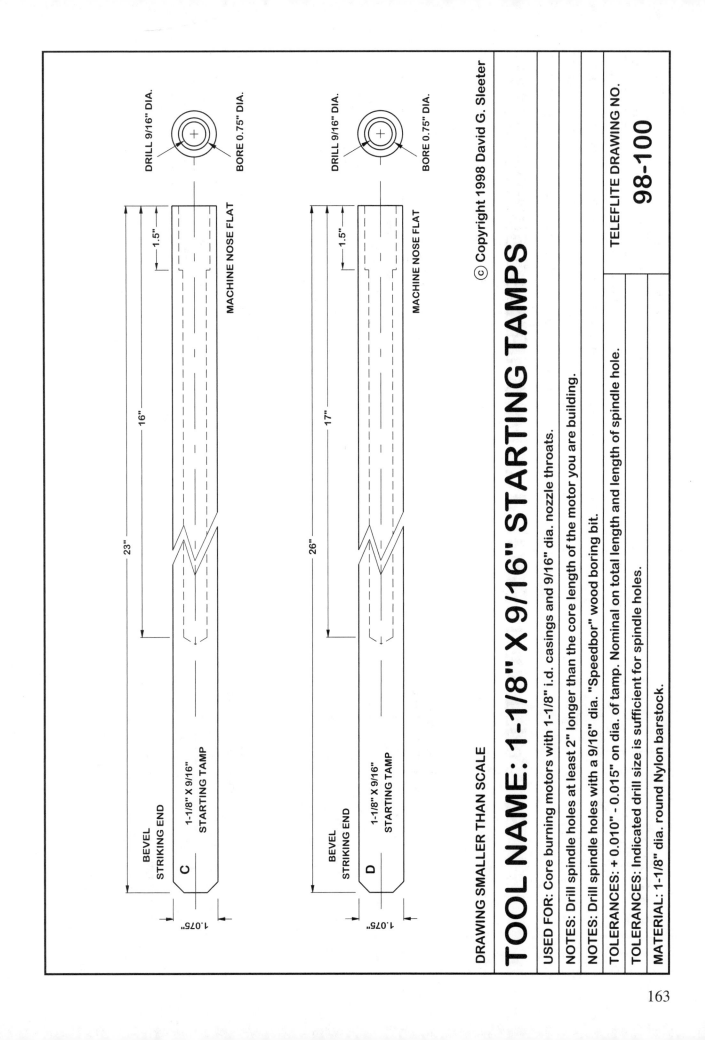

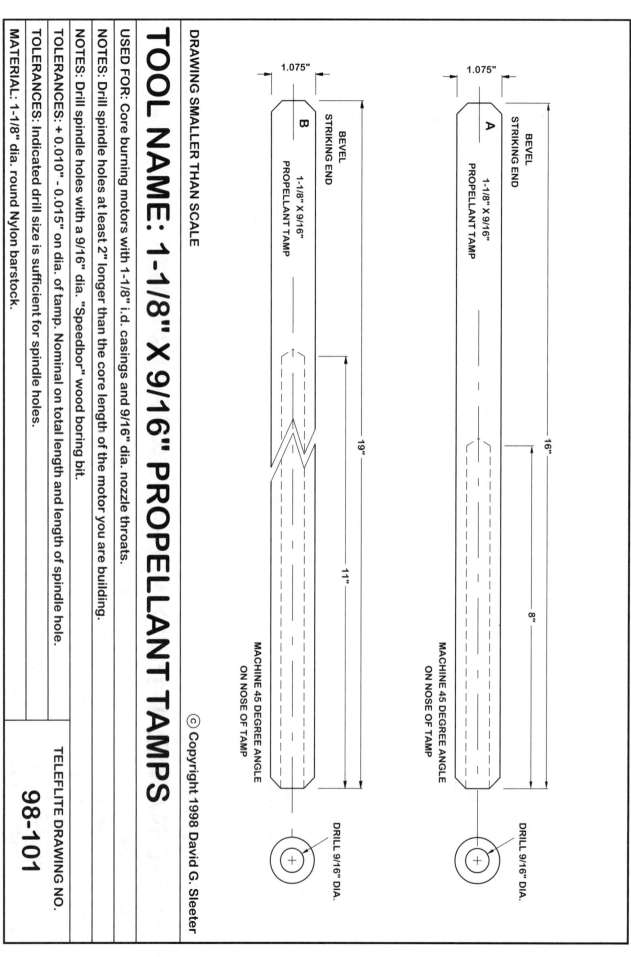

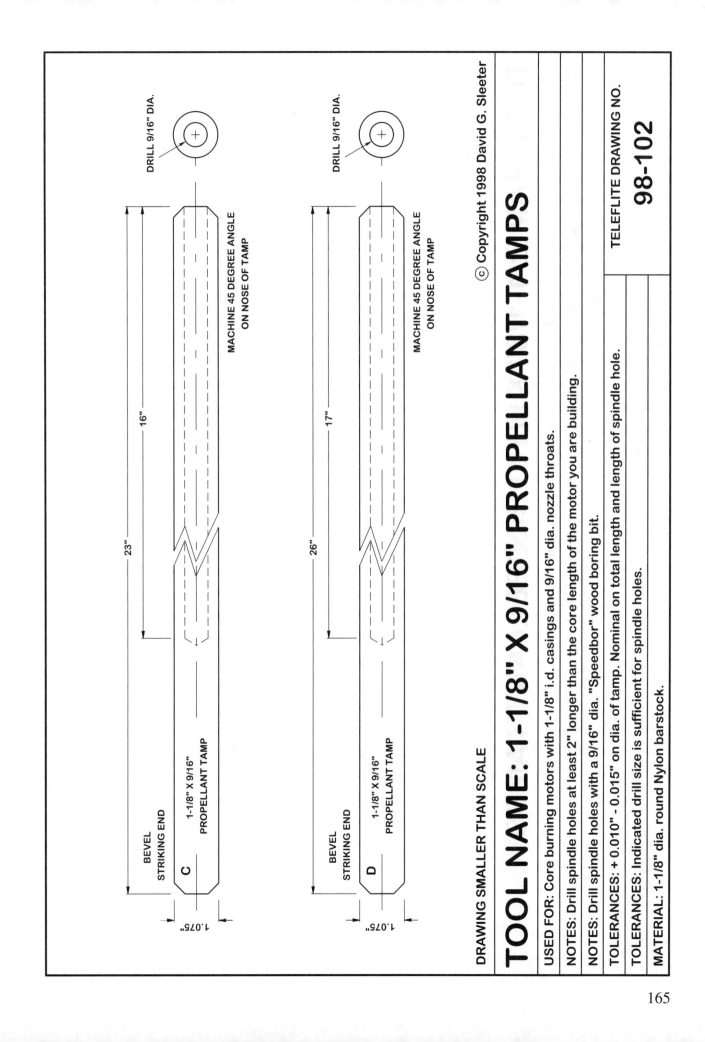

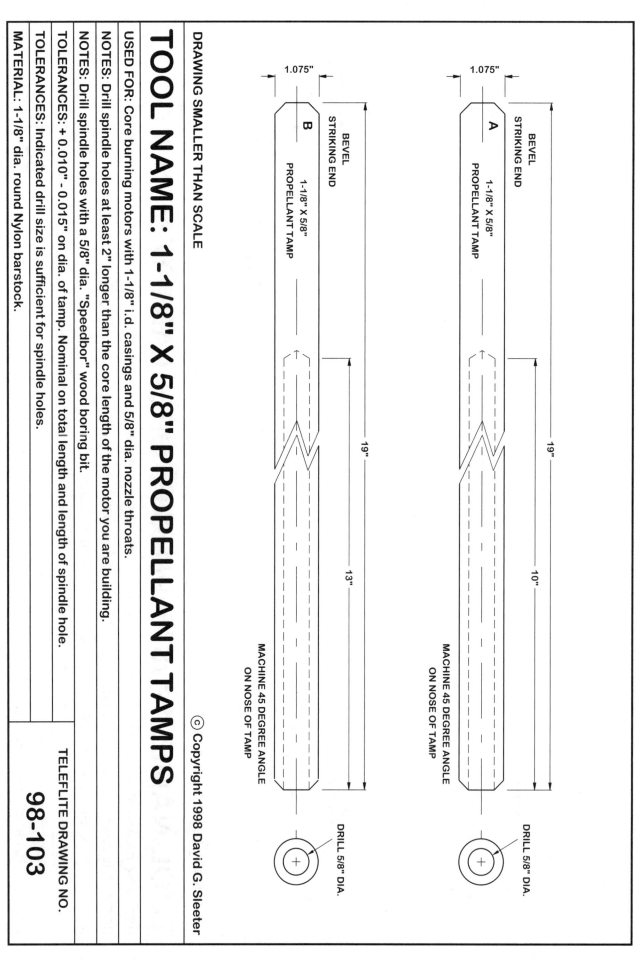

DRAWING SMALLER THAN SCALE

TOOL NAME: 1-1/8" X 5/8" PROPELLANT TAMPS

© Copyright 1998 David G. Sleeter

USED FOR: Core burning motors with 1-1/8" i.d. casings and 5/8" dia. nozzle throats.
NOTES: Drill spindle holes at least 2" longer than the core length of the motor you are building.
NOTES: Drill spindle holes with a 5/8" dia. "Speedbor" wood boring bit.
TOLERANCES: + 0.010" - 0.015" on dia. of tamp. Nominal on total length and length of spindle hole.
TOLERANCES: Indicated drill size is sufficient for spindle holes.
MATERIAL: 1-1/8" dia. round Nylon barstock.

TELEFLITE DRAWING NO.

98-103

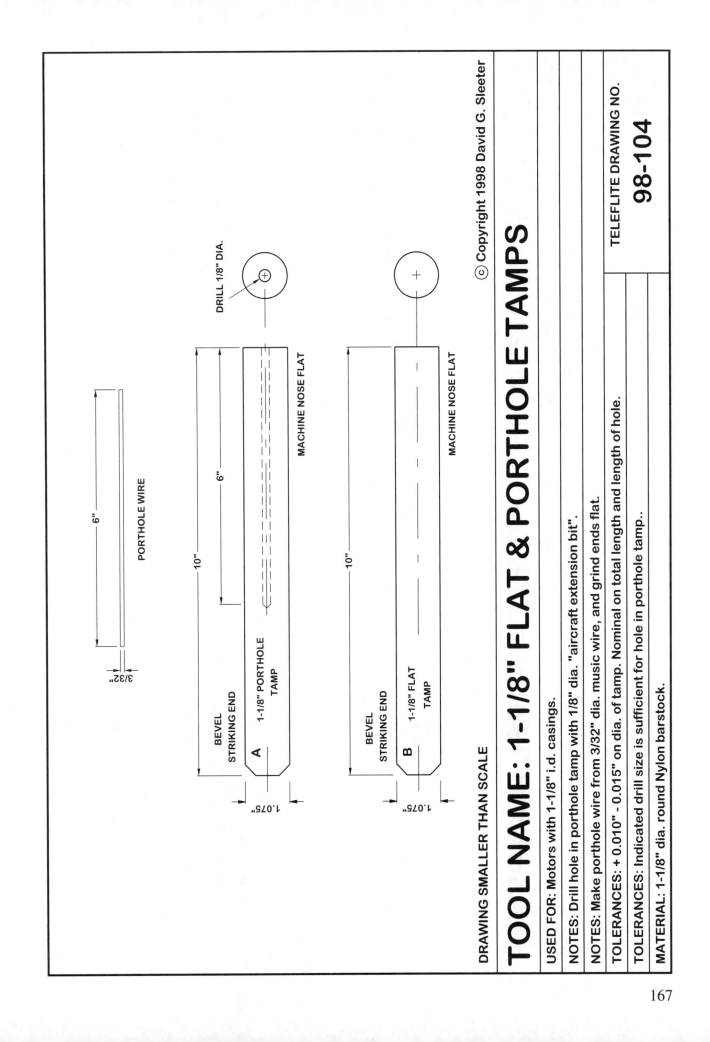

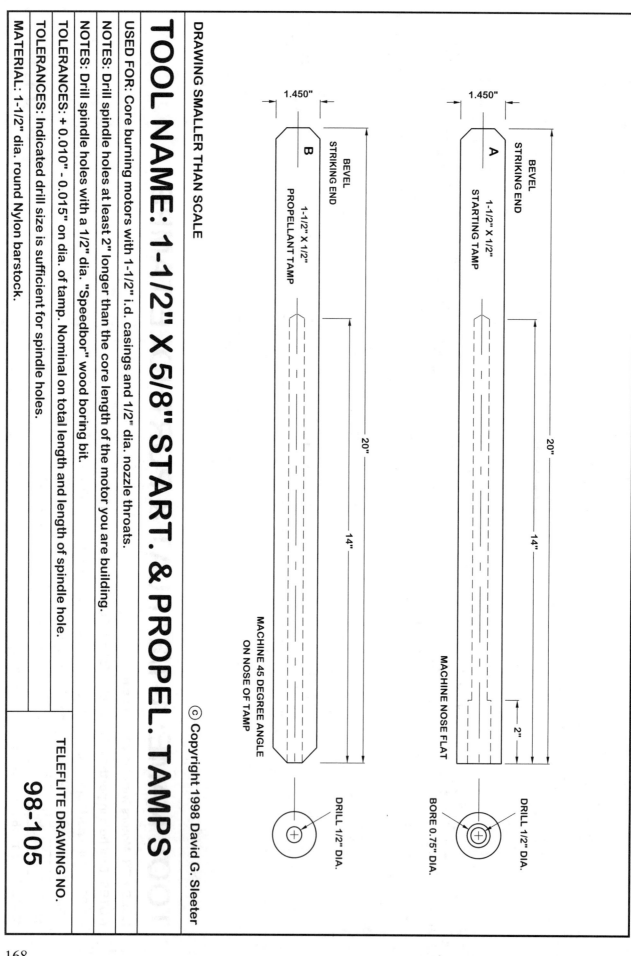

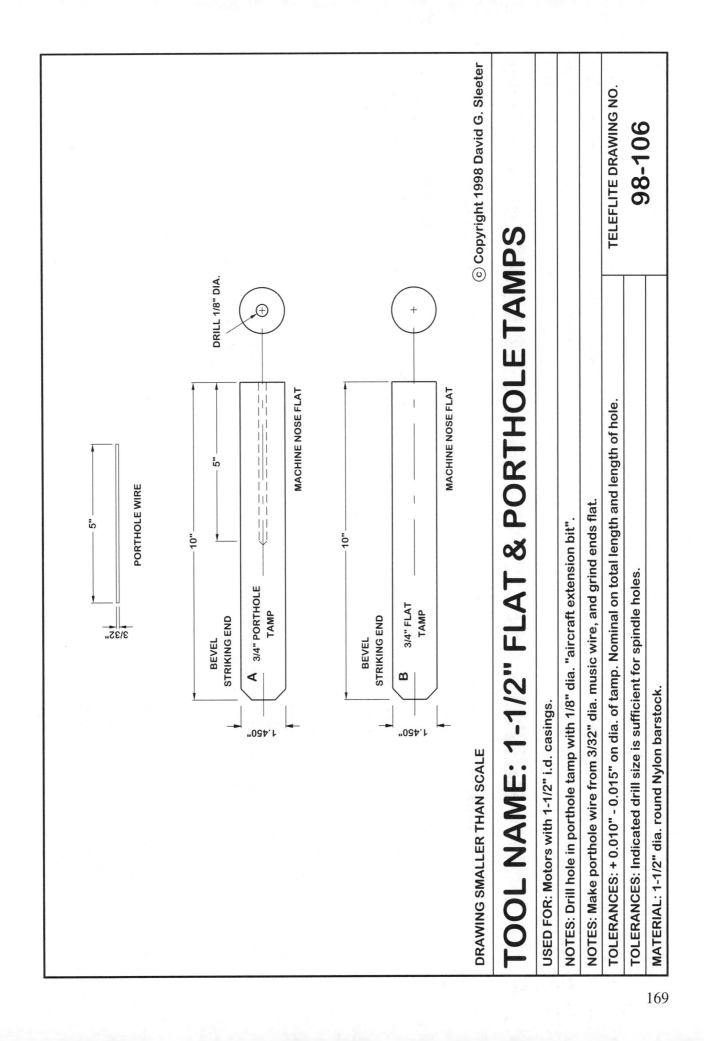

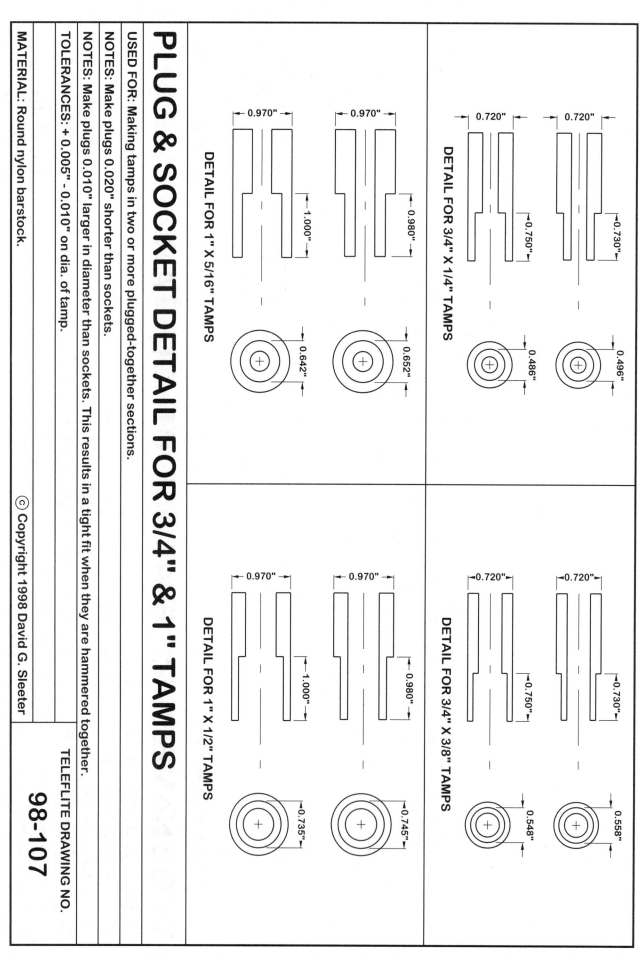

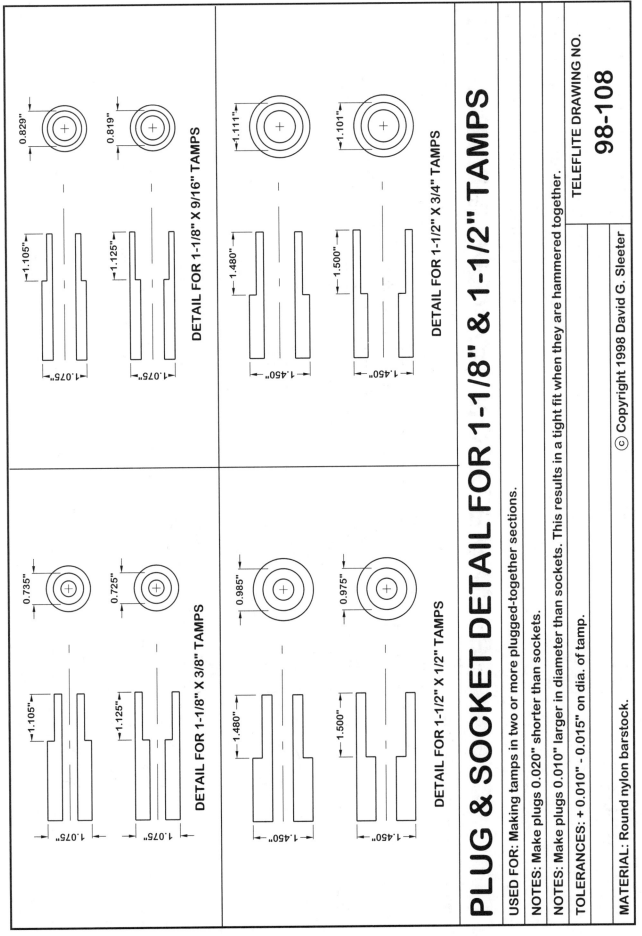

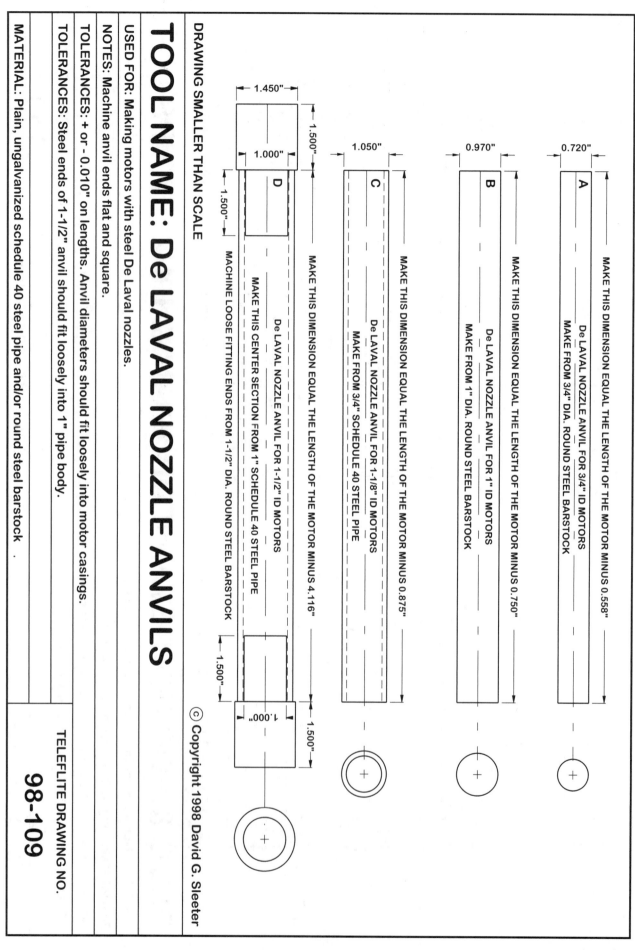

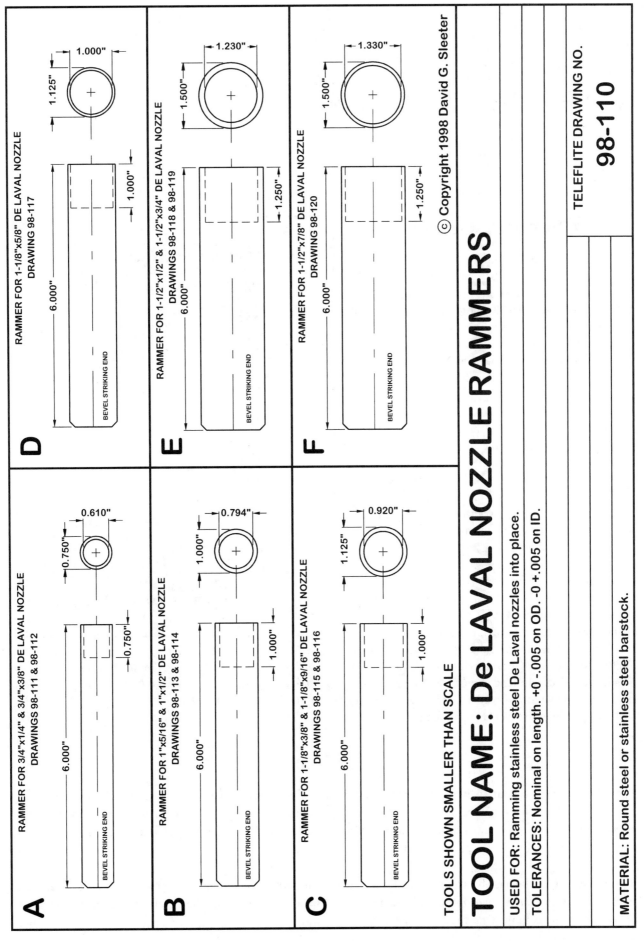

6. THE CHEMICALS

Chemical Grades

There are five propellant formulas in this book, and four of them are made with potassium nitrate or sodium nitrate, one or two carbon fuels, sulfur, and an acetone-soluble gum. The **KS** propellant is a mixture of potassium nitrate and sugar. This chapter explains what these chemicals are, where they come from, how to buy them, and in some cases, how to make them yourself, or extract them from common garden products.

A common mistake made by people new to amateur rocketry is to assume that they have to buy the chemicals from a laboratory chemical dealer. Another mistake is to fail to specify what *purity* they need. Chemical dealers often specialize in the highly purified chemicals used by high tech industries and testing labs. If you don't specify *exactly* what you need, they *might* assume that you want a lab grade chemical, and quote you a price of $30 per pound.

Chemicals are manufactured in *many* grades of purity. **Chemically Pure (C.P.)**, **U.S. Pharmaceutical (U.S.P.)**, and **Reagent Grade** are the purest, but you do *not* need chemicals of this quality to make the rocket propellants in this book. If you specify either **Technical Grade**, **Commercial Grade**, or **Pyro Grade**, the prices will range from $2 to $10 per pound. If you can find what you need at a fertilizer store, the cost will plummet to 50 cents per pound or less.

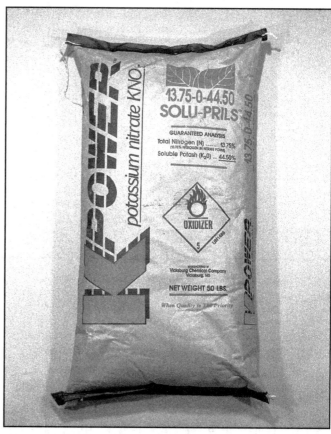

Figure 6-1. A 50 lb. bag of potassium nitrate fertilizer. This brand is manufactured by the Vicksburg Chemical Co. of Vicksburg, Mississippi. This is what I used to develop some of the propellants for this book.

Oxidizers

The solid fuel rocket industry uses *many* oxidizers. But most are expensive and hard to find, and they can't be safely mixed with sulfur or the simple carbon fuels that make the propellants in *this* book so ideal for amateurs. These *homemade* propellants are made with **potassium nitrate** or **sodium nitrate**, and I've specifically chosen them, because they are readily available, inexpensive, and *completely* safe to handle and store.

Should you have any doubts about their safety, I'll explain here that potassium nitrate and sodium nitrate are *so* chemically stable that it is *absolutely impossible* to make them burn or explode on their own. When heated to a temperature of 800° Fahrenheit, they melt; then glow red hot, and simply boil away.

● Potassium Nitrate Fertilizer

Potassium nitrate (chemical formula KNO_3), also called **saltpeter**, **nitrate of potash**, and **niter of potash**, comes in many forms, and the cheapest is the *fertilizer* grade. *Figure 6-1* shows a 50 lb. bag of the nitrate that I used to develop the KNO_3-based propellants in this book. I paid $14.00 for it at a fertilizer store here in Southern California. Potassium nitrate fertilizer is made by many companies, and *this* brand, called "K-POWER", is made by the **Vicksburg Chemical Company** of **Vicksburg, Mississippi**.

Figure 6-2. This 3-number code on the outside of the bag indicates that the product is potassium nitrate.

All fertilizer bags display a three-number code, and different fertilizers have different codes. Together, the numbers indicate how much nitrogen, phosphoric acid, and soluble potash the fertilizer supplies when it biodegrades in the soil. *Figure 6-2* shows the code on the bag of K-POWER, and you can see that it says "**13.75 - 0 - 44.50**". This is the code for potassium nitrate. Some manufacturers round off the numbers, but if the middle number is a zero, and the other two are close to 14 and 44, then the product inside is potassium nitrate. If the numbers are different, the product is *not* potassium nitrate.

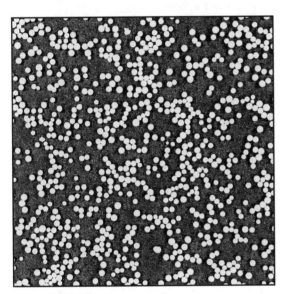

Figure 6-3. K-POWER *potassium nitrate fertilizer prills shown actual size.*

Potassium nitrate fertilizer is sold by the dealers listed under **FERTILIZERS** in *The Yellow Pages*. It comes in a granular form like table salt, and two beaded, or "prilled", types (large and small) that look like little, white beads of buckshot. The K-POWER nitrate in *Figure 6-1* is the *small* prilled kind, and *Figure 6-3* is an actual-size photo of these K-POWER prills, sprinkled onto a piece of black paper. Any of these forms will work, but I like the prilled kind, because it doesn't form lumps, and it doesn't cake up in storage.

● Potassium Nitrate Stump Remover

If you can't find potassium nitrate fertilizer, look for **stump remover**. Stump remover is used to rot out tree stumps. You'll find it at fertilizer and hardware stores, nurseries, and the garden shops in places like Kmart, Walmart, and Home Depot. The process is simple. You drill holes in the stump, and fill them with a mixture of stump remover and water. The chemical accelerates the rotting process, and in a few months, the stump decomposes to a point where it can be broken up and removed with a shovel. There are many brands on the market, and *most* of them are 99% pure potassium nitrate. *Figure 6-4* is a photo of two such types that I bought locally. One is packaged by the **Charles H. Lilly Co.**, and the other by **Cooke Laboratory Products**, both of **Portland, Oregon**. Inside each box is a chemical so similar in appearance to the K-POWER prills that, for all practical purposes, it may as well be the same product. And, of course, the number codes on the boxes *verify* that the chemical is potassium nitrate. **Important note.** As of the year, 2004, the Teleflite Corporation publishes information about alternate brands of stump remover on its Internet website at **www.teleflite.com** and **www.amateur-rocketry.com**.

● Potassium Nitrate from a Chemical Dealer

If you can't find potassium nitrate in the forms described above, try the commercial chemical dealers. Look in *The Yellow Pages* of nearby cities under headings like **CHEMICALS, RETAIL**, or **LABORATORY EQUIPMENT & SUPPLIES**, and *don't* buy it from the first place you find. Chemical prices vary *greatly*, so take your time, shop around, and specify one of the cheaper grades. *Some* dealers won't sell it to the public; usually due to fears about product liability. If you meet one of these people, don't argue; just move on. Some dealers require a minimum purchase, like $100 or 100 lbs. In this case, you'll have to decide if you want to buy that much, or you might think about sharing the cost with a friend. A *great* resource for chemicals is *American Fireworks News*. It is mailed each month to fireworks-makers throughout the world, and many chemical dealers advertise in the classified ad section in the back. At the time of this writing you can reach *AFN* at **HC 67, Box 30, Dingman's Ferry, PA 18328**, or through their Internet website at **www.fireworksnews.com**.

As of the year, 2004, I know for a fact that potassium nitrate is difficult or impossible to find in Kansas, Nebraska, Michigan, and northern Florida. If you live in one of these areas, and the situation doesn't change, you'll have to order it from one of the dealers in *AFN*, or expand your search to other states.

If you order the nitrate from a distant source, you can have it shipped to you. But the Department of Transportation recently established a new set of rules regarding the *commercial* shipment of what *they* consider to be hazardous materials. The special packaging and fees can drive up the delivered price considerably. It might be more practical to find a friend who's traveling out of state, and ask them to buy a bag for you on their way home.

Figure 6-4. As of the year, 2000, these two brands of Stump Remover are 99% pure potassium nitrate. Notice the 13-0-44 number codes.

● Can I Make Homemade Potassium Nitrate?

It is theoretically possible to make homemade potassium nitrate by reacting ammonium nitrate with potassium hydroxide in a cold water solution. Ammonium nitrate fertilizer is readily available, and you can buy potassium hydroxide from a chemical dealer. But potassium hydroxide is *extremely* hygroscopic, and the cheaper grades contain up to 15% water, acquired during the manufacturing process. Fort technical reasons, it isn't practical to remove the water with the kitchen-chemistry techniques described in this book. If you knew the amount of water the hydroxide contained, you could make an appropriate weight correction. But the amount of water varies, and it's never stated on the package. It is therefore impractical to make potassium nitrate by this method.

● Extracted Potassium Nitrate

If you can't buy potassium nitrate, *as of this writing* you can extract it from at least two, common garden products. Both are made by **Dexol Industries** of **Torrance, CA**. Though *neither* would work as a rocket propellant or an oxidizer on its own, *both* products contain usable amounts of potassium nitrate, which you can extract by the following methods.

● Dexol Stump Remover

You can buy **Dexol Stump Remover** (*Figure 6-5*) at nurseries and garden shops. Though isn't *pure* potassium nitrate, you can see from the label (*Figure 6-6*), that it *contains* potassium nitrate. The nitrate is soluble in water, and the other ingredients are not. If you dissolve one pound of the stump remover in water, filter the liquid, and evaporate the water, you'll extract about 200 gms. of usable potassium nitrate. To do so, proceed as follows.

POTASSIUM NITRATE EXTRACTED from DEXOL STUMP REMOVER
(to make approx. 200 gms.)

MATERIALS LIST

1. One lb. of Dexol Stump Remover
2. Sixteen ounces (2 cups) of boiling water
3. A one quart Pyrex measuring cup
4. A small pot and a spoon to stir with
5. Two coffee filters
6. A large kitchen sieve
7. Four 1x2 wooden boards
8. A sheet of plastic, preferably black.
   (A 30" x 36", black trash bag works nicely.)

Figure 6-6. The back of a Dexol Stump Remover label indicates that the product contains potassium nitrate.

1. Place 1 lb. of Dexol Stump Remover in the Pyrex measuring cup. Add 16 oz. (2 cups) of boiling water, and stir for 3 minutes (*Figure 6-7*).

2. Place the 2 coffee filters (a double layer) in the kitchen sieve, and place the sieve over the pot. Pour the hot solution from **step 1** into the filters, and let it drain for 5 minutes. Some pressing with the spoon (*Figure 6-8*) will help squeeze the last of the liquid through. Then discard the material that remains in the filters.

3. Place the plastic sheet on a level surface outdoors, and elevate the 4 edges with the wooden boards. Pour the filtered solution from **step 2** onto the plastic, and let it dry in the sun (*Figure 6-9*). This will get it *almost* dry, but not quite. To finish the job, separate the sun-dried crystals from the plastic, spread them on a cookie sheet, and dry them in a kitchen oven for 15 minutes at 250° Fahrenheit.

4. Granulate the oven-dried crystals by rubbing them through the kitchen sieve (*Figure 6-10*). For scaled up recipes with larger quantities, use the 18" x 18" window screen sieve described on **page 85**. The potassium nitrate made by this method is similar to the granulated potassium nitrate fertilizer described on **page 175**.

*Figure 6-5. Dexol Stump Remover. Though it's not pure potassium nitrate, it **contains** potassium nitrate, which you can recover by following the instructions on this page.*

● **Dexol Gopher Gassers**

You can buy **Dexol Gopher Gassers** (*Figure 6-11*) at nurseries, garden shops, and hardware stores. When ignited and placed in a gopher hole, they burn *very slowly*, and produce a large amount of poisonous, sulfur dioxide gas that spreads through the gopher's burrow and kills the gopher. At the time of this writing, the chemical composition that does this contains 45% potassium nitrate, 45% sulfur, 8% charcoal, and 2% dextrin (*Figure 6-12*). You will *immediately* notice that it contains the *same* ingredients used in black powder, *but in the wrong proportions*. This is an *excellent* example of how a change in the weights of the chemicals in a formula can *greatly* alter the formula's behavior. A box of 6 gopher gassers yields 115 gms. of composition, containing about 52 gms. of potassium nitrate.

The nitrate is soluble in water, but the sulfur and the charcoal are not. If you dissolve the composition from 6 gopher gassers in water, filter the liquid, and evaporate the water, you'll extract about 35 gms. of usable potassium nitrate. A small amount of dextrin is listed on the ingredient label, but it doesn't appear to a significant extent in the finished product. It *might* be absorbed by the charcoal.

Because of their cost, gopher gassers are *not* an economical source for large amounts of the potassium nitrate, but for *small* quantities they will get you by. To extract 35 gms. of potassium nitrate from 6 Dexol Gopher Gassers, proceed as follows.

Figure 6-7. A cup of boiling water is added to 1 pound of Dexol Stump Remover, and the mixture is stirred.

Figure 6-8. The hot solution is poured through two coffee filters.

Figure 6-9. The extracted potassium nitrate crystals are dried in the sun.

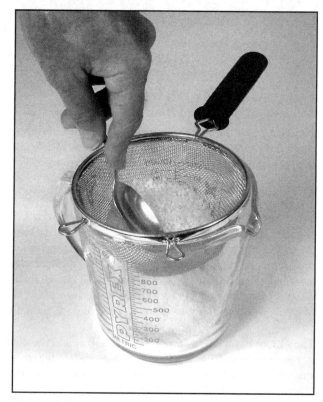

Figure 6-10. The dried potassium nitrate is granulated by rubbing it through a kitchen sieve.

Figure 6-11. A package of Dexol Gopher Gassers.

Figure 6-12. Note that the ingredients include 45% potassium nitrate.

*Figure 6-13. Each gopher gasser comes with a fuse. You can remove these fuses, and use them for the time delay burn rate test described on **page 294**.*

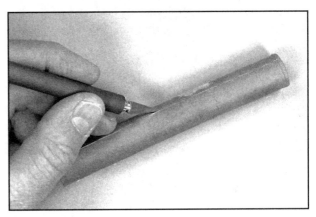

Figure 6-14. The cardboard tube is slit lengthwise.

POTASSIUM NITRATE EXTRACTED from DEXOL GOPHER GASSERS
(to make 35 gms.)

MATERIALS LIST

1. Six Dexol Gopher Gassers
2. Eight ounces of boiling water
3. A one pint Pyrex measuring cup
4. A small pot and a spoon to stir with
5. Two coffee filters
6. A large kitchen sieve
7. A large cookie sheet
8. A sheet of thin plastic, preferably black

1. Each gopher gasser has a fuse (*Figure 6-13*). Pull out the fuses, and save them for the **time delay burn rate test** described on **page 294**. Tear off the paper wrappers, and with a single edge razor blade or an X-acto knife, slit the cardboard tubes lengthwise (*Figure 6-14*). Then peel away the layers of cardboard, and expose the dark, gray composition inside (*Figure 6-15*).

179

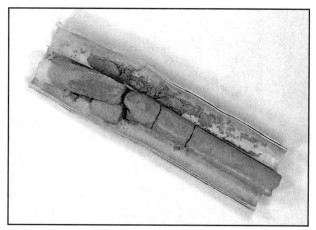

Figure 6-15. The cardboard tube is peeled away.

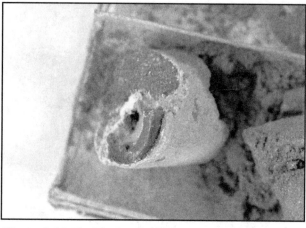

Figure 6-16. The black composition, and the light gray cap where the fuse passed through are both discarded..

Figure 6-17. Boiling water is added to the composition, and the mixture is stirred.

Figure 6-18. The hot solution is poured through two coffee filters.

At the end of each gasser you'll see a light gray clay cap with a hole in it where the fuse passed through (*Figure 6-16*). These caps are painted on one end with a flammable black material that might be dangerous if mixed with other propellant chemicals. **Do not use this black material. Discard the black material *and* the clay cap by soaking in water.** At the opposite end of the gasser is another clay cap, but it's buried in the composition, and it might be hard to see. Find it by crumbling the composition away with your fingers, and discard this cap too.

2. Place the composition from 6 gopher gassers into the Pyrex measuring cup, and crush it up as thoroughly as you can with a screwdriver handle. Add 8 ounces (one cup) of boiling water, and stir for 3 minutes (*Figure 6-17*). As you work, crush and break up any lumps with the bowl of the spoon.

3. Place the 2 coffee filters (a double layer) into the kitchen sieve, and place the sieve over the pot. Pour and scrape the hot solution from **step 2** into the filters, and let it to drain for 5 minutes. Then with the bowl of a spoon (*Figure 6-18*), squeeze out as much of the remaining liquid as you can, and discard the material that remains in the filters.

4. Place the cookie sheet on a level surface outdoors, and cover it with the sheet of plastic. Pour the filtered solution from **step 3** onto the plastic, and let it to dry in the sun (*Figure 6-19*). This will get the nitrate *almost* dry, but not quite.

5. Separate the sun-dried crystals from the plastic. Place them back on the cookie sheet (this time *without* the plastic), and dry them in a kitchen oven for 15 minutes at 250° Fahrenheit. Then granulate the oven-dried crystals by rubbing them through the kitchen sieve (*Figure 6-10*). The potassium nitrate made by this method is similar to the granulated potassium nitrate fertilizer described on **page 175**.

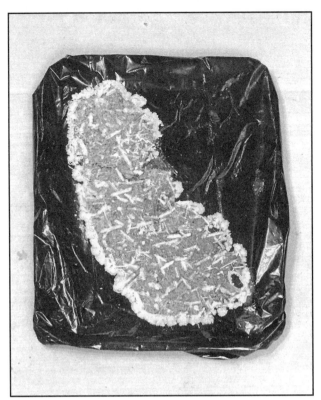
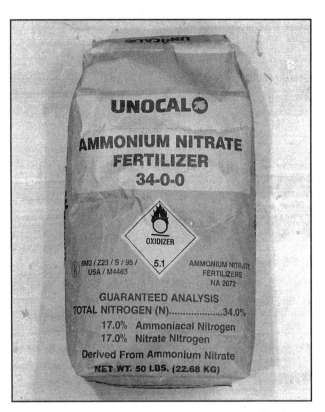

Figure 6-19. The extracted potassium nitrate crystals are dried in the sun.

Figure 6-20. A 50 lb. bag of ammonium nitrate fertilizer.

● Sodium Nitrate

Sodium nitrate (chemical formula **NaNO$_3$**), also called **nitrate of soda**, **niter of soda**, and **soda niter**, occurs naturally in large mineral deposits in South America. Because most of the world's supply used to come from Chile, it was once called **Chile saltpeter**. In the United States it is *sometimes* used as a fertilizer. But in some situations, it increases the salinity of the soil, so it has fallen out of favor with the farming industry in all but a few, widely scattered areas. As of this writing, it isn't likely that a fertilizer dealer would sell it, but it never hurts to ask.

Despite its agricultural limitations, sodium nitrate has many industrial uses, and most chemical dealers sell it. When looking for sodium nitrate, do the same things that you did to find potassium nitrate, and ask for one of the cheaper grades. Like potassium nitrate, sodium nitrate comes in both granulated and prilled forms, and either will work, as long as the product is at least 99% pure.

● Homemade Sodium Nitrate

If you can't buy sodium nitrate, you can make it yourself by reacting **ammonium nitrate** with **sodium hydroxide** in a **cold water** solution. *Sodium* hydroxide is not as hygroscopic as potassium hydroxide. It doesn't absorb a significant amount of water during its manufacture, and you can remove any residual moisture by drying it in a kitchen oven.

> **But the chemical reaction, and the evaporation of the water afterwards, release powerful and caustic ammonia fumes!** *Always* **wear the appropriate safety equipment (*Figure 6-22*).** *Always* **do it** *outdoors***, and arrange to have the wind or a fan blowing the fumes away from you.**

● Ammonium Nitrate

Like potassium nitrate, **ammonium nitrate** (chemical formula **NH$_4$NO$_3$**) is sold as a fertilizer, but in most parts of the country it is easier to find. Many dealers who don't sell potassium nitrate *do* sell ammonium nitrate. It typically comes in the prilled form in 1 lb. to 50 lb. bags.

Look in *The Yellow Pages* under **FERTILIZERS**, and shop around for the lowest price. *Figure 6-20* is a photo of a bag that I bought locally. If you look at the center of the bag, you'll see the number code, "**34 - 0 - 0**". This is the code for ammonium nitrate, and once again, if the box or the bag displays this code, then you know that the product is ammonium nitrate.

Because ammonium nitrate is also an explosive, you probably wonder if it is safe to handle and store. Please be assured that it *is* safe, *as long as it hasn't been mixed with, or contaminated by anything*. In its *pure* form, it is difficult to make quantities under 50 lbs. detonate, and it takes a brisant, commercial blasting cap to set it off. If you doubt what I am saying, look at the tons of ammonium nitrate stacked around a fertilizer warehouse, and watch the casual way in which the employees toss large bags of it onto the trucks of waiting customers. In medium to small quantities, ammonium nitrate is dangerous *only* when mixed with things like oil, diesel, or gasoline. When mixed with a carbon fuel, it makes a black powder-like mixture called **ammonpulver** that, unfortunately, doesn't make a good rocket propellant. If your ammonium nitrate ever becomes contaminated, destroy it immediately by dissolving it in water.

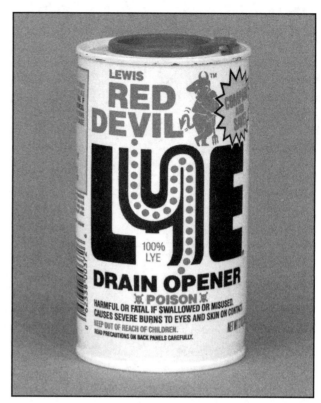

Figure 6-21. A 12 oz. can of "lye".

People new to amateur rocketry *always* want to know if they can *substitute* ammonium nitrate for potassium nitrate or sodium nitrate, and the answer is *no*! **You *cannot* use ammonium nitrate as a substitute for potassium nitrate or sodium nitrate**. Ammonium nitrate propellants operate efficiently *only* at higher chamber pressures than a cardboard tube can withstand, and they require the addition of expensive chemicals that substantially raise the cost of the finished propellant.

● Sodium Hydroxide

You can buy **sodium hydroxide** (chemical formula **NaOH**), also called **lye** or **caustic soda**, from a chemical dealer, but *first* check at a grocery store. Sodium hydroxide is sold as a cheap, drain opener under the generic name, **lye**. Look for it on the shelf with the "Draino" and the "Liquid Plumber".

Figure 6-21 is a photo of a 12 ounce can that I bought at a supermarket. At the time of this writing, a can like this costs about $3.00. For larger quantities at lower prices, look for it at farm and feed supply stores under the generic name, **caustic soda**. Farmers and ranchers use it for everything from cleaning septic tanks to stripping old paint, and as of this writing, a 50 lb. bag costs about $35.00.

> **But sodium hydroxide, and solutions of sodium hydroxide, are *very* caustic! They can cause *severe* chemical burns to the skin and the eyes, so *always* wear the appropriate safety equipment (*Figure 6-22*). Handle sodium hydroxide (lye) with *great* care. In case of contact with the skin or the eyes, flood the area immediately with water, and keep rinsing until all traces of the chemical are gone.**

When ammonium nitrate and sodium hydroxide are mixed in the right proportions in a cold water solution, the sodium ion replaces the nitrate's ammonium ion according to the following reaction:

$$NH_4NO_3 + NaOH > NaNO_3 + NH_4OH$$

As the reaction proceeds, the ammonia goes into solution in the water. When the reaction is complete, you evaporate or boil off the water-ammonia solution, and the result is pure sodium nitrate. To make 500 gms. of homemade sodium nitrate, proceed as follows.

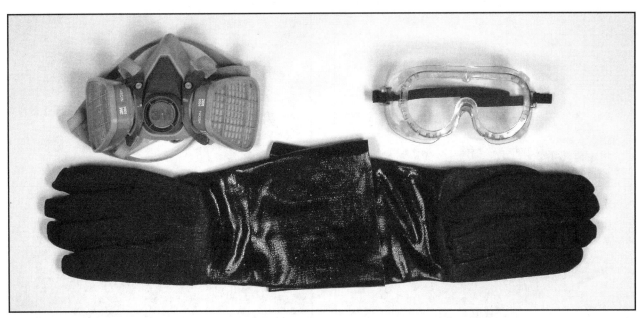

*Figure 6-22. The items shown in this photo **must** be worn when making homemade sodium nitrate. **They are not optional!** They include a dual-cartridge respirator mask with (**very important**) **ammonia-rated** cartridges, a pair of rubber chemical gloves, and a pair of chemical goggles that cover not only the front of the eyes, but the sides, top, and bottom as well.*

HOMEMADE SODIUM NITRATE
(to make 500 gms.)

MATERIALS LIST

1. A table to work on **OUTDOORS**
2. **An ammonia-rated respirator mask**
3. **A pair of chemical-rated safety goggles**
4. **A pair of rubber chemical gloves**
5. An electric fan
6. A 2 quart steel, iron, or Pyrex pot
7. A large spoon to stir with
8. A sheet of plastic, preferrably black. (A 30" x 36" black trash bag works nicely.)
9. Four 1x2 wooden boards
10. A large kitchen sieve
11. 235.0 gms. of dry sodium hydroxide
12. 476.0 gms. of dry ammonium nitrate
13. Sixteen oz. (2 cups) of *ice* water

1. **Dry the chemicals thoroughly.** Ammonium nitrate and sodium hydroxide are hygroscopic, and that means that they absorb moisture from the air. If they are damp when you weigh them, your weights will not be accurate. The reaction won't be balanced, and the end product won't be pure. Before you begin, measure out *more* of each chemical than you need. Place the two chemicals in *separate, non-aluminum* pots, and dry them for one hour in a kitchen oven at 150° Fahrenheit. Let them cool, and place them in *separate*, tightly sealed, *non-aluminum* containers. **Important note.** Sodium hydroxide reacts with aluminum, so you *cannot* use aluminum containers.

2. **Most important of all!** Put on the safety equipment shown in *Figure 6-22*. The respirator mask *must* have ammonia-rated cartridges. If the package doesn't specifically *say* that the cartridges are designed for ammonia, *then don't buy the mask*. Some of the masks sold in hardware stores have replaceable cartridges, and though they don't *come* with ammonia-rated cartridges, you might be able to special-order them. *But you'll have to ask the manufacturer.* Welding supply shops sell respirators with replaceable cartridges, and though they don't normally *stock* the ammonia-rated cartridridges, they can order them for you. The mask in the photo was made by 3M, and the cartridges are rated for ammonia and methylamine. This is the equipment that I used to develop this process. **Important note.** The cartridges will quickly deteriorate if left in open air, so store them in a zip-lock bag when not in use.

You can buy the gloves and the safety goggles at hardware stores and the big, hardware supermakets like Lowe's and Home Depot. The goggles *must* be the kind that protect not only the front of your eyes, but the sides, top, and bottom as well. These goggles are normally vented. Though they'll protect you from a chemical splash, they will *not* protect you from the eye-stinging ammonia fumes. Therefore, either put tape over the vent holes, or stand *upwind* of all ammonia sources.

Figure 6-23. As the ammonium nitrate is added, a thin scum of impurities will float to the top.

Figure 6-24. Homemade sodium nitrate crystals drying in the sun.

3. Place the 2-quart, *non-aluminum* pot or Pyrex measuring cup on a table outdoors. Either stand *upwind* of the pot, or set up an electric fan so that it blows the air across the top of the pot *in a direction away from you*. **Important note.** A water-sodium hydroxide solution reacts with and dissolves aluminum, so you *cannot* use an aluminum pot. Fill the pot with 16 oz. (2 cups) of *ice* water. To make the ice water, put a dozen ice cubes in a large measuring cup. Then fill it with water to the 2-cup line. Weigh out 235.0 gms. of dry sodium hydroxide (lye), and dissolve it *slowly and carefully* in the water. Add it a little bit at a time while stirring constantly. As the chemical dissolves, it generates heat, and the ice will melt. As the water warms up, it will release choking, eye-stinging fumes!

Warning! This is *very* nasty stuff! Do *not* breath the fumes, and don't splash it on yourself! *Important note.* **Before you proceed to step 4, *allow this hot solution to cool*.**

4. *When the solution you made in step 3 has cooled to room temperature,* weigh out *476.0 gms.* of dry ammonium nitrate. While stirring constantly, *with the wind or a fan blowing the fumes away from you*, add it *slowly and carefully* to the lye-solution you made in **step 3**. As the chemicals react, the water will bubble, releasing *large* amounts of eye-stinging ammonia fumes. **Keep your respirator on. *Stand upwind at all times*, and do *not* breath these fumes. Ammonia vapors are *very* powerful, and one good whiff can knock a person unconscious!**

As you work, a scummy foam of minor impurities will form on the surface (*Figure 6-23*), and if you are careful, you can skim it off with a tea strainer (a miniature sieve). When the last of the ammonium nitrate has been added, stir the pot *gently* for a minute or two, and let the reaction finish.

5. Lay the sheet of plastic on a level surface outdoors, and elevate the edges with the four wooden boards. Then, *standing upwind, and still wearing the respirator*, pour the solution onto the plastic sheeting, and let it dry in the sun (*Figure 6-24*). **Be careful when you do this, because the liquid contains a large amount of dissolved ammonia. It is *very* caustic and irritating to the nose, the skin, and the eyes.**

Warning! As the water starts to evaporate, the liquid will become saturated with sodium nitrate, and the ammonia will be driven off as a gas. *During this initial stage in the drying process, the ammonia fumes will be so powerful that, ten to twenty feet downwind, they will sting your eyes!* One hundred feet downwind, the ammonia odor will still be strong, so alert you're neighbors to what you are doing *ahead* of time.

Figure 6-25. A package of 4 Giant Destroyer gopher gassers.

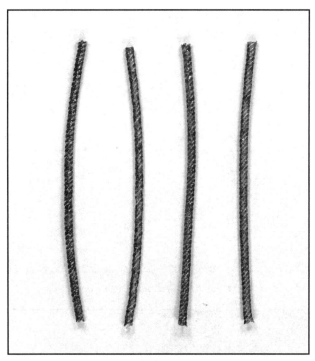

Figure 6-26. Each Giant Destroyer gopher gasser comes with a piece of green Bickford fuse.

Depending on the rate of evaporation (i.e. the temperature and humidity), in 2 to 8 hours the sodium nitrate crystals will be forming, the ammonia will be gone, and the liquid will be odorless and safe to handle. At *this* point, if you check it with pH paper (see a laboratory supply dealer), you should get a pH reading of 6 to 7. If the pH is *greater* than 7, you've got residual sodium hydroxide, and you'll have to add a few grams of ammonium nitrate to use it up.

6. Sun-drying will get the nitrate *almost* dry, but not quite. To finish the job, separate the sun-dried crystals from the plastic, spread them on a cookie sheet, and dry them in a kitchen oven for 15 minutes at 250° Fahrenheit. Then granulate the oven-dried crystals by rubbing them through the kitchen sieve (***Figure 6-10***). The sodium nitrate made by this method is similar to the granular sodium nitrate that you'd buy from a chemical dealer.

Why start with ice water? As the sodium hydroxide (the "lye") dissolves, it generates *heat*. If you start with room-temperature or warm water, the solution might boil, *and a boiling solution of sodium hydroxide is **dangerous**!* Starting with ice water cools the reaction, keeps the water from boiling, and makes the process a little safer.

Why use 2 cups of water instead of more or less? If you use *less* water, there won't be enough to absorb the ammonia. The *excess* ammonia will bubble violently, making a foaming mess that overflows the pot. Using *more* than 2 cups of water is unnecessary, and makes the sodium nitrate harder to recover. You can scale this recipe up or down by multiplying or dividing the amounts of sodium hydroxide, ammonium nitrate, and water by a constant. But the proportions should *always* be, *by weight*, 2.03 parts *dry* ammonium nitrate to 1.00 parts *dry* sodium hydroxide.

Also of note, if you work the chemical equation on page 182 with the atomic weights of the elements involved, you'll notice that my process is about 1% rich in ammonium nitrate. I've done this *intentionally* to insure that all the sodium hydroxide is used up. Residual NaOH would render the finished product alkaline. Alkalinity in the finished propellant might degrade the propellant's resin binder, and the glue that bonds the propellant grain to the motor casing wall.

● Extracted Sodium Nitrate

As of this writing, sodium nitrate is a major component in the **Giant Destroyer gopher gassers** (*Figure 6-25*) made by the **Atlas Chemical Corp.** of **Cedar Rapids, IA**. You can buy them at nurseries, garden shops, and hardware stores. Like the Dexol products, though the composition itself can't be used, it makes a good *source* for sodium nitrate. A package of four gassers contains 178 gms. of composition, from which you can extract about 70 gms. of nitrate by the following method. To extract 70 gms. of sodium nitrate from 4 Giant Destroyer gopher gassers, proceed as follows.

Figure 6-27. The cardboard tube is slit lengthwise.

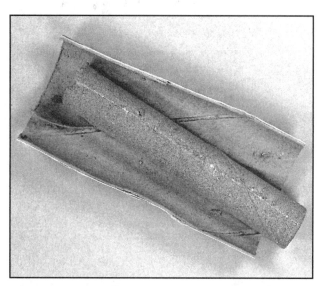

Figure 6-28. The cardboard tube is peeled away.

Figure 6-29. Boiling water is added to the composition, and the mixture is stirred for 3 minutes.

Figure 6-30. The hot solution is poured through the two coffee filters.

SODIUM NITRATE EXTRACTED from GIANT DESTROYER GOPHER GASSERS
(to make approx. 70 gms.)

MATERIALS LIST

1. Four Giant Destroyer Gopher Gassers
2. Eight ounces of boiling water
3. A one pint Pyrex measuring cup
4. A small pot and a spoon to stir with
5. Two coffee filters
6. A large kitchen sieve
7. A large cookie sheet
8. A sheet of thin plastic, preferably black

1. Open the package, and remove the 4 gopher gassers. Packed among them, you'll find 4 pieces of green "Bickford" fuse (*Figure 6-26*), which you can save and use for the **time delay burn rate test** (**page 293**). With a single edge razor blade or an Xacto knife, slit the cardboard tubes lengthwise, and peel them away from the gray composition inside (*Figures 6-27* and *6-28*).

Figure 6-31. Sodium nitrate crystals forming as the filtered solution is dried in the sun.

2. Place the composition from 4 gopher gassers into the Pyrex measuring cup, and crush it up as thoroughly as you can with a screwdriver handle. Add 8 ounces (one cup) of boiling water (*Figure 6-29*), and stir for 3 minutes. As you work, crush and break up any lumps with the bowl of the spoon.

3. Place the 2 coffee filters (a double layer) into the kitchen sieve (*Figure 6-30*), and place the sieve over the pot. Pour the hot solution from **step 2** into the filters, and let it to drain for 5 minutes. With the bowl of the spoon, squeeze out as much of the remaining liquid as you can (*Figure 6-22*). Then discard the material that remains in the filters.

4. Place the cookie sheet on a level surface outdoors, and cover it with the sheet of plastic. Pour the filtered solution from **step 3** onto the plastic, and let it to dry in the sun (*Figure 6-31*). This will get the nitrate *almost* dry, but not quite.

5. Separate the sun-dried crystals from the plastic. Break them up, place them on the cookie sheet (this time *without* the plastic), and dry them in a kitchen oven for 15 minutes at 250º Fahrenheit. Then granulate the oven-dried crystals by rubbing them through the kitchen sieve (*Figure 6-10*). The sodium nitrate made by this method is similar to the sodium nitrate fertilizer described on **page 181**.

Fuels

Rocket propellants often use things like rubber or plastic as a fuel, but these materials will *not* work with the propellants in this book. Before a fuel can take part in a combustion reaction, it has to be vaporized, and the complex molecules that make up these materials take *more* heat to vaporize than their reactions with sodium nitrate or potassium nitrate will produce. So *much* heat is consumed that the reaction self-extinguishes into a smoldering mass that never gets hot enough to really burn. Rubbers and plastics are used as the fuels in *other* solid rocket propellants, but *not* in propellants that use potassium nitrate or sodium nitrate as the oxidizer.

All the propellants in this book use some form of carbon as the fuel, and with the exception of graphite, most forms of carbon will work. The harder and denser types produce slower burning propellants, and the softer, lighter ones make propellants that burn faster. They vary considerably in their effects on a propellant's burn rate, so I'll introduce them one at a time.

● Coal

Prior to 1950, coal was the main fuel for heating homes in the eastern half of the U.S. Since 1950 it has been largely replaced by oil and natural gas. As of the year, 2000, most of the country's coal supply is used for generating electricity, or making things like steel and cement. Coal still finds enough use in *light* industry to be sold at the retail level, and like the fertilizers, you'll find it in *The Yellow Pages*, this time under the **COAL** heading.

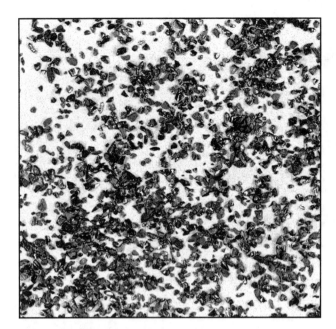

Figure 6-32. Anthracite granules shown actual size.

● Anthracite

Anthracite is a very pure, dense, and *hard* coal. It is easy to find in the East, but hard to find in the West. When used as the fuel, it makes a *slow* burning propellant, suitable for time delays and long core burners with comparatively long burn times. It comes in many sizes ranging from fist-size chunks down to particles the size of a grain of sand. Since the final step in making a propellant is to grind it into a fine powder, you should start with the smallest particles you can find.

Figure 6-33. A bag of mesquite Bar-B-Que charcoal.

As of the year, 2000, here in Southern California, the only place I've found that sells it is the **Anthracite Carbon Filter Media Company** at **734 E. Hyde Park Blvd., Inglewood, California 90302**. They manufacture water filters, and they use granulated anthracite as the filter medium. They sell it to the public, but their minimum order is one full bag. The material is not expensive (about $40.00 for 60 pounds), but they prepare it by washing it with water, and they ship it soaking wet in sealed, plastic bags.

When it arrived, I tried to dry it in my kitchen oven, and the kitchen oven worked, *but it made the house smell like a coal mine*! I solved the problem by drying it outdoors, a few pounds at a time, in an electric frying pan with the lid slightly ajar. I ordered the smallest "mesh" size they sell. It has a particle size down in the 1 mm. range, and this is what I used to develop the **KG1** propellant for this book. *Figure 6-32* is an actual-size photo of these tiny, anthracite granules. In preparing the photo for print, I was careful to adjust the camera so that the grains in the picture appear exactly as they would, had I sprinkled them onto the page in front of you.

To find anthracite where you live, look in *The Yellow Pages* under the **COAL** headings, and try to find a product who's particle size approximates what I used. Barring that, you can start with larger pieces, break them down with the homemade coal crusher (**page 86**), and sift the results through the homemade window screen sieve (**page 85**). Then see **page 206** for instructions on how to prepare the granulated anthracite for use as a rocket fuel.

● Charcoal

Wood contains a large amount of carbon, and charcoal is made by burning wood in an oxygen-starved atmosphere. The carefully controlled combustion process consumes or drives off most of the other elements, and leaves the carbon behind. The hardness and density of the wood you start with determine the hardness and density of the finished product, and these factors determine what effect the charcoal has on a propellant's burn rate.

Figure 6-34. Chunks of mesquite charcoal (shown smaller than actual size).

Figure 6-35. A bag of garden charcoal.

Generally speaking, charcoal makes propellants that burn *faster* than propellants made with coal. But within the family of charcoals, there's some variation. Hardwood charcoals make slower burning propellants, and softwood charcoals make faster burning propellants. Willow charcoal is one of the softest. It is noted for making pyrotechnic compositions that burn *very* fast, and is highly prized by the makers of fireworks. You can make charcoal at home, and science teachers might think it an interesting class project. You can make a charcoal kiln from a 5 gallon metal paint bucket, but charcoal is so easy to buy that I won't describe the process here. If you want to learn how to make charcoal, look for books on related subjects, or search the Internet.

● **Mesquite Charcoal**

Mesquite charcoal is a popular Bar-B-Que fuel, and you'll find it where Bar-B-Que supplies are sold. Mesquite charcoal is made from the hard wood of the desert mesquite bush. It usually comes in a paper bag, containing large chunks that weigh several ounces each. *Figure 6-33* is a photo of a bag that I bought at a supermarket, and *Figure 6-34* is a photo (*smaller* than scale) of some of the contents. See **page 207** for instructions on how to prepare mesquite charcoal for use as a rocket fuel.

● **Charcoal Briquets**

People new to amateur rocketry *often* ask if they can use **charcoal briquets**, and the answer is *no*. Briquets are made from cheap, stump and root wood. They contain large amounts of dirt and sand that contaminate a propellant, and make its performance unpredictable. Also, the charcoal in most briquets is premixed with small amounts of glue and nitrate oxidizers that hold the briquets together, and help keep them burning. These additives effect the charcoal's behavior in unpredictable ways, and make charcoal briquets *unsuitable* for use as a rocket fuel. When you shop for the mesquite charcoal described above, *read the labels carefully*, because *some* companies sell mesquite charcoal in this unusable, briquet form.

Figure 6-36. Granules of garden charcoal shown actual size.

Figure 6-37. Carbon black particles magnified **20,000** times by a scanning, electron microscope. The ones in this photo average 1/4 micron (ten millionths of an inch) in diameter.

● Garden Charcoal

Incredibly, a teaspoon of charcoal has the same absorbing area as a flat sheet of carbon *the size of a football field*. This is because charcoal contains a microfine structure of tiny air channels *less than 50 Angstroms across*. Please note that an "Angstrom" is *one ten millionth* of a millimeter; about the size of an atom. Through a physical phenomenon called "capillary action", these channels attract and hold moisture. In larger pieces, the tubular structure of the wood from which the charcoal is made provides an ideal environment for the growth of beneficial bacteria. For these two reasons, the addition of charcoal to soil improves the soil's quality.

Because farmers and gardeners use charcoal, you can buy it at nurseries, garden shops, and fertilizer and farm supply stores. It is a common component in potting soil, and the places that sell it call it **horticultural charcoal** or **garden charcoal**, and sometimes **number 2 charcoal** or **number 4 charcoal**, depending on the particle size.

Garden charcoal typically comes in quarter inch-size chunks in 1 to 25 pound bags at a cost (as of this writing) of 50 cents to $2 per pound. It is usually softer than mesquite charcoal, so it makes propellants that burn a little faster. *Figure 6-35* shows a bag that I bought at a nursery, and *Figure 6-36* is an actual-size photo of the contents spread out on a sheet of paper. See **page 208** for instructions on how to prepare garden charcoal for use as a rocket fuel.

● Airfloat Charcoal

You can buy **airfloat charcoal** from laboratory and pyrotechnic chemical dealers. It is already finely powdered, so it needs no preparation, But you *do* have to make sure that it's dry. You'll find the dealers who sell it under the **CHEMICALS** and **LABORATORY SUPPLIES** headings in *The Yellow Pages*, and in the classified ads in *American Fireworks News* (**page 175**). The term, "airfloat", derives from the fact that, when shaken or disturbed, a noticeable amount billows up into the air in a dirty black cloud. Because airfloat charcoal is so messy, it is a good idea to work outdoors when pouring it from one container into another.

● Activated Charcoal

Another good carbon fuel is **activated charcoal**. Activated charcoal is *regular* charcoal that has been purified to maximize its absorbancy for use in things like water filters and gas masks. Fertilizer and farm supply stores sometimes sell it, and pet shops sell it as a water filtering agent for fish aquariums. I found *six* different brands at a local pet shop, and several called themselves **activated carbon**. When I looked at them closely, they appeared to be made from something other than wood. I didn't get a chance to try them out, but I'm sure that most of them would have worked. Activated charcoal should be prepared in the same manner as the other charcoals (**Chapter 7**), the only questions being how hard it is, and what kind of wood or coal it was made from. You'll have to experiment on your own, and because of the elaborate process used to make it, you can expect to pay (as of this writing) as much as $8 per pound.

● Carbon Black

Of all the carbon fuels, **carbon black**, also called **lampblack**, makes the fastest burning propellants, and this is mainly due to its *very* small particle size. *Figure 6-37* shows a cluster of carbon black particles magnified *twenty thousand times* by a scanning electron microscope. Samples vary greatly, but the one in the photo is composed of minute balls of carbon atoms averaging 1/4 micron in diameter. Please note that 1/4 micron is **ten millionths** of an inch. Each ball is made of smaller, *loosely-agglomerated* balls that break down even farther during the milling process. The result is a finished propellant with carbon particles *mere Angstroms* in diameter.

In a *powdered* rocket propellant, their ball-bearing shape and their *very* small size allow them to fill in tiny spaces where other chemicals won't fit. This allows powdered propellants made with carbon black to be packed to a higher density than those made with other forms of carbon. During my own work and the associated motor testing, I learned that, when added to potassium nitrate or sodium nitrate-based propellants made with *other* carbon fuels, *carbon black reduces or eliminates combustion instability*. Therefore, with the exception of the **KS** propellant, I've included carbon black in *all* the propellants in this book.

Carbon black is made by burning organic resins, fat, wax, waste-oil, or natural gas in an oxygen-starved atmosphere. As with the making of charcoal, the incomplete combustion process drives off or consumes everything but the carbon. The substitution of carbon black for *other* carbon fuels accelerates a propellant's burn rate, and the extent to which it does so is governed by the carbon black's purity, and the presence of residual organic compounds. The *purest* carbon blacks are *electrically-conductive*, and the dirtier types are not. **Important note.** *If you have a choice, you should always use a pure, electrically-conductive carbon black*. In a *powdered* rocket propellant, *the presence of an electrically-conductive carbon black helps prevent the buildup of a static electric charge*.

Warning! *Carbon black is the powerful black pigment used to make black paint! It stains everything that it touches!* When working with carbon black, put on old clothes, work out-of-doors, cover the table **and the ground beneath your feet** with something washable or disposable like a sheet of plastic. Cover your shoes with the disposable, plastic "booties" sold by many industrial suppliers, and throw them away when you're finished. These "booties" are, essentially, disposable plastic gloves-for-your-feet. Carbon black floats in the air *as a fine, invisible dust. A thin coating on the soles of your shoes, barely visible to the naked eye, is enough to ruin a carpet, and you won't see the damage until the stains start to appear!* When working on a concrete slab, scrub down the slab on a regular basis. Carbon black spills clean up with spray cleaners like "409" and "Mr. Clean", but it takes repeated applications and lots of wiping. Hands wash clean with a nail brush, soap and water. Work clothes should be washed separately from other clothing in a regular laundry detergent.

Tiny amounts of residual oil cause most carbon blacks to bond firmly with skin and fabrics, making the stains hard to remove. To reduce the cleanup problem, the **Teleflite Corporation** at www.teleflite.com and www.amateur-rocketry.com, sells a specially-processed carbon black that is **electrically conductive**, **oil-free**, and **comparatively dust-free**. Teleflite's carbon black cleans up with comparative ease, and unlike most carbon blacks, when shaken or disturbed, it stays in the container, and does *not* billow up in a dirty black cloud.

You can usually buy carbon black from the chemical dealers that sell potassium nitrate and airfloat charcoal. Look in **The Yellow Pages**, or contact the **Teleflite Corporation**. Carbon black's price varies *greatly* depending on the application for which it was made. As of this writing, the carbon black used to make ink costs $20 per pound. Carbon black made for other purposes costs as little as $2 per pound, so take your time, and shop carefully. Carbon black is *also* the black colorant in automobile tires. One of my readers works at a tire factory where they give him all he can use for free. Carbon black needs no preparation other than the assurance that it is dry. For drying instructions, see **page 200**.

Figure 6-38. A box of powdered confectioner's sugar. It makes a cheap and convenient rocket fuel.

Figure 6-39. A bag of Soil Sulfur.

● Powdered Sugar

Powdered sugar (chemical formula $C_{12}H_{22}O_{11}$), also called **sucrose** or **confectioner's sugar**, is the fine, white sugar used in cake frosting. *Figure 6-38* is a photo of a box that I bought at a supermarket. The manufacturers add a small amount of corn starch, but not enough to keep it from being used as a cheap and convenient rocket fuel. Its only drawback is that a rocket propellant made with sugar and a nitrate oxidizer *will not keep*.

In a sugar-nitrate propellant, both the fuel and the oxidizer are water-soluble, crystalline chemicals. When two such chemicals remain in contact with one another, *even when almost dry*, their crystals grow slowly into and around one another. The result is that the propellant continues to self-mix *after* it's been loaded into the rocket motor. As it does so, the propellant grain shrinks, and its burn rate increases; eventually exceeding the motor's design limits. Sugar-nitrate propellants are also hygroscopic. The word, "hygroscopic", means that they attract and hold moisture, and this compounds the problem. **For these two reasons, propellants and motors made with powdered sugar and a nitrate oxidizer must be used within 48 hours of when they are made**. See **Chapter 13** for instructions on how to make a rocket propellant using powdered sugar as the fuel.

Sulfur

Sulfur (also spelled **sulphur**) is one of the basic elements that make up the universe; atomic number 16 on the Periodic Table. It readily combines with oxygen, so it makes a good fuel. When added to black powder, it doubles as a low temperature melting flux that helps the other ingredients burn faster. *Up to a point*, the addition of sulfur to black powder increases the powder's burn rate, and it makes the powder easier to ignite.

Sulfur occurs in nature as a soft, pale yellow, crystalline mineral (*not* water-soluble), which, when heated or finely powdered, smells like burned matches. For *practical* purposes, it is soluble *only* in the two *highly* poisonous chemicals, **carbon disulfide** and **sulfur monochloride**. It is *not* soluble to any useful extent in water, alcohol, or acetone. It is mined by several methods; then purified and used in everything from battery acid to antibiotics. It is sold at the retail level in four forms known as **soil sulfur**, **dusting sulfur**, **flowers of sulfur**, and **powdered sulfur**.

● **Soil Sulfur**

In the presence of oxygen and water, sulfur oxidizes to form weak sulfurous acid. Sulfurous acid dissolves the limestone in hard, alkaline soils, and it makes them softer and better for farming. The product that does this is called **soil sulfur**, and you can buy it at fertilizer and farm supply stores, and *some* garden shops. The soil sulfur that I use (*Figure 6-39*) comes in the form of tiny, flattened beads, and costs $3.95 for 5 lbs. Most soil sulfur is better than 99% pure, but before you buy it, *read the label carefully*, and make sure that *nothing* has been added to it. *Any* impurity greater than 1% will *ruin* it for making rocket propellant.

● **Dusting Sulfur**

Dusting sulfur is a finely powdered sulfur sold at fertilizer and farm supply stores, and at nurseries. Here in California, it was used for many years to control fungus infections on grapes, but new, state environmental laws have banned its use. Unless the situation changes, you can't buy dusting sulfur in California, but it *might* be available in other states.

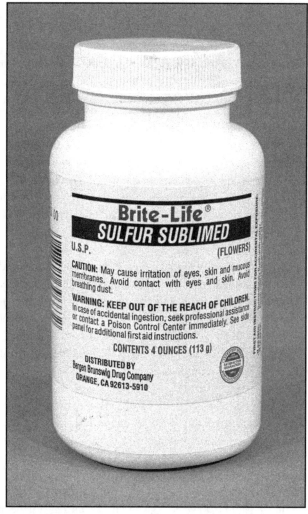

Figure 6-40. A small bottle of sublimed sulfur.

Just 10 years ago, *all* dusting sulfur was 99-1/2% pure. In recent years, some of the manufacturers have added a wetting agent that makes it better for farming, but *ruins* if for making rocket propellant. Before you buy it, look at the package carefully. If the label calls it "wettable sulfur", it is no good. If the label says 90%, 96%, or even 98% sulfur, it is no good. If it lists *any* "inert" or "other" ingredients, it is no good. I emphasize here that it must be *at least* 99% pure sulfur.

● **Flowers of Sulfur**

As of the year, 2000, most drug stores sell **flowers of sulfur** (also called **sublimed sulfur**) in small bottles costing $2 to $3. Flowers of sulfur (*Figure 6-40*) are made by vaporizing crude sulfur in a vacuum or an inert-gas atmosphere; then condensing the vapors into a mixture of powder and fine crystals. Chemical dealers sell it in larger quantities at lower prices, and because of the process by which it is made, it is more than pure enough to use in a rocket propellant.

● **Powdered Sulfur**

Powdered sulfur (also called **flour sulfu**r) is made by a crushing and grinding process that creates a product more finely powdered than flowers of sulfur. I'm not *absolutely* sure, but dusting sulfur might be nothing more than powdered sulfur that's been repackaged and renamed for a specific, agricultural purpose. Commercial chemical dealers sell powdered sulfur, and sometimes call it **pyro grade sulfur**. To find powdered sulfur, look in *The Yellow Pages* under **CHEMICALS** or **LABORATORY SUPPLIES**, and in the classified ads in *American Fireworks News* (**page 175**). For instructions on how to prepare sulfur for use in a rocket propellant, see **page 209**.

Binders

To form a powdered propellant into a sturdy and durable propellant grain, you have to convert it into a solid stick, and you do this with a material called a **binder**. A binder can be anything that, when added to the propellant, fills in the spaces between the particles, and glues the whole mass together. Ideally, a binder should take part in the combustion reaction, so rocket makers normally choose a binder that doubles as a fuel.

● Red Gum

Red gum, also called **Accroides gum**, or **yacca gum**, is the burgundy colored resin exuded by the bark of the Australian grass tree of the genus, Xanthorrhoea. The Australians dry it, clean it, crush it into a fine, orange powder; then ship it all over the world, where it's used in everything from varnishes to fireworks. Red gum is soluble in alcohol and acetone. Its high structural strength and excellent adhesive properties, combined with its ability to take part in a combustion reaction, make it an ideal binder for powdered rocket propellants.

From the 1940s to 1993, red gum was the main binder used by the American fireworks industry. In 1994, under pressure from environmentalists, the Australian government declared the grass trees (also called "yacca trees") a protected species. They temporarily stopped all exports of red gum; then resumed them in 1995, when the gum industry agreed to a resource management program. As of this writing, the future supply of red gum is secure. However, during the shortage, the American supply ran out, and the dealers who used to sell it deleted it from their catalogues. It is currently being imported again in large quantities, but as of the year, 2001, *a few* dealers are still not aware of this fact. Hopefully, this book will help spread the word, and remedy that situation.

Most red gum comes from Kangaroo Island, off the coast of South Australia, about 70 miles from the city of Adelaide. The island is large; about 1,000 square miles. The growing environment is ideal, and there are several exporters shipping it out of the main town of Kingscote. Anyone who wants to import large quantities of red gum should contact the business community on Kangaroo Island. I've noticed in the past year that there are several Kangaroo Island websites on the Internet.

To find red gum in the U.S., look in *The Yellow Pages* under headings like **CHEMICALS, RETAIL**, or **LABORATORY EQUIPMENT & SUPPLIES**. If a nearby dealer doesn't sell it, you can order it from the **Teleflite Corporation** at www.teleflite.com and www.amateur-rocketry.com, or the chemical dealers in the classified ads in *American Fireworks News* (**page 175**). If you encounter someone who used to sell red gum, and thinks that it's unavailable, show them this book, and tell them to check with their wholesalers. And a final thought. Because the environment on Kangaroo Island is similar to some Southeastern parts of the U.S., it might be possible to grow the grass trees domestically. They take 15 to 20 years to reach harvestable size, and it might be an interesting business venture for someone already engaged in tree farming.

● Goma Laca Shellac

In 1995 one of my readers sent me a half-kilo sample of something called **goma laca shellac**. She said that it came from Spain, and that during the red gum shortage, the Spaniards developed it as a man made substitute for red gum. Like red gum, it is reddish in color, and soluble in alcohol and acetone. Its adhesive properties are about the same, and so are its effects on propellant performance.

Both red gum and goma laca shellac have an acrid odor when wet with acetone. When dry, the goma laca smells like freshly sawed pine. Otherwise, the match is so close that if she hadn't told me it was synthetic, I wouldn't have guessed. I used it in a batch of the **NG6** propellant; then built an **E** and an **H** motor, and found no significant difference in performance between these goma laca motors and the ones made with red gum.

Under a regular, light microscope, red gum granules vary in color from black to pale pink, and here and there you'll see a piece of dirt or yacca bark fiber. The goma laca is a clean, uniform, apricot color, but what's most interesting is its price. As of this writing, rumor has it that at least two U.S. fireworks companies are importing it in large quantities for their own use at about half the price of red gum. None of my contacts know anything about it, and it's apparently (at least as of 2001) considered a trade secret by the people importing it. The person who sent me the sample wants to remain anonymous, and would not divulge the name and address of the manufacturer. This being the case, I have high hopes that the publication of this book will shake something loose. If enough people ask for it, someone might get motivated enough to locate the source, and start importing it for retail sale. Finally, since it is man made, I see no reason why it couldn't be produced right here in the U.S. Goma laca shellac is apparently inexpensive to make, and the market for it is already established.

● Vinsol Resin

Vinsol resin is a by-product of the logging industry. Pine tree wood is pulverized into a mass of fiber containing up to 4% pine pitch. This crude pitch, called "gum rosin", is extracted and sent to a refinery, where it is chemically separated into several dozen products, one of which is *vinsol resin*. Among other things, vinsol resin is used as an additive in asphalt and concrete, as a modifier in phenolic resins, and as a fuel and a binder in fireworks.

Like goma laca shellac, it smells like freshly sawed pine, and it quickly dissolves in alcohol and acetone. *Unlike* goma laca or red gum, it noticeably retards a propellant's burn rate. This means that a rocket motor made with vinsol resin requires a smaller nozzle throat, or a longer propellant core than an identical motor made with red gum or goma laca shellac.

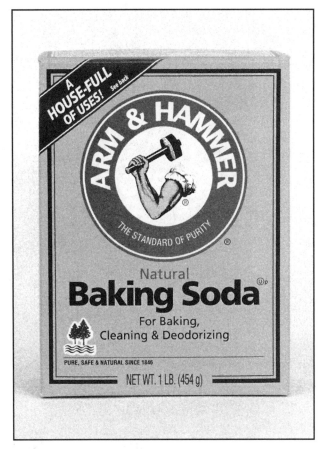

Figure 6-41. A box of baking soda. Baking soda is a cheap and effective burn rate modifier.

As of 2001, the main producer of vinsol resin is **Hercules Incorporated, 1313 N. Market St., Wilmington, Delaware 19894**. They make it in both flaked and powdered forms, but their minimum purchase is 2,500 lbs. For 50 lb. quantities, contact a company called **Chemcentral Corp.** at www.chemcentral.com. Chemcentral is the main distributor for Hercules resins, and they have offices in almost every state. For 1 lb. to 25 lb. quantities look in *The Yellow Pages* under headings like **CHEMICALS, RETAIL**, and **LABORATORY EQUIPMENT & SUPPLIES**, or contact the **Teleflite Corporation** at www.teleflite.com and www.amateur-rocketry.com, or the dealers that advertise in *American Fireworks News* (**page 175**). If you *still* can't find it, call or email Chemcentral. Tell them where you live, and ask them for the name and address of the nearest retailer that sells it.

Burn Rate Modifiers

A burn rate modifier can be any material that changes the speed at which a propellant burns. Most modifiers accelerate the burn rate. The one that I use in *this* book *slows it down*.

● Baking Soda

Baking soda's ability to slow a combustion reaction without *significantly* effecting combustion stability makes it an ideal burn rate modifier for black powder propellants. It does so by adiabatically cooling as it expands into a gas. By decreasing or increasing a propellant's baking soda content, you can speed up or slow down its burn rate, and fine-tune a rocket motor's performance. A brief reference in John Conkling's book, *The Chemistry of Pyrotechnics*, prompted me to experiment with it, and I was so happy with the results that I've included baking soda in *all* the propellants in this book. Baking soda's chemical name is **sodium bicarbonate**. Its chemical formula is **$NaHCO_3$**, and you can buy it at a supermarket (*Figure 6-41*). I'll mention it again in **Chapter 7**, but it requires no preparation. You'll be using straight from the box.

195

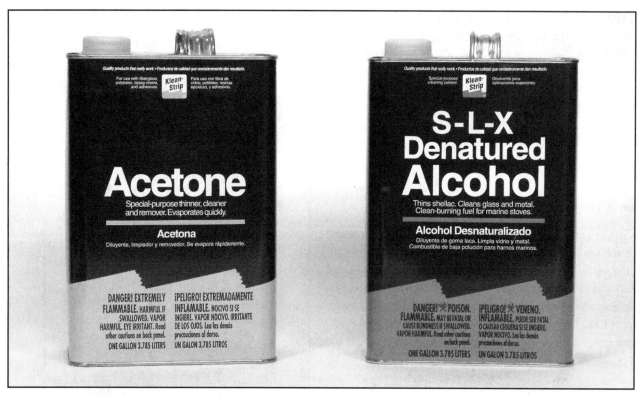

Figure 6-42. One gallon cans of acetone and denatured alcohol.

Solvents

With the exception of the **KS** propellant, all the propellants in this book contain a binder. To soften the binder and make it sticky, you need a solvent. **Acetone** and **alcohol** (*Figure 6-42*) will both work, but because acetone evaporates so fast, I like it the best.

● Acetone

Acetone is the most volatile of the two solvents. You can buy it in pint to gallon size cans at paint and hardware stores. A core burning **H** motor made with acetone will air-dry in 48 hours, and vacuum-dry in 1-1/2 hours. Acetone's only drawback is that it dissolves the buna seals found in many vacuum pumps. To vacuum-dry motors made with acetone, you'll need to verify that your pump's rubber parts are made of **Viton** (**page 47**), If they are *not* made of Viton, you'll have to replace them. Acetone will *not* harm the rubber plunger in a disposable syringe.

● Denatured Alcohol

Denatured alcohol, also called **shellac thinner**, is the *least* volatile of the two solvents. Like acetone, you can buy it in pint to gallon size cans at hardware and paint stores. It is *usually* made of ethanol (grain alcohol) with a little isopropanol mixed in to make it unfit for human consumption. **H** motors made with alcohol will air-dry in 5 to 6 days, and vacuum-dry in 6 to 8 hours. Alcohol's advantage is that it does *not* dissolve the buna seals found in many vacuum pumps. To vacuum-dry motors made with alcohol, you can use a pump containing *buna* (aka "nitrile") seals. Denatured alcohol will *not* harm the rubber plunger in a disposable syringe.

● Evaporation Rates

One of the tools in my shop is a digital analytical balance. With it, I can measure differences in weight with an accuracy of 1/10,000 of a gram. In this case I used it to compare the evaporation rates of alcohol and acetone from the 2.2 square inch surface of a 30 ml. medicine cup at a temperature of 80° Fahrenheit. The test period, as determined by a stop watch, was

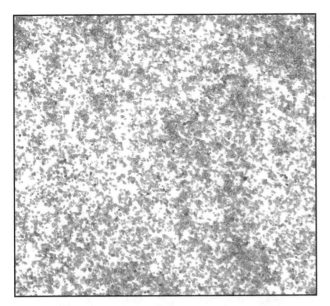

Figure 6-43. The grog that I use shown actual size.

5 minutes, and the changes in weight were: **acetone = 416 mg.**, **alcohol = 102 mg.**, and **water** (used for comparison) **= 19 mg**. The weather was humid, so the evaporation of the water was retarded, but for the purpose of making estimates, you could say that **alcohol evaporates 5 times as fast as water**, and **acetone evaporates 22 times as fast**. It was a simple experiment to perform, and it illustrates why propellants made with acetone are such an improvement over the old water-packed formulas.

Figure 6-44. Particles of grog magnified just 20 times by a scanning electron microscope. The sharp angles and edges lock the clay particles together, and prevent them from forming into layers.

IMPORTANT WARNING. Acetone and alcohol are highly flammable. Do not smoke in the presence of acetone or alcohol. When working indoors, make sure that the room is well ventilated, and do *not* allow the vapors to accumulate in a closed area.

Nozzle Clay Materials

When mixed together in the right proportions (**page 209**), the following materials form the dense, flameproof ceramic used to make the rocket motor nozzles and bulkheads.

● Powdered Clay

The main ingredient is **dry, powdered clay**, and *any* powdered clay will work. Brick and building supply dealers sell a product called **fireclay**, also called **mortar clay**. Brick layers add it to their mortar to make it sticky. Some places mix it with water, and some sell it dry. You will need the *dry* form. If the local brick yards don't have it, look in **The Yellow Pages** under headings like **CERAMICS, EQUIPMENT & SUPPLIES**. Potters and sculptors who mix their own clays buy the raw materials in the dry, powdered form. The dealers that sell to them have powdered clay, but *finding a dealer can be confusing for the following reason.*

In the past 30 years, *thousands* of small businesses have appeared that call themselves "ceramics shops". In some towns, they have a storefront in every mini-mall, while their ads dominate the **CERAMICS** column in the phone book. Unfortunately, they know **little** about *real* ceramics. They confine their activities to giving china painting classes, and teaching the use of premanufactured products. They almost *never* sell powdered clay, and they often don't know where to buy it.

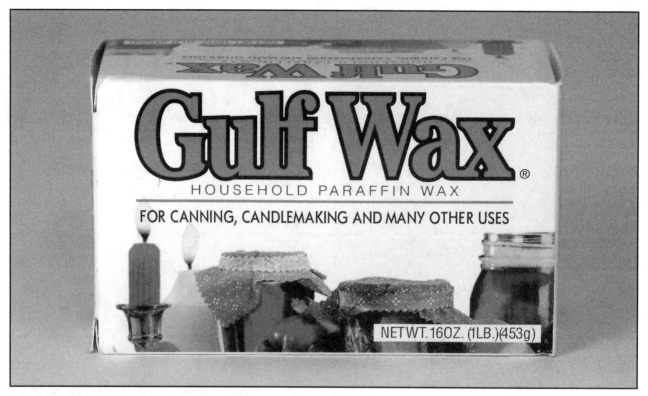

Figure 6-45. A package of "Gulf Wax" paraffin.

Look instead for the places that sell the *raw, ceramic materials*. If you live in a small town, you probably won't find one. If you live in a big city, you might find three or four. If you have no luck, call some of the china painting shops. Tell them what you need, and ask them where to buy it. If they don't know, call the art department at the nearest college or university. Ask the ceramics teacher where he buys *his* supplies. When other queries have failed, this one often works. When you've located a source, you'll have many clays to choose from, so pick one of the medium to high firing types. Rocket exhaust is *very* hot (several thousand degrees), and the more heat resistant clays work the best. Cost and quantity are the only other concerns.

● Grog

Grog is man-made sand. It is made by crushing and sieving freshly fired pottery. Potters and sculptors add it to their clay to give it a gritty, rustic texture called "tooth". You can buy it in assorted mesh sizes, but the exact size isn't critical. **Figure 6-43** shows the material that I use, *actual-size*, spread out on a sheet of paper. The place where I buy it calls it "medium-fine", but another dealer might call it something else. If the grog you buy is a little coarser or a little finer, it doesn't matter. If it's even *close* to what I use, it will work. Grog is sometimes called **firesand**, and you can buy it from the same places that sell the clay. Many years ago it was used in making fireplaces, but you *probably* won't find it at a brick yard, since its use has been all but abandoned by the building trades.

Clay molecules are *flat*, and a gram of clay contains *billions* of flat particles that form into layers when rammed against a nozzle mold. During rocket motor operation, the supersonic exhaust jet catches under these layers and rips them out, destroying the nozzle's shape, and throwing the rocket off course. Each particle of grog (*Figure 6-44*) is a masterwork of sharp edges and angles. When mixed with the clay, they lock the structure together, and keep the layers from forming. Without the grog, many of your nozzles will fail. *With* the grog, the failure rate should be zero.

● Sand

As a substitute for grog, some of my readers have used clean **sand** with marginal success. But natural sand grains have been rounded by wind and water, and they aren't as effective at strengthening the ceramic structure. Also, natural sand contains minerals that shatter when exposed to the heat of a rocket's exhaust. I personally feel that grog works *much* better, and I encourage you to use grog, even if you have to go out of your way to find it.

● **Paraffin Wax**

The model rocket industry makes rocket motor nozzles out of clay and grog alone. But this *only* works because they ram the mixture at a *very* high pressure with hydraulic presses. To achieve the same density when hand-tamping, you have to mix a lubricant with the clay to help it pack together. The people who make fireworks often use motor oil, and at first the oil seems to work. But many years ago I discovered that, during storage, the oil soaks into the paper casings, and it *weakens* them. This causes the nozzles to blow out when the motors are fired. I solved the problem by switching to **paraffin wax**, which lubricates the clay, but remains solid at room temperature, and *doesn't* weaken the paper.

You can buy paraffin at a supermarket, where it is sold as a sealant for home canning. Here in Southern California a popular brand is "Gulf Wax" (*Figure 6-45*), but other brands, and even old candles, will work just as well. Candles are made of paraffin, and you can cut off what you need with a pocket knife. For large quantities of paraffin, look in *The Yellow Pages* under **HOBBIES** and **CRAFTS**. Craft shops sell candle making supplies, and they *usually* sell large blocks of paraffin at a very reasonable price.

● **Additional Notes on Oxidizers**

When sun-drying either potassium nitrate or sodium nitrate, if the weather is *not* hot and dry; if it is cool and/or humid, residual moisture under the crystals (i.e. between the plastic and the crystals) will be difficult to remove. When this happens, you can speed the drying process with a heat lamp, but be careful not to melt the plastic.

Also of note. Oxidizers that are slightly *less* than 99% pure *might* work, but the impurities will retard the propellant's burn rate, and you'll have to compensate by lowering the propellant's baking soda content. Learning if an impure oxidizer works, or how to *make* it work, will be a process of trial and error.

Very Important. When making homemade sodium nitrate (**pages 183** and **184**), unless you live in a remote area, *limit your production to 1 or 2 pounds at a time*. The obvious concern is the large amount of gaseous ammonia released by the process. The powerful ammonia fumes are obnoxious and potentially dangerous. When making larger quantities, stay *away* from residential areas. Work *downwind* of any habitable buildings or structures where the fumes might accumulate, and keep yourself *upwind* of the fumes at all times.

Regarding ammonia's long-term enrivonmental concerns, **there aren't any!** Ammonia is *quickly* absorbed by the environment, and it *quickly* degrades into water and soluble nitrates, aka **food for the plants!** It does *not* pose a long-term environmental hazard. In many parts of the country it is used as a *fertilizer!* The fields-to-be-treated are first wet down with sprinklers. Then pure, *anhydrous* ammonia (the compressed and liquified gas) carried in thousand-gallon, pressurized tanks is injected into the soil with a device that looks like a giant, tractor-towed rake. The ammonia is *instantly* absorbed by the wet soil, *where it stays until its nitrogen is absorbed by the crop.* Amazingly, you can walk a field just an hour after a treatment, and there is *no* ammonia odor.

● **Additional Notes on Fuels**

Bituminous coal is easier to find than anthracite, but it is high in residual organic compounds (i.e. oil). Like the rubbers and the plastics mentioned on **page 187**, it will *not* work as a fuel with potassium nitrate or sodium nitrate.

7. CHEMICAL PREPARATION

Before the chemicals can be used, they *must* be correctly prepared, and this chapter explains what to do. Experience has taught me that you get the best results if you powder the fuels and the oxidizers *before* you mix them together. A powder mill is the best tool for doing this. If you don't have a powder mill, you can powder the oxidizers by precipitation with alcohol. I'll explain both methods in the pages ahead, but before you proceed, make sure that you've read **Chapter 6**. If you *haven't* read Chapter 6, *go back, and read it now*. If you see something in *this* chapter that you don't recognize, you can refer to **Chapter 6** for an explanation of what it is and where to buy it.

Drying

Before you begin, all of the chemicals must be thoroughly dry. You can place **potassium nitrate** and **sodium nitrate** in *non-aluminum* pots (i.e. steel, iron, Pyrex, enamel, etc.), and dry them in a kitchen oven for 1 to 2 hours at 250° Fahrenheit. When they are not mixed with anything, both chemicals are *very* stable, and there is *no* hazard in doing this. An aluminum pot will work, but these chemicals will etch the surface of the metal, and ruin the finish.

You can dry **charcoal** in a kitchen oven, but if it gets too hot, it will ignite and burn like it does in a Bar-B-Que. You'll have to keep the temperature at 225° to 250° Fahrenheit, and you can expect a noticeable odor.

You can dry **coal** and **carbon black** at 225° to 250° Fahrenheit, but the odor is so unpleasant that you'll want to do it outdoors. An electric frying pan with a cover works fine, but you'll have to place the cover askew, and open its vent to let the steam escape. If you don't, the water will recondense, and drip back into the pan.

Baking soda and **powdered sugar** are dried before they are packaged, and no further drying is necessary.

Sulfur, **red gum**, **vinsol resin**, and **goma laca shellac** either melt or give off unpleasant odors at low temperatures, so spread them on a sheet of plastic, and dry them in the sun. Sun-drying isn't as effective as oven-drying, but these chemicals are used in such small quantities that tiny, residual amounts of moisture shouldn't be a problem.

Oxidizers

● Milled Potassium Nitrate and Milled Sodium Nitrate (4 Hour)

The simplest way to powder an oxidizer is to process it in a powder mill. You can start with any of the forms described in **Chapter 6**; i.e. the fertilizer, the stump remover, the lab grade material, or the homemade or extracted versions. *Any* of them will work. To prepare a batch of milled potassium nitrate or milled sodium nitrate, proceed as follows.

1. Per the instructions in **Chapter 3**, set up a powder mill, and refer to *Figure 3-10* to determine how many brass pellets, and what weight of nitrate to use. To calculate the amount of nitrate mathematically, multiply the weight (in pounds) of the pellets that the mill uses by 125 gms. Place the indicated number of pellets, and the proper weight of nitrate, into the mill's jar. Tighten down the lid, place the jar on the mill's motorized base, turn the mill on, and run it for **exactly 4 hours**.

Figure 7-1. A plastic kitchen colander and a plastic bucket are used to sift the milled nitrate from the brass pellets.

Figure 7-2. Milled potassium nitrate (4 hour) magnified 200 times by a scanning electron microscope. Particles range from about 10 to 150 microns in size.

EXAMPLE 7-1:

Your mill has a 1/2 gallon jar. According to Figure 3-10, it uses 4 lbs. of brass pellets. The correct amount of nitrate is therefore:

$$125 \text{ gms.} \times 4 = 500 \text{ gms.}$$

EXAMPLE 7-2:

You are using a toy rock tumbler who's jar has a capacity of 1-1/2 cups. According to Figure 3-10, it uses 12 ounces (or 0.75 lbs.) of brass pellets. The correct amount of nitrate is therefore:

$$125 \text{ gms.} \times 0.75 = 93.75 \text{ gms. (rounded off to 93.8 gms.)}$$

2. Open the mill's jar, dump the contents into a plastic colander or a stainless steel french fry basket (*Figure 7-1*), and sift the finished product into a moistureproof container. Then return the pellets to the jar. In the pages ahead, and in *all* the propellant formulas, I call this finished product either **4 hour potassium nitrate**, or **4 hour sodium nitrate**.

Figure 7-2 is a photo of 4 hour potassium nitrate magnified 200 times. The process produces a mixture of particles ranging from about 10 to 150 microns in size.

● Precipitated Potassium Nitrate

Precipitated potassium nitrate is a substitute for the milled potassium nitrate described above. Its only advantage is that you don't need a powder mill to make it. You can begin with any of the forms described in **Chapter 6**; i.e. the fertilizer, the stump remover, the lab grade material, or the extracted versions. *Any* of them will work.

Figure 7-3. 12 oz. (1-1/2 cups) of water are added to one pound of potassium nitrate. The mixture is brought to a boil, and the nitrate is stirred until thoroughly dissolved.

Figure 7-4. When the boiling solution is poured into the alcohol, it releases a white mud, or "precipitate", of fine potassium nitrate crystals.

In chemistry, the process is called "precipitating". In the field of improvised munitions, it is called "salting". It relies on the fact that water has a greater affinity for alcohol than it does for potassium nitrate. To make 375 gms. of precipitated potassium nitrate, proceed as follows.

PRECIPITATED POTASSIUM NITRATE
(to make approx. 375 gms.)

MATERIALS LIST

1. A kitchen stove top, or an electric hot plate
2. 2 pots and a spoon to stir with
3. A large kitchen sieve
4. A rubber spatula
5. 6 paper coffee filters
6. A cookie sheet
7. A sheet of thin plastic, preferably black
8. A plastic jar with a tight fitting lid
9. A handful of large nuts or bolts
10. 1 lb. (453 gms.) of potassium nitrate
11. 12 oz. (1-1/2 cups) water
12. 28 oz. (3-1/2 cups) denatured alcohol

1. Put 16 oz. (2 cups) of denatured alcohol in one of the pots (**pot #1**). **Do this outdoors, or well away from the stove to avoid any fire hazard**.

2. In the *other* pot (**pot #2**), put one pound (453 gms.) of potassium nitrate, and 12 oz. (1-1/2) cups of water. Place **pot #2** on the stove, and bring the water-nitrate mixture to a boil. Then stir the water with a spoon until the nitrate is *completely* dissolved (*Figure 7-3*).

3. Remove the boiling nitrate solution (**pot #2**) from the stove. *Before it starts to cool*, **take it outside, or well away from the stove**, and pour it *slowly* into the alcohol in **pot #1**. I say *slowly*, because the alcohol has a lower boiling point than the water. If you do it too fast, the alcohol will boil violently, and make a splattered mess all around the pot. As the alcohol mixes with the water, the water releases the potassium nitrate in the form of a thick, white mud, or "precipitate", of tiny potassium nitrate crystals (*Figure 7-4*). Because the crystals are so small, they remain suspended in the liquid for a long time. So set the pot aside, and let the contents settle for *one full hour*. Then proceed to **step 4**.

Figure 7-5. After settling, the clear top liquid is poured off, leaving a thick, white mud of potassium nitrate crystals in the bottom of the pot.

Figure 7-6. 1/2 cup of fresh alcohol is added to the mud, and the mixture is stirred for 30 seconds.

Figure 7-7. Some final pressing with the bowl of a spoon squeezes out most of the alcohol.

*Figure 7-8. The nitrate crystals from **step 5** are dumped back onto the pot.*

Figure 7-9. The wet potassium nitrate mud is spread out in the sun to dry.

Figure 7-10. The dried potassium nitrate is broken up, and placed in a **plastic** jar along with a handful of heavy nuts. Three minutes of shaking convert it into a powder.

4. After the precipitate has settled to the bottom of the pot, pour off and discard as much of the clear, top liquid as you can (*Figure 7-5*). Then add 4 oz. (1/2 cup) of *fresh* alcohol, and stir for 30 seconds (*Figure 7-6*).

5. Place 2 coffee filters (a double layer) in the sieve. Rest the sieve on the now-empty **pot #2**, and with the spatula, pour and scrape the white mud from **step 4** into the filters. Set this pot with the sieve and filters aside, and allow enough time for the liquid to thoroughly drain away. As shown in *Figure 7-7*, some final pressing with the bowl of a spoon will finish the job.

6. Discard the clear liquid in **pot #2**. Then dump the white mud from the coffee filters *back* into **pot #2**. At this point the nitrate crystals will look like mashed potatoes (*Figure 7-8*). Add another 4 oz. (1/2 cup) of denatured alcohol, and stir for another 30 seconds.

7. Repeat **steps 5 & 6**.

8. Repeat step **5** one more time. Place the cookie sheet on a level surface *outdoors*, and cover it with a sheet of black plastic. Dump the white mud from the filters onto the plastic. Then spread it out with the spatula, and let it dry in the sun (*Figure 7-9*). *When the white mud is as dry as the sun can make it*, separate it from the plastic, place it back on the cookie sheet (this time *without* the plastic), and dry it in an oven at 250° Fahrenheit for 1/2 hour. **Warning! Alcohol vapors are explosive! Do not place the nitrate in the oven until *all* traces of alcohol odor are gone!**

9. Remove the dried potassium nitrate from the oven, and let it cool. Place it in a plastic jar (**not glass!**) along with a handful of large nuts (as in "nuts and bolts") (*Figure 7-10*). Tighten the jar's lid, and shake the jar vigorously for *3 minutes*. Open the jar, dump the contents into the kitchen sieve, and sift out the nuts & bolts. Then place the finished, powdered potassium nitrate in a moistureproof container for storage.

Figure 7-11 is a photo of precipitated potassium nitrate magnified 100 times by a scanning electron microscope. This process produces a mass of tiny crystals ranging from about 50 to 200 microns in size.

Important note. *When people first hear of this process, they ask the following questions.*

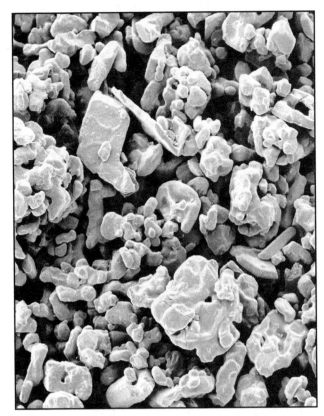

Figure 7-11. Precipitated potassium nitrate magnified 100 times by a scanning electron microscope. Particles range from about 50 to 200 microns in size.

Question. Why not just boil off the water in step 1, and avoid the trouble and expense of adding the alcohol? Answer. If you boil off the water, or let the solution dry in the sun, it has time to grow an interlocking mass of *large* crystals, and you end up with a hard, caked mess in the bottom of the pot. When you pour the solution into the alcohol, you force the nitrate to crystallize *instantly*. The large crystals don't have time to form, and the result is a mass of *tiny, disconnected* crystals that are ideal for use in a rocket propellant.

Question. Why pour the nitrate solution into the alcohol instead of the other way around? Answer. If you do it the other way around, the nitrate precipitates more slowly. It forms slightly larger crystals, and the finished product isn't as finely powdered.

Question. Why rinse the nitrate 3 times with alcohol? Answer. After the first rinse, the white, nitrate mud is wet with a solution that still contains a small amount of nitrate dissolved in water. When dried, this nitrate will recrystallize, and form the precipitate into a hard cake. Each alcohol rinse removes more of the water, and forces more of the nitrate out of solution. Three washings produces a product that, when dry, is soft enough to be powdered by the method described in **step 9**.

Figure 7-12. The homemade coal crusher is used to break up and crush large chunks of coal or charcoal prior to milling.

Figure 7-13. Use the window screen sieve (the finest of the 3 homemade sieves) for sifting the finely crushed anthracite.

Figure 7-14. 12 hour anthracite magnified 100 times by a scanning electron microscope. Particles range from about 20 to 200 microns in size.

Question. Why dry the nitrate in the sun before placing it in the oven? Answer. If you place *anything* wet with alcohol in a kitchen oven, the oven will ignite the alcohol vapors, **and the oven will explode! Do *not* place the sun-dried nitrate in an oven until *all* traces of alcohol odor are gone!**

Question. What about precipitated sodium nitrate? Answer. You can powder sodium nitrate by the same method, and because sodium nitrate is more water-soluble than potassium nitrate, you would use less water. But there is no use for precipitated sodium nitrate in this book. Sodium nitrate is used *only* in the **NG6** and **NV6** propellants. Both propellants are made in a powder mill. If you own a powder mill, you don't *need* to powder the nitrate by precipitation.

Fuels

● Milled Anthracite (12 Hour)

Because anthracite is *very* hard, you *absolutely must* mill it separately *before* you mix it with the other propellant ingredients. If you do *not* mill it separately, you will end up with a useless product, in which everything is powdered *except* the anthracite.

If you're starting with large chunks of anthracite, you'll first have to crush and sieve them down to the proper size with the coal crusher and the window screen sieve described in **Chapter 4**. If this is what you are doing, begin with **step 1**. If you are starting with the fine, granular anthracite described in **Chapter 6**, then *skip* **step 1**, and go to **step 2**.

1. Stand the large, coal crusher pipe upright on concrete or hard ground (*Figure 7-12*). Toss in a few chunks of coal, and ram down on them 5 to 10 times with the capped end of the small pipe. Dump the crushed material into the window screen sieve (*Figure 7-13*), and sift it onto a sheet of plastic. Pour the material that remains in the sieve *back* into the crusher, and repeat the process until all the material passes through the screen. Because natural coal contains a small amount of moisture, dry the granules according to the instructions on **page 200**. Then store them in a moistureproof container.

2. Per the instructions in **Chapter 3**, set up a powder mill, and refer to *Figure 7-15* to determine how many brass pellets and what weight of anthracite to use. To calculate the amount of anthracite mathematically, multiply the weight (in pounds) of the pellets that the mill uses by 62.5 gms. Place the indicated number of pellets and the proper weight of anthracite into the mill. Tighten down the lid, turn the mill on, and run it for **exactly 12 hours**.

Powder Mill Volume	Number of Pellets	Weight of Pellets	Pellet Dimensions	Weight of Carbon Fuel (Coal or Charcoal)
1-1/2 Cups Toy Tumbler	26 pellets	12 oz.	1/2" x 1/2"	46.9 grams
1-1/2 Lb.	34 pellets	1 lb.	1/2" x 1/2"	62.5 grams
3 Lb.	68 pellets	2 lbs.	1/2" x 1/2"	125.0 grams
4-1/2 Lb.	31 pellets	3 lbs.	3/4" x 3/4"	187.5 grams
6 Lb.	41 pellets	4 lbs.	3/4" x 3/4"	250.0 grams
12 Lb.	82 pellets	8 lbs.	3/4" x 3/4"	500.0 grams

Figure 7-15. Refer to this table to determine the weight of the coal or charcoal, and the size, number, and weight of brass pellets to use when making milled anthracite, milled mesquite charcoal, or milled garden charcoal.

Figure 7-16. When making milled mesquite charcoal, the 1/8" hardware cloth sieve is used to separate out anything larger than 1/8".

Figure 7-17. The window screen sieve is used to separate the 1/8" particles from the fine powder, and the powder is discarded.

3. Open the mill, dump the contents into a plastic colander or a stainless steel french fry basket, and sift the finished product into a moistureproof container (*Figure 7-19*). In the pages that follow, and in all the propellant formulas, I call this fuel **12 hour anthracite**.

Figure 7-14 is a photo of milled anthracite magnified 100 times by a scanning electron microscope. This process produces a mixture of particles ranging from about 20 to 200 microns in size.

● Milled Mesquite Charcoal (6 Hour)

Because mesquite charcoal comes in large chunks, you'll first have to crush it, and sieve it down to the proper size with the coal crusher, the 1/8" hardware cloth sieve, and the window screen sieve described in **Chapter 4**. To make a batch of milled, mesquite charcoal, proceed as follows.

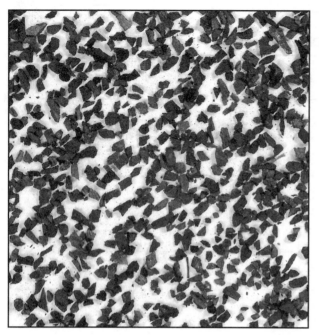

Figure 7-18. Crushed & sieved mesquite charcoal shown actual size.

Figure 7-19. The milled coal or charcoal is sifted from the pellets with a bucket and a kitchen colander.

1. Stand the large coal crusher pipe upright on concrete or hard ground (*Figure 7-12*). Toss in a few chunks of mesquite charcoal, and ram down on them 5 to 10 times with the capped end of the small pipe. Dump the crushed material into the 1/8" hardware cloth sieve, and sift it onto a sheet of plastic (*Figure 7-16*). Pour what remains in the sieve *back* into the crusher and repeat the process until all the material passes through the 1/8" sieve.

2. Filter out and discard any *powdered* charcoal by sifting the material created in **step 1** through the *window screen* sieve (*Figure 7-17*). Then dry the charcoal as described on **page 200**, and proceed to **step 3**. *Figure 7-18* shows the crushed and sieved material, actual-size, spread out on a sheet of white paper.

3. Per the instructions in **Chapter 3**, set up a powder mill, and refer to *Figure 7-15* to determine how many brass pellets and what weight of charcoal to use. To calculate the amount of charcoal mathematically, multiply the weight (in pounds) of the pellets that the mill uses by 62.5 gms. Place the indicated number of pellets and the proper weight of charcoal into the mill. Tighten down the lid, turn the mill on, and run it for **exactly 6 hours**.

4. Open the mill, dump the contents into a plastic colander or a stainless steel french fry basket, and sift the finished product into a moistureproof container (*Figure 7-19*). In the pages that follow, and in all the propellant formulas, I call this fuel **6 hour mesquite charcoal**.

● Milled Garden Charcoal (6 Hour)

Garden charcoal comes in 1/4" chunks, so you don't need the coal crusher. To make a batch of milled garden charcoal, proceed as follows.

1. Garden charcoal naturally contains moisture, so first dry it according to the instructions on **page 200**. Then, per the instructions in **Chapter 3**, set up a powder mill, and refer to *Figure 7-15* to determine how many brass pellets and what weight of charcoal to use. To calculate the amount of charcoal mathematically, multiply the weight (in pounds) of pellets that the mill uses by 62.5 gms. Place the indicated number of pellets and the proper weight of charcoal into the mill. Tighten down the lid, turn the mill on, and run it for **exactly 6 hours.**

2. Open the mill, dump the contents into a plastic colander or a stainless steel french fry basket, and sift the finished product into a moistureproof container (*Figure 7-19*). In the pages that follow, and in all the propellant formulas, I call this fuel **6 hour garden charcoal**.

● Airfloat Charcoal

Airfloat charcoal is powdered during the manufacturing process, so no milling is necessary. It *should* have been dried by the manufacturer, but *just in case it wasn't*, dry it according to the instructions on **page 200**.

● Carbon Black

Carbon black particles are *so* small that milling is not necessary. But like the other forms of carbon, it can absorb moisture during shipping and storage, so dry it according to the instructions on **page 200**.

● Powdered Sugar

Powdered confectioner's sugar is powdered during the manufacturing process, so milling is not necessary. Slight amounts of moisture can ruin it, so the manufacturers take *great* care to keep it dry at all times. Powdered confectioner's sugar requires *no* preparation. You can use it straight from the box.

Sulfur

Sulfur is so soft (number 2 on the Moh's hardness scale) that milling is not necessary. However, farm products like soil sulfur are often stored carelessly. In the process, they can become wet with dew or rainwater. If the sulfur you've purchased shows *any* sign of moisture, dry it according to the instructions on **page 200**.

Homemade Nozzle Clay

The nozzles and the bulkheads in these rocket motors are all made from a dry-rammed mixture of (**by weight**):

Powdered clay	**61%**
Grog	**31%**
Paraffin Wax	**8%**

For lack of a better name, I call this mixture **nozzle clay.** To make 1000 gms. of nozzle clay, proceed as follows.

HOMEMADE NOZZLE CLAY
(to make 1000 gms.)

MATERIALS LIST

1. A kitchen oven
2. A stovetop burner
3. 2 large pots (one of them oven-safe)
4. A table fork to mix with
5. 610 gms. of dry powdered clay
6. 310 gms. of grog
7. 80 gms. of paraffin, or old candle wax

1. Weigh out the clay and the grog. Place them in an oven-safe pot, and mix them together with the table fork (*Figure 7-20*). Place the pot in the oven, and heat the mixture at 300° Fahrenheit for 1/2 hour.

2. When the clay-grog mixture is *thoroughly* hot, place 80 gms. of paraffin or candle wax in the *other* pot. Place it on the stove, and heat it enough to melt the wax, but *not* so hot that it smokes or burns (*Figure 7-21*).

3. When the wax has melted, turn off the stove. Remove the hot clay-grog mixture from the oven, and pour the melted wax *slowly* into the hot, clay-grog mixture (*Figure 7-22*). Then mix all three ingredients together with the fork. **Warning!** The pot will be **hot**, so hold it with a thick pot holder.

Figure 7-20. Before heating, the clay and the grog are mixed together in an oven-safe pot.

Figure 7-21. The candle wax is melted in a pot on the stove.

Important note. *Step 3 will take several minutes, and should not be rushed.* Turn the mixture over frequently by reaching down, and pulling up the contents on the bottom. Also be sure to scrape the corners and the sides of the pot to pick up any excess waxy material that might have accumulated in these areas.

The ceramic materials that you heated in the oven are hot enough to keep the wax melted. When you begin, you will see dark, wet lumps of wax-saturated clay that will gradually disappear as you work. When all the material in the pot is a uniform gray with no visible lumps or wet spots, the job is done.

4. Remove the pot from the stove, take it outdoors, set it on the ground, and let it cool. Place the finished, cooled nozzle clay in a suitable container for storage.

Though the finished product doesn't *feel* waxy, it contains enough wax to lubricate the clay particles during ramming. This allows the mixture to be packed to a high density with a series of comparatively light hammer blows. And, of course, this recipe can be scaled up or down as needed.

Figure 7-22. The hot mixture of clay and grog is removed from the oven. The melted wax is added, and the three components are mixed until thoroughly blended.

The question arises as to why you don't use more or less wax. The answer is that *more* than 8% wax results in a mixture that melts and erodes away during motor operation. For the rocket motors in *this* book, 8% is optimum. Using *less* than 8% makes the nozzle clay more difficult to ram.

8. MOTOR CASINGS

The motor casings for commercial and military rocket motors are made of rigid materials like metal and fiberglass, but the casings for these homemade motors *cannot* be rigid. The materials that you ram into the casings expand outward, and the casings have to respond by squeezing back. The two opposing forces create a tight seal between the nozzle clay, the propellant, and the casing wall. To make the system work, the casings *must* be made of something resilient, *and metal and fiberglass will not work*. Cardboard is strong, inexpensive, and easy to work, and it has the necessary resilience. So *all* the motors in this book are made with *cardboard* casings.

A cardboard rocket casing is an ordinary cardboard tube that has been made for a special purpose. Cardboard tubes come in two types; **spiral** and **convolute** (aka **parallel wound**). The spiral tubes are the most common, but you *cannot* use a spiral tube to make a rocket motor. Spiral tubes are plentiful. You see them everywhere, and it would be convenient if you could make them work. But each layer of paper in a spiral tube contains a tiny spiral air channel. When a rocket motor ignites, its combustion chamber pressurizes. If the casing is made from a spiral tube, and at any time during the burn, the inside wall is exposed to the flame, the flame burns through the first layer of paper, and exposes the first air channel. The elevated pressure then forces the flame *up* the channel, where it prematurely lights the rest of the propellant, causing the motor to *explode*.

To avoid confusion, I've provided a photo of a spiral tube in *Figure 8-1*. Spiral tubes are made from multiple layers of narrow paper strips wound up like a barber pole. If the tubes you have look like the one in *Figure 8-1*, regardless of how thick they are, *they will not work*.

To properly contain the burning gases generated during a rocket motor's operation, you need a *convolute* tube. *Figure 8-2* is a photo of a convolute tube, and you can see that it lacks the spiral structure of the tube in *Figure 8-1*. A convolute tube is made by rolling a single sheet of glue-coated paper around a rod or a mandrel. The result is a tube that is made like a jelly roll with a solid wall of paper and glue, and *no* voids or air pockets.

How to Buy Convolute Tubes

Convolute tubes are not as common as the spiral kind, but they're easy to find if you know where to look. To make the motors in this book, the sizes you will need are **1/2" i.d. x 3/4" o.d., 3/4" i.d. x 1" o.d., 1" i.d. x 1-1/4" o.d., 1-1/8" i.d. x 1-1/2" o.d., and 1-1/2" i.d. x 2" o.d.**

Convolute tubes are often used in the packaging of paper and fabrics. Check with friends and businesses. If you know someone who uses a lot of computer or fax paper, or works in the garment industry, you might find a free supply. If this doesn't work, look for the companies that sell tubes to the fireworks industry. You'll see their ads in the classified sections of magazines like ***Popular Science, Soldier of Fortune, Guns, Shotgun News,*** and ***American Fireworks News*** (**page 175**). Check the **PERIODICALS** section at the library, and look for the ads offering things like chemicals, fuse, and fireworks-making supplies. Some of these companies sell convolute tubes, and even the ones who don't can often tell you where to buy them. Failing that, you can order the tubes needed to make *most* of the motors in this book from the **Teleflite Corporation** at www.teleflite.com and www.amateur-rocketry.com.

To buy the tubes in *large* quantities, you can order them directly from a manufacturer. As of the year 2000, I know of six U.S. companies that make convolute tubes in sizes and wall thicknesses suitable for rocket motors. They are:

Figure 8-1. A spiral cardboard tube. Its basic design makes it **unsuitable** for building a rocket motor.

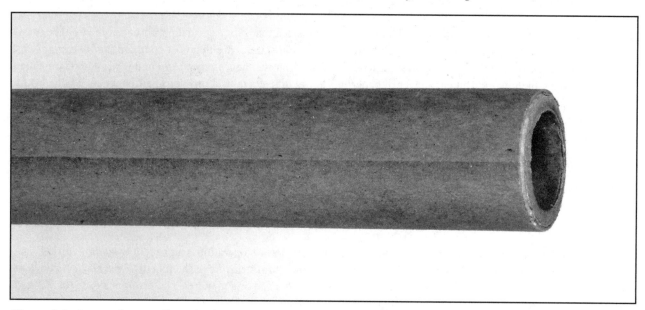

Figure 8-2. A convolute cardboard tube. Its solid wall of paper and glue make it **ideal** for building a rocket motor.

Caraustar (formerly Star Paper Tube)
1379 McDow Dr.
Rockhill, SC 29732

New England Paper Tube
173 Weeden St.
Pawtucket, RI 02860

Sonoco Products Co.
1 North Second St.
Hartsville, SC 29550

Newark Paperboard Products
10215 San Sevaine Way
Mira Loma, CA 91752

Pacific Fabric Reels
3401 Etiwanda Ave., Bldg. 811-C
Mira Loma, CA 91752

Thames River Tube Co.
64 High St.
Ashaway, RI 02804

I've spoken with each of them, and they all said that they'll make tubes for anyone who pays the setup charge, and meets their minimum order requirement. Tube manufacturers are geared toward making *large* quantities of tubes to *your* specifications. They'll accept small orders, but their minimum charge usually makes runs of fewer than 500 tubes uneconomical. If you're a member of a group like Tripoli or the RRS, you might consider finding other members to share the cost of a large order. When you buy tubes in this way, you'll pay *less* than you'd pay for the materials needed to make them yourself, and you'll pay just a fraction of what a retail tube dealer would charge.

When placing an order with a manufacturer, *tell them to use plenty of glue*. One of the difficulties in making convolute tubes is judging how much glue to use. If the machine operator uses too much, the tubes stick together as they dry. To avoid this problem, inexperienced operators sometimes reduce the amount of glue to a point where there isn't enough. In many of the industries that use convolute tubes, the amount of glue doesn't matter. In the rocket industry it is *very* important. It is therefore important that you explain your needs *ahead of time*, and ask the manufacturer to have one of his more experienced machine operators handle your production run.

The *type* of glue used is also important. Most manufacturers like **dextrin**, because it is cheap and easy to clean up. A **silicate** glue makes a stronger tube, but the post-production cleanup is difficult. Manufacturers often shy away from silicate unless you insist on it. You can expect to pay an extra cleanup fee for silicate, but if the fee is reasonable, the increase in the tube strength is worth the added cost.

Most convolute tubes, including the ones used in fireworks, are made from partially recycled kraft paper. The model rocket industry uses a *much* stronger tube made from a cream or "manila" colored paper with a *very* high virgin wood fiber content. These tubes have *several times* the burst strength and burn-through resistance of a standard fireworks tube. If you use them to build the rocket motors in this book, you'll be able to push motor performance substantially higher than the listed figures.

The problem is that, as of this writing, New England Paper Tube is the only company that makes them. They are very expensive, and people on the West Coast must pay the substantial charges involved in having them shipped all the way from Rhode Island. In August of 1995 I spoke with Newark in California. They expressed an interest in making these tubes, but *only* if they receive enough orders to justify the 10 ton minimum purchase of the special paper required. I encourage anyone who needs *large* quantities of these tubes to contact Newark. New England Paper Tube also makes the standard kraft paper fireworks tubes, and they've assured me that they can make the 1-1/2" i.d. x 2" o.d. tubes needed for the **I-65**.

Small Homemade Tubes

If you can't buy the tubes, you can make them yourself. It takes a little time, but the materials are cheap, and if you make them properly, they'll be just as good as the commercial tubes. The smaller and shorter tubes are the easiest, and the level of difficulty rises as the length and the diameter increase. Commercial tubes are made on machines that cut the paper, coat it with a water-based glue, and roll it around a mandrel. It seems logical that you could make a homemade tube in the same way. But when making *homemade* tubes, **this method will not work**.

A commercial machine makes about 20 tubes per minute. Each tube passes through the machine so quickly that the paper doesn't absorb the water from the glue until *after* the tube is rolled. *Then* the paper swells with moisture, the tube expands, and shrinks again as it dries. When making a *homemade* tube, anywhere from 1 to 3 minutes can pass from the time you apply the glue to the time that you roll tube, and the paper has plenty of time to absorb the water from the glue. The rolling is therefore done with the paper in a *premoistened and swollen condition*. A tube made in this way looks fine when you take it off the mandrel. But as it dries, the layers of paper shrink away from one another, and separate, creating wrinkles, voids and air pockets that render the tube useless.

To make a commercial-quality tube by hand, you have to minimize the amount of water in the paper during the rolling process. After struggling with the problem for several weeks, I found that I could coat the paper with glue *ahead* of time, let it dry, and lightly remoisten it just prior to rolling. This minimized the problem, but if I dampened the paper too much, or waited too long before I rolled the tube, the problem reappeared.

The best glue for making a homemade tube is common *white glue*. Once it is dry, you can resoften it with water or alcohol, and though water makes the paper swell, *alcohol does not*. But alcohol *alone* makes the glue so slippery that rolling the tube is impossible. As the alcohol evaporates, the glue becomes sticky again, but the time window before it dries is so short that you don't have enough time to work. Through trial and error I learned that, for remoistening the glue, a *mixture of alcohol and water works perfectly*. **Important note.** *Two parts alcohol to one part water works well for papers up to .010" thick, and a one-to-one mixture works best for anything thicker*. Different size tubes require different thicknesses of paper, and slightly different construction techniques. You can make a 1/2" or 3/4" i.d. tube up to 8" long from a brown paper shopping bag, and here's how to do it.

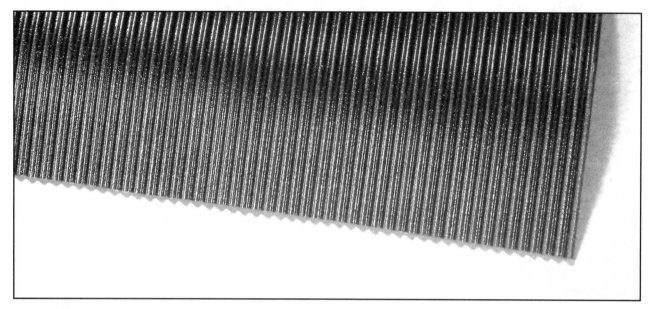

Figure 8-3. *A piece of ribbed, vinyl "hallway runner" makes a nice tube-rolling pad. You'll find it at carpet and flooring shops, and big hardware warehouses like Home Depot. The cut edge in this photo is shown actual size with 8 ribs per inch of material width.*

A 3/4" TUBE MADE FROM A SHOPPING BAG

MATERIALS LIST

1. A large, heavy duty, brown paper shopping bag
2. A smooth, round metal bar, 3/4" dia. x 12" long
3. A straight, heavy, steel block at least 8" long
4. A carpenter's framing square
5. A smooth, flat, washable surface to work on
6. A 2 ft. square piece of vinyl or rubber sheeting
7. An oven or refrigerator rack
8. A pair of scissors
9. A small, short nap paint roller with tray
10. A sponge
11. White glue (like Elmer's "Glue-All")
12. Some old newspapers
13. Waxed paper
14. Water, and some denatured alcohol

When you buy groceries at a supermarket, *most* checkers will ask if you want plastic or paper bags, so always say paper. I often shop at one of those big warehouses where you save money by bagging your own groceries. If I double-bag everything, I come home with six, large, heavy duty shopping bags, and that's enough to make *ten* of the tubes in this demonstration. If you ask your friends to save their bags for you, you'll have more paper than you can use.

The round, metal bar can be made of any metal, but it has to be *smooth*. You make the tubes by rolling the paper around the bar, and when you are finished, you have to slide them off. If the bar is rough, you won't be able to do it. Cold rolled steel, stainless steel, brass, and aluminum work fine, but hot rolled steel, with its typically rough finish, does *not* work. If the bar you find is gouged or scratched, smooth it with a file and some #120 sandpaper. If you can't find *exactly* what you need, you can make it on a metal lathe. The straight, heavy steel block (square or rectangular in shape) is used as a straight edge, and it has to be *heavy*, so that once in place, it doesn't wiggle around.

White glue comes in plastic bottles up to a gallon in size, you can buy it almost anywhere, and the most popular brand is probably Elmer's "Glue-All". Stationery stores and places like Walmart sell the small bottles, and the big hardware warehouses like Home Depot and Lowe's sell it by the gallon.

The flat, washable surface can be a Formica counter top, but most counter tops are not *perfectly* flat. An uneven surface generates uneven pressure as the tubes are rolled, and uneven pressure causes areas where the paper is poorly bonded. The vinyl/rubber sheeting acts as a resilient rolling pad, and evens out the pressure between the high spots and the low spots. For this demonstration, I bought a 27 inch-wide x 2 foot length of black, ribbed, vinyl "hallway runner" at Home Depot (*Figure 8-3*). I found it in the department where they keep the rubber floor mats. To use it, I turn it upside down, and put the ribbed side on the bottom.

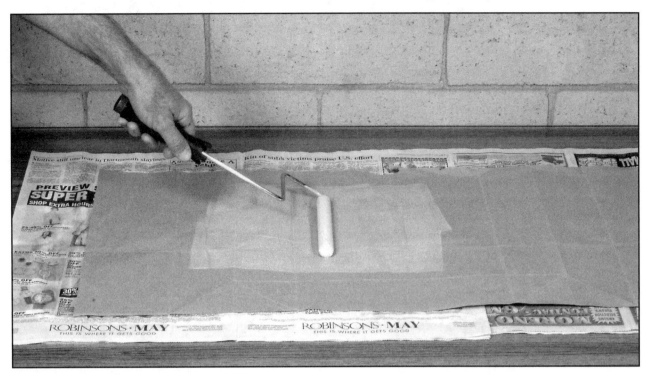

Figure 8-4. The water-thinned glue is applied to the paper with a small, short nap paint roller.

The paint roller should be the small 4" to 6-1/2" type used for trim work, and the nap should be *short*. The roller puts the glue on the paper., and if the nap is too long, it soaks up more glue than you need.

Before you begin, make up the following two solutions:

GLUE SOLUTION: 1 cup (8 oz.) of water + 1 cup (8 oz.) of white glue

and

ALCOHOL SOLUTION: 1/2 cup (4 oz.) of water + 1 cup (8 oz.) of denatured alcohol

---

The **GLUE SOLUTION** is applied to the paper ahead of time, and the **ALCOHOL SOLUTION** remoistens the glue during the rolling process. Because the balances of water to glue, and water to alcohol are important, you should keep these solutions in sealed containers so that they are not altered by evaporation. Plastic soft margarine tubs with snap-on lids work fine, and they'll keep longer if you store them in a refrigerator.

1. Cut the bottom out of a large, heavy duty shopping bag. Cut away the vertical seam in the back, and trim the paper so that the edges are square and straight. Set up an ironing board, and using a steam iron with the steam turned **ON**, and the heat set for cotton or linen (i.e. *hot*), iron out the creases as best you can. You'll end up with a sheet of paper about 16-1/2" x 36", and about .006" thick.

2. Lay the sheet on *two* layers of newspaper, and using the paint roller and a paint roller tray, coat one side with the **GLUE SOLUTION** (*Figure 8-4*). The coating needn't be thick, but it must *completely cover* the paper (i.e. no bare spots). Lift the glue-coated sheet off the newspapers, *discard the newspapers*, and set it aside to dry. When it is *thoroughly* dry, turn it over. Lay it on two *fresh* sheets of newspaper, coat the opposite side in the same manner, and let it dry again. Then inspect the both sides of the paper carefully. As the glue dries, leaves, insects, dirt, and other debris can become trapped in the glue. If you see anything in the dried glue, scrape it off with a putty knife (*Figure 8-5*). Then cut the paper into two 8-1/4"-wide x 36"-long strips.

Important Note. *As the first side dries, the paper will curl. As the second side dries, the paper will straighten out again.*

Figure 8-5. Small bits of debris are scraped away with a putty knife.

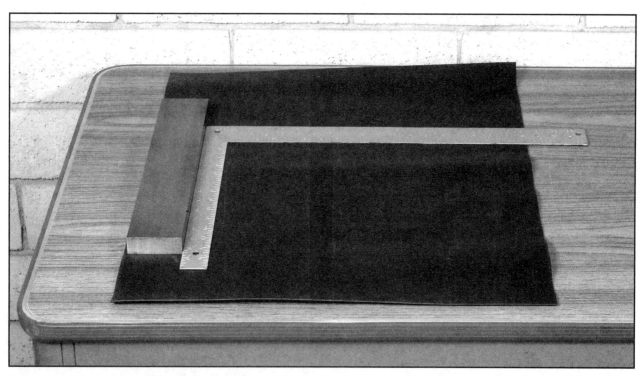

Figure 8-6. The rolling setup for small tubes. A vinyl pad forms the work surface. A heavy metal block and a carpenter's square are used to align the paper.

3. Lay the vinyl pad (the upside down hallway runner) on a flat, washable surface. Lay the steel block on the vinyl, and lay the carpenter's square against the block (*Figure 8-6*).

4. Pick up one of the paper strips. Bevel the corners of one end (cut them at a 45° angle with a pair of scissors). Align the paper's edge with the inside edge of the carpenter's square, and as shown in *Figure 8-7*, hold it firmly in place with a heavy weight (i.e. a can of V8 juice). *Without wiggling the paper or disturbing it's position*, remove the carpenter's square.

***Figure 8-7**. A strip of paper is aligned with the carpenter's square, and held in place with a heavy weight; in this case a can of V8 Juice.*

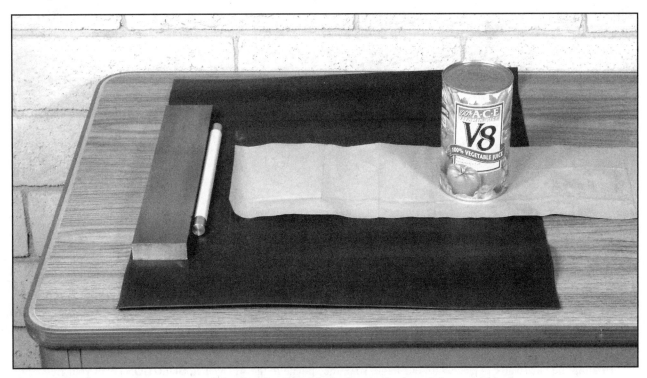

***Figure 8-8**. The waxed paper-covered bar is aligned with the forward edge of the metal block.*

5. Cut a piece of waxed paper 5" wide x 10" long. Roll it around the bar as tightly as possible, and align the bar with the forward edge of the steel block (***Figure 8-8***). The waxed paper keeps the bar from sticking to the tube during the rolling process, and the steel block insures that the bar is square with the edge of the paper.

6. Pressing firmly down on the bar *so that it doesn't wiggle*, roll it forward onto the paper (***Figure 8-9***). Without wiggling the bar, lift up the beveled end of the paper. Roll it over the top of the bar, and tuck it under the bar as tightly as possible (***Figure 8-10***). You will find that long fingernails and the beveled corners of the paper are helpful when doing this.

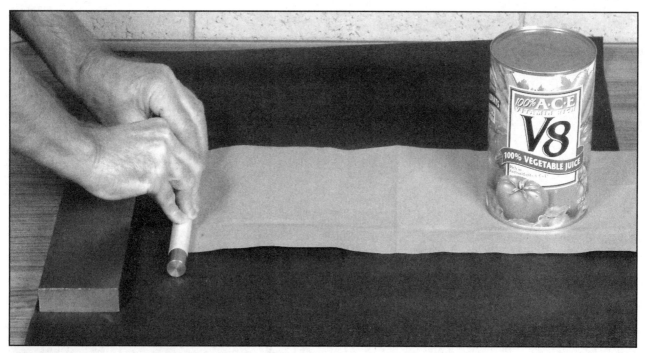

Figure 8-9. The bar is rolled forward onto the paper.

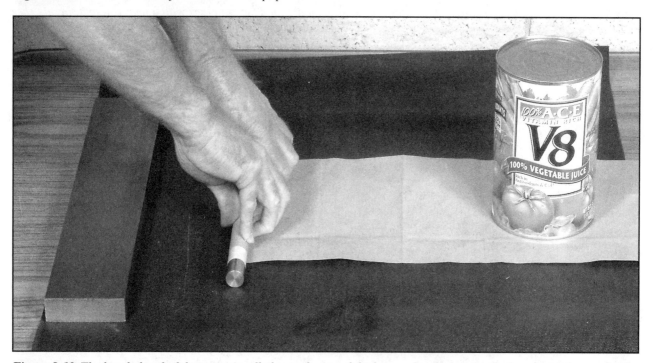

Figure 8-10. The beveled end of the paper is rolled over the top of the bar, and tucked under the bar as tightly as possible.

7. Holding the bar firmly in place with one hand, soak the sponge in the **ALCOHOL SOLUTION**. With your free hand, squeeze the sponge about 3/4 dry. Dampen about 6" of paper in front of the bar, taking care to squeeze the sponge as far under the bar as possible (*Figure 8-11*). The amount of moisture needed isn't much; just enough to make the glue glossy and sticky again. *Without delay*, roll the bar forward over the dampened paper, pressing down firmly as you go (*Figure 8-12*).

8. When you reach the end of the dampened area, remove the weight (the juice can). Lift the bar-with-paper *off* the pad, and take it back to the starting point. Resoak the sponge. Wring it out again. Dampen another 12" of paper, and roll the bar forward again. Continue dampening and rolling in this manner to within 1/2" of the paper's end.

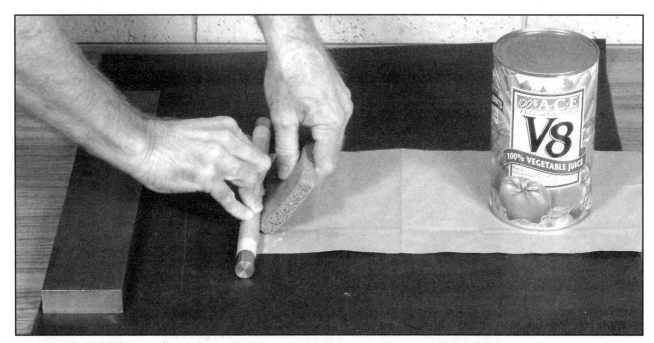

Figure 8-11. The first 6" of paper are dampened.

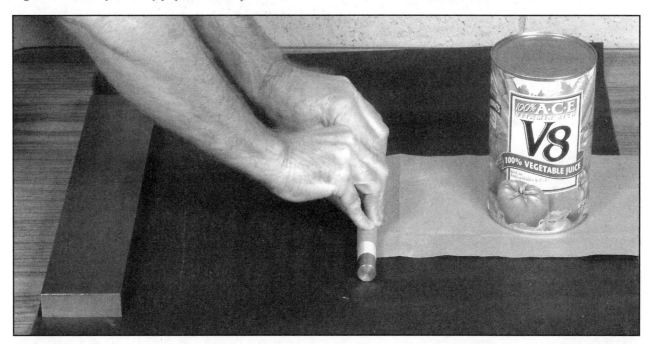

Figure 8-12. The bar is rolled forward over the dampened paper.

9. Squeeze the dampened sponge *between* the end of the paper and the tube, dampening *both* surfaces in the process (*Figure 8-13*). *Without delay,* roll the bar forward, and rock it back and forth over the end of the paper until the end is firmly bonded.

At this point the tube is almost finished, *but not quite.* You will probably find that a 3/4" i.d. tube with a 1/8" thick wall takes *more* than the 36" of paper that you've already used, and you'll have to keep adding paper until the outside diameter equals *one inch.* To do so, proceed with **step 10**.

10. Place the bar with the partially completed tube against the metal block. Cut a piece of preglued paper strip 8-1/4" wide x 12" long, and dampen its *entire surface. Without delay*, slide one end of the paper under the bar until its end *touches* the partially completed tube (*Figure 8-14*). Roll the bar forward onto the paper, and rock it back and forth until the end is firmly bonded. Continue rolling until the outside diameter equals 1.00", as measured by an accurate pair of calipers. Then cut off the excess paper, and repeat **step 9**.

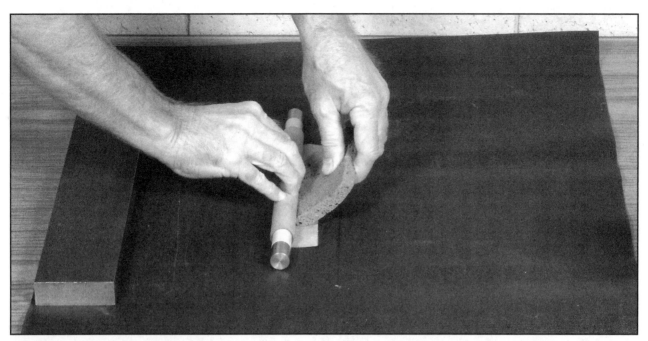

Figure 8-13. The sponge dampens the end of the paper and the surface of the tube beneath it..

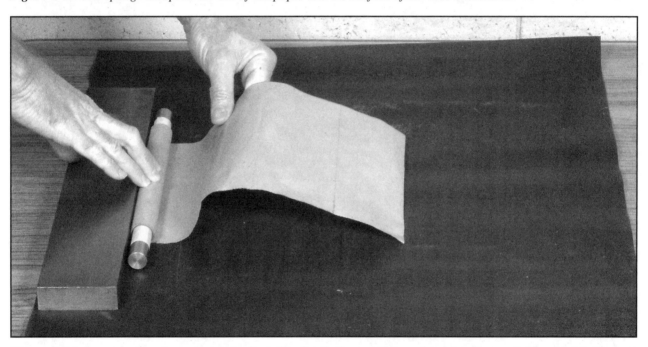

Figure 8-14. A new preglued paper strip is slipped under the bar until its end touches the partially completed tube.

11. Slide the finished tube off the bar, and inspect the inside of each end. The inside end of the paper should be firmly bonded to the inside of the tube. If it is *loose*, fill a small, disposable syringe (**page 48**) with the **ALCOHOL SOLUTION**, and squirt it under the end of the paper (*Figure 8-15*). Allow the excess to run out the opposite end of the tube and back into the solution container. *Without delay*, slide the metal bar *without the waxed paper* back into the tube, and roll it back and forth in rolling pin fashion (*Figure 8-16*) until the end of the paper is firmly glued in place. When doing this, you will find that the inside of the tube is slippery, and you'll have about 30 seconds to work before it gets sticky again. When you are finished, *without delay,* slide the finished tube off the bar, and let it dry on a rack indoors for 48 to 72 hours (*Figure 8-17*).

12. When the tube is *thoroughly* dry, cut it to the length required for the motor you are making (a bandsaw with a 4 tooth-per-inch woodcutting blade works best), and sand off the resultant paper fuzz with a belt sander or #120 sandpaper. Then give the ends of the tube a visual inspection. If you've done everything right, it should look like the one in *Figure 8-18*: a solid wall of paper and glue with no voids or air pockets.

Figure 8-15. A disposable syringe injects the water-alcohol solution under the loose end of the paper.

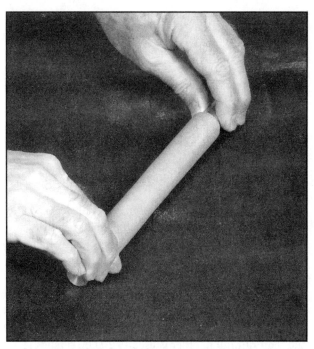

Figure 8-16. Rolling the bar back and forth bonds the end of the paper to the inside of the finished tube.

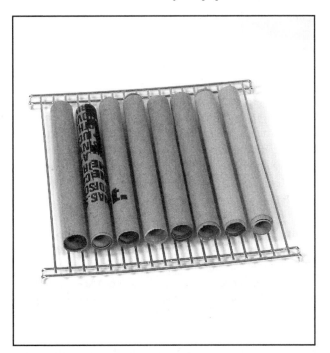

Figure 8-17. Small, finished tubes should be air dried indoors on a rack for 48 to 72 hours. Note in the next photo that I've "jazzed up" the tube's appearance by putting the shopping bag label on the outside.

Figure 8-18. The cut and sanded end of a properly made tube should reveal a solid wall of paper and glue with no voids or air pockets.

How to Buy Paper

Shopping bags are nice because they are *free*. But ironing out the wrinkles is a chore, and for just a few dollars, you can buy the paper used to make the bags. It's called **60 pound kraft**, or **60 pound postal wrapping paper**, and you can buy it at an office supply store. As of the year, 2000, a 125 square foot roll costs about $3, and that's enough to make 3 dozen of the tubes just described.

For $20, a nearby paper dealer sold me a 1650 square foot roll, and that's enough to make *600* tubes at a cost of less than 4 cents per tube. Of course the glue is expensive when you buy it in small bottles. But like the other supplies, the price drops dramatically when you buy more. As of 2003, a gallon of white glue costs $11.

● Kraft Linerboard

Sixty pound kraft paper works fine for short tubes with walls up to 1/8" thick, but not for anything larger. Sixty pound kraft is only .006" thick. In tubes made from paper this thin, the normal variations in paper thickness, plus the variations in the thickness of the glue, make up a significant percentage of the wall thickness. As you increase the number of wraps, the variations accumulate, and the tube's outside diameter becomes irregular. The tube in the previous example took 18 wraps to make. At 18 wraps, the variations are acceptable. At anything over 20, they become troublesome.

When working with paper this thin, you will also find it hard to roll a sheet more than 8" wide without making it wrinkle. This limits the tubes made from 60 pound kraft to a maximum length of 8". Larger and longer tubes need thicker paper. The *best* paper for making large tubes is **kraft liner board**, but before you start looking for it, you need to understand how the paper industry works.

The paper industry is split into three levels. At the top are the **paper mills**. Paper mills are gigantic factories that buy scrap paper and wood chips by the trainload, convert them into pulp, and then into rolls of paper weighing *up to 6 tons each*. They make paper *only* when someone orders it. They do *not* keep paper in stock. They do *not* sell paper in small quantities, and their minimum order is usually 20,000 pounds (10 tons). Paper mills produce **prime quality paper** and **B-grade paper**. Prime quality paper is made *exactly* to the customer's specifications. Prime quality kraft liner board is the *best* paper for making hand-rolled rocket casings, but it's almost *impossible* to buy in small quantities. The mills make it *only* for the customers who order it. They *rarely* have any left over, and even when they do, they *don't* sell it to the public.

B-grade paper, also called **odd lot** or **job lot** paper, is paper that's "a little out of spec."; what another industry might call a "factory second". It might be from the beginning of a mill run before the machine was properly adjusted. The color or the texture might be wrong, or the thickness might vary. And rather than throw it away, the mill sells it to a "paper converter".

Paper converters form the middle tier of the industry. Paper converters buy *a few* prime quality papers, but *only the most popular types*, which they repackage and sell to retailers like the store where I bought the 60 lb. kraft. They deal mostly in large rolls of *B-grade paper*, which they buy from the mills, and sell at a discount to nearby industries that don't care about color, texture, or thickness. With rare exceptions, paper converters take orders by phone, load them onto trucks, ship them to *commercial* destinations, and sell *only* to customers with established accounts. They do *not* normally deal with the public, and though they may have exactly what you want, they will probably ask you to buy it through a retailer.

At the bottom of the industry are the **retailers**, also called **paper dealers**, and the paper dealers are the only members of the industry who maintain stores where you can walk in and buy something. In most parts of the country you'll find a **PAPER DEALERS** column in *The Yellow Pages*. The problem is that kraft linerboard is an *industrial* material used *primarily* to make cardboard boxes. It is *not* used by the public, so most dealers don't sell it, and even the ones who do don't keep it in stock. This doesn't mean that they can't get it for you. It means that it's a *special order* item, and will be subject to the minimum dollar amount imposed by the converter where they buy it.

Because paper dealers don't normally sell kraft linerboard, they will *not* have samples to show you, and they *won't* be able to help you select what you need. The paper converters have the samples *and* the paper, but they *usually* won't sell it to the public, so here's how to proceed.

Paper converters tend to locate in large cities near the industries that buy from them. Check *The Yellow Pages* of nearby cities under the **PAPER CONVERTERS** or **PAPER CONVERTING** headings. When you find a list of converters, pick up the phone, and start calling them. In the example ahead, we'll pretend that you need some .015" thick linerboard for making **I-65** casings. **I-65** motors are 15" to 16" long, and we'll assume that you plan to experiment with longer motors too, so you'll need some paper that is 24" wide.

1. *First* ask each converter if they sell kraft linerboard. If they say yes, ask them if they have any that's approx. .015" thick. If they say yes again, ask them what their minimum purchase is (i.e. the dollar amount), and make a note of what they say.

2. When you've spoken with all of them, call back the ones whose minimum order is affordable, who sound like they have what you want, and tell them that you want to see some samples. They *should* be happy to show them to you. They'll probably want you to make an appointment, and they might want some more information about you. At that point you can explain that you're a private party, that you *know* they don't sell to the public, and that if they have what you want, you'll order it through a dealer. Also note that paper converters often deal with customers over long distances. If a personal visit is difficult, they *might* be able to send you a sample by mail.

Important note. *When making the tubes for these rocket motors, you need the strongest paper you can find. Kraft linerboard varies* **tremendously** *in strength, depending on what kind of fiber it is made from. It is* **not** *something you can order sight-unseen, and before you buy it, you* **absolutely must** *look at a sample.* Linerboard made from 100% *newly* processed wood fiber is called **virgin kraft linerboard**. Because the fibers are long and unbroken, it is *very* strong and *very* hard to tear. **Virgin kraft linerboard** is the *best* paper for making rocket casings.

Linerboard made from 100% *recycled* fiber is called **100% recycled kraft linerboard**. Because the fibers have been chopped up and broken in the recycling process, it is weak, and it tears very easily. **100% recycled kraft linerboard** is the *worst* paper for making rocket casings. Virgin kraft linerboard is *hard* to find. Recycled kraft linerboard is *easy* to find, and most linerboards are a made from a *mixture* of virgin and recycled fibers.

Important note. *These tubes are made by rolling the paper around a metal bar. When you go to look at the samples, take the bar with you, and bring along a micrometer or a vernier caliper, so you can verify the paper's thickness.*

Important note. *Paper that is too thick or too rigid will kink when rolled around the bar, and paper that is too thin will amplify the problems of wrinkling, telescoping, and uneven wall thickness. As a general rule, I like a paper whose thickness equals* **1%** *of the i.d. of the motor casing. Given a choice, I'd make an 0.75" i.d. casing from paper that's .0075" thick, and I'd make a 1.5" i.d. casing from paper that's 0.015" thick. If my choices were limited, I might vary from these figures by plus or minus 25 percent.*

3. When you arrive at a converter's office, ask to see samples of kraft linerboard *approximately* .015" thick. Tell them that you need a roll at least 24" wide, and tell them that you need a very *strong* paper. When they bring out the samples, roll each one around the bar, and pick the toughest paper they have that wraps around the bar *without kinking*.

At this point *don't be surprised if the paper they show you is not the thickness you specified.* As I said before, paper converters sell mostly to industries that don't care about texture and thickness. During my own odyssey, I quickly learned that, when I asked for paper that was 0.015" thick, my request was interpreted to mean anything from 0.010" to 0.020".

Also, the paper industry uses several standards for measuring thickness. When you say 0.015", they *might* think you mean "15 mil", and 15 mil paper can actually be *thinner*. When the samples they showed me were *not* the thickness I wanted, I asked if I could walk through the warehouse and pick something out. *Taking the micrometer with me,* and measuring the paper as I went, I found and bought some paper that was suitable. But they *didn't* have a sample, because they were very busy at the time. They hadn't kept their samples up to date, and what they displayed in the office represented *less than half* of what they actually had for sale.

4. When you've seen the best they have to offer, take a sample with you. Write down the details (like the name, the thickness, the width, the stock number, etc.), and go to the next converter. When you've seen the samples of several converters, *pick the one that is best*, and ask the converter if he'll sell it to you. If he says yes, you're done. If he says *no*, ask him if he knows a dealer who can order it for you. If he does, your search is over. If he doesn't, start calling paper dealers. When they answer the phone, say approximately the following. *"Hi. My name is So-and-So. I need to buy some paper from United Paper Converting over in Gotham City. I know exactly what I want, but they don't sell to the public, and they've asked me to order it through a dealer. Can you order it for me?"*

5. When you've found a dealer who can help you, give them the details, show them the sample, *and call the converter yourself* to verify that all the specifications are correct; i.e. *no mistakes*. Because it's a special order, they'll usually want payment in advance, and delivery will take about a week.

6. If you are serious about making homemade casings, when you've found the paper you want, you should buy as much as you can for the following reason. The paper you've chosen is probably a B-grade paper. The converter may have only a few rolls, and when they are gone, he will *not* be able to order more. He *might* get something like it again, but you can't count on it. And even if he does, there's no way to predict when.

● Tagboard

If you can't find kraft linerboard, look for **tagboard**. Tagboard is used for making price tags and file folders. Many years ago it was also used for making dress patterns in the garment industry. Most tagboard is as strong as good linerboard, and though it's more expensive, it is often easier to buy. It comes in several colors (manila and white are the most common), and you can buy it in 24" x 36" sheets from distributors and retailers who sell paper to the printing industry. Look in *The Yellow Pages*, or search the Internet for sites selling "tagboard" or "tag board".

Tagboard comes in standard weights and thicknesses that vary a little depending on the manufacturer. Of the companies I spoke with, the *approximate* figures were: **100 pound tag = .007" thick, 125 pound tag = .009" thick, 150 pound tag = .011" thick, 175 pound tag = .013" thick, and 200 pound tag = .015" thick**. They also said that they could special order it in 24" wide rolls. But the smallest rolls were 40" in diameter, *and they weighed 1,000 lbs.*!

● A Caveat and a Suggestion

The preceding information is based on my own experience shopping for paper in Southern California. There are dozens of dealers and converters within a 100 mile radius of where I live. Even so, I found it *very* difficult to buy linerboard in small quantities. Of all the converters I spoke with, all but two had $200 to $300 minimums. After many phone calls, I located and bought *just two* 42 lb. "counter rolls". But a month later, when I tried to buy more, they were no longer available. I also found that, here in Southern California, *many* of the independent dealers, who used to sell small quantities of tagboard at reasonable prices, have recently been bought out by *big* companies that *also* have $300 minimums. By calling a tagboard manufacturer, I was able to get the name of a nearby *retailer*, who sold me 100 sheets for $40. But at 40 cents per sheet, the cost of a homemade tube was about equal to the retail price of a commercial tube.

The problem is *not* that people in the paper industry are inherently greedy, but that the paper industry is a high volume, *low profit* business. Converters buy the paper in rolls weighing up to 6 tons each, and the average converter has to sell $200 worth to make $20. If a customer wants one or two small rolls, the converter has to get a *large* roll from the warehouse, load it onto a machine, set up the machine, cut and roll the paper, and return the large roll to storage. This can occupy a fork lift driver and a machine operator for half-an-hour. By the time the converter has paid their wages and his other expenses, he's lucky if he's made $5. Most of the converters I spoke with said they'd be *happy* to sell me small rolls if I'd order them *20 to 50 at a time*. This leads me to think that someone who lives near a converter might make a business out of buying small rolls of linerboard or tagboard in large quantities, and reselling them to the people who need them. To anyone thinking of doing this, I suggest the following. *Start with paper that's .012" thick*. Rocket builders can make casings up through 1" i.d. from 60 lb. kraft, but they need thicker paper for anything larger. The .012" liner board is ideal for casings from 1-1/8" through 1-1/2" i.d., so the market for the .012" paper should be the greatest.

Sell the paper in 65 lb. rolls. When shipping by United Parcel, as the weight of a package increases, there's a rapid *decrease* in the cost-per-pound for anything up to 70 lbs. Beyond 70 lbs., the cost-per-pound rises. A 69 lb. roll is the most economical to ship, but paper can increase in weight by several percent in damp weather. Making the rolls weigh 65 lbs. allows for a weather-related weight increase, and keeps them safely under the 70 lb. limit. *Advertise in American Fireworks News and on the Internet*. AFN is distributed to rocket and fireworks makers throughout the country, and there are several chat groups on the Internet that talk about rocketry and pyrotechnics.

● Red Rosin Paper

If you can't find linerboard or tagboard, a *less* desirable alternative is something called **red rosin paper**, and most roofing suppliers sell it. Red rosin paper is about 0.012" thick, and comes in 3 ft. wide x 500 square foot rolls. The problem is that red rosin paper is made from 100% recycled fiber, and *only the best grade* will work for rocket casings. The best grade in Southern California is called **4 pound red rosin paper**, and that means that it weighs 4 lbs. per 100 square feet. A 500 square foot roll should weigh *at least 20 lbs. If it doesn't, **don't buy it!***

When you shop for red rosin paper, take along a bathroom scale. When you find the paper in question, weigh yourself alone. Then weigh yourself holding the roll of paper. If the difference in weight isn't **at least** 20 lbs., ***don't buy the paper!*** Before I knew better, I bought something called "standard grade" red rosin paper. The roll weighed less than 15 lbs., and the tubes that I made from it were too light and pithy to be of any use.

Here in Southern California I found some 4-pound paper in the paint department at Home Depot. Painters use it as a cheap, disposable drop cloth. As of this writing, it is called **Ratan Red Rosin-Sized Sheathing**, and I've successfully used it for tubes up to 1-1/8" i.d. by 18" long with wall thicknesses up to 3/16". At the time of this writing, it is made by

Salinas Valley Wax Paper Inc. of Salinas, CA, and distributed through Home Depot in the Western half of the U.S. Salinas does *not* sell to the public, and the paper is *not* sold in the eastern part of the U.S. at this time, but similar products might be available. As of the year, 2003, you can expect to pay $10 to $15 per roll. Red rosin paper is *usually* made from **kraft chipboard**, and "chipboard" is *usually* made from ***all the scraps that the paper mills want to get rid of!*** It is *very* soft. It tears easily, and though the tubes made from the 4 pound paper are not very good, they are at least usable.

Large Homemade Tubes

You can make a small tube by following the instructions on **page 214**, but a large tube requires a different setup. In the example below, I'll make a 1-1/8" i.d. tube for an **NG6-129-G37** motor from 150 pound tagboard. Its .011" thickness is ideal for a tube of this size, and the finished tube will be at least as strong as a standard fireworks tube. To make a 1-1/8" i.d. tube from 150 pound tagboard, proceed as follows.

A 1-1/8" TUBE MADE FROM 150 POUND TAGBOARD

MATERIALS LIST

1. A 24" x 36" sheet of 150 lb. tagboard
2. A smooth, round metal bar, 1-1/8" dia. x 16" long
3. A heavy steel or aluminum block at least 12" long
4. A dry wall square
5. A smooth, flat, washable surface to work on
6. A 30" square piece of vinyl or rubber sheeting
7. An oven or refrigerator rack
8. A pair of scissors
9. A 9", medium (3/8") nap paint roller
10. A sponge
11. White glue
12. A roll of plastic sheeting, 36" wide.
13. Waxed paper
14. Water, and some denatured alcohol

With these larger sheets of paper, you'll need a larger surface to work on when applying the glue. A concrete slab covered with a sheet of plastic works fine, and most hardware and paint stores sell the plastic in 3 foot-wide rolls. The paint roller should be the standard 9" type, and the nap should be medium (3/8"). The steel or aluminum block should be *very* large and *very* heavy, and you'll also need a "dry wall square". A dry wall square is a large, aluminum T-square with a 4-foot blade. You can buy one at a hardware store.

Before you begin, make up the following two solutions:

GLUE SOLUTION: 1 pint (16 oz.) of water + 1 pint (16 oz.) of white glue

and

ALCOHOL SOLUTION: 1 cup (8 oz.) of water + 1 cup (8 oz.) of denatured alcohol

1. Cut a 3 foot-wide sheet of plastic sheeting 6 feet long, and lay it on a concrete slab (*Figure 8-19*). Lay the sheet of tagboard on the plastic, and using the paint roller, coat one side with the **GLUE SOLUTION** (*Figure 8-20*). The coating needn't be thick, but it must *completely cover* the paper (i.e. no bare spots). Lift the glue-coated sheet off the plastic, and let it dry. When it is *thoroughly* dry, turn it over, and place it back on the plastic. Then coat the *opposite* side, and let it dry again.

2. When *both* sides are dry, inspect each side carefully. As the glue dries, leaves, insects, and dirt can become trapped in the glue. Also, glue that overruns the edges of the paper adheres to the plastic; then dries, flakes off, and transfers to the next sheet of paper. Before you proceed, scrape away any debris with a putty knife (*Figure 8-21*). Then cut the sheet lengthwise into two 12" x 36" strips.

3. Lay the vinyl pad on a flat, washable surface. Lay the large metal block on the vinyl, and lay the dry wall square against the block (*Figure 8-22*).

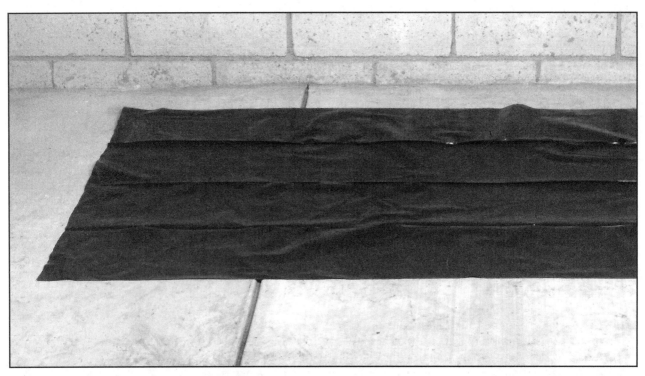

Figure 8-19. Large sheets of paper are glued on a 3 foot-wide sheet of plastic.

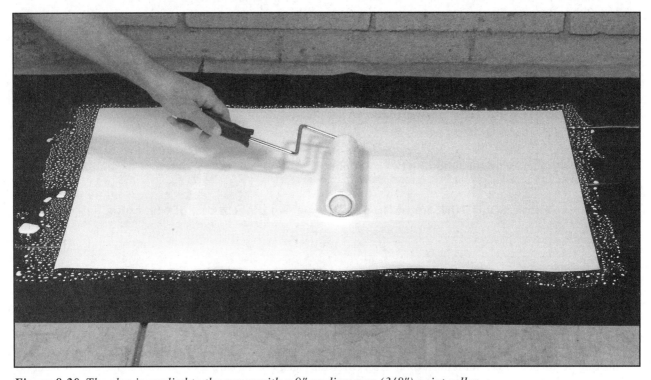

Figure 8-20. The glue is applied to the paper with a 9" medium nap (3/8") paint roller.

5. Pick up one of the strips. Bevel the corners of one end (snip them at a 45° angle with a pair of scissors), and orient the paper so that the beveled end rests on the steel block. Align the paper's edge with the inside edge of the dry wall square, and hold the paper firmly in place with a heavy weight (***Figure 8-23***). *Without wiggling the paper or disturbing it's position*, remove the dry wall square.

6. Cut a piece of waxed paper 14" to 15" long. Roll it as tightly around the metal bar as possible, and align the bar with the forward edge of the metal block (***Figure 8-24***). The waxed paper keeps the bar from sticking to the tube during the rolling process, and the metal block insures that the bar is square with the edge of the paper.

Figure 8-21. Bits of dried glue and other debris are scraped away with a putty knife.

Figure 8-22. The rolling setup for large tubes. A vinyl pad forms the work surface. A large metal block and a dry wall square are used to align the paper.

7. Pressing firmly down on the bar so that it doesn't wiggle, roll the end of the paper over the top of the bar (***Figure 8-25***), and tuck it under the bar as tightly as you can (***Figure 8-26***). You'll find that long fingernails and the beveled corners of the paper are helpful when doing this.

   Important note. *You'll notice here that I use a different technique than I do when working with the smaller bar and the thinner paper, With this larger bar, I find it easier to start with the paper **under** the bar with its end sandwiched between the bar and the metal block. Since either method will work, I suggest that you try both, and adopt the one that works best **for you**.*

227

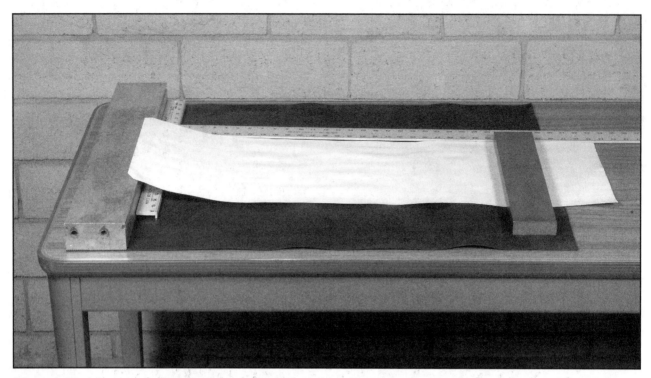

Figure 8-23. A heavy weight (in this case the steel block from the small tube setup) holds the paper firmly in place.

Figure 8-24. The bar is laid on top of the end of the paper, and aligned with the metal block.

8. Holding the bar firmly in place with one hand, soak the sponge in the **ALCOHOL SOLUTION**, and with your free hand, squeeze it about 3/4 dry. Dampen about 12" of the paper in front of the bar, and squeeze the sponge as far under the bar as possible (*Figure 8-27*). The amount of moisture needed isn't much; just enough to make the glue glossy and sticky again. *Without delay,* roll the bar forward over the dampened paper, pressing down *firmly* as you go.

9. When you reach the end of the dampened area, remove the weight, lift the bar-with-paper *off* the pad, and take it back to the starting point. Resoak the sponge. Wring it out again, dampen another 12", and roll the bar forward again. Continue dampening and rolling to within 1/2" of the paper's end.

Figure 8-25. The end of the paper is rolled over the top of the bar.

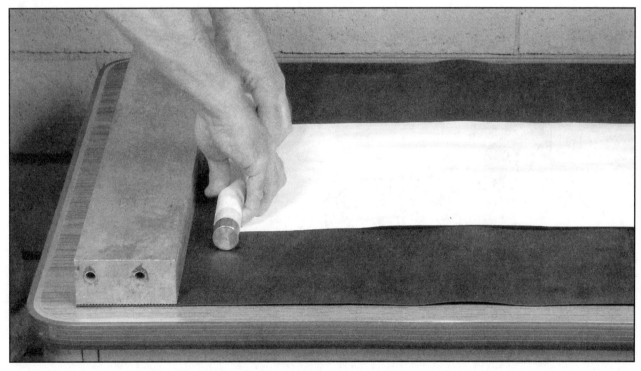

Figure 8-26. The end of the paper is tucked under the bar.

10. Squeeze the damp sponge *between* the end of the paper and the tube, dampening *both* surfaces in the process (*Figure 8-28*). *Without delay*, roll the bar forward, and rock it back and forth over the end of the paper until the end is firmly bonded. Then set the bar with the partially rolled casing aside, and proceed to **step 11**.

11. Replace the drywall square. Pick up the *second* strip of tagboard. Bevel the corners of one end. Lay it on the pad with its end about 3" from the metal block. Align its edge with the dry wall square, and just as you did with the first sheet, hold it firmly in place with a heavy weight. *Without wiggling the paper or disturbing it's position*, remove the dry wall square, and place the bar with the partially rolled tube back against the starting block.

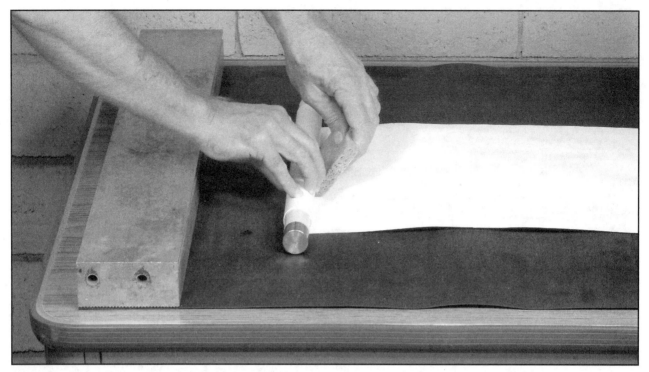

Figure 8-27. The first 18" of paper are dampened.

Figure 8-28. The sponge dampens the end of the paper and the surface of the tube beneath it.

12. Dampen the sponge with the **ALCOHOL SOLUTION**. Wring it about 3/4 dry, and dampen the first 12" of paper. Roll the bar forward onto the end of the paper, and rock it back and forth until the end is firmly bonded. *Without delay*, continue rolling and dampening until the outside diameter of the tube equals 1.50", as measured by an accurate pair of calipers. Then cut off the excess paper, and repeat **step 10**.

13. Slide the finished tube off the bar, and inspect the inside of each end. The beveled end of the paper should be firmly bonded to the inside of the tube. If it is *loose*, fill a small, disposable syringe (**page 48**) with the **ALCOHOL SOLUTION**, and squirt it under the end of the paper (*Figure 8-15*). Allow the excess to run out the opposite end of the tube, and

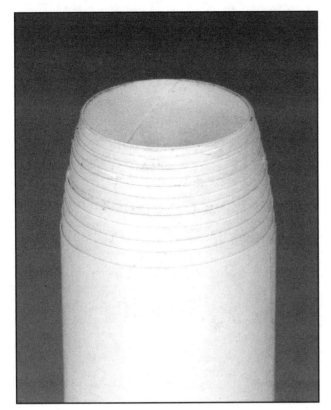 

Figure 8-29. A bad case of telescoping.   *Figure 8-30.* A near-perfect tube.

back into the solution container. *Without delay*, slide the round bar *without the waxed paper* back into the tube, and roll it back and forth in rolling pin fashion (*Figure 8-16*), until the end of the paper is firmly glued in place. When doing this, you will find that the inside of the tube is slippery, and you'll have about 30 seconds to work before it gets sticky again. When you are finished, *without delay*, slide the tube off the bar, and let it dry for a week on a rack indoors.

When rolling tubes larger than 3/4" i.d., your biggest problem will be **telescoping**. Telescoping is the tendency of the paper to spiral off to one side. The square and the metal block insure that the paper is straight when you start, but during the rolling process, irregularities accumulate, and the paper eventually spirals in one direction or the other. There is little you can do to prevent this, and the extent to which it happens is determined partly by skill and experience, and partly by chance. *Figures 8-29* and *8-30* show two 1-1/8" tubes made with 150 pound tagboard. *Figure 8-29* shows a bad case of telescoping, and in *Figure 8-30*, you can see that I got lucky. In *Figure 8-30* I lightly brushed the edges of the paper with a black marking pen to show how closely they are aligned. A typical tube lies somewhere in between.

The **G-37** motor design specifies an 0.188" casing wall thickness. But paper is a squishy, organic material, and a handmade tube will vary from this figure considerably. This is due to expansion of the paper from moisture, uneven pressure while rolling, and variations in the thickness of the glue. Basic arithmetic says that a tube made with .011" paper should take 17 wraps, but the one in the photos took just 13. When you apply the glue, the paper absorbs some of the glue, and expands. And it *never* expands evenly. You started with paper that was .011" thick, but measurements taken after the glue has dried will vary from .012" to .013", and this added thickness means that the finished tube will take fewer wraps of paper than expected.

Important note. When rolling very *long* tubes, cut the starting end of the paper at a slight angle (just a few degrees). This will make one edge slightly longer than the other, and when you start to roll, the longest edge will tuck under the bar first. Starting at that long edge, roll the bar forward a little at a time as you "tuck your way" to the opposite edge (the short edge).

Another important note. In the previous demonstrations, I told you to bond the starting and finishing ends of the paper by rocking the bar back and forth. When working with a stiff and springy paper like tagboard, it can take up to a *minute* of this rocking motion to achieve a strong and permanent bond.

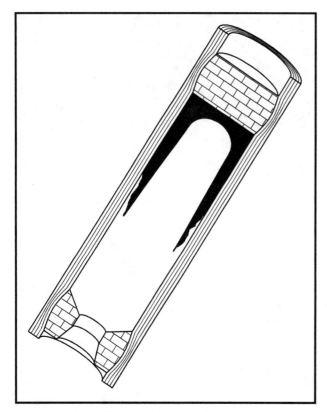

Figure 8-31. As the last of the propellant is consumed, the flame works its way under the edge of the remaining propellant.

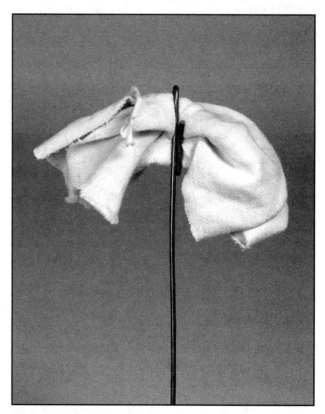

Figure 8-32. A piece of rag is crimped to the end of a straightened coat hanger.

And a final note. In the previous demonstrations, I've specified exact percentages for the glue and alcohol solutions, because these are the percentages that work for *me*. When making your own tubes, start with these percentages. Then feel free to experiment. You will *probably* find that, for a specific paper, adjusting the amount of water in each solution results in something that works even *better*.

Important glue note. Those of you familiar with woodworking might wonder if you can use aliphatic woodworking glue (like the "Tightbond" brand). In the course of my work, I experimented with aliphatic glue, and quickly found that, unlike white glue, which is flexible when dry, dried aliphatic glue is rigid and brittle. The tubes that I made with aliphatic glue were correspondingly too rigid and brittle to be of any use.

DRYING

Paper tubes take a long time to dry, **and the process cannot be rushed**. Putting them in a kitchen oven will *ruin* them by causing the layers of paper to separate. Paper tubes should be air dried *indoors* on an oven or refrigerator rack. ***If you place them on a flat surface, they will warp***. The rate at which they dry will vary with the wall thickness, the density of the paper, how much glue you used, and the weather. It is impossible to quote an exact time, but in "normal weather", plan on *at least three days* for each 1/8" of wall thickness. If you want to know how long a tube takes to dry, weigh it right after you've made it. Then weigh it again each day thereafter. When it stops losing weight, it is dry.

GLUE COATING

The motors in this book are all core burners. To allow for removal of the core spindle, all the cores are tapered. As a result, a point occurs during motor operation, when the flame reaches the casing wall near the nozzle, *before* the last of the propellant toward the *front* of the motor is consumed. If there are any problems with the bond between the propellant and the casing, the flame works its way under the edge of the propellant (***Figure 8-31***). This causes a sudden increase in chamber pressure and thrust during the final moments of the motor's burn, and you can minimize the problem by coating the inside of the tubes with glue. You can use yellow contact cement or a 50/50 mixture of white glue and water. Contact cement is the glue used to bond Formica counter tops, and you can buy it at a hardware store. To glue-coat a batch of tubes, proceed as follows.

Figure 8-33. The rag is dipped into the glue.

Figure 8-34. As the rag is swabbed up and down inside the tube, excess glue drips back into the container.

Straighten out a coat hanger, and crimp a small piece of clean rag in one end (*Figure 8-32*). Place a few layers of newspaper on the floor. Place the open container of glue on the newspapers, and dip the rag into the glue (*Figure 8-33*). Slip one of the tubes *over* the coat hanger, and holding the hanger by its upper end, swab the rag up and down until the inside of the tube is completely covered with glue. If you hold the tube directly over the container, any excess glue will drip back in (*Figure 8-34*).

Lay more sheets of newspaper next to a wall, and as shown in *Figure 8-35*, lean the finished tubes on-end until the glue is dry (at least an hour). The bottom end of each tube will stick to the paper, but you'll cut this part away when you trim the tubes to length. With a little practice and the right size rag, you can work *very* quickly, and coat a dozen tubes in 5 minutes.

Important note about contact cement. As this book goes to the printer, some of the manufacturers of contact cement have started making a "nonflammable" version. As of the year, 2004, I see it in all the hardware stores. In my experiments with just one brand, I found that, when dry, this nonflammable contact cement does *not* resoften with acetone, so I'm guessing that it will *not* work for this application.

Figure 8-35. A batch of glue-coated tubes leaned against a wall to dry.

9. WORKING WITH PROPELLANTS

● First and Most Important!

Before you proceed, if you haven't read **Chapter 1**, go back and read it now. It describes important safety procedures that you *must* follow when preparing or handling the propellants described in this book. Also pay attention to the bold warning at the beginning of **Chapter 1**. To emphasize its importance, I'll repeat that warning here.

> **Do *not* add or substitute other chemicals either to-or-for the chemicals described in this book. Specifically do *not* use POTASSIUM CHLORATE, SODIUM CHLORATE, POTASSIUM PERCHLORATE, SODIUM PERCHLORATE, POTASSIUM PERMANGANATE, AMMONIUM PERCHLORATE, PHOSPHOROUS, MATCH HEADS, ALUMINUM, MAGNESIUM, or ANY METALS AT ALL! *All* of these substances will form dangerous and explosive mixtures when combined with the other ingredients described herein, *AND OTHER CHEMICALS WILL FORM DANGEROUS MIXTURES AS WELL.* When mixing and handling propellants, be sure to ground yourself and all containers and utensils according to the instructions on page 14.**
>
> *If you decide to experiment on your own outside the bounds of the instructions provided in this book, consult with a professional chemist before you proceed.*

Propellant Names

To identify the propellants in this book, and to tell them apart, I've given them each a name. The names are **KG1**, **KG3**, **NG6**, **NV6**, and **KS**. The first character in each name identifies the propellant's oxidizer. Because **K** is the atomic symbol for potassium, a **K** indicates that the oxidizer is **potassium nitrate**. Because **N** is the first letter in the atomic symbol for sodium, an **N** indicates that the oxidizer is **sodium nitrate**.

The second character identifies the propellant's binder. A **G** means that the binder is either **red gum** or **goma laca shellac**, and a **V** stands for **vinsol resin**. The third character identifies the propellant's fuel. In the course of the work, I experimented with six forms of carbon, and I assigned each one a number from 1 through 6. Beginning with the hardest, and ending with the softest they are: **anthracite coal** (1), **bituminous coal** (2), **mesquite charcoal** (3), **garden charcoal** (4), **airfloat charcoal** (5), and **carbon black** (6). The letter, **S**, stands for **sugar**.

As the work progressed, I learned that, due to its high impurity content, bituminous coal does *not* make a good fuel. I learned that carbon black is the *only* carbon fuel that works well in combination with sodium nitrate and/or vinsol resin. I learned that potassium nitrate-based propellants made with *only* carbon black exhibit troublesome case bonding problems. And I found that mesquite charcoal, garden charcoal, and airfloat charcoal are so similar in their effects on performance that I could compensate for any small differences between them with a burn rate modifier. I therefore lump them together, and I use mesquite charcoal (number 3) to represent the group.

Together, these considerations eliminate all possible combinations but the five listed above. Please note that this naming system is something I created myself. It applies *only* to the propellants in this book, and any similarity to the names of propellants used by the rocket industry is purely coincidental.

Figure 9-1. A half gallon mill jar with 4 lbs. of 3/4" brass pellets.

Figure 9-2. The chemicals have been added, and the mixture is ready to be milled.

The Propellant Mixing Procedure

Important note. *With the exception of the **KS** propellant, all the propellants in this book should be mixed in a powder mill set up according to the instructions in **Chapter 3***. I've chosen the **KG3** propellant for this demonstration, and now I'll describe the process that I use to make it. I'll use a **Lortone Model QT-6 mill** with a **1/2 gallon jar** and **4 lbs. of brass pellets**, and I'll make a **500 gram batch**. The formula for the **KG3** propellant is **by weight**:

4 hour potassium nitrate	72.1%
6 hour mesquite charcoal	7.3%
Carbon black	4.0%
Red gum	7.0%
Powdered sulfur	9.6%
Total	100%

For a 500 gram batch, the actual weights are:

4 hour potassium nitrate	360.5 gms.
6 hour mesquite charcoal	36.5 gms.
Carbon black	20.0 gms.
Red gum	35.0 gms.
Powdered sulfur	48.0 gms.
Total	500.0 gms.

Load the mill's half gallon jar with *4 lbs.* of brass pellets (*Figure 9-1*). Then weigh out the indicated amounts of each chemical, and add them in on top of the pellets (*Figure 9-2*). **Important note.** *The charcoal and the carbon black are light and airy. If you dump the potassium nitrate in on top of them, they will billow up, and splash out onto the workbench. Therefore, put the potassium nitrate in first. Then add the charcoal, the carbon black, the red gum, and the sulfur.*

Figure 9-3. The mill jar on the drive base, and ready to go.

Figure 9-4. The small appliance timer is set for 12 hours.

Figure 9-5. The finished propellant is sifted through a stainless steel french fry basket into a plastic bucket.

Figure 9-6. Small amounts of propellant remaining in the jar are brushed out with a paintbrush.

Place the mill in a sandbag containment (**pages 16** & **17**). Tighten the jar's lid, and place the jar on the mill's drive base (*Figure 9-3*). Then set the timer for *12 hours* (*Figure 9-4*). Please note that I say 12 hours, because *that* is the milling time that I used to develop these propellants. When making your own propellants, you can reduce the time to as little as 4 hours, and compensate by the methods described in **Chapters 11** through **15**. **Important note.** *Rock tumbler motors have a low starting torque, and you'll have to give the jar a gentle push to get it moving.* At the end of 12 hours, the timer will automatically turn off the motor, and stop the mill.

Important safety note. *Never stand over the containment, reach into the containment, or try to remove the jar from the mill while the mill is running. Before you do **anything** in a containment with a running mill, turn off the mill, and unplug the motor.*

Figure 9-7. The propellant goes into the bucket, and the pellets stay in the french fry basket.

Figure 9-8. The finished propellant is transferred to a zip lock plastic bag, and stored in a tin cookie box.

To remove the propellant from the mill, **ground yourself, and all utensils and containers according to the instructions on page 14**. Then unplug the motor. Lift the jar off the drive base, place it on the ground, and open the lid. Place a stainless steel french fry basket (or a colander) on top of a plastic bucket (*Figure 9-5*). Then *slowly* dump the contents of the jar into the basket.

I say *slowly,* because the propellant is as fine as baby powder. If you dump it too quickly, it will billow up and spill on the ground. When you do this, a small amount of propellant will remain stuck to the inside of the jar, but you can dust it loose with a paintbrush. *Lay the paintbrush on the ground for a minute to drain away its static charge.* Then brush the remaining propellant into the basket (*Figure 9-6*). By gently shaking the bucket and the basket, sift the propellant into the bucket. The pellets will remain in the basket (*Figure 9-7*), and from there you can put them back in the mill.

Lay a zip-lock plastic bag on the ground for a minute to drain away its static charge. Then transfer the finished propellant to the bag, and place it inside of a lightweight, metal box for storage (*Figure 9-8*). Please note that, though the ingredients and the milling times may differ, with the exception of the **KS** propellant, all the propellants in this book should be mixed in this way.

Propellant Burn Rate Control

To fine-tune a rocket motor, or adjust a propellant's burn rate, you need a **burn rate modifier**, and **baking soda** works *very* well. When you add baking soda to a propellant, it slows the propellant's burn rate by dissociating into a gas, and adiabatically cooling the reaction. Too much soda causes combustion instability (**page 246**), and the process of finding the upper limit is one of trial and error. To do so, pick a propellant formula, and build motors with increasing amounts of baking soda until they start to "chuff" (**page 246**). Then back off to the last amount that *didn't* cause chuffing, and use that as the upper limit.

To add the baking soda, weigh out the recommended amount of propellant, multiply that weight by the percentage of soda you've decided upon, and stir it in *before* you load the motor. You can mix it with the propellant during the milling process, but if you guess wrong, you'll have to throw out the whole batch. It is more practical to add it at the last minute. That way, you can customize each motor, and make fine adjustments as you go.

When building rocket motors, you'll be thinking in *percentages* of baking soda. To calculate the proper *weight* of baking soda, multiply the desired percentage by the motor's estimated propellant weight. This will make a little *more* propellant / soda mixture than you actually need, but having extra is better than running short.

EXAMPLE 9-1:

You're perfecting a KG1-71-F40. The estimated propellant weight is approximately 113 gms., and based on previous experiments, you've decided to add 10% of that weight in baking soda. The actual weight in soda should be:

0.10 x 113 gms. = 11.3 gms.

You should then weigh out 113 gms. of propellant, and add 11.3 gms. of baking soda.

The KG1-71-F40 was designed to peak at approx. 20 lbs. of thrust. If your test motor performs poorly, and peaks at a thrust of (for example) only 15 lbs., you would then build *another* motor, and reduce the baking soda to perhaps 8% or 9%.

EXAMPLE 9-2:

The motor in Example 9-1 performed poorly. To improve its performance, you'll build another motor with 8% baking soda. You'll start again with 113 gms. of propellant., but the desired amount of soda is now 8% of that weight. The actual weight in baking soda should be:

0.08 x 113 gms. = 9.04 gms. (rounded off to 9.0 gms.)

You should then weigh out 113 gms. of propellant, and add 9.0 gms. of baking soda. Assuming that you did everything correctly, this motor would perform noticeably better than the one in Example 9-1.

Important note: *When building a **new** motor design, make the **first** motor with the amount of soda recommended in the motor drawing's "**BAKING SODA NOTE**". Then decrease the amount one or two percent at a time, until the motor's performance is up to the specs. listed in the drawing.* If you seriously *underestimate* the correct amount of soda, the propellant will burn too fast, the chamber pressure will exceed the motor's design limits, and the motor will *explode*. Reducing the soda one, careful step at a time minimizes the chance of making such an error.

Adding the Binder Solvent

To soften a propellant's binder and make it sticky, you mix in a measured amount of solvent. The solvent can be **acetone** or **alcohol**, but whichever you chose, **the amount must be correct**. If you don't use enough, you compromise the propellant's strength and density, *and the propellant burns too fast*.

If you use too much solvent, the excess liquid traps air between the powder grains. This air compresses when the propellant is rammed, and shortly thereafter expands again, causing the propellant grain to crack and crumble apart. The amount of solvent needed to produce a solid propellant grain of the *proper* density is called **the solvent requirement**. It can vary from 1 cc. per 20 gms. of propellant to more than 3 cc, and the correct amount is determined by **the pellet test (page 240)**.

Very Important note. *Using the proper amount of solvent (i.e. the solvent requirement) is absolutely critical to a motor's performance.* When mixing a propellant for the first time, **you must perform the pellet test**. If you make a change in any of the ingredients, or even the *proportions* of those ingredients, **you must perform the pellet test**. If you shorten or lengthen the milling time, **you must perform the pellet test**. Any time you mix a *new* propellant, or alter a known propellant *in any way*, **you must perform the pellet test**. *To reemphasize*, the solvent requirement, as determined by **the pellet test**, *is absolutely critical* to a motor's performance, and skipping the pellet test is *not* an option.

Important note. *When building a motor, or performing the pellet test, the amount of solvent **must** be measured accurately with a disposable syringe.* See **page 48** for details on how to use a syringe. Under most circumstances, I recommend that you use acetone as the binder solvent. Alcohol works fine, but motors made with alcohol take longer to dry.

To calculate the amount of solvent needed for a particular motor, add the weight of the propellant (in gms.) to the amount of baking soda you've added (in gms.). The total of the two will be the actual weight of the material in the mixing cup. Then divide this number by 20, and multiply the resultant figure by **the solvent requirement**, as determined by **the pellet test**. To put it arithmetically,

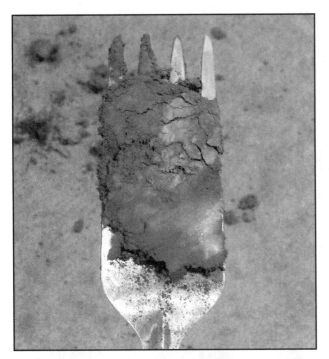 

Figure 9-9. The Fork Test. If the solvent hasn't been evenly distributed, the material stuck to the fork looks lumpy and blotchy with large and noticeable dark and light patches.

Figure 9-10. The Fork Test. When the material stuck to the fork takes on a reasonably uniform gray or black appearance, the distribution of the solvent is complete.

The amount of solvent in cc. = (gms. of material in mixing cup / 20) x the solvent requirement in cc.

EXAMPLE 9-3:

You're building the KG1-71-F40 in Example 9-2. The total amount of material in the mixing cup is (113 gms. propellant + 9 gms. baking soda) 122 gms., and we'll pretend that the solvent requirement (as determined by the pellet test) is 1.5 cc. per 20 gms. of propellant.

(122/20) x 1.5 cc. = 6.1 x 1.5 cc. = 9.15 cc. (rounded off to 9.2 cc.)

You should therefore add 9.2 cc. of solvent to the 122 gms. of material in the mixing cup.

EXAMPLE 9-4:

You're building a KG3-116-G102, and the test motor took 155.4 gms. of propellant (quoted from the motor drawing). To be sure that you don't run short, you're starting with 170.0 gms. We'll pretend that you've added 8.5 gms. of baking soda (5%) for a total of 178.5 gms. of material in the mixing cup. And we'll also pretend that the solvent requirement (as determined by the pellet test) is 1.8 cc. per 20 gms. of KG3 propellant.

(178.5/20) x 1.8 cc. = 8.925 x 1.8 cc. = 16.065 cc. (rounded off to 16.1 cc.)

You should therefore add 16.1 cc. of solvent to the 178.5 gms. of material in the mixing cup.

Important note. *The solvent must be mixed in thoroughly.* To insure that it is, I developed a quick procedure that I call **the fork test**. To perform this test, weigh out the desired amount of propellant, place it in a polyethylene mixing cup, and add the required amount of baking soda. Then add the desired amount of solvent, and mix it in with a table fork (the kind you eat with).

The proper motion when doing this is to alternately stir and knead the powder against the side of the cup with the back of the fork while rotating the cup a fraction of a turn after every 1 to 2 strokes. After two dozen strokes, draw the fork away from the cup with a sliding motion. As you do so, some of the propellant will stick to the back of the fork, creating a smooth

*Figure 9-11. The things you will need to perform **The Pellet Test**.*

inspection surface. If the solvent is *not* thoroughly distributed, the surface will look blotchy, and the color will be uneven (*Figure 9-9*). Keep mixing and examining the back of the fork (typically 30 to 60 seconds). When the inspection surface takes on a reasonably uniform black or gray appearance (*Figure 9-10*), the job is done.

● The Pellet Test

To determine the correct amount of solvent for each propellant, I developed a procedure that I call **the pellet test**. To perform the test, you'll need a nylon tamp and an improvised pellet mold. To prepare for the pellet test, proceed as follows.

MATERIALS LIST

1. One 3/4" x 5" long brass pipe nipple with as smooth an inside finish as possible
2. One 3/4" galvanized or plain cast iron floor flange
3. One piece of round stainless steel, 1" long, and machined to slide smoothly through the brass nipple
4. A nylon tamp, 8" long with flat ends, and machined to slide smoothly through the brass nipple
5. A medium-sized hammer (A 20 oz. carpenter's hammer works fine)
6. An anvil, a heavy block of metal, or a concrete floor to pound on
7. A 3 cc. to 6 cc. disposable syringe *with a dull-sanded needle* (any size in this range will work)
8. A small polyethylene mixing cup and a small polyethylene funnel
9. A table fork (the kind you eat with)
10. Acetone or denatured alcohol (I prefer acetone)
11. 20.0 grams of the propellant to be tested (weigh it out, and place it in the mixing cup)

Figure 9-11 shows most of these items laid out and ready to use. You'll find the brass pipe nipple and the galvanized floor flange at a hardware store. When buying the flange, try to pick one that isn't warped. When shopping for the brass nipple, examine each one carefully, and try to pick one that's free of any burrs or roughness on the inside. If you can't find one in this condition, you can smooth and deburr the inside with an improvised hone. To make an improvised hone, proceed as follows.

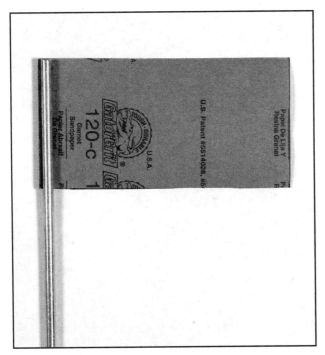

Figure 9-12. The sandpaper is glued to a 1/4" dia. steel rod with Krazy Glue. Glue it as shown, like a flag pointing right.

Figure 9-13. The sandpaper and steel wool are coiled around the rod like a jelly roll.

Figure 9-14. With the drill running, the hone is worked in and out of the brass nipple, smoothing its inside surface.

MATERIALS LIST

1. One piece of 1/4" dia. steel rod, 12" long
2. One piece of #120 sandpaper, 3" x 6"
3. One small piece of fine steel wool
4. Krazy Glue (or similar instant cyanoacrylate cement)

1. Cut the sandpaper to size, and glue the rod to the *back* of the sandpaper with the Krazy Glue (*Figure 9-12*). Glue it as shown like a flag pointing right, and place a weight on the rod to hold it firmly to the paper until the glue has hardened.

2. Spread the steel wool into a broad, flat pad, and lay it on the back of the sandpaper. Coil the sandpaper *and* the steel wood around the rod like a jelly roll (*Figure 9-13*), and be sure to coil it in the direction shown.

*Figure 9-15. **The Pellet Test.** The pipe fitting-pellet mold is placed over the round piece of stainless steel.*

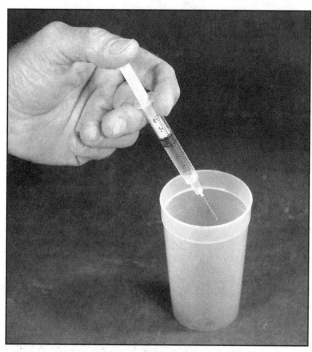

*Figure 9-16. **The Pellet Test.** 1.6 cc. of acetone are added to the propellant in the mixing cup.*

*Figure 9-17. **The Pellet Test.** The dampened propellant is poured into the pellet mold.*

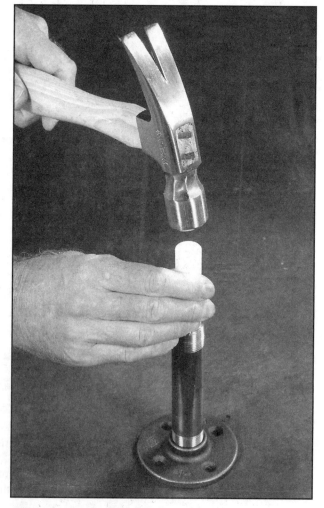

*Figure 9-18. **The Pellet Test** Eight to ten solid hammer blows compress the propellant into a dense and hard pellet.*

To smooth the inside of the brass nipple, place the nipple in a vise with a cloth rag to protect it from the vise jaws. Mount the hone in an electric hand drill, and with the drill running, work the hone in and out of the pipe (*Figure 9-14*) until the inside surface takes on a smooth, satin finish. For an even smoother job, repeat the honing with #220 sandpaper.

● **Assembling the Pellet Mold**

1. Screw the galvanized floor flange *tightly* onto one end of the brass nipple, and place the assembly on a flat surface. If the flange doesn't sit level, mount the nipple in a metal lathe, and true up the face of the flange with a face cut.

2. Place the short piece of stainless steel on an anvil, a heavy block of steel, or a concrete floor, and place the pipe nipple-flange assembly over the piece of steel (*Figure 9-15*).

● **Performing the Pellet Test**

1. I'll use the **NG6** propellant for this example. Weigh out **20.0 gms.** of **NG6** propellant, and place it in a polyethylene mixing cup. Fill a 3 cc. syringe with **1.6 cc.** of acetone, and squirt it into the cup on top of the propellant (*Figure 9-16*). *Without delay*, mix it in with a table fork, and use **the fork test** to determine when the acetone is thoroughly distributed. The entire process shouldn't take more than 30 seconds. You can buy the disposable syringe at a veterinary supply shop, but before you use it, be sure to dull the tip of the needle (**pages 48 & 49**).

2. *Without delay (**don't let the solvent evaporate!**)*, pour the dampened propellant into the brass pipe pellet mold (*Figure 9-17*). You can poke it through the funnel with a bamboo Bar-B-Que skewer. Insert the nylon tamp, and starting with a few light taps to squeeze out any excess air, hammer down on the tamp with 8 to 10 strong, solid blows (*Figure 9-18*). The material will at first feel soft under the hammer; then harden up as it reaches a state of full compaction.

3. On a drill press table, place a pair of wooden or metal blocks with a folded rag between them. *Without removing the tamp*, lift the pellet mold off the work surface (the short piece of steel will probably fall out), and place it on the blocks. Using the nose of the drill chuck as a press arbor, use the drill's quill feed lever to push down on the tamp, and force the pellet out of the mold and onto the rag (*Figure 9-19*). Please note that you may have to manually remove the pellet from the end of the tamp by twisting it gently with your fingers.

4. Place the pellet on a sheet of paper, and label it "**1.6 cc.**"

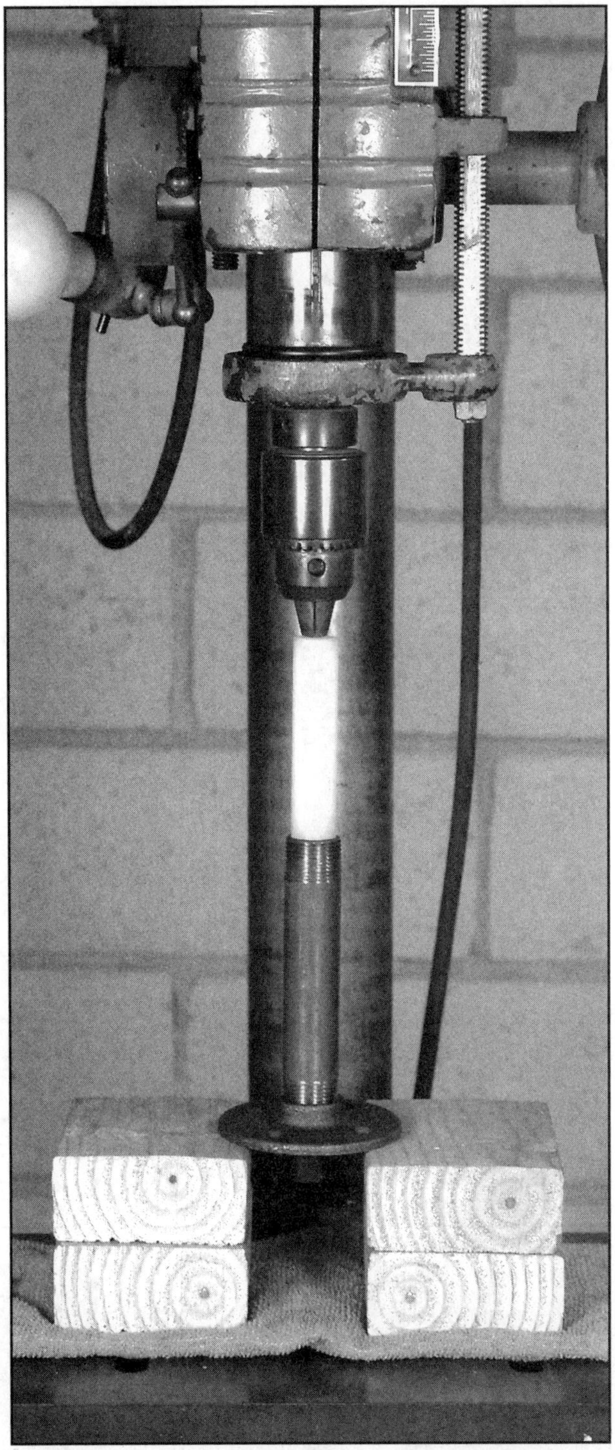

Figure 9-19. The Pellet Test. The drill press's quill-feed lever pushes the finished pellet out onto the rag.

5. Repeat **steps 1** through **4** five more times, and increase the amount of acetone each time by **0.2 cc**. When you're finished, you'll have a row of six test pellets made with acetone in the amounts of **1.6, 1.8, 2.0, 2.2, 2.4,** and **2.6** cc's (*Figure 9-20*).

● **Evaluating the Pellet Test**

To learn which amount of solvent is correct, carefully examine the pellets in *Figures 9-21* through *9-26*. In *this* example you'll notice the smooth finish on the pellets made with **1.6, 1.8,** and **2.0 cc.** of acetone. On the pellet made with **2.2 cc.** you'll see a single crack. On the **2.4** and **2.6 cc.** pellets you'll see a *noticeable* amount of cracking. The ring-shaped cracks were formed when compressed air, trapped between the powder grains, expanded as the pellets came out of the mold. The air was trapped by an *excess* amount of acetone.

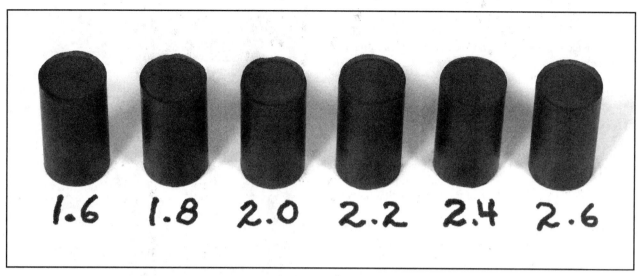

Figure 9-20. Six finished test pellets made with (from left to right) **1.6**, **1.8**, **2.0**, **2.2**, **2.4**, and **2.6 cc.** of acetone.

Figure 9-21. A pellet made with **1.6 cc.** of acetone. Note the smooth finish.

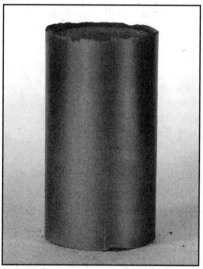

Figure 9-22. A pellet made with **1.8 cc.** of acetone. Note the smooth finish.

Figure 9-23. A pellet made with **2.0 cc.** of acetone. Note the smooth finish.

Figure 9-24. A pellet made with **2.2 cc.** of acetone. Note the single crack just below the center.

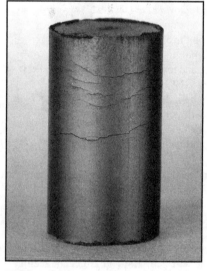

Figure 9-25. A pellet made with **2.4 cc.** of acetone. Note the multiple cracks.

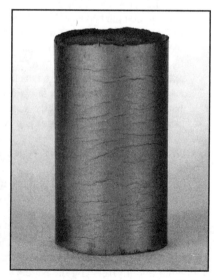

Figure 9-26. A pellet made with **2.6 cc.** of acetone. Note the **extensive** cracking.

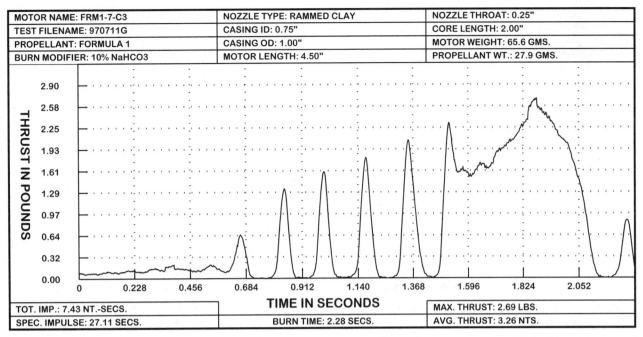

Figure 9-27. The thrust-time curve of a **Formula 1** core burner. It chuffs badly at a frequency of 5.9 cycles per second.

The *correct* amount of acetone is the *maximum* amount that *doesn't* trap air and cause cracking. In *this* example, **2.2, 2.4,** and **2.6 cc.** are *too much* acetone, and **2.0 cc. is correct**. Based on the results of *this* test, you would know that, when building rocket motors with the **NG6** propellant you've just tested, you should use **2.0 cc.** of acetone for every **20 gms.** of propellant.

The Logic Behind the Propellant Formulas

With the exception of the **KS** propellant, all the propellants in this book are modified versions of black powder. To show you the logic behind their makeup, I'm going to walk you through the *approximate* process that I used to develop the **KG1** propellant. My own work involved a large amount of trial and error. In my early tests, I *didn't* use baking soda, and I spent more time learning what *doesn't* work than what does. What I describe here are only the steps that lead to success, and to simplify the demonstration, I've *assumed* the addition of 10% baking soda to each motor's propellant. If you stay with the formulas in this book, you won't need this information. If you wish to experiment on your own, it may be of some help. Failing that, it will at least illustrate how small changes in a propellant formula can *greatly* effect a rocket motor's performance.

The standard formula for black powder is (**by weight**):

Potassium nitrate	**75%**
Carbon	**15%**
Sulfur	**10%**

This is a **stoichiometric** mixture (or close to it). The word, "stoichiometric", means that the proportions are such that, during a chemical reaction, *all* the ingredients are used up, leaving nothing behind. It burns furiously, but without a binder to hold it together, it can't be used in a rocket motor. When I started working with red gum, my first goal was to find out how much I needed to form the powder into a dense and solid propellant grain.

For the first experiment I used charcoal as the carbon fuel, and I substituted red gum for some of the charcoal in amounts varying from 14% down to 3%. I made a pellet mold like the one in *Figure 9-15*, and using alcohol as the solvent, I made a series of test pellets to see how the gum would effect their structural strength. I learned immediately that using gum in amounts greater than 9% made the powder so doughy that the alcohol wouldn't distribute properly. Further experiments taught me that reducing the amount below 4% compromised pellet strength. A batch made with 5% gum mixed easily and evenly, and produced a pellet that was dense and hard, and looked ideal. I therefore adopted a 5% gum content as my initial standard, and I changed the makeup of the classic black powder mixture to **75% potassium nitrate, 10% carbon, 5% red gum**, and **10% sulfur** (**Formula 1** on the **page 246**).

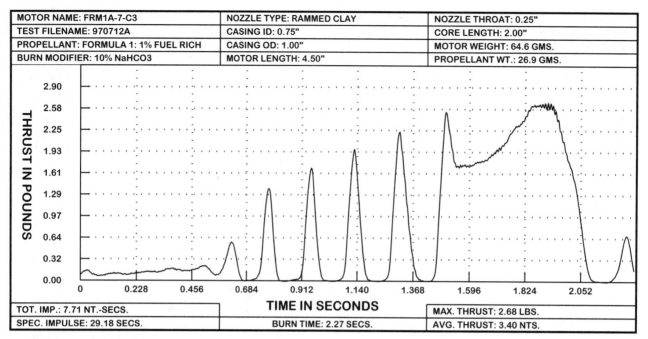

*Figure 9-28. The thrust-time curve of a **Formula 1** motor made **1% fuel-rich**.*

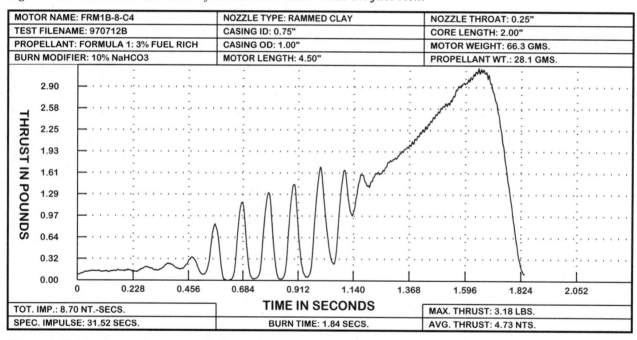

*Figure 9-29. The thrust-time curve of a **Formula 1** motor made **3% fuel-rich**.*

◆ FORMULA 1

4 hour potassium nitrate	75%
12 hour anthracite	10%
Red gum	5%
Powdered sulfur	10%

My next task was to test the mixture in a working rocket motor, but I knew ahead of time that combustion instability might be a problem. **Combustion instability** is the tendency of a propellant to burn roughly, or to ignite, self extinguish, and reignite in a repetitive and rhythmic pattern. This rhythmic burning is commonly called **chuffing**, and it makes the motor "chug" like a steam engine. Red gum is a complex organic material, and such materials tend to promote unstable combustion. Harder forms of carbon have the same effect, and because anthracite was the hardest of the carbon fuels available, I chose 12 hour anthracite (**page 206**) as the fuel for the experiments that followed. I reasoned that if I could make a stable-burning propellant with red gum and anthracite, then propellants made with red gum and softer forms of

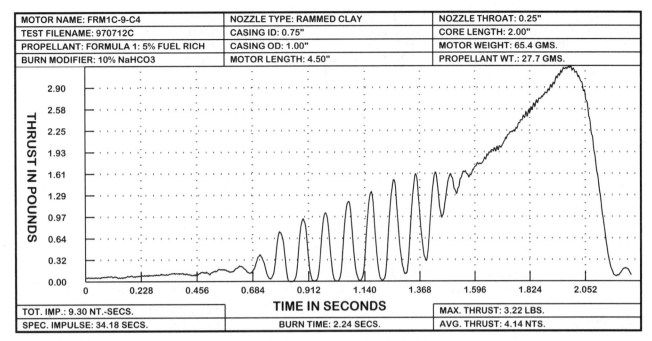

*Figure 9-30. The thrust-time curve of a **Formula 1** motor made 5% fuel-rich.*

carbon would work automatically. I substituted 12 hour anthracite for the 6 hour garden charcoal, named the propellant **Formula 1**, and used it to build and test a 3/4" i.d. core burner. ***Figure 9-27*** illustrates what happened. As you can see, the motor chuffed badly at 5.9 cycles per second.

It is well known in propellant engineering that you can often reduce combustion instability by increasing the amount of fuel beyond what is needed for a stoichiometric reaction. Propellants made in this way are called **fuel-rich**, and they have the added benefit that they protect the inside of the rocket motor during operation. Small amounts of left over oxidizer in the combustion gases can eat away at a motor's internal parts, and a fuel-rich propellant uses up the excess oxidizer. Therefore, making a propellant a little fuel-rich is almost always a good thing to do.

To see what the effect of a fuel-rich mixture would be, I tested three motors using *extra* anthracite in the amounts of 1%, 3%, and 5%. A positive outcome would be an *increase* in the chuffing frequency (*i.e. faster* chuffing) with a corresponding *decrease* in the chuffing amplitude (i.e. the *height* of the chuffs). ***Figures 9-28**, **9-29**,* and ***9-30*** show what happened. There was a barely perceptible improvement with 1%, a measurable improvement with 3%, and a slightly greater improvement with 5%. Above 5% the benefits dropped off, and more extensive testing lead me to adopt a 4% increase as the best. **Important concept**. *When you add something to a formula without taking an equal amount of something away, the percentages of all the ingredients change*, and you calculate the *new* percentages in the following way.

1. Instead of writing **Formula 1** in percentages, write it in gms. It then becomes:

4 hour potassium nitrate	75 gms.
12 hour anthracite	10 gms.
Red gum	5 gms.
Powdered sulfur	10 gms.
Total	100 gms.

2. When you add the extra 4% anthracite (4 gms. in this case), the formula changes to:

4 hour potassium nitrate	75 gms.
12 hour anthracite	14 gms.
Red gum	5 gms.
Powdered sulfur	10 gms.
Total	104 gms.

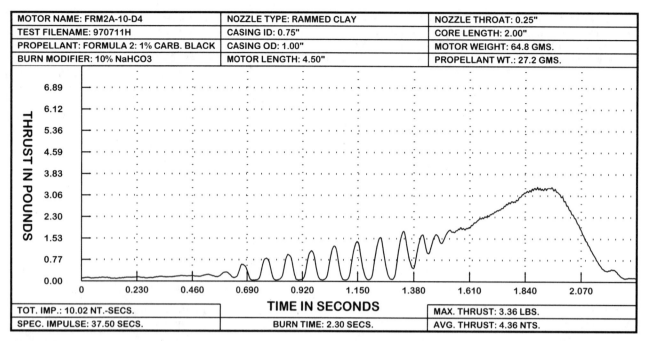

*Figure 9-31. The thrust-time curve of a motor made with a modified version of **Formula 2**, containing **12.3% anthracite** and **1% carbon black**.*

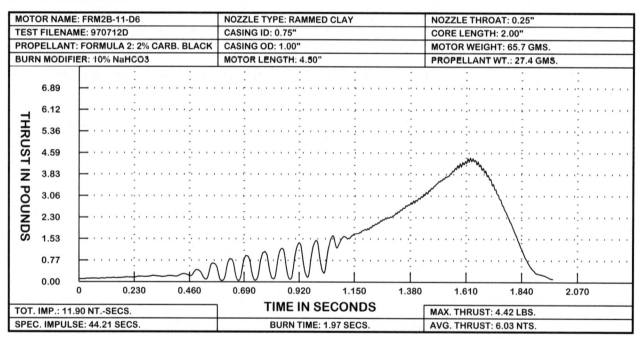

*Figure 9-32. The thrust-time curve of a motor made with a modified version of **Formula 2**, containing **11.3% anthracite** and **2% carbon black**.*

3. To calculate the *new* percentages, divide the amount of each ingredient (in gms.) into the *new* total of 104 gms. The arithmetic goes like this: **Potassium nitrate = 75/104 = 0.721, or 72.1%. 12 hour anthracite = 14/104 = 0.135, or 13.5%. Red gum = 5/104 = 0.048, or 4.8%. Powdered sulfur = 10/104 = 0.096, or 9.6%**. When converted into these percentages, the formula in **step 2** becomes:

4 hour potassium nitrate	72.1%
12 hour anthracite	13.5%
Red gum	4.8%
Powdered sulfur	9.6%
Total	100%

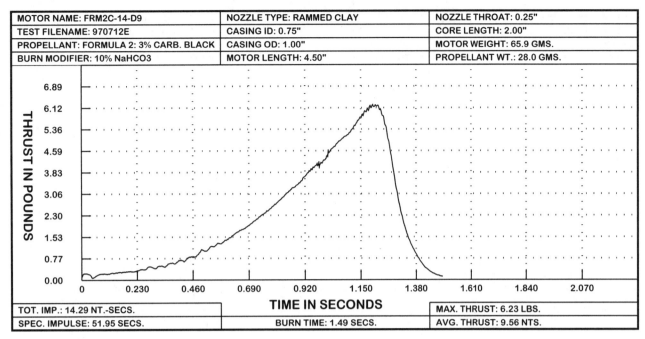

*Figure 9-33. The thrust-time curve of a motor made with a modified version of **Formula 2**, containing **10.3% anthracite** and **3% carbon black**.*

Now, if you look at the new percentages, you'll notice that, when the percentages were recalculated, the red gum content dropped to 4.8%. To get it back to the 5% required for good structural strength, I *added* 0.2% red gum, and balanced the mixture by *removing* 0.2% of the anthracite. The new formula became:

◆ **FORMULA 2**

4 hour potassium nitrate	72.1%
12 hour anthracite	13.3%
Red gum	5.0%
Powdered sulfur	9.6%
Total	100%

A motor made with **Formula 2** produced a thrust-time curve similar to the one generated by a 4% fuel-rich **Formula 1**, and I had to find a way to eliminate the rest of the combustion instability. Knowing that softer forms of carbon generate *less* instability than the harder forms, I substituted 4 hour garden charcoal for some of the anthracite. This eventually worked, but I had to use so *much* charcoal that the mixture could no longer be considered an anthracite-based formula.

Thinking that carbon black might work better, I replaced the anthracite with carbon black in the amounts of 1%, 2%, 3%, and 4%. *Figures 9-31*, *9-32*, *9-33*, and *9-34* show what happened. 1% carbon black produced a small improvement. 2% carbon black produced a dramatic improvement, and 3% carbon black reduced combustion instability to an insignificant ripple at the beginning of the motor's burn. At 4% carbon black, combustion remained stable, and overall performance improved as well. I rewrote **Formula 2** to include 4% carbon black, and the result was:

◆ **FORMULA 3**

4 hour potassium nitrate	72.1%
12 hour anthracite	9.3%
Carbon black	4.0%
Red gum	5.0%
Powdered sulfur	9.6%
Total	100%

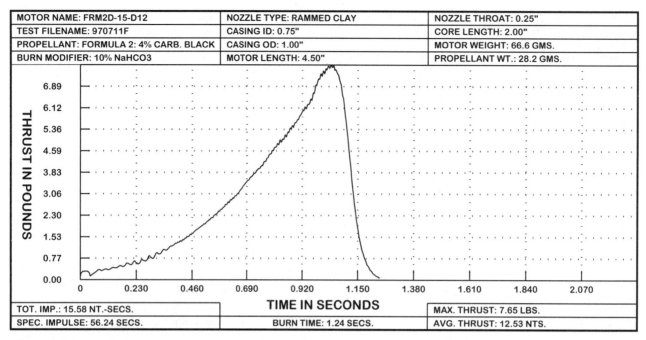

*Figure 9-34. The thrust-time curve of a motor made with a modified version of **Formula 2** containing 9.3% anthracite and 4% carbon black. I renamed this propellant **Formula 3**.*

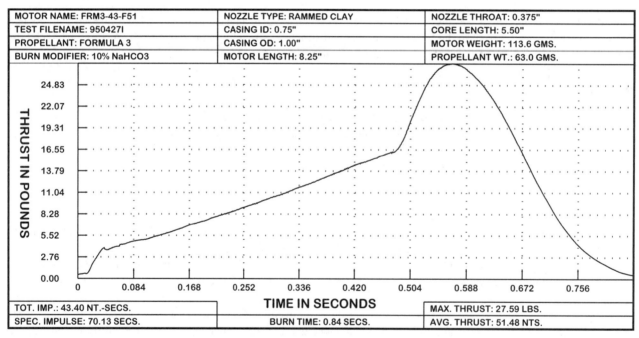

*Figure 9-35. Some of the motors made with the **Formula 3** proportions experienced a sudden increase in chamber pressure and thrust near the end of their burns.*

In further experiments with coal, charcoal, and carbon black-based propellants, I learned that I could get away with less carbon black and/or less extra fuel in *some* of the motor designs, but not all. If I returned to the amounts used in **Formula 3**, combustion instability vanished, and I therefore adopted the **Formula 3** proportions for *all* my propellants. During the next eighteen months, I built and tested hundreds of motors using these proportions, and though most performed well, a significant percentage produced a sudden increase in chamber pressure and thrust during the final moments of their burns. ***Figure 9-35*** illustrates the problem. As you can see, the motor works correctly for the first 0.5 seconds. Then something goes wrong, and the performance parameters take a sudden leap upward. In some of the motors the effect was large. In others it was small.

Through a process of elimination, and a series of conversations with a friend, a professional rocket engineer named Rick Loehr, I narrowed the cause down to a failure in the bond between the propellant grain and the casing wall. Because the propellant cores in these motors are slightly tapered, a point is reached during operation when the flame reaches the casing wall *before* the

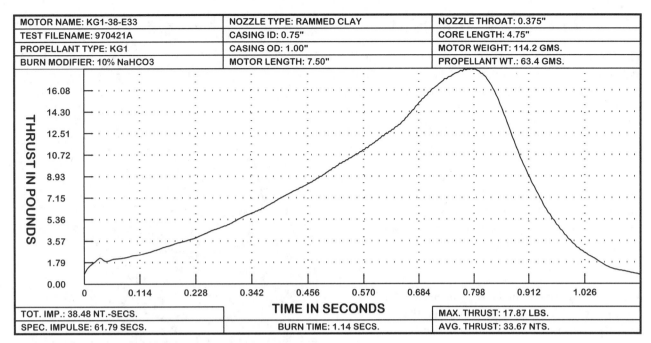

Figure 9-36. The KG1 propellant (with 7% red gum) produces a proper thrust-time curve.

propellant is completely consumed. At this point, if the bond between the propellant and the casing is weak, the flame works its way under the thin edge of the remaining propellant, and the amount of propellant burning at any moment suddenly increases. In my first attempt to fix the problem, I coated the insides of the casings with a 50-50 mixture of aliphatic glue and water. The acetone in the propellant resoftened the glue, and made it bond with the propellant as it dried. This helped a lot, but it didn't solve the problem completely. In frustration, I began poring over the thrust-time curves of all the motors I'd tested.

As I did so, I noticed that the motors made with sodium nitrate and carbon black were free of the problem. Motors made with charcoal were the worst, and motors made with anthracite fell somewhere in between. Because charcoal is *very* porous, anthracite is *slightly* porous, and carbon black particles are *not* porous, I reasoned that the answer lay in the amount of gum I was using, and the porosity of the carbon fuel. I guessed that the charcoal-based motors were the worst, because the charcoal was soaking up most of the gum, leaving nothing to bond with the casing wall. In propellants made with carbon black, the gum remained as a coating on the *outside* of the carbon particles. In my final and *successful* attempt to solve the problem, I *decreased* the anthracite by 2%, *increased* the red gum by 2%, and created:

◆ THE KG1 PROPELLANT

4 hour potassium nitrate	72.1%
12 hour anthracite	7.3%
Carbon black	4.0%
Red gum	7.0%
Powdered sulfur	9.6%
Total	100%

With the new proportions, I rebuilt and retested the motors in which the problem had been the greatest, and though most of the motors worked properly, a vestige of the problem remained in a few. In these motors I tried to increase the gum content to 8%, but this made the propellant unduly sticky and difficult to mix. As a final solution, I adopted the **KG1** proportions for *all* the black powder propellants, and began coating the insides of *all* the casings with glue. The result was the series of *successful* motor designs published in this book (*Figure 9-36*).

If you forget about the process, and look strictly at the formulas, all the propellants are easy to understand. You pick a carbon fuel, and start with the classic 75-15-10 black powder mixture. Replace about 1/2 of the carbon with red gum, goma laca shellac, or vinsol resin. Then enrich the whole mixture with an extra 4% carbon black, and you're there. The proportions in the sodium nitrate-based propellants are different, but that's *only* because the atomic weight of sodium is different than the atomic weight of potassium. Atom-for-atom, the amounts are the same.

The KS Propellant

The **KS** propellant is a simple mixture of powdered sugar and powdered potassium nitrate. It has no binder, and you don't need a powder mill to make it. You mix the ingredients by shaking them together in a plastic soft margarine tub, and this makes it ideal for settings where time is limited and attention spans are short.

With the homemade test equipment described in **Chapter 19**, a teacher can mix the propellant, build a rocket motor, fire the motor, and have the class analyze the data all in the span of an hour or two. A complete study of how changes in a propellant or a motor design effect a motor's performance can be finished in a week.

The main *disadvantage* of the **KS** propellant is that *it doesn't keep*. For the reasons discussed on **page 192**, a powdered propellant made with potassium nitrate and sugar continues to self-mix, even *after* it's been loaded into a rocket motor. As it does so, the propellant grain shrinks, and its burn rate increases, eventually exceeding the motor's design limits.

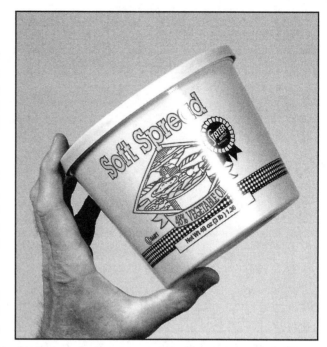

Figure 9-37. The KS propellant is mixed by manually shaking the ingredients for 2 minutes in a plastic soft margarine tub.

Also, potassium nitrate-sugar propellants are *very* hygroscopic. The word, "hygroscopic", means that they readily attract and hold moisture, and this compounds the problem. For these two reasons, **both the KS propellant, and all motors that contain the KS propellant, must be used within 48 hours of when they are made.**

Another disadvantage is that, because the sugar is sticky (even when it *appears* to by dry), removing the core spindle becomes increasingly difficult as motor size increases. My own experience has been that 1/4" dia. spindles come out easily. 3/8" dia. spindles require some muscle, and spindles larger than 1/2" dia. are impossible to remove without damaging the motor. Short of experimenting with nonstick coatings on the spindle, this limits you to a maximum motor size of **E** or maybe **F**. Finally, because there is no binder in the propellant, the propellant grains are fragile, and the motors must be handled carefully to avoid damage. To make 200 gms. of the **KS** propellant, proceed as follows.

You can use either **4 hour milled potassium nitrate** or **precipitated potassium nitrate** (**pages 200 - 205**). The formula for the **KS** propellant is **by weight**:

Powdered potassium nitrate	**60%**
Powdered sugar	**40%**

For a 200 gram batch, this translates into:

Powdered potassium nitrate	**120.0 gms.**
Powdered sugar	**80.0 gms.**

Weigh out 120.0 gms. of either 4 hour potassium nitrate or precipitated potassium nitrate, and 80.0 gms. of powdered sugar, and break up any lumps with a screwdriver handle. Place them together in a large, plastic, soft margarine tub. Snap on the lid, and shake the tub for **2 minutes** (*Figure 9-37*). **Important note.** As you work, shake it in varying directions to thoroughly mix the ingredients inside. When you're finished, wait a minute for the dust inside to settle. Then open the lid.

To use the propellant, weigh out the amount needed for the motor you are building. Place it in a mixing cup. Add the desired percentage of baking soda, and mix it in thoroughly with a table fork. Then build the motor. When you are finished, be sure to use up *all* the propellant, fire *all* the motors within 48 hours, and dispose of any extra propellant by dissolving it in water. Please note that I am *not* the inventor of the potassium nitrate-sugar propellant. Rocket amateurs have been using nitrate-sugar propellants in one form or another since the 1940s.

10. THE MOTOR BUILDING PROCESS

WARNING! Use the following motor-building method *ONLY* with the *homemade* propellants described in this book *made according to the instructions in this book*. *NEVER* add other chemicals to the propellants described in this book, and *NEVER* substitute other chemicals for the chemicals recommended in this book. *NEVER* use this building method with commercially-made black powder, commercially-made explosives and propellants, or *ANY* other chemical mixtures.

The use of this building method with ANYTHING but the homemade propellants made according to the instructions in this book could result in an explosion!

An Illustrated Outline

Most of the motors in this book have rammed clay nozzles, and the instructions for making these motors begin on **page 274**. A series of photos will show you what to do at each step along the way. But before you begin, read the following outline, and look over the drawings on **pages 254 - 273**. They provide an *inside* view of the building process that will help you to better understand how the various parts of a motor are formed. Shop drawings for the required tools begin on **page 88**, and the actual motor designs begin on **page 328**. Instructions for building motors with the more sophisticated De Laval nozzles begin on **page 319**.

1. Cut a dry, glue-coated (**page 232**) convolute motor casing to length, and smooth the ends with sandpaper.

2. Place the nozzle mold and the core spindle on a concrete floor or a heavy block of steel. Clamp the casing into the casing retainer, and place it over the nozzle mold-core spindle assembly (*Figure 10-1*).

3. Pour half of the nozzle clay into the casing (*Figure 10-2*). Insert the starting tamp, and ram the clay into place (*Figure 10-3*). Pour in the rest of the clay (*Figure 10-4*). Insert the propellant tamp, and ram the remainder of the clay into place (*Figure 10-5*).

4. Weigh out the amount of propellant needed for one motor. Mix in the desired amount of baking soda, and stir for 30 seconds.

5. Mix in the correct amount of solvent. This shouldn't take more than 60 seconds.

6. *Without delay*, put in the first dose of dampened propellant (*Figure 10-6*). Insert the propellant tamp, and ram the propellant into place (*Figure 10-7*).

7. Repeat **step 6** until the proper amount of propellant has been loaded (*Figures 10-8* through *10-13*), and use a flat-ended tamp to ram in the final dose (*Figure 10-13*). With the flat-ended tamp, ram in the amount of clay indicated for the forward bulkhead. Then lift the motor off the nozzle mold. Pull out the core spindle. Release the motor from the casing retainer, and lay it on a rack to dry.

Important note. *When ramming in propellant or nozzle clay, start with a few light taps to squeeze out any excess air. Then complete the job with 8 to 10 solid hammer blows.* A 20 oz. carpenter's hammer or a 20 oz. ball-peen hammer works well for the 1/2" and 3/4" i.d. motors. A 2 lb. "single jack" is about right for the 1" i.d. motors, and a 6 lb. sledge hammer on a short handle is best for the 1-1/8" and 1-1/2" i.d. motors.

When building motors with the **KS** propellant, because it uses no binder solvent, eliminate **step 5**. **Steps 1** through **7** produce a "plugged" motor with no time delay or ejection charge. To make a motor with a time delay and an ejection charge, finish loading the propellant, and proceed to **step 8**.

8. Select a *slow* burning propellant, and further retard its burn rate with baking soda. Calculate the amount needed to produce the desired delay time, and dampen it with the correct amount of solvent. With a flat-ended tamp, ram it into the front of the motor on top of the propellant grain (*Figures 10-14* and *10-15*).

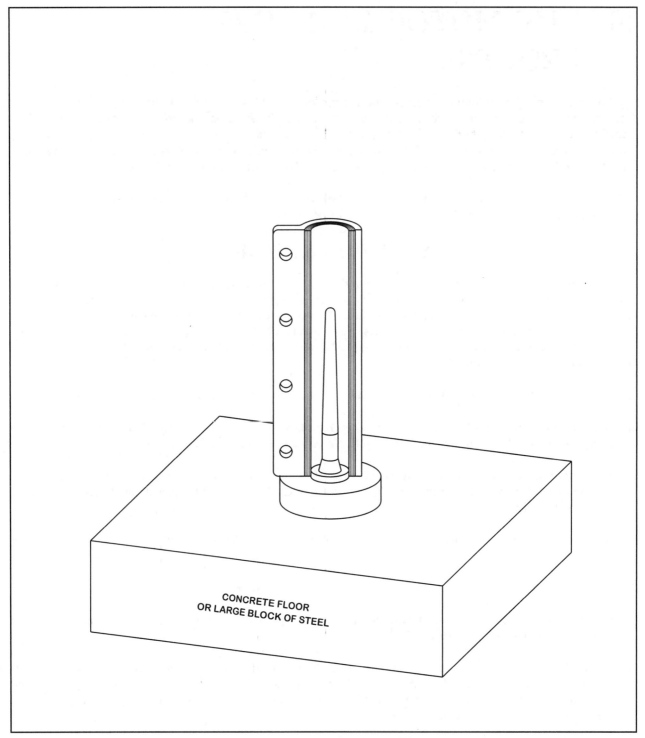

Figure 10-1. The motor casing (inside the casing retainer) is placed over the nozzle mold - core spindle assembly.

9. Place a punched cardboard disk on top of the time delay (*Figure 10-16*). Insert the porthole rod into the hole in the disk (*Figure 10-17*). Then with a porthole tamp, ram in the amount of clay indicated for the forward bulkhead (*Figures 10-18* and *10-19*).

10. Remove the porthole rod, the nozzle mold, and the core spindle. Release the motor from the casing retainer, and lay it on a rack to dry.

Steps 8 through **10** add a time delay and a porthole that can be used for second stage ignition, or for igniting a parachute ejection charge (*Figure 10-20*). *Figure 10-21* illustrates the two motor types just described.

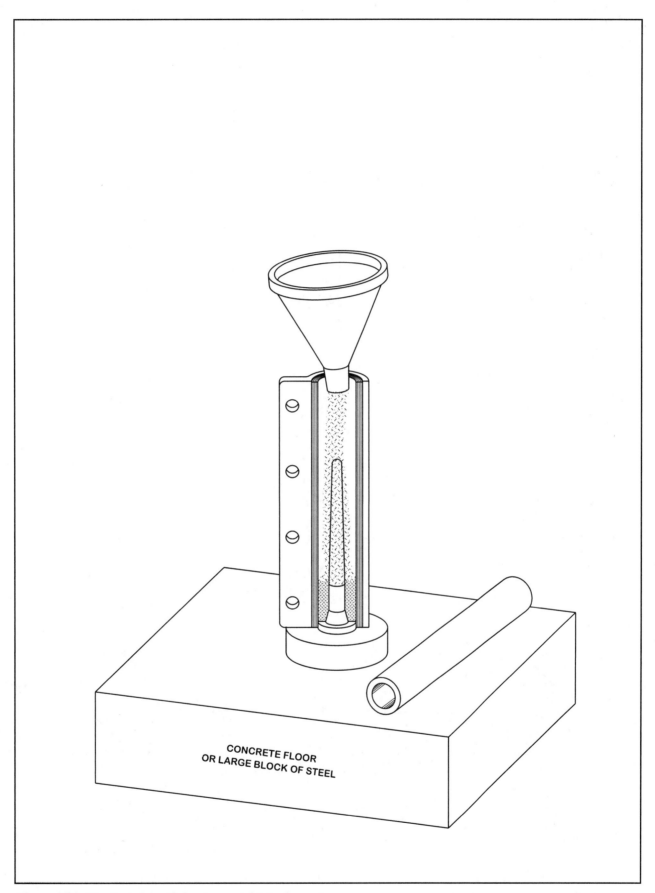

Figure 10-2. The first dose of clay is poured into the casing.

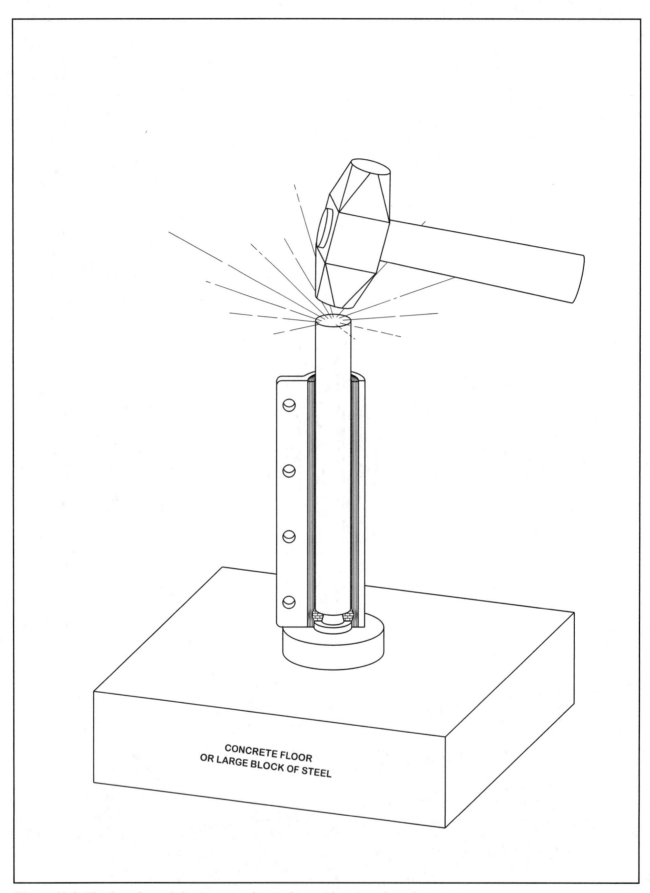

Figure 10-3. *The first dose of clay is rammed into place with a ring-shaped starting tamp.*

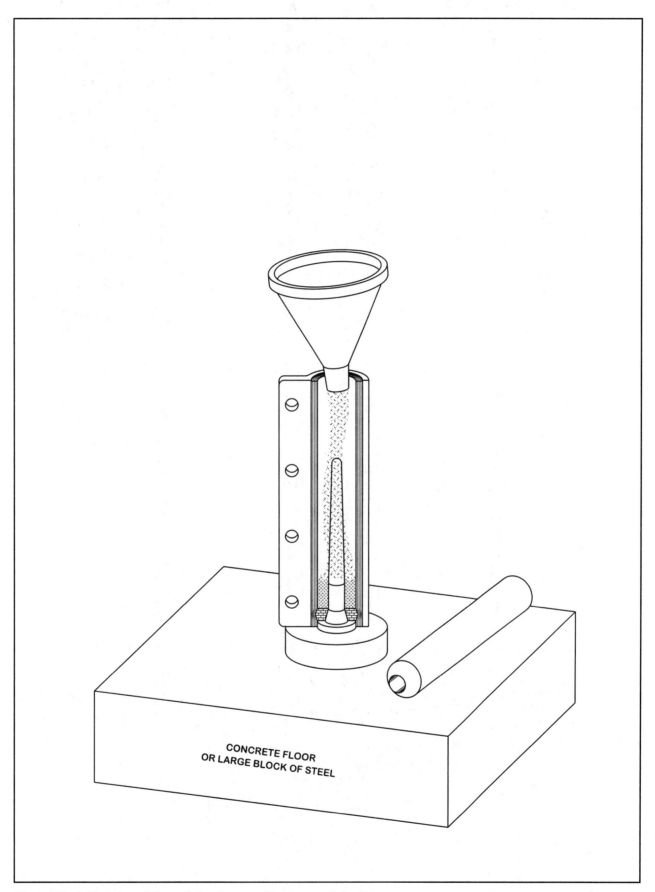

Figure 10-4. The second dose of clay is poured in on top of the first.

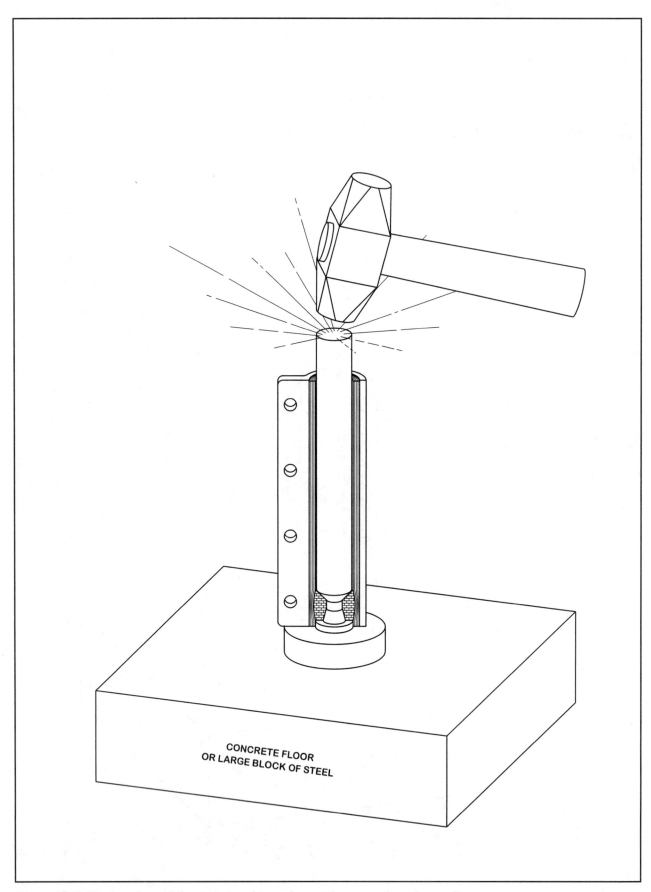

Figure 10-5. The last dose of clay is rammed into place with a cone-shaped propellant tamp.

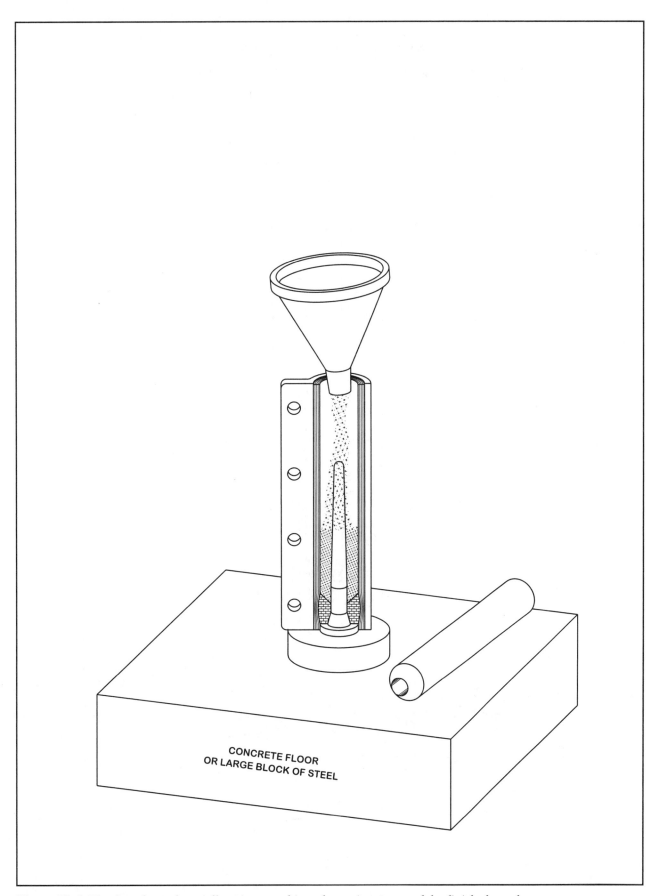

Figure 10-6. *The first dose of propellant is poured into the casing on top of the finished nozzle.*

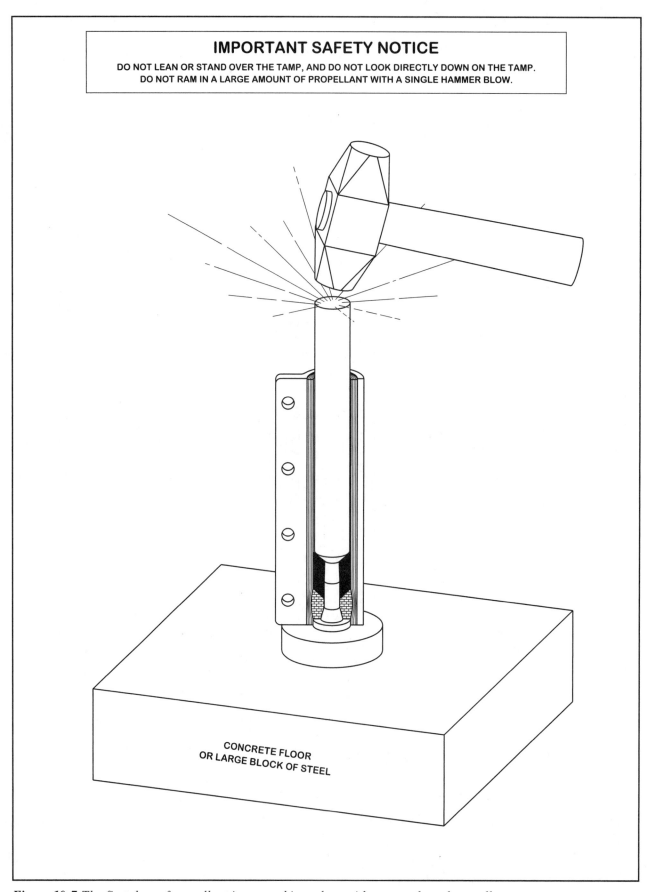

Figure 10-7. The first dose of propellant is rammed into place with a cone-shaped propellant tamp.

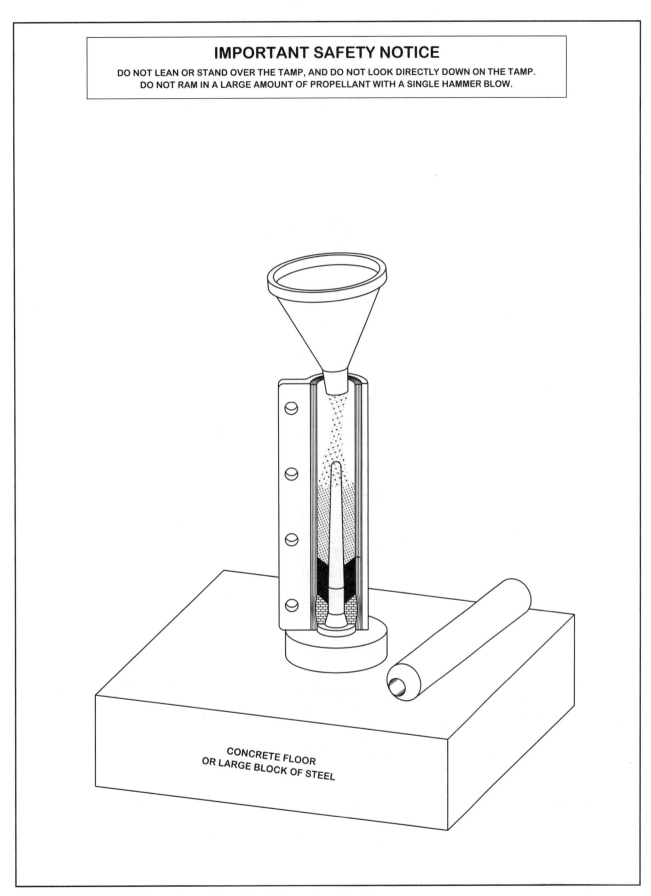

Figure 10-8. A second dose of propellant is added on top of the first.

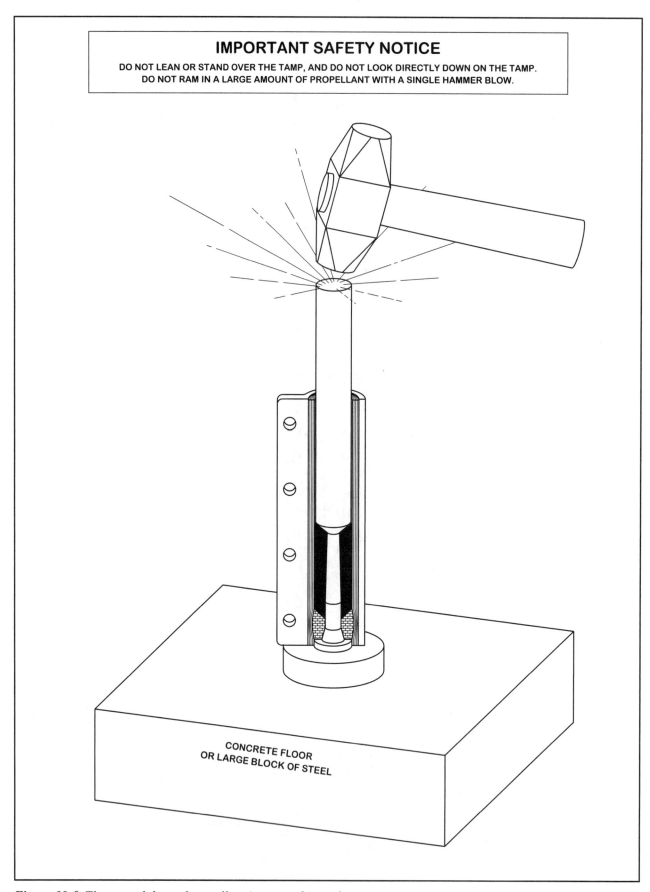

Figure 10-9. *The second dose of propellant is rammed into place.*

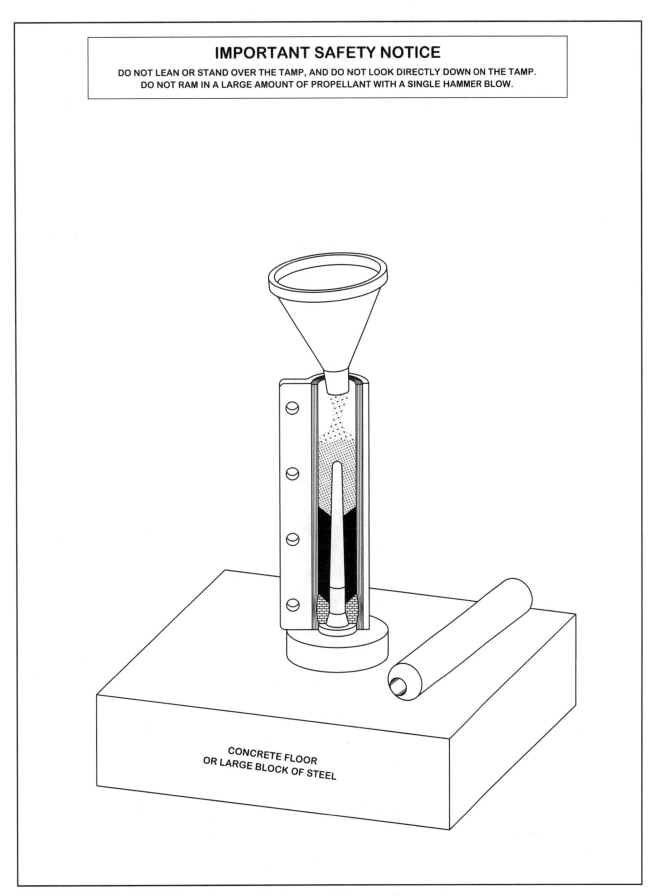

Figure 10-10. A third dose of propellant is added on top of the second.

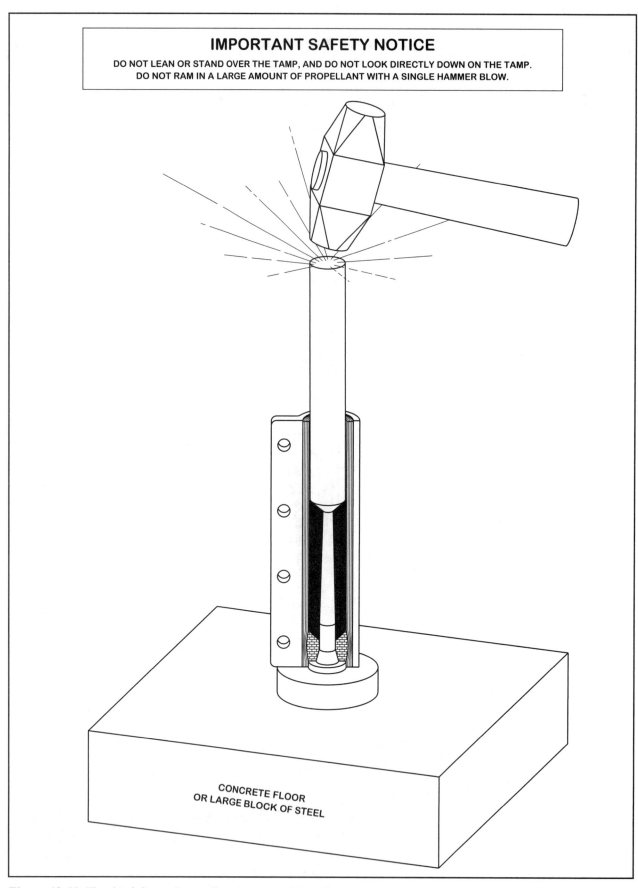

Figure 10-11. The third dose of propellant is rammed into place.

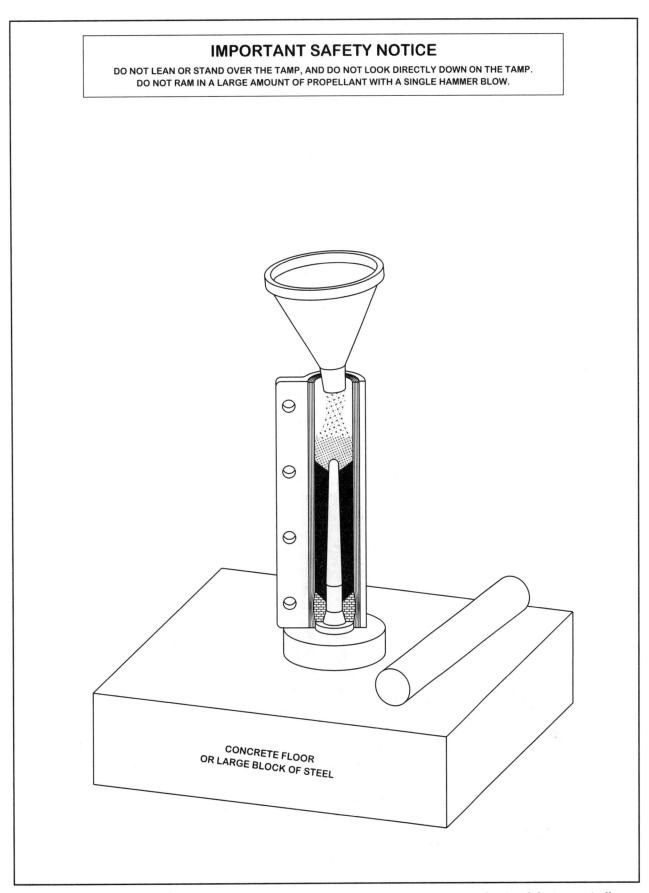

Figure 10-12. A fourth and final dose of propellant is poured into the casing. It covers the tip of the core spindle.

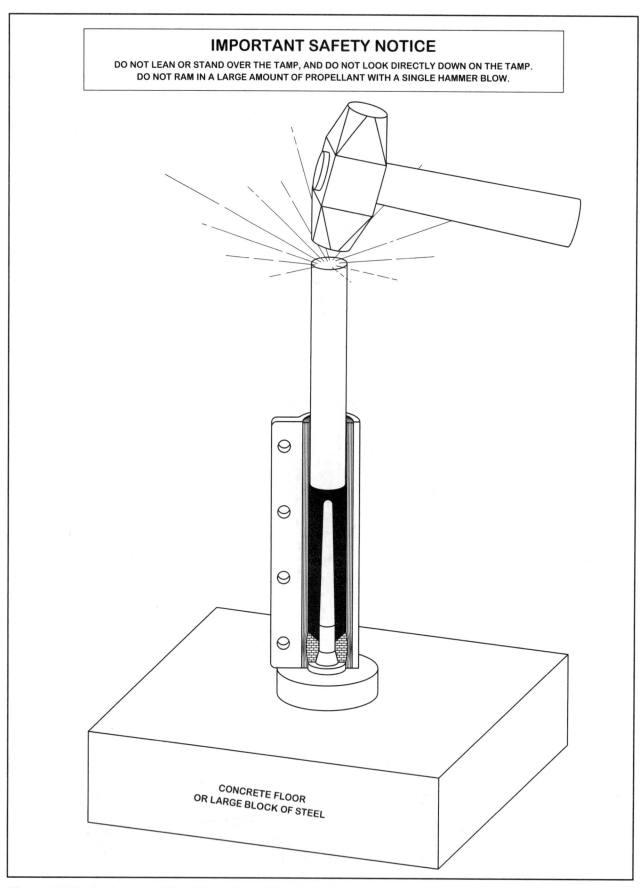

Figure 10-13. *The fourth and final dose of propellant is rammed into place with a flat-ended tamp.*

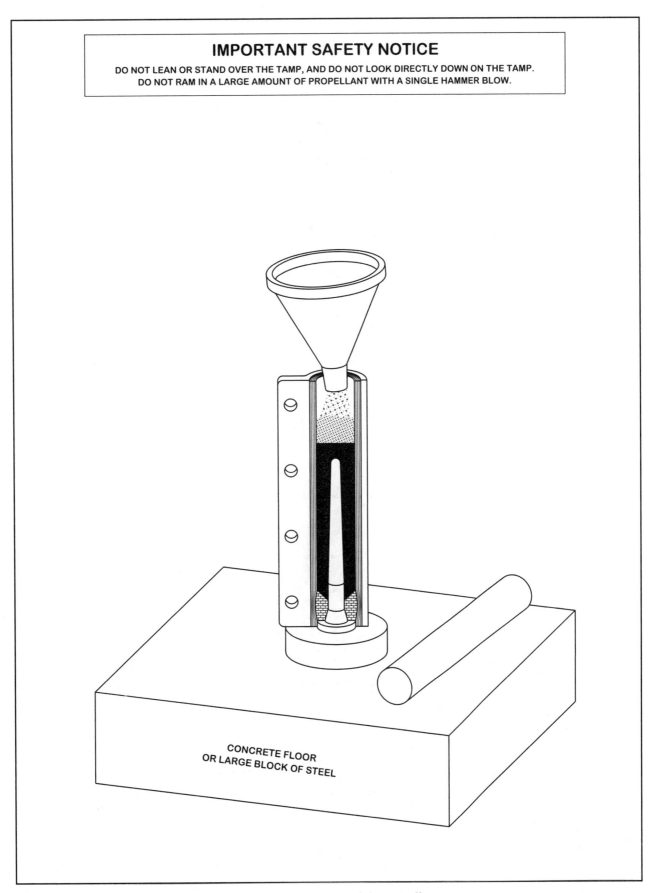

Figure 10-14. The time delay is poured into the casing on top of the propellant.

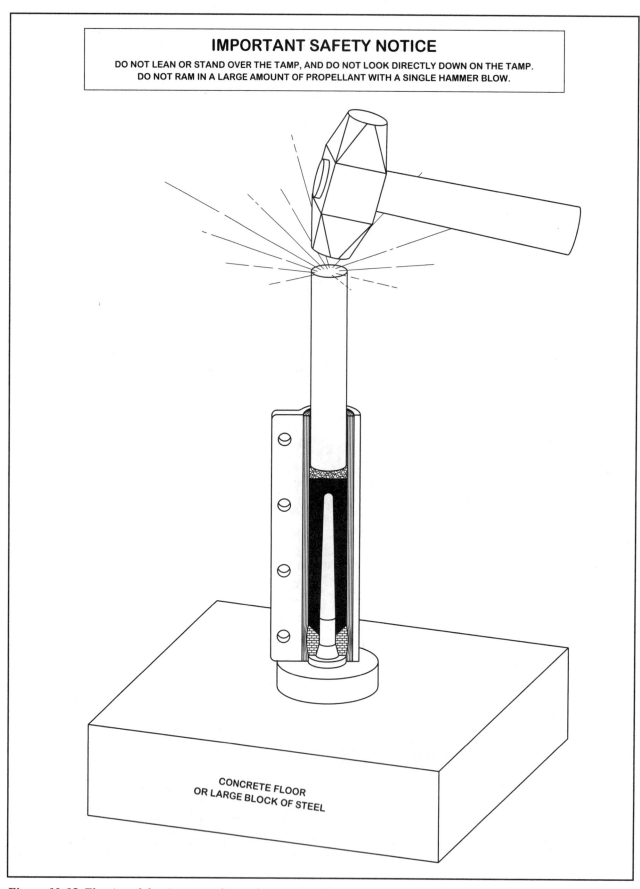

Figure 10-15. The time delay is rammed into place with a flat-ended tamp.

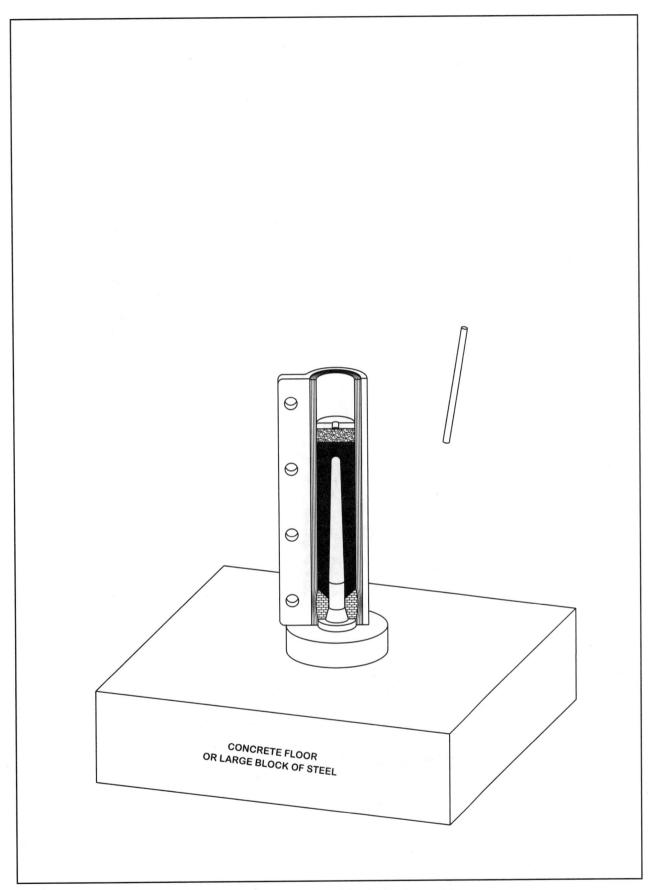

Figure 10-16. A punched cardboard disk is placed on top of the finished time delay.

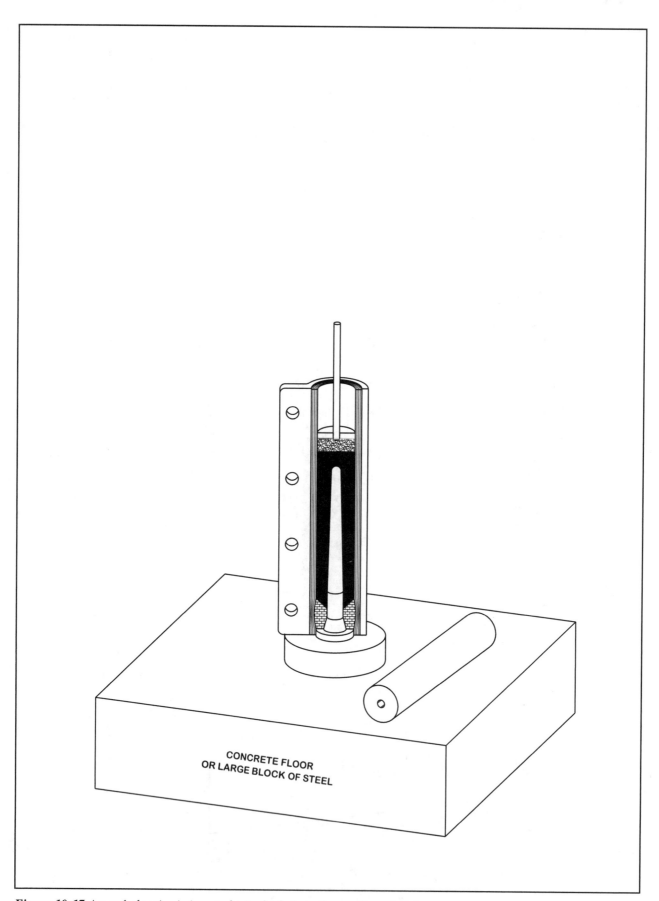

Figure 10-17. A porthole wire is inserted into the hole in the cardboard disk.

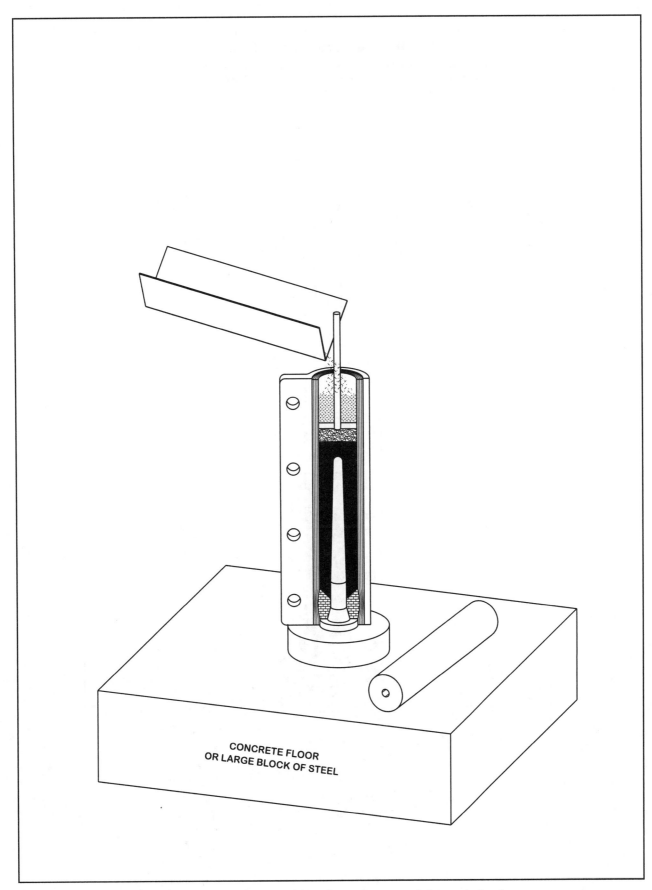

Figure 10-18. The forward bulkhead clay is poured into the casing around the porthole wire.

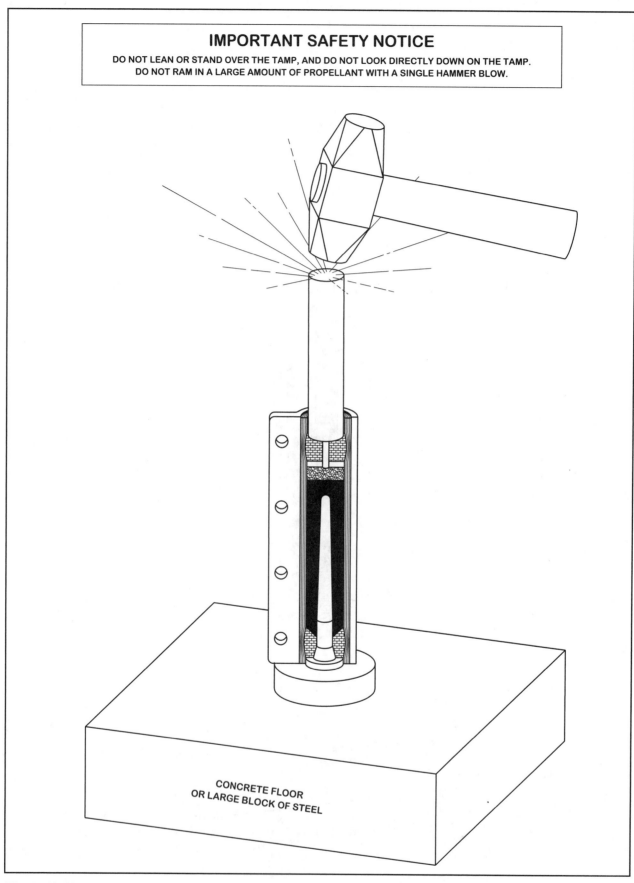

Figure 10-19. The forward bulkhead clay is rammed into place with a porthole tamp.

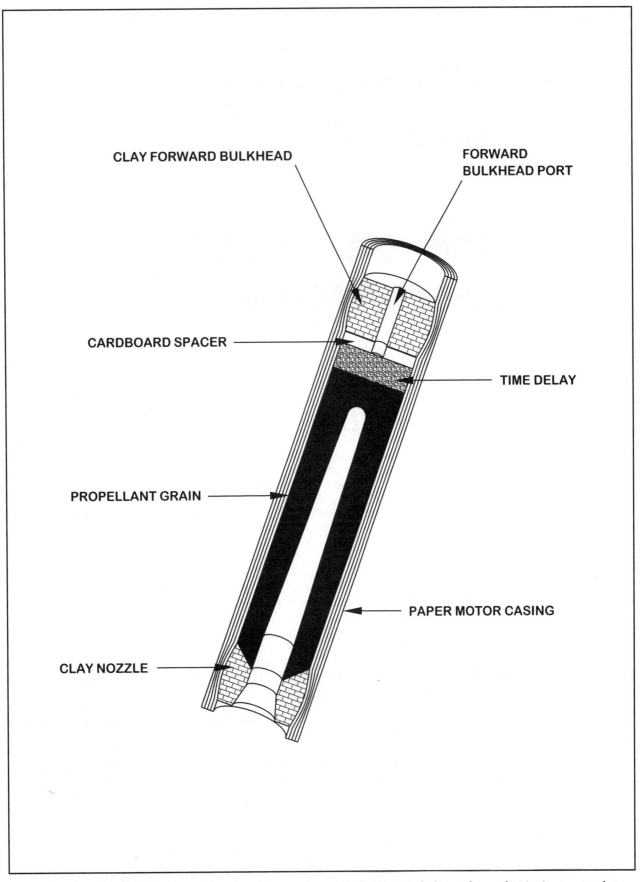

Figure 10-20. *The finished motor with a time delay and a porthole. The porthole can be used to ignite a second stage motor or a parachute ejection charge.*

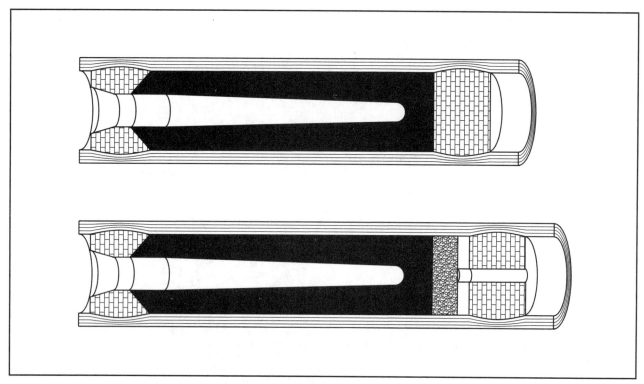

Figure 10-21. A plugged rocket motor (above), and a motor with a time delay and a porthole (below).

Your First Rocket Motor

When learning to make rocket motors, it is wise to start with a small one. In the following demonstration I'll show you how to make a **KG3-26-E31** (*Figure 10-22*) with an 8 second time delay, and for this example, I'll *assume* the addition of 10% baking soda to its propellant. Note that the actual test motor, made with *my* chemicals, required just 5% soda to make it work as expected. It might take you half an hour to complete your first motor, but with experience you'll gain speed, and you'll soon be able to make a motor like this in 10 minutes. The motor drawing and the list of required tools are on **page 356**. The motor's estimated performance specs. are on **page 357.**

● Getting Started

A **KG3-26-E31** with an 8 second time delay should be about 6-1/2" long, but you'll start with a 7" casing, and you'll build the motor in a 7" casing retainer. The 7" retainer is a standard piece of tooling, and when you are finished, you'll cut the *front* of the motor back to the proper length. I know that this wastes some casing material, but the trouble and cost of making a custom retainer for this particular motor are not worth the small savings that would result. This is something you'll have to consider when making your own tools.

1. Cut a dry, glue coated (**page 232**), 3/4" i.d. x 1" o.d., convolute motor casing to a length of 7" (*Figure 10-23*), and smooth the ends with sandpaper or a belt sander (*Figure 10-24*).

2. Cut a sheet of heavy construction paper 7" wide, and wrap it around the outside of the casing (*Figure 10-25*) until the wrapped casing fits tightly in the casing retainer when the clamping bolts are tight.

The inside of a casing retainer, when pinched shut, should always be a little bigger than the *outside* of the dry motor casing. This is because cardboard tubes not only vary in size, but because they expand and contract with changes in humidity. A casing made to fit a retainer perfectly in dry weather would be too large on a rainy day. The paper-wrap allows you to adjust for minor variations in the casing's diameter.

Cutting the first sheet of paper is a matter of trial and error. Begin with a shorter piece than you think you will need. Wrap it around the casing. Insert the wrapped casing into the casing retainer, and tighten it shut. If the casing is loose, try a longer sheet of paper and more wraps. Keep experimenting until you've achieved a tight fit. Then keep the piece of paper that works as a pattern, and cut several dozen like it. You can use each one 4 to 5 times before it wears out. Thereafter, the act of wrapping the paper around a casing will take only a few seconds. Construction paper works fine for the small motors. If one sheet isn't enough, add a second or third sheet as needed. For the big motors, try the black building paper sold at hardware stores, or the red rosin paper described on **page 224**.

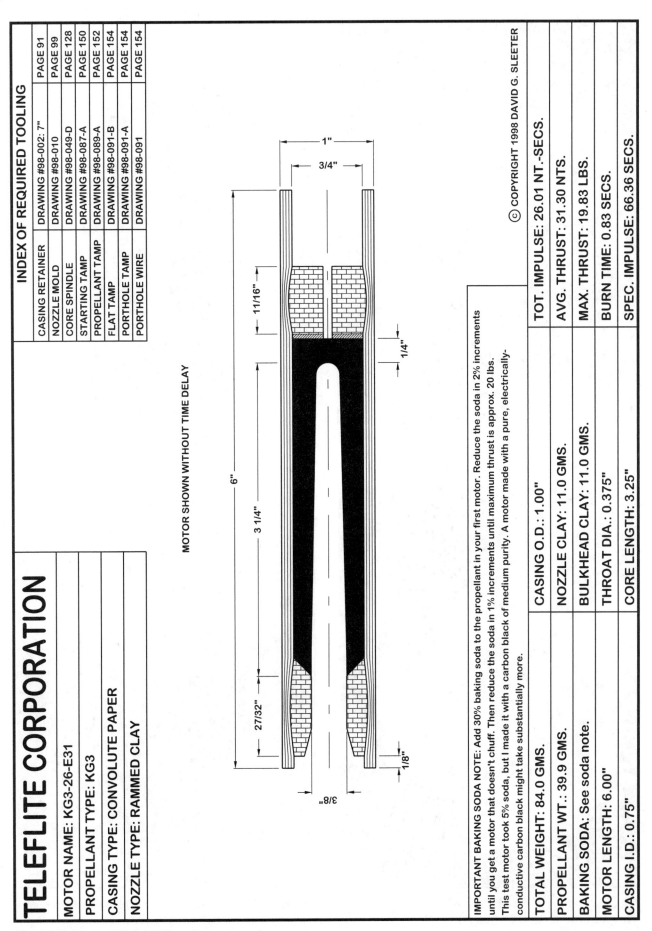

Figure 10-22. A KG3-26-E31 rocket motor.

Figure 10-23. A dry, glue-coated convolute motor casing is cut to length. You can use a handsaw, but a bandsaw with a wood-cutting blade works faster.

Figure 10-24. The ends of the casing are sanded smooth. You can do this by hand, but a belt sander is faster.

Figure 10-25. A sheet of thick construction paper is wrapped around the outside of the casing.

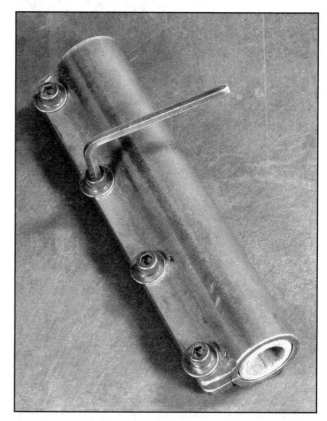

Figure 10-26. The casing retainer's clamping bolts tighten it snugly over the outside of the casing.

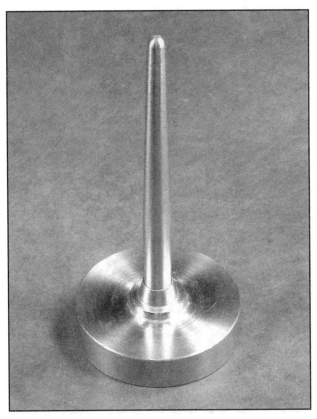

Figure 10-27. The nozzle mold and core spindle are placed on a concrete floor or a heavy block of steel (in this case an old cast iron surface plate).

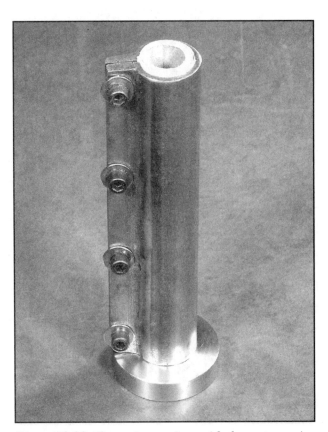

Figure 10-28. The casing retainer with the motor casing clamped inside is placed over the nozzle mold-core spindle assembly.

Figure 10-29. Half of the nozzle clay is poured into the top of the casing.

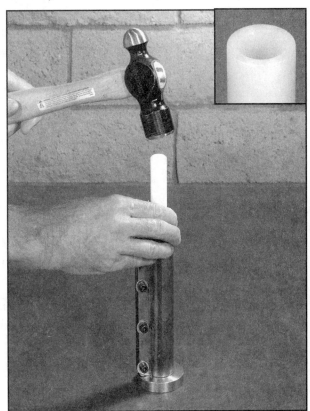

Figure 10-30. The first half of the clay is rammed into place with a starting tamp. The insert at the top right-hand corner shows the starting tamp's narrow, ring-shaped tip.

277

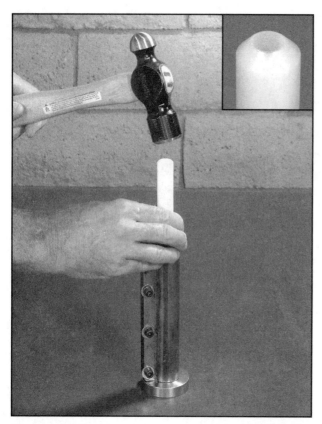 

Figure 10-31. *The remaining nozzle clay is rammed into place with a propellant tamp. The insert at the top right-hand corner shows the propellant tamp's cone-shaped tip.*

Figure 10-32. *An ultrafine point "Sharpie" marking pen marks the depth of the base of the core spindle on a bamboo Bar-B-Que skewer.*

3. Insert the paper-wrapped casing into the casing retainer, and tighten the clamping bolts (*Figure 10-26*). To minimize wear to the retainer's bolt threads and flanges, place washers under the bolt heads, and lubricate the bolts with bee's wax or "Door Ease". You can buy these products at a hardware store.

4. Place the nozzle mold and the core spindle on a concrete floor or a heavy block of steel (*Figure 10-27*). Then place the casing (inside the casing retainer) *over* the nozzle mold-core spindle assembly (*Figure 10-28*).

You'll notice in these photos that I am *not* working on a block of steel. A few years ago I bought an old 24" x 36", cast iron surface plate mounted on tall, cast-iron legs. Its surface was damaged and no longer accurate, so the owner let it go for $50. It is so massive that I use it as a combination anvil-and-workbench.

● Forming the Nozzle

1. Weigh out **11.0 gms.** of nozzle clay, and pour *half* of it into the casing with the aid of a funnel (*Figure 10-29*). Then with the starting tamp, ram the clay into place with 8 to 10 solid hammer blows (*Figure 10-30*).

A starting tamp has an extra large spindle hole, and a flat, ring-shaped nose. Its *only* purpose is to ram in the first dose of nozzle clay. It concentrates the force of the hammer into a narrow ring around the inside edge of the motor casing, and insures that the clay in this area is rammed in tight. If you *don't* use a starting tamp, the clay in this area will be *poorly* compacted, and the outer edge of the nozzle's exit taper could tear out during motor operation. In extreme cases, this can divert a motor's exhaust enough to throw a rocket off course.

2. Pour in the rest of the clay, and ram it into place with the propellant tamp and 8 to 10 solid hammer blows (*Figure 10-31*).

A propellant tamp has a 45 degree cone-shaped nose, and a spindle hole just big enough to fit smoothly over the fattest part of the core spindle. The cone forms the nozzle's convergent taper, and when ramming in the propellant, it expands the propellant outward, forcing it tightly against the casing wall.

 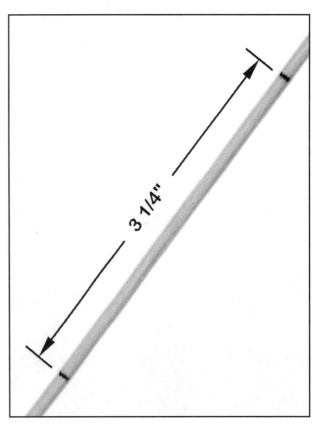

Figure 10-33. The same pen and skewer mark the depth of the core spindle's tip.

Figure 10-34. The distance between the two marks equals the height of the core spindle inside the motor casing, which equals the length of the motor's propellant core. For the **KG3-26-E31**, the distance should be 3-1/4".

Many rocket makers would say that the convergent taper should be 60 degrees, but this would require a tamp with a 60 degree cone on its tip. I worked with a 60 degree tamp for several days, but the leading edge around the spindle hole was so sharp and fragile that it quickly broke away. A flat-ended tamp would be the most durable, but the lack of *any* taper would compromise motor performance. The 45 degree angle is a practical compromise.

Now it's time to take a measurement. The **KG3-26-E31** has a 3-1/4"-long propellant core, and when *I* make this motor, its nozzle takes 11.0 gms. of clay. If I load *less* than 11.0 gms., the top of the clay nozzle throat ends up *lower* on the base of the spindle. The throat is *too short*, and the core is *too long*. If I load *more* than 11.0 gms., the end of the throat falls *higher* on the spindle. The throat is *too long*, and the core is *too short*.

Variations in the density of the clay and the squishiness of the cardboard casing might cause *your* motor to take a little more or less clay than what I use. If 11.0 gms. is correct, the distance from the base of the spindle to its tip will be *exactly* 3-1/4", and you can measure that distance in the following way.

Insert the pointed end of a bamboo Bar-B-Que skewer into the casing, and maneuver it around until the point touches the *bottom* of the clay cone right at the base of the core spindle. Then lean the top of the skewer to one side, and mark the point where it touches the front edge of the casing (*Figure 10-32*).

Now lift the skewer up, and place its point on the *tip* of the core spindle. Sight across the top edge of the casing, and mark the skewer again (*Figure 10-33*). Lift the skewer out, and measure the distance between the marks (*Figure 10-34*). If the distance is 3-1/4", then 11.0 gms. of clay is correct. If the distance is *greater* than 3-1/4", you'll have to ram in more clay, and accurately measure how much you've added.

To do so, weigh out 5.0 gms. of clay. Sprinkle a small portion into the casing, and ram it into place. Keep doing this, and keep taking measurements until the distance between the marks is 3-1/4". Then weigh the clay that you have left, and *subtract* that amount from the 5.0 gms. that you started with. If the result (for example) is 0.7 gms., then you've *added* 0.7 gms. of clay to the 11.0 gms. you originally loaded. The total will be 11.7 gms., and *forever thereafter,* you will know that, when building a **KG3-26-E31** rocket motor, you'll need *exactly* 11.7 gms. of nozzle clay.

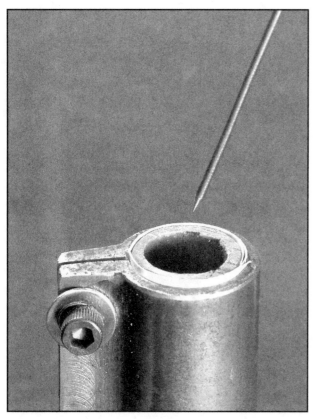

Figure 10-35. The sharpened end of a piece of music wire is used to chip away excess clay inside the motor casing.

Figure 10-36. The loosened clay is dumped into a clean container, then weighed. The weight of the clay removed is subtracted from the weight of the clay originally loaded.

Figure 10-37. The solvent is distributed by stirring and kneading the dampened propellant against the side of the mixing cup with a table fork.

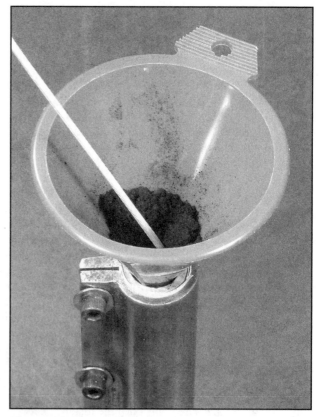

Figure 10-38. A bamboo Bar-B-Que skewer pokes the dampened propellant down into the motor casing.

If the distance between the marks is *less* than 3-1/4", you'll have to *remove* some clay. To do so, sharpen the end of a piece of music wire (aka "piano wire" or "model airplane landing gear wire") (*Figure 10-35*). Insert the pointed end into the casing. With a series of downward stabbing motions, work your way once around the nozzle clay near the casing wall, chipping away at the clay as you go. Pick up the entire casing retainer-nozzle mold assembly. Turn it over, and dump the loosened clay into a clean container (*Figure 10-36*).

Insert the propellant tamp, and redistribute the clay in the casing with 4 to 5 solid hammer blows. Make a *new* measurement, and keep removing clay in this manner until the distance between the marks is 3-1/4".

Then weigh the clay you removed. If, for example, you removed 0.4 gms. of clay, then *subtract* 0.4 gms. from the 11.0 gms. you originally loaded. The new total will be 10.6 gms., and *forever thereafter* you will know that, when building a **KG3-26-E31**, you'll need *exactly* 10.6 gms. of nozzle clay.

I *know* that this procedure is cumbersome and time consuming, but you only have to do it *once* for each new motor design. When you've established the correct amount of clay for a particular motor, you can use that amount for *all* the motors that follow.

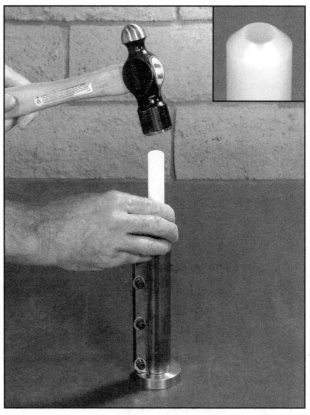

Figure 10-39. The first dose of propellant is rammed into place with a cone-shaped propellant tamp. The insert at the top right-hand corner shows the propellant tamp's cone-shaped tip.

● Loading the Propellant

Important note. *Before you can load the propellant, you have to dampen it with acetone*. **The dampened propellant dries quickly, so once you've started the loading process, you must finish it without delay**. For this demonstration, we'll pretend that you've already done **the pellet test (page 240)**, and learned that the solvent requirement for *your* **KG3** propellant is 1.8 cc. per 20 gms.

The actual test motor chosen for this example took 39.9 gms. of propellant, but the *next* motor might take more or less. Due to the squishiness of the cardboard casings, there will *always* be some variation. To insure that you don't fall short, you should dampen *more* propellant than you think you will need. In this example, I'd advise 45 gms.

1. Weigh out 45.0 gms. of **KG3** propellant, and place it in a polyethylene mixing cup. Add 4.5 gms. (10%) baking soda, and stir it with a fork for 30 seconds. The total amount of material in the mixing cup will now be 49.5 gms.

Important note. *When choosing a mixing cup, the propellant plus the baking soda should fill the cup 1/4 to 1/2 full*. If you fill the cup more than halfway, it makes the mixing process difficult. If you fill it less than 1/4 full, the acetone evaporates too fast.

2. Calculate the amount of acetone required as follows. There are 49.5 gms. of material in the mixing cup, and the solvent requirement is 1.8 cc. for every 20 gms.

Amount of acetone in cc. = (Weight of material in mixing cup / 20) x the solvent requirement in cc.

Amount of acetone in cc. = (49.5 / 20) x 1.8 = 2.475 x 1.8 = 4.455 rounded off to 4.5 cc.

Draw 4.5 cc. of acetone into a 6 cc. or 12 cc. disposable syringe. Squirt it into the mixing cup on top of the propellant, and mix it in with a table fork (the kind you eat with). The proper motion when doing this is to alternately stir, and then knead the powder against the side of the cup with the back of the fork (*Figure 10-37*), while rotating the cup a fraction of a turn after every 1 to 2 strokes. Use **the fork test (page 239)** to determine when the solvent is thoroughly distributed. The entire process shouldn't take more than 30 seconds.

***Figure 10-40**. A **teaspoon** and a **Tablespoon** shown actual size. These are the two spoons that I use for measuring rocket propellant during the propellant-loading process.*

Figure 10-41. When loading a *1/2" i.d.* rocket motor, this is the amount of propellant that I use for each dose. It's about **half a teaspoon** (shown actual size).

Figure 10-42. When loading a *3/4" i.d.* rocket motor, this is the amount of propellant that I use for each dose. It's about **one teaspoon** (shown actual size).

Figure 10-43. When loading a *1" i.d.* rocket motor, this is the amount of propellant that I use for each dose. I call it a **rounded teaspoon** (shown actual size).

Figure 10-44. When loading a *1-1/8" i.d.* rocket motor, this is the amount of propellant that I use for each dose. I call it a **rounded Tablespoon** (shown actual size).

Figure 10-45. When loading a *1-1/2" i.d.* rocket motor, this is the amount of propellant that I use for each dose. I call it a **heaping tablespoon** (shown actual size).

3. *Without delay*, put one teaspoon (***Figure 10-42***) of dampened propellant into the funnel, and poke it down into the casing with a bamboo Bar-B-Que skewer (***Figure 10-38***). Then ram it into place with the *cone-tipped* propellant tamp (***Figure 10-39***). Begin with a few light taps to squeeze out any excess air. Then finish the job with 8 to 10 solid hammer blows. **Important note.** *"Without delay" does **not** mean that you have to rush, but if the phone rings, **don't answer it**.*

Important note. *When hand-tamping a powdered rocket propellant, the state of compaction decreases with the distance from the face of the tamp. It is therefore important to load the propellant in a series of small doses. If you make the doses too large, each one will be properly compacted* only *near the face of the tamp. The material farther away will be* poorly compacted, *and the motor will perform erratically.*

Figure 10-46. The tip of a flat-ended tamp.

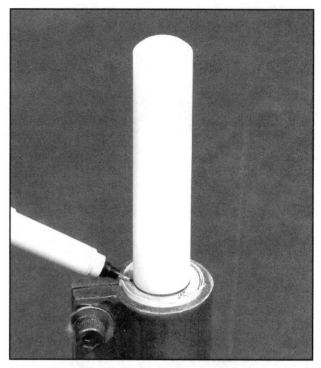

Figure 10-47. The flat-ended tamp sits on top of the core spindle. An ultrafine-point "Sharpie" marking pen scribes a ring where it enters the front of the casing.

The size of each dose does *not* have to be exact, and the following method of measurement is more than adequate. Like most people, I have 2 kinds of spoons in the kitchen drawer. *Figure 10-40* is an actual-size photo of these spoons. The small one is called a **teaspoon**, and I use for eating soup or stirring my coffee. The big one is called a **Tablespoon**, and I use it for all the things for which a teaspoon is too small. I also use these spoons to measure rocket propellant, and *Figures 10-41* through *10-45* show the size of the propellant doses that I use when building the rocket motors in this book. As in some of the other photos, I've adjusted the camera to show everything actual size. If you use these approximate amounts, propellant density and compaction will not be a problem.

4. *Without delay*, repeat **step 3** until the surface of the rammed propellant falls just below the tip of the core spindle.

Now it's time to take another measurement. Insert the *flat-ended* tamp (*Figure 10-46*) into the top of the motor so that it rests on the tip of the core spindle. With an ultrafine-point "Sharpie" marking pen, scribe a ring around the tamp where it enters the front of the casing (*Figure 10-47*). Per the instructions in **Example 2-4** on **page 29**, calculate the distance from the tip of the core spindle to the front end of the propellant grain, and make a *second* mark *exactly* this distance *below* the first (*Figure 10-48*). In this example, the distance should be **0.25"** or **1/4"**. Like the clay measurement, this **measurement-and-calculation will take a few minutes, but you only have to do it *once* for each new motor design.**

5. *Without delay*, add a small amount of the dampened propellant to the front of the motor, and ram it in with the flat-ended tamp. Then repeat the process until the *second* mark on the tamp lines up with the front edge of the casing (*Figure 10-49*).

6. **Important note. This final step is very important, so don't forget to do it.** If you look down into the top of the motor, you will *usually* see a thin film of propellant stuck to the inside of the casing wall. **Before you can ram in a time delay or the forward bulkhead, you have to remove this excess propellant.** To do so, scrape *downward* with the front edge of the flat-ended tamp. Work you way completely around the casing wall, and when you are finished, ram the scrapings into the propellant below with a few solid hammer blows. With the completion of this final task, the propellant-loading process is finished.

At this point, if you're building a **plugged** motor (one *without* a time delay or a porthole), use the *flat-ended* tamp to ram in an amount of clay *equal to the amount that you used for the nozzle*, and the motor will be finished. To add a time delay and a porthole, proceed as follows.

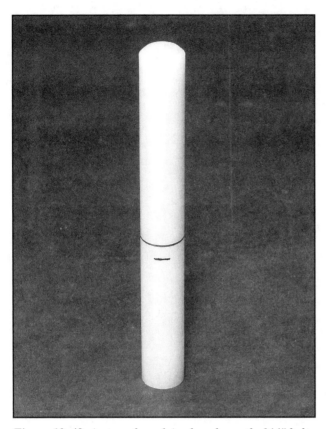

Figure 10-48. A second mark is placed exactly 1/4" below the ring-mark.

Figure 10-49. Small amounts of propellant are rammed into place until the second mark lines up with the front of the motor casing.

● Adding a Time Delay

You can use any of the propellants in this book as a time delay, but the slower-burning propellants work the best. I'll use the **KG1** for the time delay in this example. You want the delay to burn as slowly as possible, and you can slow a propellant's burn rate by adding baking soda. For the following demonstration, we'll pretend that you've already tested the **KG1,** and found that at atmospheric pressure, it burns stably with the addition of up to 15% baking soda. We'll pretend that you've done **the pellet test (page 240)**, and found that the solvent requirement is 1.5 cc. per 20 gms. of propellant. We'll also pretend that you've already performed **the time delay burn rate test (page 293)**, and found that the burn rate at atmospheric pressure (with 15% baking soda) is 0.063" per second.

1. Place 10.0 gms. of **KG1** propellant in a *small* mixing cup. Add 1.5 gms. (15%) baking soda, and stir it with a fork for 30 seconds. Then, per **Example 2-3** on **page 29**, calculate the length needed to produce an 8 second time delay. Since the length of the delay equals the propellant's burn rate at atmospheric pressure times the desired delay time:

> **Length of time delay = 0.063 x 8 = 0.504", or about 1/2".**

2. On the flat-ended tamp, make a *third* mark (*Figure 10-50*) 1/2" below the *second* mark that you made in **step 4** on **page 284**.

3. With a 3 cc. disposable syringe, add the proper amount of acetone to the mixture in the cup. In this case the cup contains (10.0 + 1.5) 11.5 gms. The solvent requirement for *your* **KG1** propellant is 1.5 cc. per 20 gms. of propellant, so:

> **(11.5/20) x 1.5 = 0.575 x 1.5 = 0.863 rounded off to 0.9 cc.**
> **You should therefore add 0.9 cc. of acetone to the mixture in the cup**

Now stir and knead the acetone into the mixture until it is evenly distributed as determined by **the fork test**.

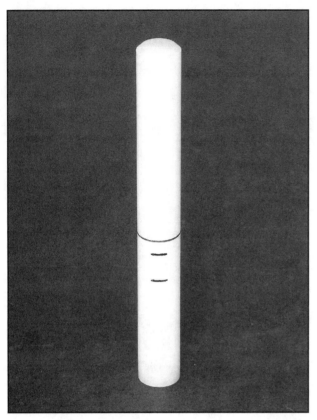

Figure 10-50. A third mark is placed exactly 1/2" below the second mark.

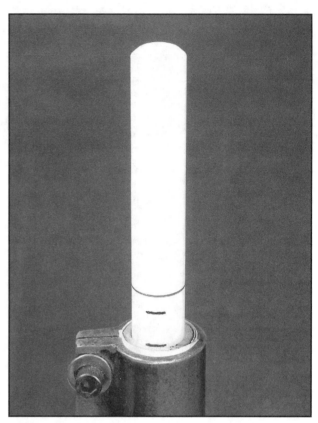

Figure 10-51. Small amounts of time delay mixture are rammed into the front of the motor until the third mark is level with the front of the motor casing.

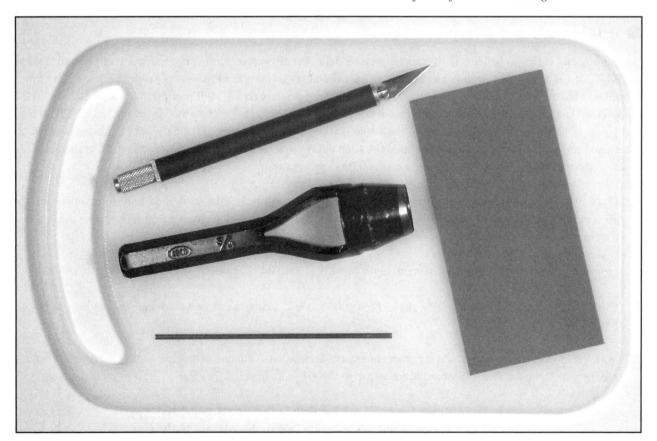

*Figure 10-52. An **X-acto knife** or an **arch punch**, a short length of **3/32" dia. music wire**, a scrap of picture framer's **mat board**, and a small plastic kitchen cutting board are the things you will need to form a porthole in the forward bulkhead.*

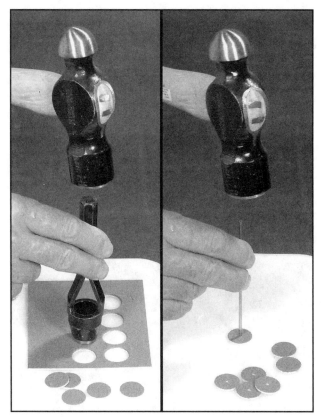

Figure 10-53. 3/4" dia. disks are cut from a piece of mat board, and 3/32" dia. holes are punched in their centers.

Figure 10-54. One of the punched disks is placed in the front of the motor casing on top of the time delay.

4. **Without delay**, with the *flat-ended* tamp, ram in small amounts of this time delay mixture until the mark that you made in **step 2** is level with the front of the motor casing (*Figure 10-51*). Look down into the top of the casing, and with the front edge of the *flat-ended* tamp, scrape off any time-delay material that you find sticking to the casing wall. Then ram the scrapings into the material below with a few solid hammer blows. **This final scraping and cleaning of the inside casing wall is very important, so don't forget to do it!** A thin layer of delay material on the inside of the casing wall can allow the flame to burn *around* the forward bulkhead, and prematurely ignite the motor's parachute ejection charge, or damage the inside of the rocket. With the completion of this final task, the time delay is finished.

● Forming a Porthole

To transfer the fire from the time-delay inside the motor to the parachute ejection charge *outside*, you need a forward bulkhead with a porthole, and here's how to make one.

MATERIALS LIST

1. A scrap of picture framer's mat board
2. A 4" length of 3/32" dia. music wire **with the ends ground flat**
3. A 3/4" arch punch or an X-acto knife

Mat board is the thick, dense cardboard used in picture framing. It's about 0.050" thick, comes in a variety of colors and textures, and you can buy it at an art supply store. You'll find the **music wire** (also called **piano wire** or **landing gear wire**) at hobby shops that sell model airplane supplies. It is primarily used to make model airplane landing gear. It is *very* hard and springy, and the easiest way to cut it is to grind through it with the edge of a grinding wheel.

You'll be cutting a small circle from the mat board. You can make the cut with an **X-acto knife** (also sold at hobby shops), but an **arch punch** does a neater job. Hardware stores do *not* normally sell arch punches. You can buy an arch punch from an industrial tool dealer. Look in *The Yellow Pages* under headings like **MACHINE TOOLS** or **INDUSTRIAL EQUIPMENT & SUPPLIES**. The arch punch you buy should match the inside diameter of the motor casing. In this example the punch diameter should be 3/4". *Figure 10-52* shows these items laid out and ready to use.

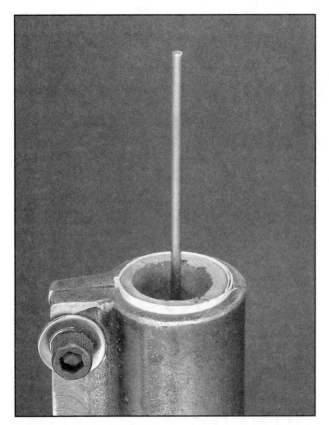
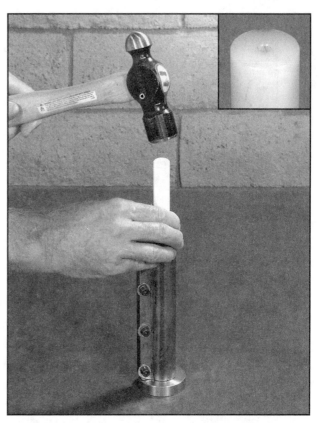

Figure 10-55. The porthole wire is inserted into the hole in the disk, so that it stands upright, and extends vertically out the front of the motor casing.

Figure 10-56. The first half of the forward bulkhead clay is rammed into place with a porthole tamp. The insert in the upper right-hand corner shows the small hole in the tamp's tip.

1. With a hammer and a 3/4" arch punch, punch 3/4" dia. disks from the mat board. With the 4" length of 3/32" dia. music wire, punch a hole in the center of each disk (*Figure 10-53*). To protect the punch and the wire from damage, do it on a plastic cutting board. The small board in the photos cost less than a dollar at Walmart. When you are finished, examine each disk carefully, and make sure that the holes are clear of any paper debris. Then insert one of the disks into the front of the motor casing, and push it down until it rests firmly on top of the time delay (*Figure 10-54*).

2. Insert the **porthole wire** (for *this* motor use **drawing #98-091** on **page 154**) into the hole in the disk, so that it stands *upright,* and extends vertically out the front of the motor casing (*Figure 10-55*).

3. Weigh out the same amount of clay that you used for the nozzle (11.0 gms. in *my* original test motor), and pour *half* of it into the front of the casing around the porthole wire. Then ram it into place with the **porthole tamp** and 8 to 10 solid hammer blows (*Figure 10-56*). A porthole tamp has a flat tip, and a central hole a little wider and deeper than the porthole wire.

Important note. *When forming the bulkhead in a 1/2" i.d. motor, load the clay in 1 dose. When forming the bulkhead in a 3/4" i.d. motor, split the total amount of clay into 2 doses. When forming the bulkhead in a 1", a 1-1/8", or a 1-1/2" i.d. motor, split the clay into 3 or 4 doses.*

4. Ram in the rest of the clay, and withdraw the porthole tamp.

5. With a pair of pliers or vise-grips, twist and pull out the porthole wire (*Figure 10-57*). Remove the nozzle mold with a twisting-pulling motion, and remove the core spindle by pulling and twisting with a T-handle or a pair of vise-grips (*Figure 10-58*). If you clamp the casing retainer flange into a vise, it makes the job easier; particularly when removing spindles from the larger motors. Loosen the casing retainer's clamping bolts. Release the motor from the casing retainer, and lay it on a rack to dry.

Figure 10-57. The porthole wire is removed by pulling and twisting with a pair of pliers.

Figure 10-58. The core spindle is removed by pulling and twisting with a T-handle (a Phillips screwdriver or a length of thick music wire) or a pair of vise-grips. A vise is of help when doing this; particularly when building the larger motors.

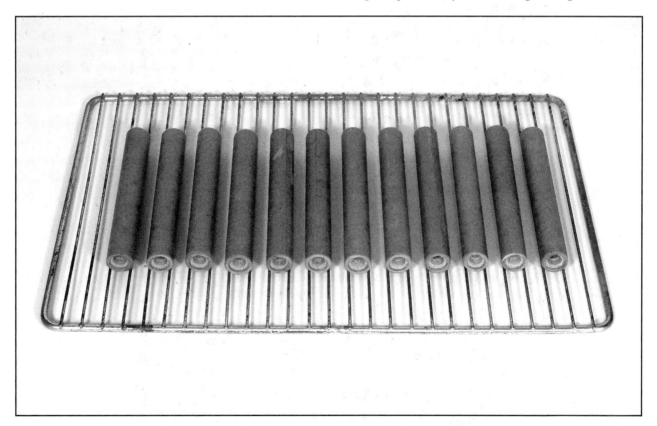

Figure 10-59. A dozen **E**-engines air-drying on an old refrigerator rack. **Keeping them up on a rack is important.** If you lay them on a flat surface, they will warp as they dry, and their propellant grains will crack.

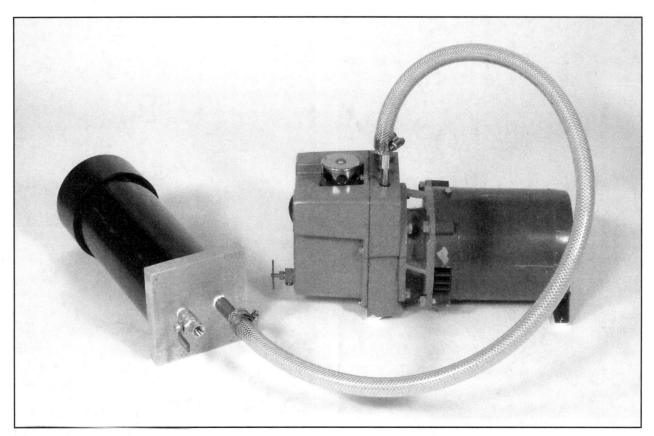

Figure 10-60. A small vacuum pump and a homemade vacuum chamber greatly speed the drying process.

● Air-Drying

Air-dry these motors on a *rack* (*Figure 10-59*). If you lay them on a flat surface, they'll warp, and their propellant grains will crack. Warm, dry air speeds the drying process, and cold, humid air slows it down. In "normal" weather (75° Fahrenheit, and average humidity), 1/2" i.d. core burners take 12 to 18 hours. 3/4" i.d. core burners take 24 to 30 hours. 1" i.d. core burners take 30 to 48 hours. 1-1/8" i.d. core burners take 48 to 60 hours, and 1-1/2" i.d. core burners take 72 to 84 hours. These drying times apply *only* to motors made with *acetone*. Motors made with alcohol will take much longer.

For end burners, or motors with *very* long time delays, *double* these times. To actually measure how long a motor takes to dry, weigh it the moment it's finished, and weigh it every 4 hours thereafter. When it stops losing weight, it is dry.

Important note. *When using alcohol as the binder solvent, multiply the above times by 6.* That is, figure 4 days for a 1/2" i.d. core burner, 7 days for a 3/4" i.d. core burner, 9 days for a 1" i.d. core burner, 1-1/2 to 2 weeks for a 1-1/8" i.d. core burner, and 2 to 3 weeks for a 1-1/2" i.d. core burner.

● Vacuum-Drying

You can speed the drying process by *10 to 30 times* with a vacuum pump and a homemade vacuum chamber. **Pages 46 - 47** explain how to shop for a vacuum pump. Instructions for making a homemade vacuum chamber begin on **page 81**, and *Figure 10-60* is a photo of a typical system that you might put together by following these instructions.

The pump in the photo is connected to the chamber with a piece of 5/8" i.d. x 7/8" o.d. vinyl tubing. If you make the hose bib on the chamber lid the same size as the bib on the pump, you can connect the two with a single piece of hose, and avoid the need for adapters. You can buy the tubing at most hardware stores; it comes in both thick-wall and thin-wall varieties. The thin-wall tubing collapses under a vacuum, so be sure to buy the thick-wall kind.

Auto parts stores sell heavy duty fuel and water hose that works, but before you buy something, squeeze it with your fingers. If it seems like it would collapse under a vacuum, look for something sturdier. If you can't find a suitable hose at a hardware store or an auto parts store, try a vacuum pump dealer. If they don't sell vacuum hose, they can usually direct you to someone who does.

Figure 10-61. A vacuum pump's sight glass tells you when the oil level is correct.

Figure 10-62. A small petcock valve makes the job of changing the pump's oil easier.

To use the system, make sure that all the hose connections, and both surfaces of the lid's rubber gasket are lightly coated with silicone vacuum grease (sold by vacuum equipment dealers and laboratory/chemical dealers). Place the motors-to-be-dried into the chamber, and stick on the lid. Close the lid's ball valve, and turn on the pump. The moment the pump starts, the lid will be sucked down tight. The vacuum alone holds the lid on, and clamps are not needed.

Vacuum pumps sound loud and "bubbly" when pumping air. Silent running, or a sharp clicking or "rattling" sound, is the normal sign that the pump has achieved a good vacuum. At this point, set a timer for the recommended drying time, and go eat lunch. The smallest motors will be dry when you return, and the largest (if made with acetone) will take 2 hours.

Important note. *When vacuum-drying motors made with acetone, the following drying times apply.* 1/2" i.d. core-burners take 20 minutes. 3/4" i.d. core-burners take 30 minutes. 1" i.d. core-burners take 45 minutes. 1-1/8" i.d. core-burners take 1 hour, and 1-1/2" i.d. core-burners take 1-1/2 to 2 hours. For end-burners, or core-burners with very long time delays, double these times. *When vacuum drying motors made with alcohol, multiply all of the above times by 6.*

Important note. *Vacuum pumps fitted with standard buna (aka nitrile) rubber seals can be used with alcohol,* **but not with acetone**. If you're drying motors made with acetone, be sure that the pump's seals art made of Viton. If they are *not* made of Viton, have them replaced. See **page 47** for details.

When the motors are dry, turn off the pump, and open the valve on the vacuum chamber's lid. This lets air into the chamber, and when the chamber reaches atmospheric pressure, the lid will come off. If you don't plan to use the motors immediately, place them in a zip lock, plastic bag for storage.

● **Vacuum Pump Maintenance**

When using a vacuum pump for the first time, make sure that it contains the recommended amount of clean, fresh **vacuum pump oil**. Most pumps have a clear sight glass where you can see if the oil level is correct (*Figure 10-61*). As the motors dry, small amounts of acetone or alcohol are mixed with the oil, and the oil in the sight glass becomes cloudy. Also, small amounts of moisture from the paper motor casings condense out, and sink to the bottom of the oil reservoir. As the water accumulates, the level in the sight glass rises. It is therefore important to drain the water often, and change the oil at least once a month; more frequently if you run the pump every day.

These tasks can be awkward on a pump fitted with a drain plug, and you can make the job easier by replacing the plug with a small, petcock valve (*Figure 10-62*). You can buy the valve and the fittings at a hardware store.

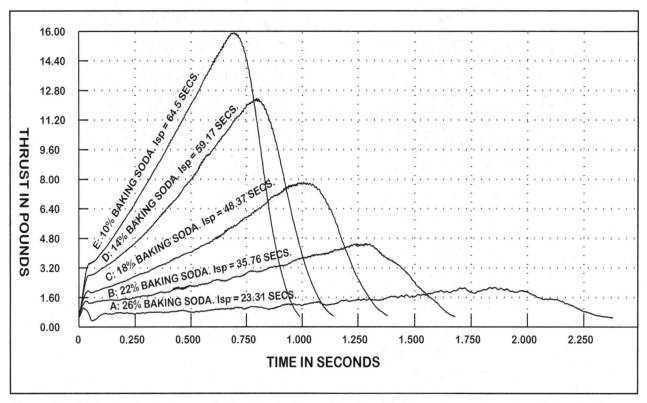

Figure 10-63. The thrust-time curves of five identical motors made with decreasing amounts of baking soda. You can see the dramatic effect that changes in a propellant's baking soda content have on motor performance. The propellants containing more than 10% soda all performed poorly.

Important note. *You cannot use regular motor oil in a vacuum pump, because it won't produce a good vacuum.* Vacuum pumps need a highly refined **vacuum pump oil** with an extremely low vapor pressure; the vapor pressure of motor oil is too high. You can buy **vacuum pump oil** from the same places that sell vacuum pumps, and you'll also find it in the Grainger catalogue (**see page 47**).

● The Effect of Baking Soda on Motor Performance

In the preceding demonstration, I *assumed* the addition of 10% baking soda to the propellant. **In actual practice, when building a new and untested design, you should add 30% baking soda to the propellant in the first motor.** This amount of soda will almost certainly make the motor chuff and perform poorly. Assuming that it does, add 28% soda to the propellant in the second motor. In the tests that follow, decrease the soda in 2% increments until you get a motor that *doesn't* chuff. Then decrease the soda in 1% increments until you get a motor whose maximum thrust approximates the maximum thrust printed in the motor drawing.

Figure 10-63 illustrates the thrust-time curves of five identical motors made with **26%**, **22%**, **18%**, **14%,** and **10%** soda. I've labeled each curve accordingly, and you can see the dramatic effect that the soda has on the motors' performance.

● The Effect of Core Length on Motor Performance

The length of a rocket motor's propellant core also has a profound effect on performance, because it directly effects the amount of propellant burning at any given moment. *Figure 10-64* illustrates the thrust-time curves of four rocket motors with core lengths of **4-1/2"**, **4"**, **3-1/2"**, and **3"**. I've labeled each curve accordingly, and *again* you can see the dramatic effect that the changes have on motor performance.

As the cores become shorter, there is a *very* slight increase in the burn time, but not enough to offset an overall *decrease* in the maximum thrust, the total impulse, and the specific impulse. Conversely, making the cores longer causes an *increase* in these parameters, but be careful when you do this. As maximum thrust, total impulse, and specific impulse rise, so does the chamber pressure. If you make the core too long, you'll exceed the motor's design limits, *and the motor will explode*.

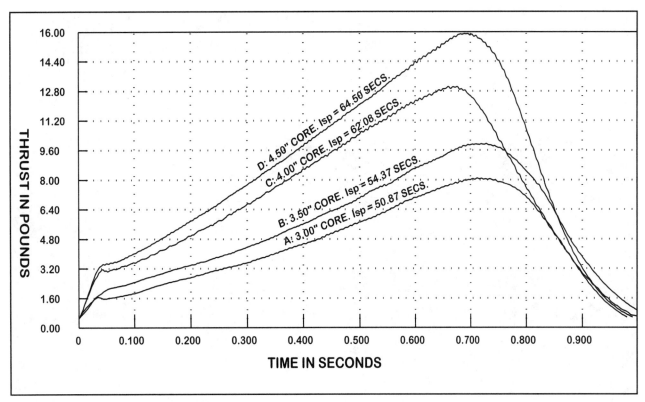

Figure 10-64. *The thrust-time curves of four rocket motors with core lengths of 4-1/2", 4", 3-1/2", and 3" show how the length of a motor's propellant core effects its performance. As you can see, the motors made with the shorter cores performed poorly.*

● The Time Delay Burn Rate Test

Before you can use a propellant as a time delay, you have to know how fast it burns at atmospheric pressure, and you can determine that speed with **the time delay burn rate test**. I used the **KG1** propellant as the time delay in the previous demonstration (**page 285**), so I'll use it again here.

1. Cut a 1/2" i.d. x 3/4" o.d. convolute motor casing to a length of 5". Wrap it with the correct amount of construction paper, and clamp it into a 3/4" i.d. casing retainer. In this case, you can use an improvised retainer made from a piece of 3/4" schedule 40 plastic pipe and a series of hose clamps (see **page 65**).

2. Place the casing (in the casing retainer) upright on a heavy block of steel or a concrete floor. With a 1/2" flat tamp (**drawing #98-084-B on page 147**), ram in 3.0 gms. of nozzle clay (***Figure 10-65***). A square of masking tape stuck to the bottom of the casing keeps the loose clay from falling out until ramming is complete. When ramming is finished, insert the tamp in on top of the clay, and scribe a ring at the point where it enters the front of the casing (***Figure 10-66***). Remove the tamp, and place a *second* mark *exactly* 3.0" below the ring (***Figure 10-67***).

3. Weigh out 20.0 gms. of **KG1** propellant, and place it in a mixing cup. Add 3.0 gms. (15%) of baking soda, and mix it with a fork for 30 seconds. The total amount of material in the cup will now be 23.0 gms.

4. Calculate the amount of acetone required as follows. There are 23.0 gms. of material in the mixing cup, and we'll pretend that the solvent requirement is 1.5 cc. for every 20 gms. of propellant

> **Amount of acetone in cc. = (23.0/20) x 1.5 = 1.15 x 1.5 = 1.725 rounded off to 1.7 cc.**

Draw 1.7 cc. of acetone into a 3 cc. or 6 cc. syringe, and squirt it into the cup on top of the propellant. Mix it in with a table fork, and use **the fork test** to determine when it is thoroughly distributed.

5. **Without delay**, put a **half-teaspoon** of the dampened propellant into a funnel, and poke it down into the casing with a bamboo Bar-B-Que skewer. Them ram it into place with the 1/2" flat-ended tamp. Repeat this process until the

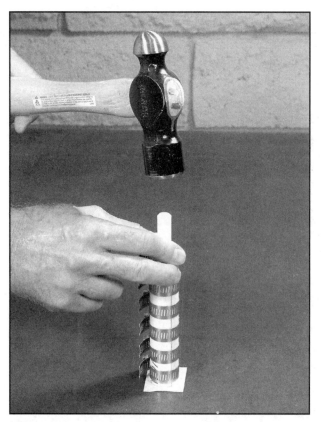 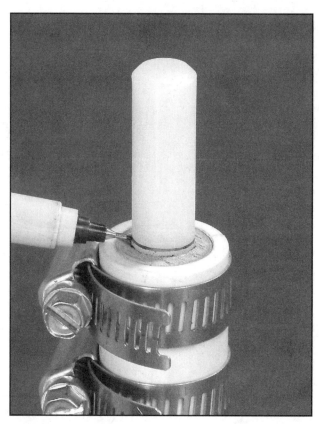

Figure 10-65. *3.0 gms. of nozzle clay are rammed into a 5" long x 1/2" i.d. motor casing. Note the square of masking tape stuck to the bottom of the casing. It holds in the loose clay until ramming is complete.*

Figure 10-66. *The tamp is reinserted into the casing until its end rests on the rammed clay. An ultrafine-point "Sharpie" marking pen scribes a ring where it enters the front of the casing.*

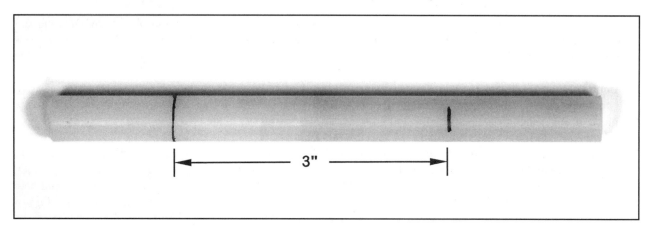

Figure 10-67. *A second mark is made exactly 3.0" below the ring.*

second mark on the tamp is level with the front of the casing (***Figure 10-68***). Then release the casing from the casing retainer, and vacuum dry it for 45 minutes, or let it air dry on a rack for 24 hours, .

To perform the Time Delay Burn Rate Test, proceed as follows:

1. Saw the excess cardboard casing back to within 1/4" of the end of the propellant (***Figure 10-69***). Then paint the end of the propellant (***Figure 10-70***) with a small amount of the black powder igniter syrup described on **pages 459 - 460**, and let it thoroughly dry.

2. **In an area where there's nothing that might catch fire**, stand the test sample upright in a bed of oil-based modeling clay, and tape a fuse or an electric igniter on the end of the sample, so that it touches the dried black powder syrup on the end of the propellant (***Figure 10-71***).

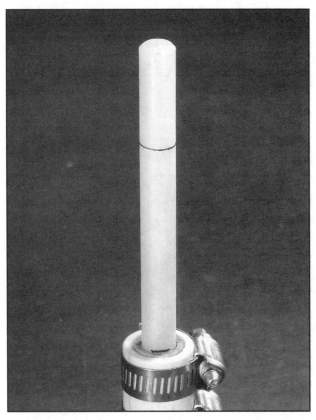

Figure 10-68. Dampened **KG1** propellant with 15% baking soda is rammed into place until the "3-inch" mark is level with the front of the casing.

Figure 10-69. When the sample is thoroughly dry, the excess cardboard is sawed back to within 1/4" of the end of the propellant. A handsaw with a miter box works fine.

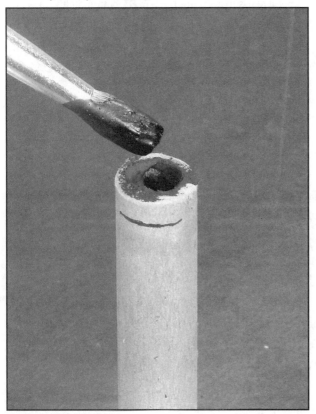

Figure 10-70. The end of the dried time delay is painted with a small amount of the black powder igniter syrup described on **pages 457 - 458**.

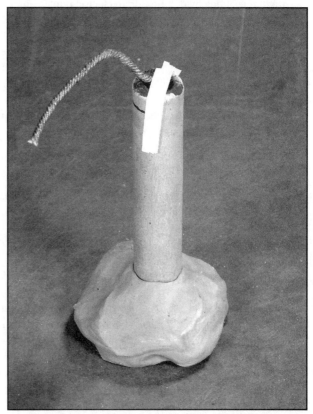

Figure 10-71. The sample is placed upright in a bed of modeling clay. A fuse or igniter is taped on the end, so that it touches the dried black powder on the end of the sample.

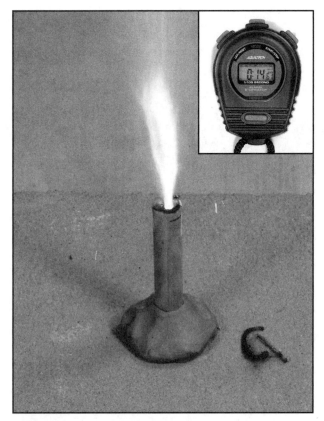 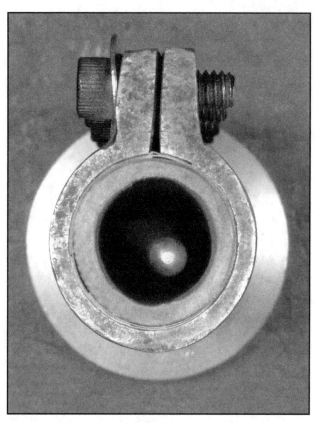

Figure 10-72. The test sample is ignited, and a stop watch measures how long it takes to burn 3 inches.

Figure 10-73. An uneven distribution of the nozzle clay forces the tip of a wobbly core spindle off-center.

3. Light the sample with a fuse or an electric igniter, and the moment it starts to burn, hit the **START** button on a stop watch (*Figure 10-72*). The moment that it burns out, hit the **STOP** button, and you'll know exactly how long the **KG1** propellant with 15% baking soda takes to burn 3 inches at atmospheric pressure. As the burn proceeds, watch and listen carefully. If you see or hear any signs of combustion instability (i.e. "chuffing"), reduce the amount of baking soda, and repeat the test with lesser amounts of soda until you achieve a stable burn. If, for example, stable burning is achieved with 12% soda, then use the KG1 propellant with 12% soda as your time delay.

To find the burn rate of the successful mixture in inches-per-second, divide the time you measured in **step 3** by *three*, and take the reciprocal of the answer.

> *EXAMPLE 10-1:*
>
> *We'll pretend that the successful mixture took 48 seconds to burn 3 inches. Dividing that figure by 3 yields a figure of 16 seconds per inch, and the reciprocal of 16 seconds per inch is approx. 0.063 inches per second. That is:*
>
> *48/3 = 16, and 1/16 = 0.0625, rounded off to 0.063*
>
> *So at atmospheric pressure, the time delay tested in this example burns at a rate of 0.063 inches per second.*

Since the mixture's burn rate at atmospheric pressure is the same as the rate at which it burns as a time delay, you can use this information to calculate the amount needed to produce a time delay of any desired length.

Solving Problems

● A Tilted Core Spindle

When experimenting with core spindles of various lengths, you'll find it *most* convenient to leave the nozzle molds and the spindles unattached. To make separation easier, the fit between the spindle and the nozzle mold should be slightly loose. If the fit is too tight, small amounts of clay trapped between the two during the loading process will cause them

Figure 10-74. One or two set screws in the nozzle former's base prevent the spindle from wobbling.

Figure 10-75. Set screws create small burrs and scars on the base of the spindle that can cause it to jam in place.

to jam together. This looseness allows the spindle to wobble around a bit, and the wobble can be a nuisance when building motors with larger and longer propellant cores.

When ramming in the nozzle clay, an uneven distribution of the clay will force a wobbly spindle out of alignment, and shove its tip off center (*Figure 10-73*). You can solve the problem by locking the spindle and the nozzle mold together with one or two set screws (*Figure 10-74*). But each time a screw is tightened, it raises a little burr on the spindle's base (*Figure 10-75*), and these burrs cause the spindle to jam in the nozzle mold's spindle hole. Until you're *certain* that you're going to use a particular spindle with a particular nozzle mold, it is unwise to lock them together in this manner.

Until you are *ready* to lock them together, you can use the following tricks to insure that the spindle stays properly aligned. The first is to load the nozzle clay in small doses, and pay close attention to its distribution. The first dose rammed in with the starting tamp isn't a problem, because it falls below the spindle's base. But close attention to how you put in the rest of the clay will keep the spindle straight.

Figure 10-76 shows a motor with part of the nozzle clay already in place, and the spindle tilted to the right. To fix the problem, lean the motor to the right as you pour in the next dose of clay (*Figure 10-77*). Then gently move it back to the vertical, and ram the clay into place. As you do so, the spindle will be forced to the *left*. If the spindle moves *too far* to the left, add the next dose of clay to the *left* side, and force the spindle back toward the center. By adding the clay in small doses, and carefully choosing where you put it, you can keep the spindle straight, and its tip properly centered.

If the spindle is tilted, and the nozzle is already in place, you can straighten it up by chipping away clay from one side, and moving it to the other. Use the sharpened music wire shown in *Figure 10-35* to chip away at the clay opposite the tilt (*Figure 10-78*). Then tip the entire motor-casing retainer assembly *below the horizontal*, and gently tap the nozzle mold on the block of steel (*Figure 10-79*). This will make the loosened clay fall to the opposite side. Then gently return the motor to the vertical, and ram the clay into place.

As with the other procedures, this will take some time at first, but with experience, you'll gain speed. When you've finalized a motor design, and dedicated a set of tools to making *just that motor*, then you can install the set screws, and eliminate the problem.

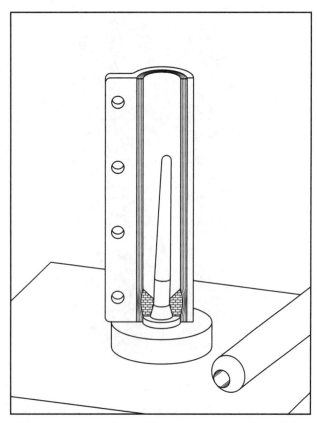

Figure 10-76. A motor with a tilted core spindle. Part of the nozzle clay has been rammed into place, but an uneven distribution has forced the spindle to the right.

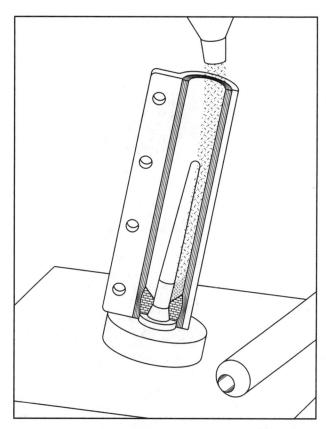

Figure 10-77. Leaning the motor to the right, and placing the next dose of clay on the right, forces the spindle back to the left when the clay is rammed into place.

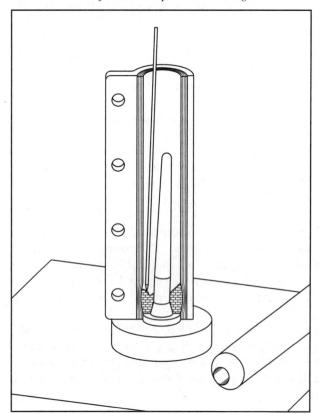

Figure 10-78. A sharpened piece of music wire chips the clay away from one side of the nozzle.

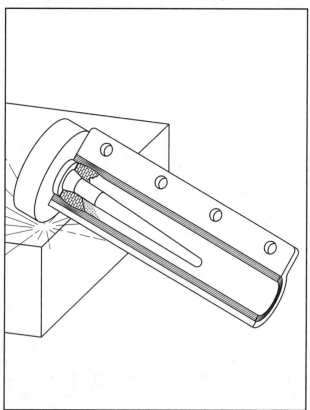

Figure 10-79. The loosened clay is moved to the opposite side by tilting the motor below the horizontal, and tapping the nozzle mold gently on the edge of the block of steel.

 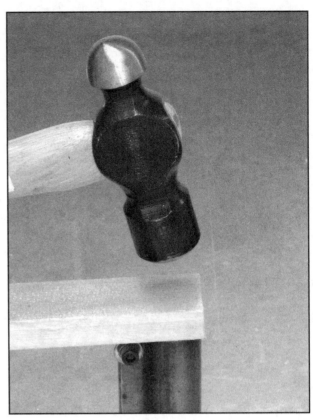

Figure 10-80. The casing retainer in this photo was not sufficiently tightened around the motor casing, and it's worked its way upward during the loading process.

Figure 10-81. The temporary fix is to tap the retainer back down by gently hammering on its upper end. A block of wood protects it from damage.

● A Slipped Casing Retainer

If the casing retainer isn't tight enough around the motor casing, it will slide upward during the loading process. *Figure 10-80* illustrates the problem. The obvious solution is to wrap more paper around the casing when you build the next motor. *Figure 10-81* shows the temporary fix, which is to simply tap the retainer back down by hammering gently on its upper end. A block of wood protects it from damage.

● Rapid Solvent Evaporation

An **E31** goes together so quickly (10 minutes) that solvent evaporation isn't a problem, but this is *not* the case with the larger motors. An **H160** takes about 30 minutes to load, and the acetone evaporates too fast to allow such a long working time. There are four solutions to the problem, and the first and most obvious is to switch the solvent to alcohol. But the trouble with alcohol is the *great* increase in the drying time.

The second and most practical fix is to split the propellant into two equal batches, and mix the soda into both of them. Then dampen and load each batch separately. When loading an **H160**, which takes about 460 gms., I weigh out *two* 250 gm. batches, and dampen just the first batch. *Only* when I've used up the first batch, do I dampen the second batch and complete the job. In this way, I limit my working time with each batch to 15 minutes.

To reduce the problem further, I use a large, broad container for mixing in the acetone (*Figure 10-82*). But I *transfer* the dampened propellant to a tall, thin cup immediately thereafter (*Figure 10-83*). By working from a tall, thin container, the propellant at the bottom is protected from evaporation by the propellant at the top. As I scoop the propellant out, I am careful to work from the top down, insuring that the propellant near the bottom stays damp.

The final trick is to increase the solvent content by 10% to 15%. This fourth fix is useful when making *very small* motors like the **KG3-7-C11**. This little motor takes just 11.3 gms. of propellant (less than 1/2 oz.), and the amount of propellant being mixed is so small that a significant portion of the acetone evaporates during the mixing process. Increasing the acetone content with such a tiny motor insures that the propellant stays damp enough to produce a good finished product.

Figure 10-82. When working with larger amounts of propellant, soft margarine tubs (1 lb. to 3 lbs.) are ideal for mixing in the solvent.

Figure 10-83. Loading the propellant from taller, thinner containers, like these tall, thin polyethylene (plastic) glasses, helps keep the propellant damp.

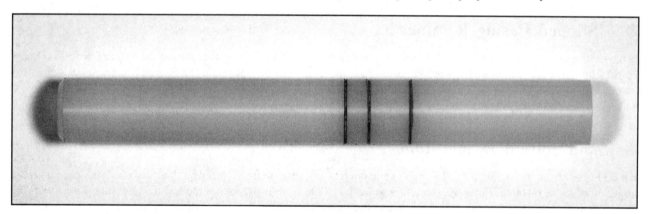

Figure 10-84. When you've decided to use a particular tamp for making a particular motor, you can replace the temporary pen marks with permanent, machined grooves. A lathe with a standard 60 degree thread-cutting tool works fine. For this photograph, I've rubbed some ink into the grooves to make them show up better for the camera.

As a general rule, if the loading time is less than 10 minutes, acetone evaporation isn't a problem. For loading times of 11 to 15 minutes, working from a tall, thin container keeps the propellant sufficiently damp. For loading times greater than 16 minutes, split the propellant into *two* batches. Then dampen and load each one *separately*. Switching the solvent to alcohol is the ultimate solution, but the trade-off is the great increase in the drying time, which may or may not be a problem, depending on how soon you want to use the motors.

● Using Up Leftover Propellant

When you've finished building a rocket motor, you will *always* have some leftover propellant. To use it again, let it thoroughly dry. Then place it in a sealed container, and write on the container exactly what's inside. After making a dozen **KG3-26-E31**s, you'll end up with maybe 100 gms. of **KG3** propellant containing (for example) 10% baking soda. Dry it out, put it in a zip-lock plastic bag, and label it "**KG3 + 10% BAKING SODA**".

The next time you make **KG3** motors, render it back into a powder by running it for a few minutes in the powder mill. This reconstituted propellant burns quite a bit faster than the original, but you can compensate by adding more baking soda. In this way, everything gets used, nothing is wasted, and you don't have to dispose of unwanted propellant.

Dedicating Your Tools

When you've finalized a motor design, you can record the exact amounts of clay, propellant, baking soda, and acetone needed to make it, and you can dedicate a set of tools to making *only that motor*. Dedicating your tools means selecting a particular core spindle, and mounting it semi-permanently to the nozzle mold with set screws. And it also means replacing the pen marks on your tamps with permanent machined grooves (*Figure 10-84*).

To make these grooves, place the tamps back in a metal lathe, and with a sharp pointed lathe tool, cut permanent, ring-shaped grooves where the pen marks used to be. These *permanent* marks will allow you to *quickly* make multiple copies of the original motor without having to take measurements.

Multistage Ignition

The most reliable device for firing a second stage motor is an onboard, electronic timer coupled to an electric igniter, but I won't discuss the technology here. For further information, contact the **Tripoli Rocketry Association**, or search the Internet for websites related to high power rocketry. Most of Tripoli's members are involved in high power rocketry, and the larger rockets flown by these people often use electronic timers.

Without this equipment, you'll have to rely on the interesting and primitive method known as **pyrotechnic staging**. Pyrotechnic staging derives its name from the fact that you actually transfer the flame from the booster motor into the core of the second stage motor, and here's how to do it.

● Preparing a Booster for Pyrotechnic Staging

MATERIALS LIST

1. A 3/4" arch punch
2. A 1/4" gasket punch or paper punch
3. A 1/4" dia. plastic drinking straw
4. A scrap of corrugated cardboard
5. Some 5 minute epoxy glue
6. A small plastic kitchen cutting board

Industrial tool dealers sell arch punches and gasket punches, and you can buy paper punches at places like Office Depot and Staples. Hardware stores sell the 5-minute epoxy, and supermarkets sell the drinking straws. Flexible straws are approximately 1/4" dia. Standard, straight drinking straws are a little thinner, and you might have to experiment with different size straws and punches to find a perfect match. If the straws are *exactly* 1/4" dia., you can use a standard 1/4" paper punch. *Figure 10-85* shows these items laid out and ready to use.

1. By following the previous instructions, build a **KG3-26-E31** with a forward bulkhead and a porthole, but *without* a time delay, and dry it thoroughly. If you make the motor with acetone, air-drying will take 24 hours, and vacuum-drying will take 30 minutes.

2. Punch a 3/4" dia. disk from the cardboard (in this case a scrap from a corrugated cardboard box), and punch a 1/4" dia. hole in the center of the disk (*Figure 10-86*). Insert the disk into the front of the motor, and push it firmly down against the forward bulkhead (*Figure 10-87*). Roughen one end of the drinking straw with sandpaper. Insert the roughened end into the hole, and press the straw firmly into place (*Figure 10-88*).

Important note. *Straws come in various sizes, so* **buy the straws first.** *Then buy a punch that makes a hole into which the straws fit tightly.* For the larger motors, use the 1/4" dia. flexible drinking straws. For the little 1/2" i.d. motors, use the thin, red, plastic "Sip-N-Stir" combination straw-coffee stirrers sold my Smart & Final Iris.

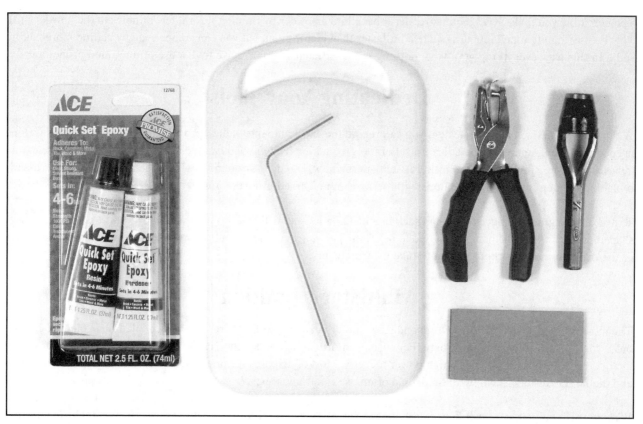

Figure 10-85. These are the things you'll need to prepare a booster motor for pyrotechnic staging. In this example I'm using a 1/4" dia. drinking straw and a 1/4" dia. paper punch. Other straws come in different sizes. The diameter of the punch for the straw should match the diameter of the straw, and the diameter of the arch punch should match the i.d. of the motor casing.

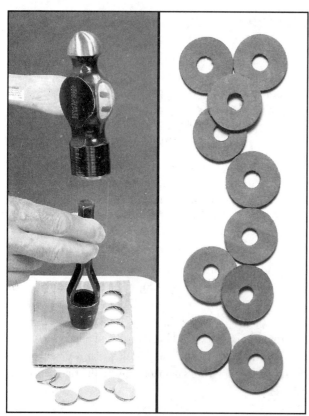

Figure 10-86. 3/4" dia. disks are punched from a scrap of corrugated cardboard, and 1/4" dia. holes are punched in their centers.

Figure 10-87. One of the punched disks is pushed firmly down against the motor's forward bulkhead. In this photo you can see the motor's porthole through the hole in the disk.

Figure 10-88. A 1/4" dia. drinking straw is inserted into the hole in the disk so that it extends straight out the front of the motor. The fit between the straw and the hole should be snug.

Figure 10-89. A 1/8" to 1/4" thick layer of 5-minute epoxy is poured on top of the disk around the straw. The epoxy seals the straw permanently in place.

3. Stand the motor on end. Mix up a small amount of 5-minute epoxy, and pour enough into the front of the motor to form a layer between 1/8" and 1/4" thick around the straw (*Figure 10-89*). When the epoxy has hardened, the booster motor is ready.

● Preparing a Second Stage Motor for Pyrotechnic Staging

These homemade propellants are slow to ignite, so you should always augment second-stage ignition with a **starter-pellet** made from a powder that is *fast* burning and easy to light. Commercial black powder works the best. To make and use these pellets, proceed as follows.

MATERIALS LIST

1. A drinking straw or a length of brass tubing
2. A 4" square piece of 1/4" thick balsa wood
3. A small amount of commercial black powder
4. A bottle of 70% rubbing alcohol
5. A tube of Duco cement
6. A table fork and a putty knife

Figure 10-90 shows these items laid out and ready to use. Each pellet is cemented to the front of the second stage motor's propellant core. You should make these pellets a diameter that will *almost* fit snugly. To form the pellets, you'll use the drinking straw or a piece of brass tubing as a "cookie cutter", so choose a piece who's inside diameter is a little *smaller* than the i.d. of the motor core's front end.

Gun shops sell the black powder. Supermarkets sell the drinking straws, and hobby shops sell the brass tubing. Hobby shops and hardware stores have the Duco cement. You use the cement to glue the pellets in place, and I've chosen *Duco* cement, because, unlike *most* cements, *Duco* cement is flammable when dry. Hobby shops sell the balsa wood, and drug stores sell the rubbing alcohol. But be sure to buy the 70% kind. To make a pellet with the necessary structural strength, you need some water in the alcohol, and the 91% disinfectant alcohol doesn't have enough water.

Figure 10-90. The things you will need to prepare an upper stage motor for pyrotechnic staging. From left-to-right they are a table fork, a putty knife, some commercial black powder, some 70% rubbing alcohol, and a 4-inch square of 1/4" thick balsa wood. You'll also need an Xacto knife to cut a 2-inch square hole from the center of the balsa wood.

Figure 10-91. A frame made of 4" square by 1/4"-thick balsa wood with a 2"-square hole cut from its center is placed on a sheet of aluminum foil.

Figure 10-92. A stiff paste of black powder and rubbing alcohol is spread into the frame with the putty knife. The frame should be filled and leveled to its full depth of 1/4".

Figure 10-93. As the tube or the straw is lifted from the paste, a pellet of the paste should stay in the straw, and be removed from the paste, An empty hole should be left behind.

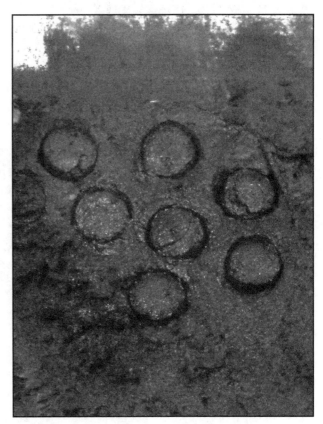

Figure 10-94. If the pellets stay in the paste (as in this photo), the paste is too wet, and you'll have to let it dry for a while.

Figure 10-95. A pile of dried igniter pellets, and the leftover scraps from the frame. The leftovers can be dried and used again by remoistening with fresh rubbing alcohol.

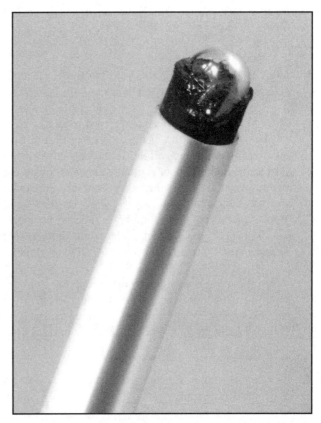

Figure 10-96. A dried igniter pellet is placed in the end of a straw or a piece of tubing. The pellet's end is coated with Duco cement.

305

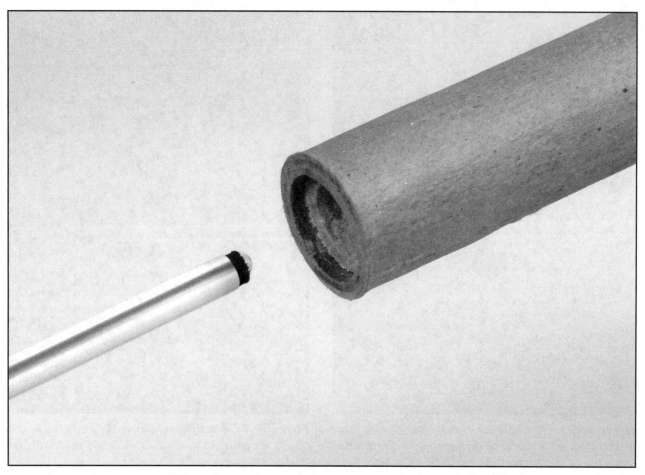

Figure 10-97. The plastic drinking straw inserts the igniter pellet into the motor's core as far as the straw will go. Note that, in this photograph, the bubble on the end of the pellet resulted from the wet glue's response to the heat of the photo studio lights.

1. Lay a sheet of aluminum foil on a flat work surface. Cut a 2" square from the center of the balsa wood, and lay this square balsa wood frame on the foil (*Figure 10-91*).

2. Place **30 gms.** of commercial black powder in a mixing cup, and with a table fork, work in small amounts of 70% rubbing alcohol until the mixture is the consistency of a stiff paste. If the powder is coarse, give the granules plenty of time to dissolve. Mix in enough alcohol to thoroughly wet the granules. Then cover the mixing cup with aluminum foil, and let it sit for an hour. An hour is usually enough time for the granules to soften to a point where you can mash them into a paste. When the paste is the right consistency, spread it into the balsa wood frame with the putty knife (*Figure 10-92*).

3. With the drinking straw, or a brass tube of the appropriate size, push downward into the paste, and lift the tube out. As you do so, a pellet of the paste should come out *with* the tube (*Figure 10-93*). If it *doesn't* come out (*Figure 10-94*), the paste is too wet, and you'll have to let it dry for a few minutes.

   When the paste is sufficiently dry, punch out as many pellets as you can, and set them on a paper towel to dry. To remove each pellet from the cutter tube, blow gently on the opposite end as you would on a pea shooter, aiming at the towel below. When finished, you'll have a small pile of igniter pellets and some left over paste inside the balsa frame (*Figure 10-95*). When the leftover paste is dry, store it in a zip lock plastic bag. You can then remoisten and reuse it as needed.

4. To mount a pellet in the core of a second stage motor, place it back in the end of the cutter-tube, so that a short portion is exposed, and apply a small amount of Duco cement to the pellet's exposed end (*Figure 10-96*).

   Without delay, insert the tube with the pellet *as far into the core as it will go* (*Figure 10-97*). Then reach *through* the tube with a bamboo Bar-B-Que skewer, *push the pellet the rest of the way to the front of the core*, and withdraw the tube and the skewer. When the Duco cement is dry, the second stage motor is ready.

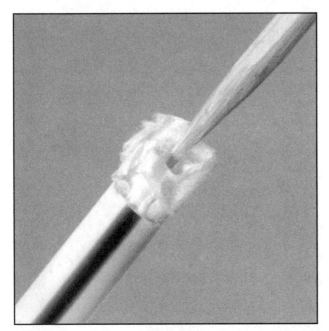

Figure 10-98. A tiny wad of Kleenex keeps the loose powder from flying forward during deceleration.

● Preparing for a Two Stage Rocket Flight

To make this staging system work, you have to mount the two rocket motors so that the straw in the front of the booster reaches *almost* to the igniter pellet in the core of the second stage motor. When building a two stage rocket, you'll have to plan things accordingly. To prepare for a two stage flight, proceed as follows.

1. Cut the straw in the front of the booster motor *back* until its end is about 1" from the second stage igniter pellet when the two stages are plugged together. Then partially fill the straw with **FFF** or **FFFF** commercial black powder. For the small cocktail straws I use about 1/4 gm. For the large straws maybe 1/2 gm. To clarify, you want *just enough* to fill the motor's porthole, plus a little bit in the straw above; that's *just enough* to blow sparks up into the core of the second stage motor. **Important note.** *I've tried Pyrodex P, but it doesn't catch fire as fast as the black powder, and it's not as reliable.*

2. **Important note.** *When a rocket's booster stops firing, the rocket immediately decelerates, and the negative G forces will throw the powder in the straw forward, preventing its ignition.* To hold the powder in place, wad up a tiny piece of Kleenex tissue, and insert it into the straw on top of the powder (*Figure 10-98*).

   Push it into place with a Bar-B-Que skewer, *but don't make it too tight.* The forces of deceleration are rather mild. It should be tight enough to hold the powder in place, but *not* so tight that it causes the flame to blow back into the empty booster.

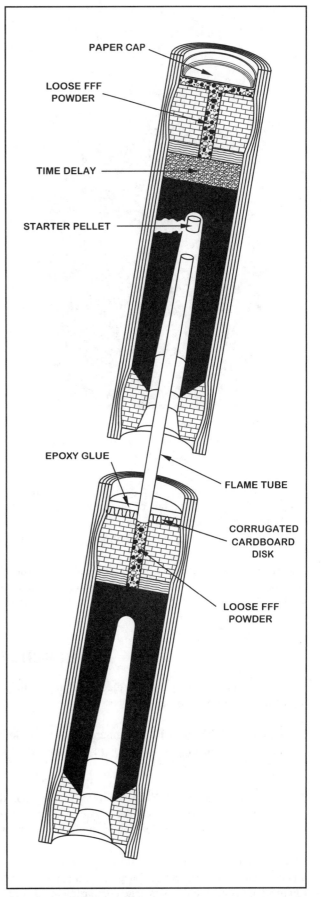

Figure 10-99. The proper motor configuration for pyrotechnic staging.

307

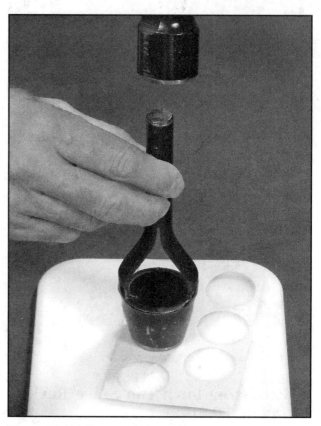

*Figure 10-100. Commercial black powder or Pyrodex P fills the motor's porthole, and forms a **very** thin layer over the front of the forward bulkhead.*

Figure 10-101. A paper disk is cut or punched a little larger than the outside diameter of the motor casing.

3. Assemble the rocket for launch, insert the igniter, cross your fingers, and push the firing button. If you've done everything right, the last of the propellant in the booster motor will light the black powder powder in the porthole, and send a puff of flame through the straw to the second stage igniter pellet, which will light the second stage motor. The second stage motor will then fire, and push the empty booster away.

Figure 10-99 illustrates the proper configuration. There's a bit of artistry involved, and it might take some practice to make the system work. I advise you to perform your first experiments with inexpensive rockets, or two stage stick rockets. Plan on the fact that your first attempts might fail, and don't put a lot of money or effort into making these test rockets look nice.

Parachute Ejection

To create a motor with a parachute ejection charge, build a **KG3-26-E31** with a time delay and a forward bulkhead with a porthole. Then proceed as follows.

1. Pour enough **FFF** or **FFFF** commercial black powder into the front of the motor to fill the porthole, and form a *very* thin layer of loose powder on top of the forward bulkhead (*Figure 10-100*). The layer shouldn't be more than 1 or 2 powder grains thick.

2. Cut or punch a disk of stiff paper *slightly larger* than the outside diameter of the motor casing (*Figure 10-101*), and push the disk down on top of the powder with the porthole tamp (*Figure 10-102*). *Figure 10-103* illustrates the result. The loose ejection powder is held in place with what looks like an upside-down, paper bottle cap.

The amount of powder needed will vary according to the size of the motor and the empty volume in the rocket's nose cone and parachute compartment. If you use too little, the nose cone will fail to come off, and the parachute will not deploy. Using too much can damage the parachute compartment and even burn the parachute. As I said in **Chapter 2**, the most practical approach is to set up a dummy body tube *on the ground,* complete with a parachute and nose cone. Then, starting with a very *small* amount of powder, electrically ignite increasing amounts until the desired effect is achieved.

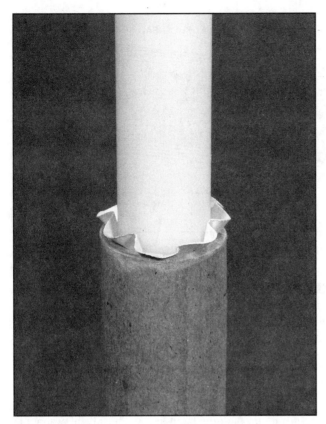

Figure 10-102. The paper disk is pushed in on top of the powder with a flat tamp or a porthole tamp.

Figure 10-103. The result is an "upside-down bottle cap" of paper that holds the loose ejection powder in place.

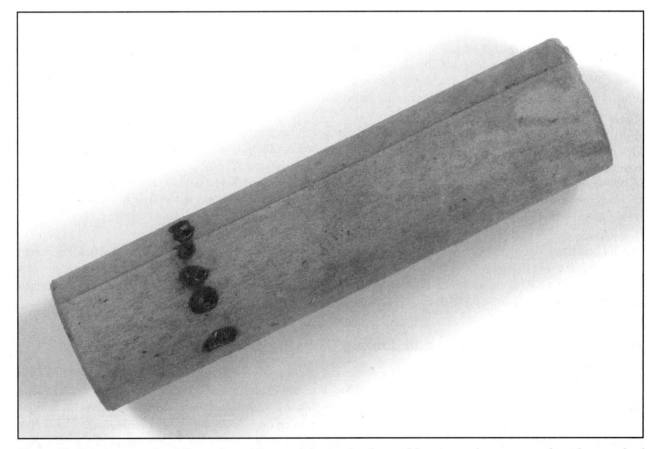

Figure 10-104. An example of "burn-through" around the nozzle of an end-burning rocket motor made with a standard fireworks-grade casing.

309

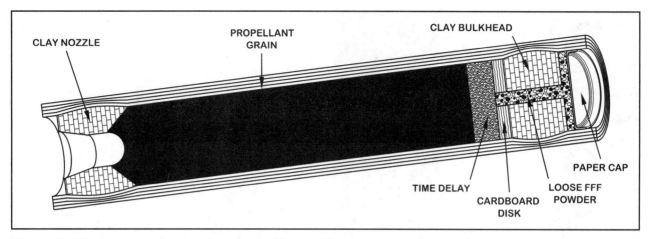

Figure 10-105. A cross section drawing of an end burner illustrates its various components

End Burners

When I started this book, I wanted to include a series of end burning rocket motors, but I quickly discovered that my standard fireworks-grade tubes don't have sufficient burn-through resistance to support the end burning design. Unlike a core burner, the inside of an end burner's casing is exposed to intense heat and flame for the duration of the burn. The ring-shaped area where ignition first takes place suffers the longest exposure, and in most of my attempts, it burned through in 3 or 4 seconds (*Figure 10-104*).

I also encountered a difficult engineering problem (at least with *this* motor building method). As the solvent evaporates from an end burning propellant grain, it carries a small amount of red gum in solution along with it. The gum migrates *away* from the motor's center, and *toward* the outer surface, leaving the center of the grain slightly deficient in gum. This causes the center to burn at a different rate than the outside, and the circular burning surface becomes increasingly conical as the burn proceeds. In small end burners with short propellant grains, it isn't a problem. In the larger and longer motors, it causes a steady increase in the area of the burning surface, and a corresponding increase in chamber pressure and thrust. If the burn time is long, the chamber pressure eventually exceeds the motor's design limit, and the motor explodes. With the pure KG6 propellant and some hand rolled casings, I constructed a few working motors with burn times of up to 8 seconds, but nothing I consider reliable enough to publish.

At the time that I stopped my work on end burners, I'd planned to counter the progressive increase in chamber pressure by adding baking soda to the forward part of the propellant grain. For example, in a 6-inch-long propellant grain, the first inch or two of the grain near the nozzle might be pure KG6, and the *forward* part of the grain would contain a small amount of baking soda. The actual percentage of soda, and the position of the change would be determined by trial and error. By tweaking the the number of changes, the positions of the changes, and the soda percentages, it might be possible to produce a long propellant grain with a reasonably constant chamber pressure and thrust.

When I began to *seriously* work on end burners, the publication of this book was three years overdue, the book was already more than 500 pages in length, and I had almost 3,000 people waiting for its completion. The urgency of that situation prevented me from taking the additional time needed to develop this idea and the actual motor designs. But I *did* manage to develop the tooling and the basic construction procedure for anyone who wants to experiment on their own.

● Construction Technique and Tooling

To begin with, you'll need a top quality cardboard motor casing. As of this writing, you can order the high-strength casings used by the model rocket industry (**pages 212 - 213**) from **New England Paper Tube, 173 Weeden St., Pawtucket, RI 02862**. But the setup charge makes it impractical to order fewer than 500 of any one size. If you're seriously interested in motor building, and can find some friends to share the cost of a large order, the *wholesale* price-per-casing, direct from **New England**, will be comparable to the *retail* price of a standard convolute fireworks tube. At the time of this writing, **New England** has a website at **http://www.nepapertube.com**. You can roll the casings yourself by following the instructions in **Chapter 8**, but to make a truly *top-quality* casing, you'll have to use the strongest tag board or liner board you can find.

Figure 10-106. *A set of tools for experimenting with end burners. The set in this photo is made of 304 stainless steel.*

Figure 10-105 illustrates an end-burning rocket motor. The method of construction is the same as that used for core burners, except that the core spindle extends into the propellant *a distance equal to the diameter of the nozzle throat*. For example, if the nozzle throat is 3/16" dia., then the spindle extends just 3/16" into the propellant. If the throat is 1/4" dia., then the spindle extends just 1/4" into the propellant. Because the distance is so short, it is important that it be fairly exact. If you make it much longer than required, the motor's initial pressure surge could exceed the design limits of the casing. It is therefore *very* important to check the height of the spindle above the nozzle clay before you load the propellant.

To produce an end burner with enough thrust to lift a rocket in vertical flight, you need a very *fast* burning propellant, and the one that I used for my own work is the one that I originally named the **KG6**. In my initial experiments I slowed its burn rate with baking soda, and I tried to use it in core burners, but troublesome case-bonding problems forced me to abandon the idea. *Without* the baking soda, its bonding properties are better, and its burn rate is sufficient for a reasonably good end burner. The **KG6** formula is:

◆ THE KG6 PROPELLANT

4 hour potassium nitrate	72.1%
Carbon black	11.3%
Red gum	7.0%
Powdered sulfur	9.6%
Total	100%

The performance of a core burner is adjusted by altering the propellant's baking soda content, or the length of the motor's core. Because an end burner's *has no core*, and because you *can't* add baking soda, your only option is to alter the diameter of the nozzle throat. To do this, I developed the special set of tools shown in *Figure 10-106*. It amounts to a flat base with a set of interchangeable core spindles. In *these* tools, the spindle forms not only the motor's nozzle throat, but its exit taper as well. Since a black powder motor's nozzle exit should be roughly twice the area of its throat, I made the diameter of each spindle's base equal to 1.414 (the square root of 2) times the spindle's diameter. The tools in the photo are used to build 1-1/8" i.d. motors, but in Chapter 5 I've provided tooling designs for other-size end burners as well.

When a motor's propellant formula and grain geometry are fixed, enlarging the nozzle throat reduces the chamber pressure and thrust. Making the throat smaller causes chamber pressure and thrust to *rise*. The nozzle throat diameter in a commercially manufactured, black powder end burner is about *one quarter* of the inside diameter of the motor casing.

That is, the nozzle throat in a 1" i.d. motor is 1/4" dia., and the nozzle throat in a 3/4" i.d. motor is 3/16" dia. If you make the **KG6** with a *cheaper* carbon black high in organics, it will *probably* burn *slower* than the commercial propellant, and you might have to make your nozzle throats correspondingly smaller. For example, a 1-1/8" i.d. end burner might work best with a 1/4" dia. throat, and the 1" and 3/4" i.d. versions of the same motor might require throats of 7/32" dia. and 5/32" dia. respectively. Some trial and error will be necessary.

When experimenting with end burners, for obvious safety reasons, you should always make your first motor with a nozzle throat *larger* than what you think you will need. Then gradually reduce the diameter until you achieve the performance you want. Cardboard motor casings are limited in the amount of pressure they can withstand. Because chamber pressure is directly related to motor thrust, you can avoid exceeding a casing's design limits by keeping the maximum thrust below a potentially hazardous "red line" value.

For standard fireworks-grade casings, I try to keep the maximum thrust *less than* 200 lbs. per square inch of nozzle throat area. You can calculate the maximum safe thrust for a nozzle throat of any diameter with the standard equation for the area of a circle, $A = \pi r^2$, where **A** is the circle's area in square inches, π is the number, **3.142**, and **r** is the circle's radius in inches. Since the radius of a circle equals half its diameter, you must first rewrite the equation as $A = \pi(d/2)^2$, where **d** is the diameter of the nozzle throat. Since motor thrust is a direct function of nozzle throat area, for any given nozzle, the maximum safe motor thrust equals the area of its throat (in square inches) times 200 lbs. For a fireworks grade casing, **maximum safe motor thrust = $\pi(d/2)^2$ x 200 lbs**.

EXAMPLE 10-2:

A motor's nozzle throat is 0.250" dia. For a standard fireworks grade casing:

Maximum safe motor thrust = 3.142 x (0.250/2)^2 x 200 lbs. = 0.049 x 200 lbs. = 9.8 lbs.

Rounded to approx. 10 lbs.

EXAMPLE 10-3:

A motor's nozzle throat is 0.500" dia. For a standard fireworks grade casing:

Maximum safe motor thrust = 3.142 x (0.500/2)^2 x 200 lbs. = 0.196 x 200 lbs. = 39.2 lbs.

Rounded to approx. 40 lbs.

For the high strength casings mentioned on **page 310,** you can increase these figures, but you'll have to establish the appropriate safety limits by experiment.

Boats, cars, and airplanes require motors with *very* long burn times and *very low* thrust. You can lower the thrust in a homemade end burner by making its nozzle throat smaller, and switching to a slower burning propellant. I'll leave it to *you* to design the smaller spindles, and I'll simply remind you to keep the maximum thrust at or below 200 lbs. per square inch of nozzle throat area. Keep the exit taper angle at 15 degrees, and make the base diameter of each exit taper cone about 1.4 times the diameter of the nozzle throat. For motors with *very* long burn times, it would also be advisable to use stainless steel De Laval nozzles.

When making end burners, a standard, cone-shaped propellant tamp forms the nozzle's convergent taper, and loads the first dose of propellant. But the main bulk of the propellant is loaded with a *flat tamp*. **Figures 10-107** through **10-112** illustrate the unique propellant-loading process. Shop drawings for the special spindles and bases are located in **Chapter 5** on **pages 115 - 120.**

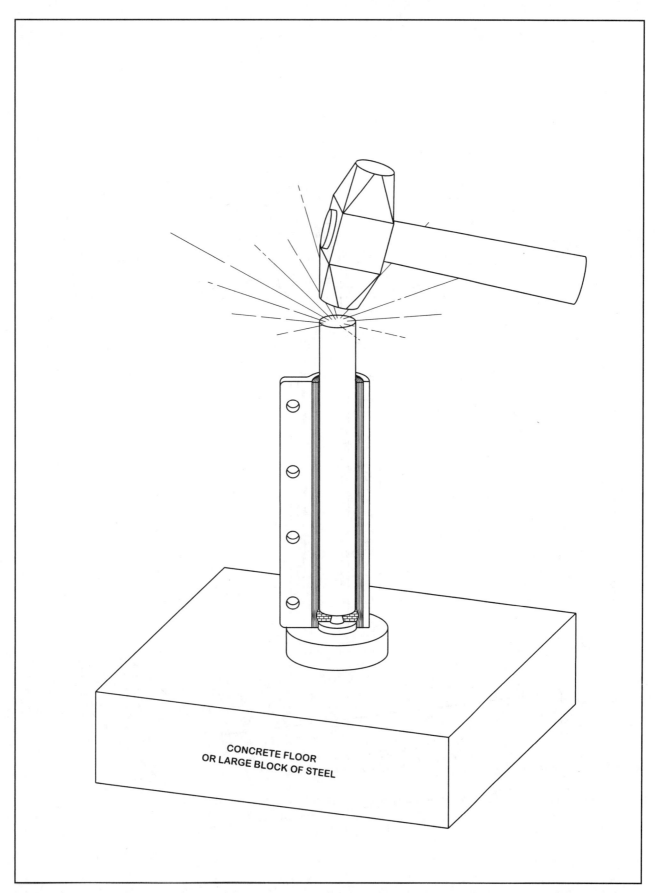

Figure 10-107. The first dose of nozzle clay is rammed into place with a starting tamp and 8 to 10 solid hammer blows.

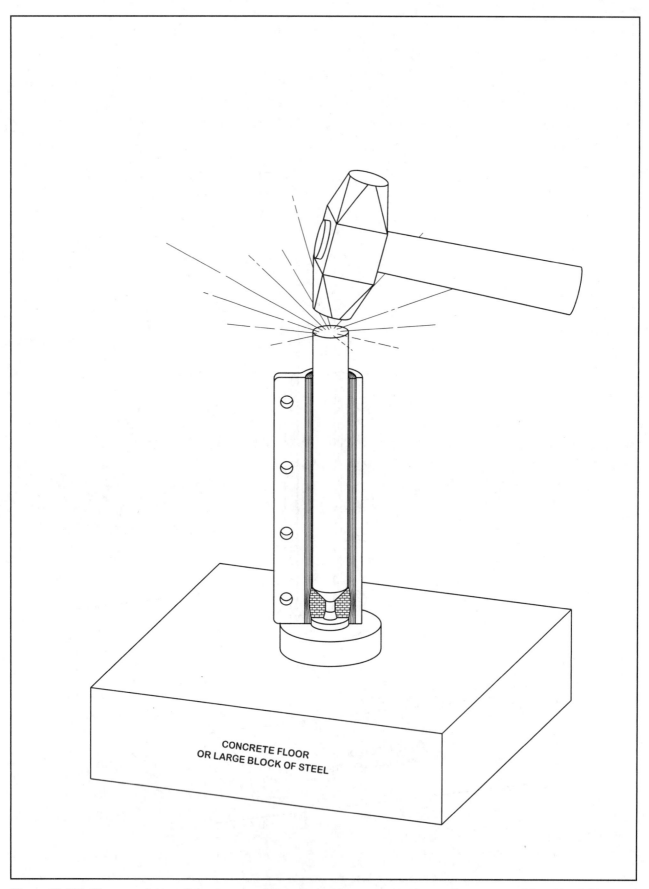

Figure 10-108. *The second dose of clay is rammed into place with a cone-tipped propellant tamp and 8 to 10 solid hammer blows. The cone-shaped tip forms the nozzle's convergent taper.*

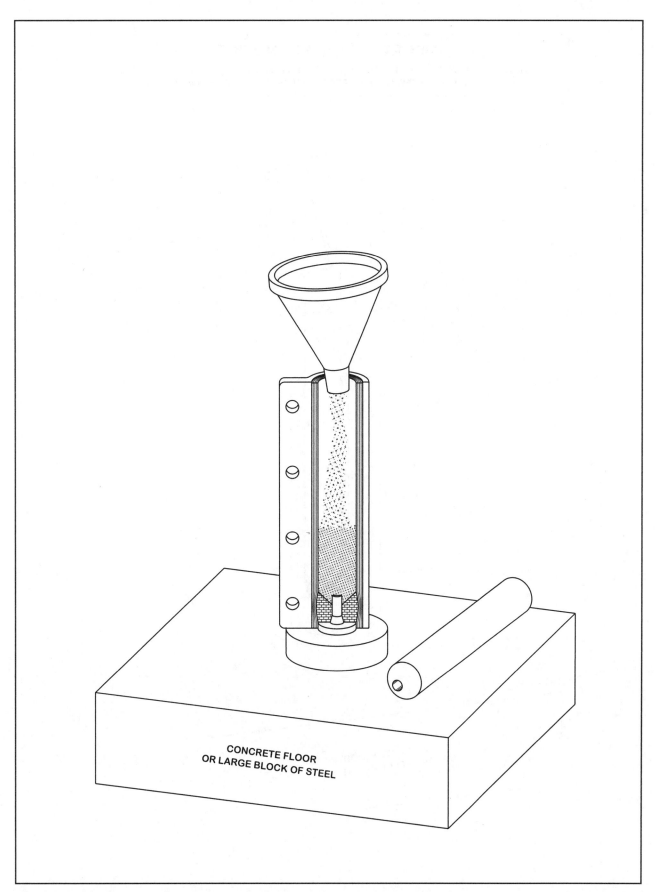

Figure 10-109. *The first dose of propellant is poured in on top of the finished nozzle.*

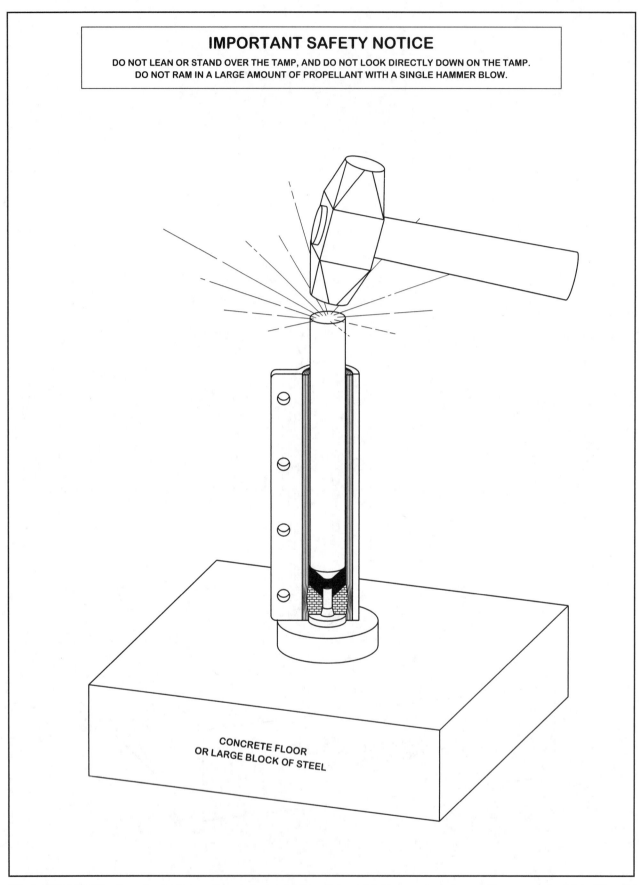

Figure 10-110. The first dose of propellant is rammed into place with a cone-tipped propellant tamp and 8 to 10 solid hammer blows.

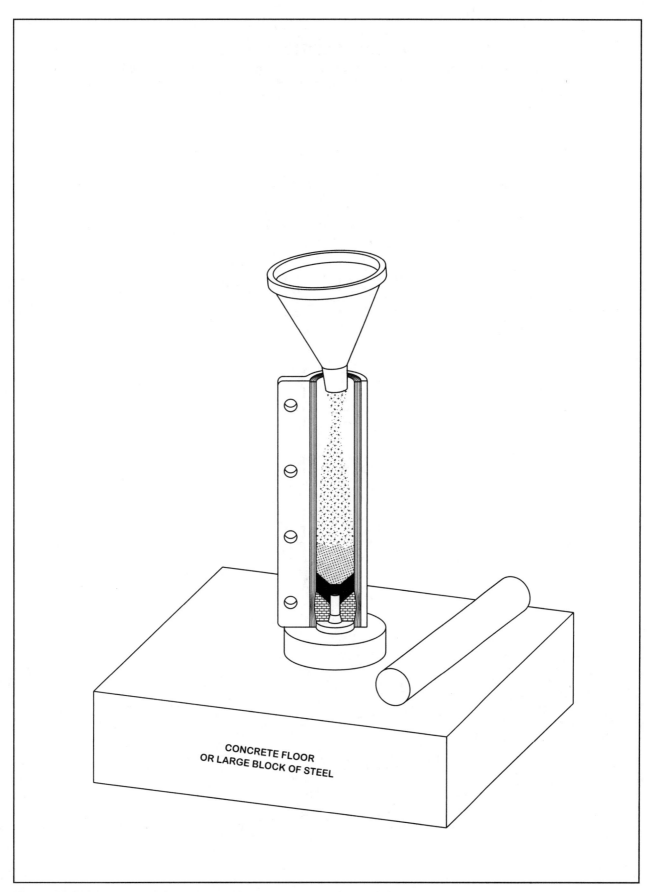

Figure 10-111. *The second dose of propellant is poured in on top of the first.*

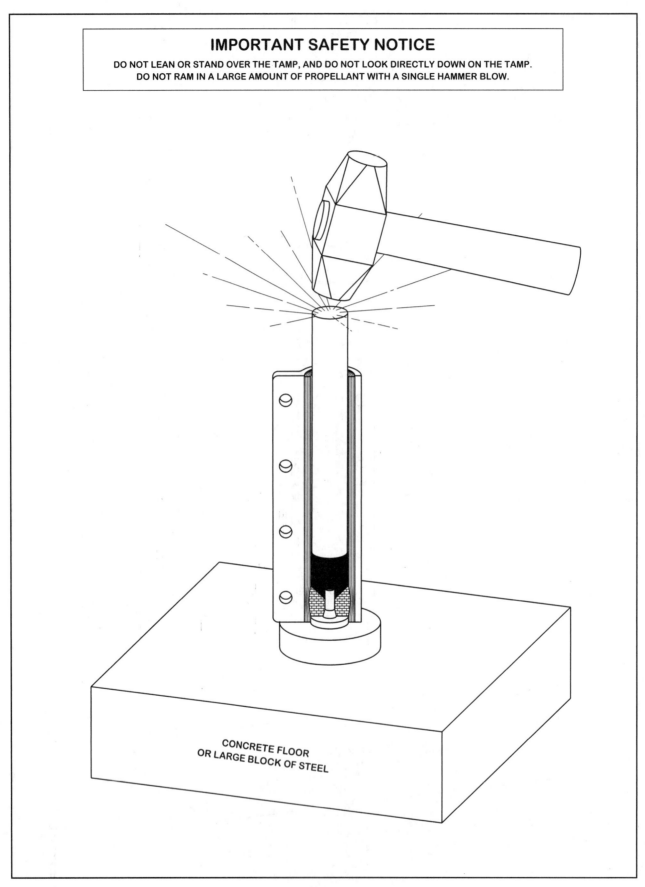

Figure 10-112. *The second dose of propellant, and all subsequent doses, are rammed into place with a flat tamp and 8 to 10 solid hammer blows. The flat tamp rams in the rest of the propellant. The time delay and the forward bulkhead are formed in the same manner used when making a core burner.*

Figure 10-113. Before-firing and after-firing photos illustrate a case of excessive nozzle erosion.

Improving Performance With De Laval Nozzles

Clay nozzles and standard casings work well for rocket motors with a specific impulse of up to 80 seconds. Above 80 seconds, the casings start to burst, and the higher velocity exhaust begins to erode the nozzle clay (*Figure 10-113*). For these reasons, I've *intentionally* kept the I_{sp} of the motors in this book *low*. If you want to improve the I_{sp}, you can do so by switching to De Laval nozzles, and high-strength motor casings. Note that, in the above-photos, I've *simulated* the high velocity erosion effect in a motor made with a *standard* casing by adding extra wax to the nozzle clay.

A De Laval nozzle accelerates the motor's exhaust more efficiently than a clay nozzle, and because it is made of steel, it eliminates the problem of nozzle erosion. Black powder has a theoretical maximum I_{sp} of 120 to 140 seconds. The better-grade **New England** casings are *very* strong, and when used in combination with De Laval nozzles, some of the propellants in this book will yield I_{sp}s of 100 seconds or more. The following text and photos explain how to install a De Laval nozzle in a cardboard tube. Shop drawings for the De Laval nozzles begin on **page 323**, and the drawings for the specialized tools are located on **pages 172** and **173**. Each nozzle has a flange that centers it in the casing (*Figure 10-114*), and the clay that holds the nozzle in place is rammed against this flange. The following example illustrates the installation of a De Laval nozzle with a 5/8" dia. nozzle throat in a 1-1/8" i.d. cardboard casing.

1. Cut a dry, glue coated, convolute casing to length. Clean up the ends with a belt sander, and mount it in a casing retainer per the instructions on **pages 274 - 278**.

2. Place the casing (in the retainer) *upside down* on a concrete floor or a heavy block of steel, and insert the specially-machined bar that I call **the anvil** (*Figure 10-115*). The anvil's purpose is to align the nozzle, and hold it in position as you ram it into place. The length of the anvil determines the nozzle's exact location in the motor casing. Each unique motor design requires its own unique anvil, and the anvil in this example is made from a piece of 3/4" schedule 40 steel pipe with the ends machined flat.

3. Place the DeLaval nozzle *with the flanged end down* into the motor casing, and press it *firmly down* onto the end of the anvil (*Figure 10-116*). If the fit is tight, you can tap it into place with a hammer and a block of wood.

4. Fill the ring-shaped area around the nozzle with loose nozzle clay (*Figure 10-117*). A black rubber stopper (i.e. a "cork") inserted into the nozzle throat keeps the excess clay from falling into the motor casing. You can buy rubber stoppers in various sizes at most hardware stores. Insert the tool that I call **the rammer** into the end of the casing, and hammer the clay firmly into place (*Figure 10-118*). Note that the nose of the rammer (*Figure 10-119*) is machined to fit the ring-shaped space around the nozzle.

Figure 10-114. A De Laval nozzle has a flange that centers it in the motor casing, and provides a surface against which to ram the clay that holds it in place.

Figure 10-115. The anvil is inserted into the motor casing. Its flat, machined end aligns the nozzle, and holds it in position as the clay is rammed into place.

Also note that, in this example, I'm using a special tool called a "dead-blow" hammer. It's a 2-1/2 pounder, and I bought it at Harbor Freight. Dead-blow hammers are hollow, and filled with a loose load of lead shot that flies to the inside of the face of the hammer on impact. This keeps the hammer from bouncing, and it transmits the full force of the blow to the object being struck. Their other advantage is that they are made of high-impact plastic, and they won't mar the striking end of the steel rammer. Dead-blow hammers are *very* effective, and you might want to experiment with them for other motor-building tasks.

5. Repeat **step 4** until the ring-shaped area around the nozzle is filled with tightly packed clay (***Figure 10-120***). Then lift the casing (in the casing retainer) off the floor. Remove the anvil, and dump out any loose clay.

6. Insert the core spindle into the nozzle mold base. Place the casing (in the casing retainer) *upright* over the core spindle - nozzle mold assembly (***Figure 10-121***), and load the propellant and the other motor components in the normal manner. Note that the cone on the nozzle mold is machined to fit the exit taper on the De Laval nozzle.

With a De Laval nozzle and a stronger casing, you can significantly raise a motor's maximum chamber pressure and thrust, but the process of learning how far you can go is one of trial and error. As you explore the upper limits of the new casings' strength, you'll most certainly explode a few motors. This being the case, be sure to conduct your tests in a place where there is **nothing flammable. Never launch an untested motor into the air**. Ignite the motors *electrically* (**never** use a fuse), and stand *a generously safe distance away* when conducting a test.

To increase a motor's specific impulse and total impulse, you must increase its maximum chamber pressure, and there are 3 ways to do this.

1. Reduce the propellant's baking soda content, and thereby increase its burn rate.

2. Increase the length of the propellant core, and thereby increase the physical surface area of propellant burning at any given moment.

Figure 10-116. The De Laval nozzle is pressed firmly down onto the end of the anvil. If the fit is tight, you can tap it into place with a hammer and a block of wood.

Figure 10-117. The ring-shaped area around the nozzle is filled with loose nozzle clay. A rubber stopper keeps excess clay from falling into the casing.

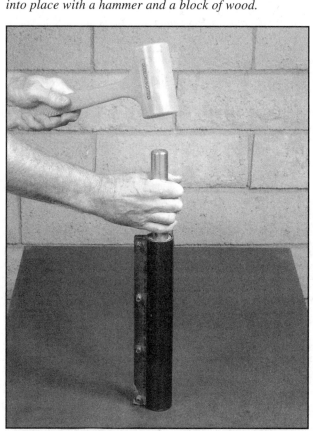

*Figure 10-118. The rammer shown in **Figure 10-119** is used to hammer the clay into place.*

Figure 10-119. The nose of the rammer is machined to fit the narrow ring-shaped area around the nozzle.

321

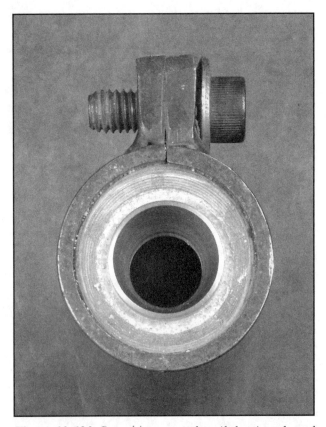

Figure 10-120. Step 4 is repeated until the ring-shaped space between the nozzle and the inside of the casing wall is full of tightly packed clay.

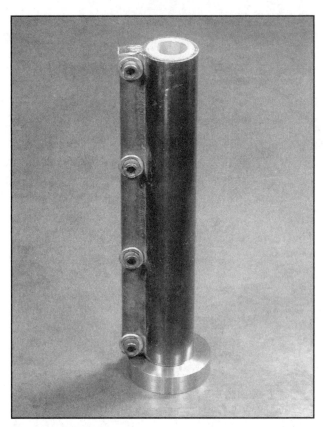

Figure 10-121. The casing (in the casing retainer) with the De Laval nozzle installed is placed upright over the core spindle-nozzle mold assembly. The propellant and the other motor components are then loaded in the normal manner.

3. Reduce the diameter of the nozzle throat, *but for the homemade propellants described in this book*, keep the diameter of the nozzle exit equal to *approximately* 1.4 times the diameter of the throat.

As a motor's chamber pressure rises, its maximum thrust increases, its burn time shortens, and its total impulse and specific impulse increase. A stainless steel De Laval nozzle erodes out .001" to .002" per firing, so you can reuse a De Laval nozzle at least a dozen times before you see any measurable decrease in performance. When performance begins to fall, you can compensate with a reduction in the propellant's baking soda content.

To recover a used De Laval nozzle from a spent motor casing, soak the casing in water for a few hours. Peel away the paper, and crumble away the clay with your fingers. Then clean the charred stainless steel with steel wool, hot water, and a strong detergent.

To provide you with some De Laval experience before you go off and experiment on your own, I've included 3 De Laval motor designs in this book. They are the **KG1-193-H144** (**pages 346 - 347**), the **KG3-140-G139** (**pages 366 - 367**), and the **NG6-227-H137** (**pages 416 - 417**). They are all designed to be made with standard *fireworks-grade* casings. If you want to improve their performance, then build them with the high-strength casings described on **page 319**.

An Important and Helpful Cleanup Note

Though I don't show it in the photos (because it would get in the way), you can make post-motor-construction cleanup easier by building your motors in a shallow, sheet metal muffin pan or cake pan. Place the pan on the work surface. Then place the nozzle mold-core spindle assembly *in the pan*, and build the motor *in the pan*. The pan will catch the small amounts of propellant and clay that normally spill during motor construction, and your work area will stay noticeably cleaner.

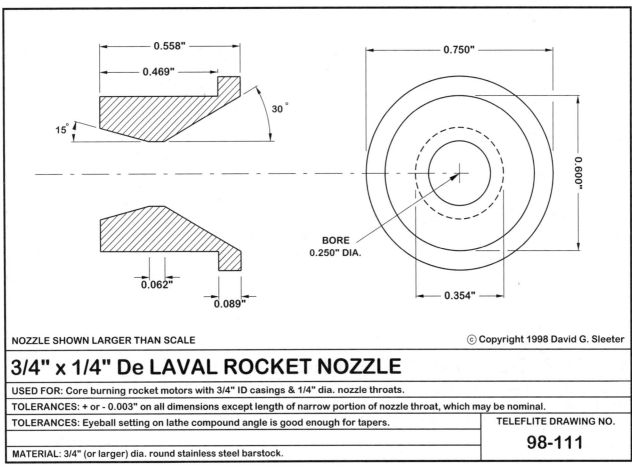

3/4" x 1/4" De LAVAL ROCKET NOZZLE

USED FOR: Core burning rocket motors with 3/4" ID casings & 1/4" dia. nozzle throats.
TOLERANCES: + or - 0.003" on all dimensions except length of narrow portion of nozzle throat, which may be nominal.
TOLERANCES: Eyeball setting on lathe compound angle is good enough for tapers.
MATERIAL: 3/4" (or larger) dia. round stainless steel barstock.

TELEFLITE DRAWING NO.

98-111

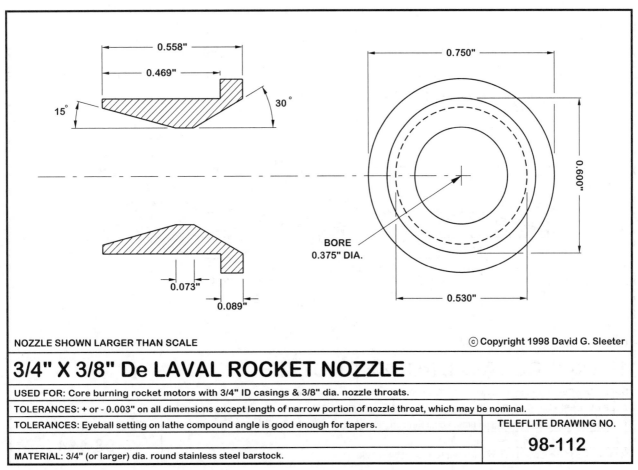

3/4" X 3/8" De LAVAL ROCKET NOZZLE

USED FOR: Core burning rocket motors with 3/4" ID casings & 3/8" dia. nozzle throats.
TOLERANCES: + or - 0.003" on all dimensions except length of narrow portion of nozzle throat, which may be nominal.
TOLERANCES: Eyeball setting on lathe compound angle is good enough for tapers.
MATERIAL: 3/4" (or larger) dia. round stainless steel barstock.

TELEFLITE DRAWING NO.

98-112

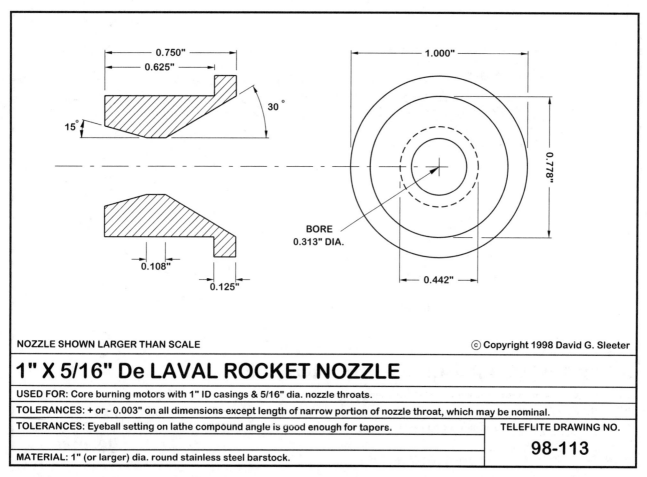

1" X 5/16" De LAVAL ROCKET NOZZLE

USED FOR: Core burning motors with 1" ID casings & 5/16" dia. nozzle throats.

TOLERANCES: + or - 0.003" on all dimensions except length of narrow portion of nozzle throat, which may be nominal.

TOLERANCES: Eyeball setting on lathe compound angle is good enough for tapers.

MATERIAL: 1" (or larger) dia. round stainless steel barstock.

TELEFLITE DRAWING NO. **98-113**

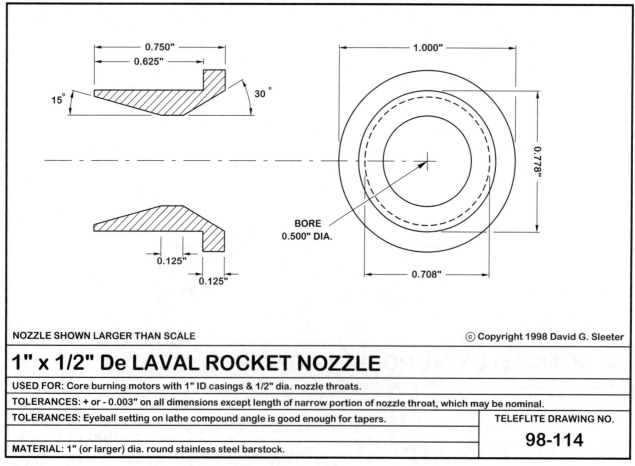

1" x 1/2" De LAVAL ROCKET NOZZLE

USED FOR: Core burning motors with 1" ID casings & 1/2" dia. nozzle throats.

TOLERANCES: + or - 0.003" on all dimensions except length of narrow portion of nozzle throat, which may be nominal.

TOLERANCES: Eyeball setting on lathe compound angle is good enough for tapers.

MATERIAL: 1" (or larger) dia. round stainless steel barstock.

TELEFLITE DRAWING NO. **98-114**

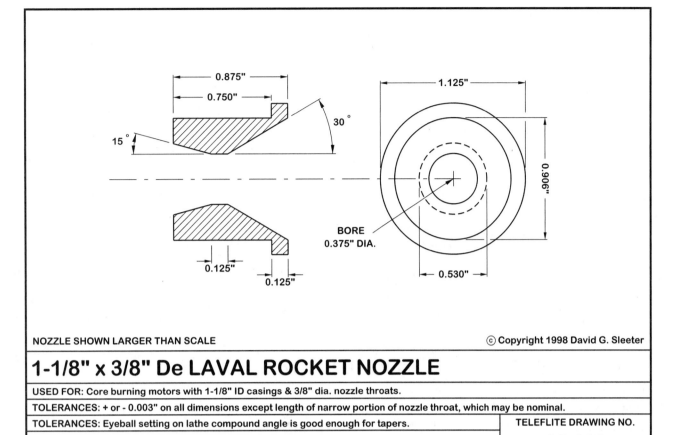

1-1/8" x 3/8" De LAVAL ROCKET NOZZLE

USED FOR: Core burning motors with 1-1/8" ID casings & 3/8" dia. nozzle throats.

TOLERANCES: + or - 0.003" on all dimensions except length of narrow portion of nozzle throat, which may be nominal.

TOLERANCES: Eyeball setting on lathe compound angle is good enough for tapers.

MATERIAL: 1-1/8" (or larger) dia. round stainless steel barstock.

TELEFLITE DRAWING NO.
98-115

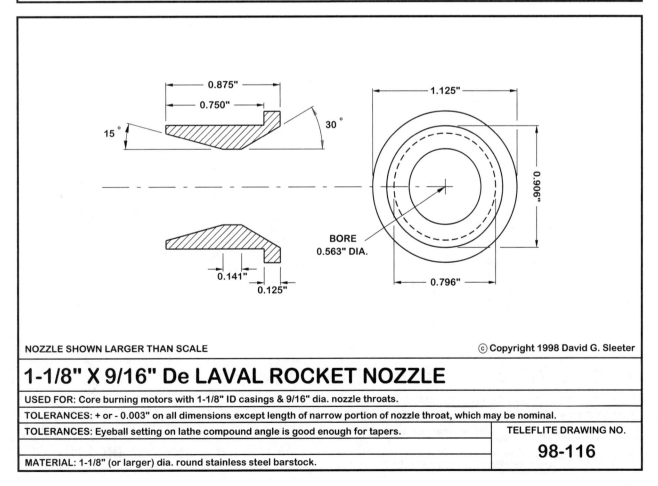

1-1/8" X 9/16" De LAVAL ROCKET NOZZLE

USED FOR: Core burning motors with 1-1/8" ID casings & 9/16" dia. nozzle throats.

TOLERANCES: + or - 0.003" on all dimensions except length of narrow portion of nozzle throat, which may be nominal.

TOLERANCES: Eyeball setting on lathe compound angle is good enough for tapers.

MATERIAL: 1-1/8" (or larger) dia. round stainless steel barstock.

TELEFLITE DRAWING NO.
98-116

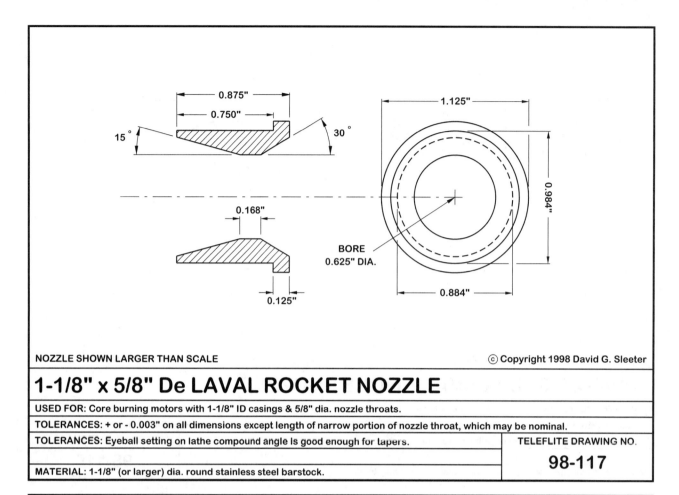

NOZZLE SHOWN LARGER THAN SCALE © Copyright 1998 David G. Sleeter

1-1/8" x 5/8" De LAVAL ROCKET NOZZLE

USED FOR: Core burning motors with 1-1/8" ID casings & 5/8" dia. nozzle throats.

TOLERANCES: + or - 0.003" on all dimensions except length of narrow portion of nozzle throat, which may be nominal.

TOLERANCES: Eyeball setting on lathe compound angle is good enough for tapers.

TELEFLITE DRAWING NO.
98-117

MATERIAL: 1-1/8" (or larger) dia. round stainless steel barstock.

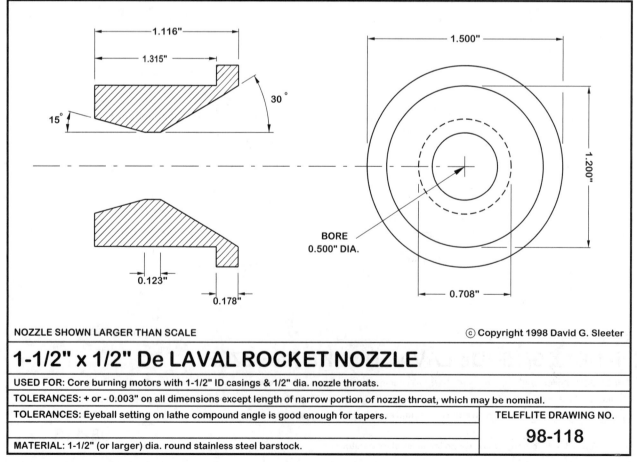

NOZZLE SHOWN LARGER THAN SCALE © Copyright 1998 David G. Sleeter

1-1/2" x 1/2" De LAVAL ROCKET NOZZLE

USED FOR: Core burning motors with 1-1/2" ID casings & 1/2" dia. nozzle throats.

TOLERANCES: + or - 0.003" on all dimensions except length of narrow portion of nozzle throat, which may be nominal.

TOLERANCES: Eyeball setting on lathe compound angle is good enough for tapers.

TELEFLITE DRAWING NO.
98-118

MATERIAL: 1-1/2" (or larger) dia. round stainless steel barstock.

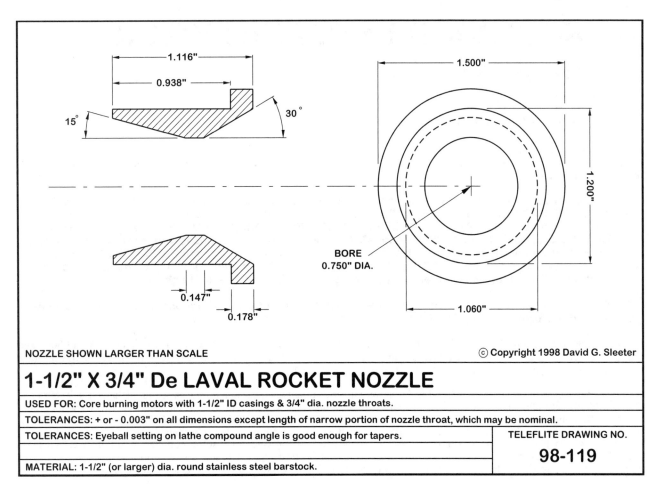

1-1/2" X 3/4" De LAVAL ROCKET NOZZLE

USED FOR: Core burning motors with 1-1/2" ID casings & 3/4" dia. nozzle throats.

TOLERANCES: + or - 0.003" on all dimensions except length of narrow portion of nozzle throat, which may be nominal.

TOLERANCES: Eyeball setting on lathe compound angle is good enough for tapers.

MATERIAL: 1-1/2" (or larger) dia. round stainless steel barstock.

TELEFLITE DRAWING NO. **98-119**

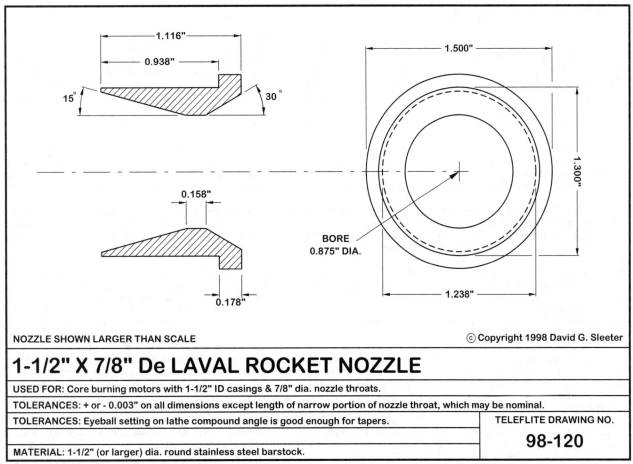

1-1/2" X 7/8" De LAVAL ROCKET NOZZLE

USED FOR: Core burning motors with 1-1/2" ID casings & 7/8" dia. nozzle throats.

TOLERANCES: + or - 0.003" on all dimensions except length of narrow portion of nozzle throat, which may be nominal.

TOLERANCES: Eyeball setting on lathe compound angle is good enough for tapers.

MATERIAL: 1-1/2" (or larger) dia. round stainless steel barstock.

TELEFLITE DRAWING NO. **98-120**

11. KG1 PROPELLANT AND MOTOR DESIGNS

● First and Most Important!

Before you proceed, if you haven't read **Chapter 1**, go back and read it now. It contains important information on the safe handling of this rocket propellant. Also pay attention to the bold warning at the beginning of **Chapter 1**. To emphasize its importance, I'll repeat that warning here.

> Do *not* add or substitute other chemicals either to-or-for the chemicals described in this book. Specifically do *not* use POTASSIUM CHLORATE, SODIUM CHLORATE, POTASSIUM PERCHLORATE, SODIUM PERCHLORATE, POTASSIUM PERMANGANATE, AMMONIUM PERCHLORATE, PHOSPHOROUS, MATCH HEADS, ALUMINUM, MAGNESIUM, or ANY METALS AT ALL! *All* of these substances will form dangerous and explosive mixtures when combined with the other ingredients described herein, *AND OTHER CHEMICALS WILL FORM DANGEROUS MIXTURES AS WELL*. When mixing and handling propellants, be sure to ground yourself and all containers and utensils according to the instructions on page 14.
>
> *If you decide to experiment on your own outside the bounds of the instructions provided in this book, consult with a professional chemist before you proceed.*

The **KG1** propellant is a milled mixture of **4 hour milled potassium nitrate**, **12 hour milled anthracite**, **carbon black**, **red gum**, and **sulfur**. If you haven't read **Chapters 5** and **6**, go back and read them now. Together they explain how to buy or make these chemicals, and how to prepare them for use in a rocket propellant.

If all other parameters are equal, then the **KG1** is the slower burning of the two potassium nitrate-based propellants. But its burn rate can vary *greatly* depending on the exact nature and purity of its chemicals. The quality of the carbon black is particularly important. A carbon black high in residual organics (i.e. oil) makes a slow burning propellant, and a pure, electrically conductive carbon black makes a fast burning propellant. The *exact* burn rate can be adjusted by substituting slower burning anthracite for some of the fast burning carbon black, or by adding or subtracting baking soda. When compressed into a solid propellant grain, the KG1's density is **25 to 30 gms. per cubic inch**. To prepare the **KG1** propellant, set up a powder mill (**Chapter 3**). Then mix the propellant (**Chapter 9**), and use the figures below to determine the exact weights of the chemicals needed for the desired batch size.

◆ THE KG1 PROPELLANT

4 hr. milled potassium nitrate	72.1%	721.0 gms.	360.5 gms.	180.2 gms.	90.0 gms.
12 hr. milled anthracite	7.3%	73.0 gms.	36.5 gms.	18.3 gms.	9.2 gms.
Carbon black	4.0%	40.0 gms.	20.0 gms.	10.0 gms.	5.0 gms.
Red gum	7.0%	70.0 gms.	35.0 gms.	17.5 gms.	8.8 gms.
Sulfur	9.6%	96.0 gms.	48.0 gms.	24.0 gms.	12.0 gms.
Total batch size	**100%**	**1000.0 gms.**	**500.0 gms.**	**250.0 gms.**	**125.0 gms.**

Before you can use the propellant, *you must perform the pellet test* (**pages 238 - 245**). The pellet test determines the exact amount of solvent needed to form 20 grams of propellant into a dense and solid propellant grain. The correct amount of solvent is called **the solvent requirement**. *The solvent requirement is critical, and cannot be guessed at*. The pellet test is time consuming, *but you only do it twice*. Perform the *first* pellet test with the **KG1** propellant *plus 30% baking soda*, and do it *before* you make your first motor.

To build your *first* motor with the **KG1** propellant, weigh out the amount indicated in the motor drawing (rounding off the numbers makes the math easier). Then add 30% of that weight in baking soda, and mix it in thoroughly with a fork. If, for example, you've weighed out 100 grams of propellant, add 30 grams of baking soda for a total of 130 grams. Mix in the proper amount of solvent as determined by your first pellet test. Then build and dry the motor (**Chapter 10**).

Per the instructions on **page 494**, do a rudimentary test by firing the motor (with the motor pointing *down* and the exhaust aimed *upward*) from a pipe in the ground. If this first motor "chuffs" (chugs like a steam engine), build a *second* motor, and reduce the baking soda to 28%. Keep building, testing, and reducing the amount of baking soda in 2% increments until you get a motor that burns smoothly and doesn't chuff. At first the chuffing may be slow. As you approach a condition of combustion stability, the frequency of the chuffs will increase (the chuffing sound will get *faster*). When you're *very* near stability, the amplitude of the chuffs will decrease (the chuffs will sound *fainter*).

Now set up a thrust-testing stand (**page 470**), and test the motor for maximum thrust (**page 475**). In the tests that follow, reduce the soda in 1% increments until you get a motor whose maximum thrust approximates the maximum thrust printed in the motor drawing. At that point you'll have a working motor with the maximum safe chamber pressure and thrust for that particular design. As you approach a final, successful soda content, *repeat the pellet test*, and use the solvent requirement indicated by this *second* test to zero-in the *final* design. Keep accurate notes. Write down the final solvent requirement and the amount of baking soda that works the best, and *forever thereafter* use those figures to build copies of your final, successful motor.

Important note. You can also alter the chamber pressure and thrust by changing the length of the motor's core. A *longer* core will *increase* the chamber pressure and thrust, and a *shorter* core will *reduce* the chamber pressure and thrust. But, of course, changing the length of the core will also change the amount of propellant required, and the physical size of the motor.

● Improving Motor Performance

Now you have a working rocket motor, but some of the material inside the motor isn't rocket propellant. It's *baking soda*, and baking soda is *dead weight*. It doesn't contribute anything to the motor's total impulse. You added the soda to control chamber pressure and thrust, but you can *also* control those parameters by altering the balance between the carbon black and the anthracite. And unlike baking soda, the anthracite is a *fuel*. It takes part in the reaction, and it *contributes* to the motor's total impulse.

In the following hypothetical illustration, I'll call the original motor a **KG1-13-D18**. I'll stipulate that its maximum thrust is 9 lbs., and I'll stipulate that it takes 10% baking soda to make it work as intended. To *improve* the motor's performance, I'll have to get rid of some of the soda. Of course when I do that, the propellant's burn rate will increase. But I can *reverse* the increase, and *compensate* for the loss of the soda by replacing some of the carbon black with slower burning anthracite. **Important note**. *Each time you change a propellant's formula, you must repeat the pellet test*. Skipping the pellet test is *not* an option. Possible replacement formulas are:

KG1X1	KG1X2	KG1X3
4 hr. mill. potassium nitrate 72.1%	4 hr. mill. potassium nitrate 72.1%	4 hr. mill. potassium nitrate 72.1%
12 hr. mill. anthracite 8.3%	12 hr. mill. anthracite 9.3%	12 hr. mill. anthracite 10.3%
Carbon black 3.0%	Carbon black 2.0%	Carbon black 1.0%
Red gum 7.0%	Red gum 7.0%	Red gum 7.0%
Sulfur... 9.6%	Sulfur 9.6%	Sulfur 9.6%

In the formulas above, though the balance between the anthracite and the carbon black changes (in 1% increments), the *total* amount of carbon fuel remains the same (11.3%). If I build a motor with the KG1X1 propellant, and add 10% baking soda, because the anthracite burns *slower* than the carbon black it replaced, the motor's maximum thrust will be *less* than 9 pounds. To restore the maximum thrust to its original level, I'll keep building, testing, and reducing the soda in 1% increments until the maximum thrust is back up to 9 lbs. I'll call this second motor a **KG1X1** (named after its propellant), and I'll pretend here that it took 8% soda to make it work.

If the **KG1X1** with 8% soda burns smoothly and doesn't chuff, I'll build an identical motor with the KG1X2 propellant. Of course when I do that, the maximum thrust will drop again, so I'll keep building, testing, and *again* reducing the soda in 1% increments until the maximum thrust equals 9 lbs. I'll call this third motor a **KG1X2**, and for the purpose of this illustration, I'll pretend that it took 6% baking soda to make it work. If the **KG1X2** burns smoothly and doesn't chuff, I'll repeat the experiment with the KG1X3 propellant. If the **KG1X2** chuffs and *doesn't* burn smoothly, I'll know that the KG1X2 propellant *doesn't* work; that I need *more* than 2.0% carbon black to achieve combustion stability, and that the KG1X1 propellant with 8% baking soda works the best.

If I now build a chart recorder (**page 478**), and generate thrust-time curves for the **KG1-13-D18** and **KG1X1** motors, I'll see that the **KG1X1** has a slightly longer burn time, and a greater total impulse. The **KG1X1** might turn out to be a KG1-*15*-D18, and I'll adopt the KG1X1 propellant with 8% baking soda for all the motors that follow. When you perform these experiments yourself, you'll eventually reach a point where the motors begin to chuff. The amount of carbon black needed to maintain combustion stability will vary according to its purity, and the process of finding the minimum will be one of trial and error.

Important note. *The KG1 motors in this book were all made with propellants milled for 12 hours*. When making your own propellants, you can reduce the milling time to as little as 4 hours, and compensate by adjusting the amount of baking soda. The process of finding the minimum milling time is one of trial and error, *and each time you change the milling time, you must repeat* **the pellet test**.

The KG1 propellant exhibits good case bonding properties with the **KG1-38-E33** and the **KG1-149-G104** showing just a vestige of the effect discussed on **pages 248 - 249**. All the **KG1** motors were made with standard fireworks grade casings, and all but the **KG1-193-H144** were made with clay nozzles. By switching to stainless steel De Laval nozzles and the high strength casings manufactured (as of this writing) by **New England Paper Tube** (**pages 163 - 164**), you can increase the motors' maximum chamber pressure and thrust, and substantially improve their performance. Shop drawings for De Laval nozzles begin on the **page 323**.

With each motor design, I've included the thrust-time curve and motor performance data generated by the test equipment and software described on **pages 468 - 469**, plus two flight performance estimates generated by **Rogers' Aeroscience Alt4** rocket performance prediction software. The "minimum rocket" is what you might build if you were trying to break an altitude or speed record. The lift-off weight is minimal. The body diameter is equal to the o.d. of the motor, and the parachute compartment is small; approx. 5 body diameter's in length. The "average rocket" is more like what you would build for scientific research or general sport flying. The body diameter is one of the industry standards approx. twice the diameter of the motor, and there's ample space and weight allowance for a large parachute and a small scientific payload.

In each case the lift-off weight is *very* approximate, and based on an educated guess. The *extremely* accurate performance specs. *are the actual numbers generated by the Alt-4 program*. They are based on the rocket's stated length, diameter, and lift-off weight, and the *tested* motor's thrust-time data. *In the real world*, variations in motor performance, the size and shape of the rocket's nosecone and fins, and even the smoothness of its paint-job will cause the rocket's performance to vary from the stated figures.

● Important G-Force Advisory

Static testing perfects a motor's performance, but it *doesn't* predict how the hardened propellant grain will respond to high G-forces during an actual rocket flight. Baking soda is *soft*, and the carbon fuels used in these propellants vary *greatly* in hardness. So variations in the soda content and the type of carbon fuel will effect the strength of the finished propellant grain. If a soft propellant grain collapses at maximum acceleration, the motor will self-destruct, and the rocket will be lost in flight. Initial flight testing of a new motor design should *always* be conducted with this possibility in mind, particularly when testing the motor in a small, high-acceleration airframe. It is always advisable to conduct your early tests in a "throwaway" rocket, i.e. something cheap that you don't mind losing. When you're sure that you've got a successful motor-airframe combination, *then* invest in something nicer.

TELEFLITE CORPORATION

MOTOR NAME: KG1-6-C6
PROPELLANT TYPE: KG1
CASING TYPE: CONVOLUTE PAPER
NOZZLE TYPE: RAMMED CLAY

INDEX OF REQUIRED TOOLING

CASING RETAINER	DRAWING #98-001.5"	PAGE 90
NOZZLE MOLD	DRAWING #98-006	PAGE 95
CORE SPINDLE	DRAWING #98-035-C	PAGE 121
STARTING TAMP	USE PROPELLANT TAMP	
PROPELLANT TAMP	DRAWING # 98-082-A	PAGE 145
FLAT TAMP	DRAWING #98-084-B	PAGE 147
PORTHOLE TAMP	DRAWING #98-084-A	PAGE 147
PORTHOLE WIRE	DRAWING #98-084	PAGE 147

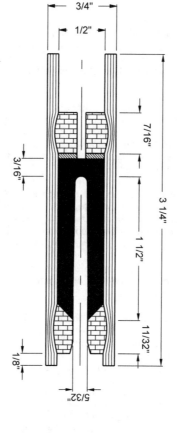

MOTOR SHOWN WITHOUT TIME DELAY

© COPYRIGHT 1998 DAVID G. SLEETER

IMPORTANT BAKING SODA NOTE: Add 30% baking soda to the propellant in your first motor. Reduce the soda to the propellant until you get a motor that doesn't chuff. Then reduce the soda in 1% increments until maximum thrust is approx. 3-1/2 lbs. This test motor took 10% soda, but I made it with a carbon black of medium purity. A motor made with a pure, electrically-conductive carbon black might take substantially more.

TOTAL WEIGHT: 28.4 GMS.	**CASING O.D.:** 0.75"	**TOT. IMPULSE:** 6.05 NT.-SECS.
PROPELLANT WT.: 10.5 GMS.	**NOZZLE CLAY:** 3.3 GMS.	**AVG. THRUST:** 6.60 NTS.
BAKING SODA: See baking soda note.	**BULKHEAD CLAY:** 3.3 GMS.	**MAX. THRUST:** 3.45 LBS.
MOTOR LENGTH: 3.25"	**THROAT DIA.:** 0.1563"	**BURN TIME:** 0.92 SECS.
CASING I.D.: 0.50"	**CORE LENGTH:** 1.50"	**SPEC. IMPULSE:** 58.65 SECS.

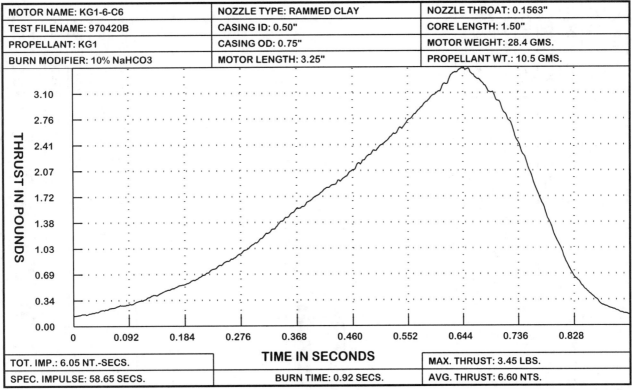

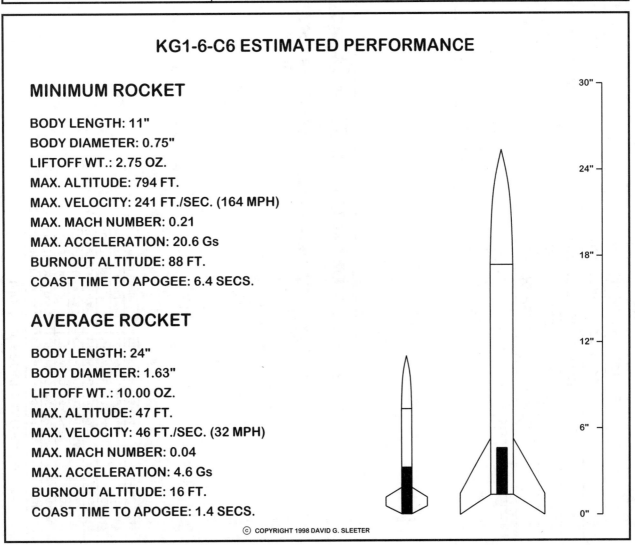

TELEFLITE CORPORATION

MOTOR NAME:	KG1-12-D18
PROPELLANT TYPE:	KG1
CASING TYPE:	CONVOLUTE PAPER
NOZZLE TYPE:	RAMMED CLAY

MOTOR SHOWN WITHOUT TIME DELAY

INDEX OF REQUIRED TOOLING

CASING RETAINER	DRAWING #98-001: 7"	PAGE 90
NOZZLE MOLD	DRAWING #98-007	PAGE 96
CORE SPINDLE	DRAWING #98-039-F	PAGE 123
STARTING TAMP	USE PROPELLANT TAMP	
PROPELLANT TAMP	DRAWING #98-082-B	PAGE 145
FLAT TAMP	DRAWING #98-084-B	PAGE 147
PORTHOLE TAMP	DRAWING #98-084-A	PAGE 147
PORTHOLE WIRE	DRAWING #98-084	PAGE 147

IMPORTANT BAKING SODA NOTE: Add 30% baking soda to the propellant in your first motor. Reduce the soda in 1% increments until you get a motor that doesn't chuff. Then reduce the soda in 2% increments until maximum thrust is approx. 9 lbs. This test motor took 10% soda, but I made it with a carbon black of medium purity. A motor made with a pure, electrically-conductive carbon black might take substantially more.

CASING I.D.: 0.50"	CORE LENGTH: 3.75"
MOTOR LENGTH: 5.50"	THROAT DIA.: 0.25"
BAKING SODA: See baking soda note.	BULKHEAD CLAY: 3.3 GMS.
PROPELLANT WT.: 22.4 GMS.	NOZZLE CLAY: 3.3 GMS.
TOTAL WEIGHT: 46.6 GMS.	CASING O.D.: 0.75"
TOT. IMPULSE: 12.91 NT.-SECS.	AVG. THRUST: 18.08 NTS.
MAX. THRUST: 8.89 LBS.	BURN TIME: 0.71 SECS.
SPEC. IMPULSE: 58.65 SECS.	

© COPYRIGHT 1998 DAVID G. SLEETER

MOTOR NAME: KG1-12-D18	NOZZLE TYPE: RAMMED CLAY	NOZZLE THROAT: 0.25"
TEST FILENAME: 970420G	CASING ID: 0.50"	CORE LENGTH: 3.75"
PROPELLANT: KG1	CASING OD: 0.75"	MOTOR WEIGHT: 46.6 GMS.
BURN MODIFIER: 10% NaHCO3	MOTOR LENGTH: 5.50"	PROPELLANT WT.: 22.4 GMS.

TOT. IMP.: 12.91 NT.-SECS.		MAX. THRUST: 8.89LBS.
SPEC. IMPULSE: 58.67 SECS.	BURN TIME: 0.71 SECS.	AVG. THRUST: 18.08 NTS.

KG1-12-D18 ESTIMATED PERFORMANCE

MINIMUM ROCKET

BODY LENGTH: 13"
BODY DIAMETER: 0.75"
LIFTOFF WT.: 3.4 OZ.
MAX. ALTITUDE: 1,929 FT.
MAX. VELOCITY: 459 FT./SEC. (313 MPH)
MAX. MACH NUMBER: 0.41
MAX. ACCELERATION: 47.9 Gs
BURNOUT ALTITUDE: 148 FT.
COAST TIME TO APOGEE: 9.6 SECS.

AVERAGE ROCKET

BODY LENGTH: 24"
BODY DIAMETER: 1.63"
LIFTOFF WT.: 10.6 OZ.
MAX. ALTITUDE: 252 FT.
MAX. VELOCITY: 123 FT./SEC. (84 MPH)
MAX. MACH NUMBER: 0.11
MAX. ACCELERATION: 13.1 Gs
BURNOUT ALTITUDE: 39.6 FT.
COAST TIME TO APOGEE: 3.6 SECS.

© COPYRIGHT 1998 DAVID G. SLEETER

TELEFLITE CORPORATION

MOTOR NAME:	KG1-18-D12
PROPELLANT TYPE:	KG1
CASING TYPE:	CONVOLUTE PAPER
NOZZLE TYPE:	RAMMED CLAY

MOTOR SHOWN WITHOUT TIME DELAY

Dimensions shown on motor diagram: 1", 3/4", 11/16", 9/32", 4 3/4", 2 1/4", 19/32", 1/8", 1/4"

© COPYRIGHT 1998 DAVID G. SLEETER

INDEX OF REQUIRED TOOLING

CASING RETAINER	DRAWING #98-002: 7"	PAGE 91
NOZZLE MOLD	DRAWING #98-008	PAGE 97
CORE SPINDLE	DRAWING #98-044-F	PAGE 126
STARTING TAMP	DRAWING #98-085-A	PAGE 148
PROPELLANT TAMP	DRAWING #98-086-A	PAGE 149
FLAT TAMP	DRAWING #98-091-B	PAGE 154
PORTHOLE TAMP	DRAWING #98-091-A	PAGE 154
PORTHOLE WIRE	DRAWING #98-091	PAGE 154

IMPORTANT BAKING SODA NOTE: Add 30% baking soda to the propellant in your first motor. Reduce the soda to the point that doesn't chuff. Then reduce the soda in 1% increments until you get a motor that doesn't chuff. Then reduce the soda in 1% increments until maximum thrust is approx. 9 lbs. This test motor took 10% soda, but I made it with a carbon black of medium purity. A motor made with a pure, electrically-conductive carbon black might take substantially more.

TOTAL WEIGHT: 71.3 GMS.	CASING O.D.: 1.00"	TOT. IMPULSE: 18.17 NT.-SECS.	
PROPELLANT WT.: 32.1 GMS.	NOZZLE CLAY: 11.0 GMS.	AVG. THRUST: 12.44 NTS.	
BAKING SODA: See baking soda note.	BULKHEAD CLAY: 11.0 GMS.	MAX. THRUST: 8.47 LBS.	
MOTOR LENGTH: 4.75"	THROAT DIA.: 0.25"	BURN TIME: 1.46 SECS.	
CASING I.D.: 0.75"	CORE LENGTH: 2.25"	SPEC. IMPULSE: 57.62 SECS.	

MOTOR NAME: KG1-18-D12	NOZZLE TYPE: RAMMED CLAY	NOZZLE THROAT: 0.25"
TEST FILENAME: 970421J	CASING ID: 0.75"	CORE LENGTH: 2.25"
PROPELLANT: KG1	CASING OD: 1.00"	MOTOR WEIGHT: 71.3 GMS.
BURN MODIFIER: 10% NaHCO3	MOTOR LENGTH: 4.75"	PROPELLANT WT.: 32.1 GMS.

TOT. IMP.: 18.17 NT.-SECS.		MAX. THRUST: 8.47 LBS.
SPEC. IMPULSE: 57.62 SECS.	BURN TIME: 1.46 SECS.	AVG. THRUST: 12.44 NTS.

KG1-18-D12 ESTIMATED PERFORMANCE

MINIMUM ROCKET

BODY LENGTH: 14"
BODY DIAMETER: 1.00"
LIFTOFF WT.: 5.00 OZ.
MAX. ALTITUDE: 1,724 FT.
MAX. VELOCITY: 413 FT./SEC. (282 MPH)
MAX. MACH NUMBER: 0.37
MAX. ACCELERATION: 30.9 Gs
BURNOUT ALTITUDE: 201 FT.
COAST TIME TO APOGEE: 8.9 SECS.

AVERAGE ROCKET

BODY LENGTH: 30"
BODY DIAMETER: 2.26"
LIFTOFF WT.: 14.50 OZ.
MAX. ALTITUDE: 219 FT.
MAX. VELOCITY: 111 FT./SEC. (75 MPH)
MAX. MACH NUMBER: 0.10
MAX. ACCELERATION: 8.8 Gs
BURNOUT ALTITUDE: 51 FT.
COAST TIME TO APOGEE: 3.2 SECS.

© COPYRIGHT 1998 DAVID G. SLEETER

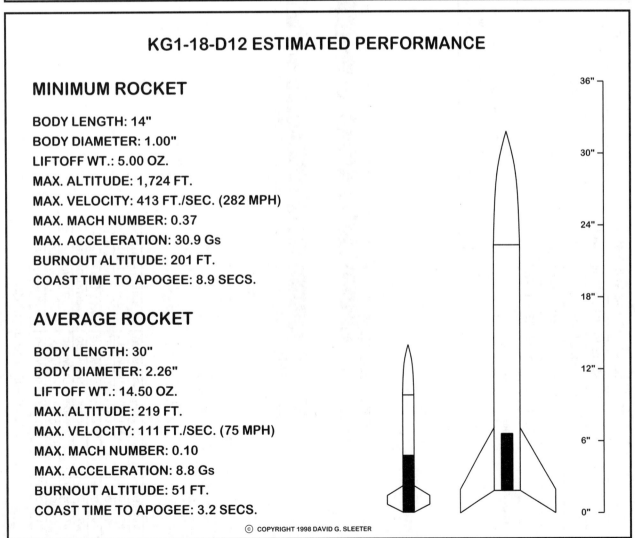

TELEFLITE CORPORATION

MOTOR NAME:	KG1-38-E33
PROPELLANT TYPE:	KG1
CASING TYPE:	CONVOLUTE PAPER
NOZZLE TYPE:	RAMMED CLAY

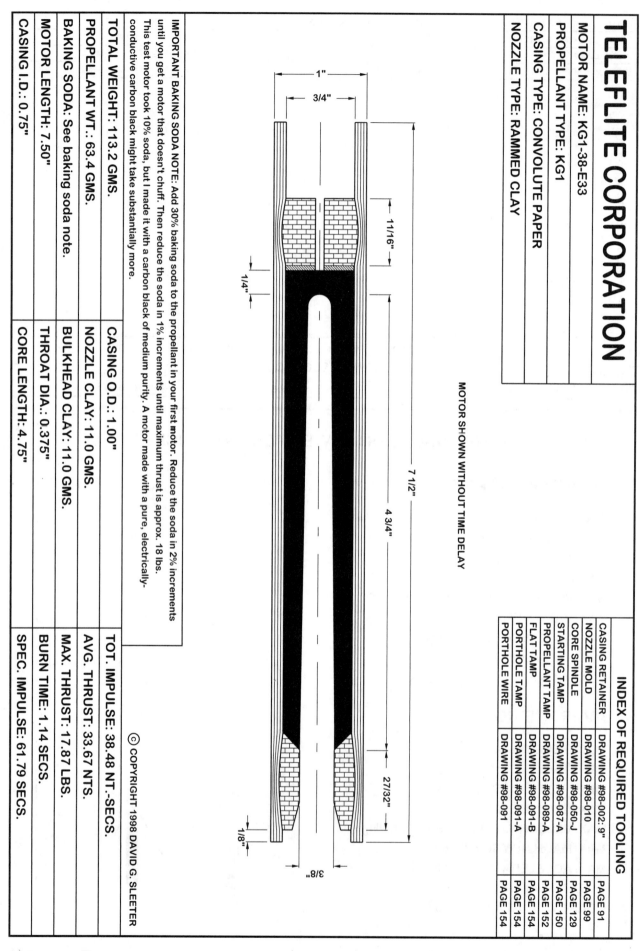

MOTOR SHOWN WITHOUT TIME DELAY

INDEX OF REQUIRED TOOLING

CASING RETAINER	DRAWING #98-002: 9"	PAGE 91
NOZZLE MOLD	DRAWING #98-010	PAGE 99
CORE SPINDLE	DRAWING #98-050-J	PAGE 129
STARTING TAMP	DRAWING #98-087-A	PAGE 150
PROPELLANT TAMP	DRAWING #98-089-A	PAGE 152
FLAT TAMP	DRAWING #98-091-B	PAGE 154
PORTHOLE TAMP	DRAWING #98-091-A	PAGE 154
PORTHOLE WIRE	DRAWING #98-091	PAGE 154

CASING I.D.:	0.75"
MOTOR LENGTH:	7.50"
BAKING SODA:	See baking soda note.
PROPELLANT WT.:	63.4 GMS.
TOTAL WEIGHT:	113.2 GMS.
CASING O.D.:	1.00"
NOZZLE CLAY:	11.0 GMS.
BULKHEAD CLAY:	11.0 GMS.
THROAT DIA.:	0.375"
CORE LENGTH:	4.75"
TOT. IMPULSE:	38.48 NT.-SECS.
AVG. THRUST:	33.67 NTS.
MAX. THRUST:	17.87 LBS.
BURN TIME:	1.14 SECS.
SPEC. IMPULSE:	61.79 SECS.

IMPORTANT BAKING SODA NOTE: Add 30% baking soda to the propellant in your first motor. Reduce the soda in 2% increments until you get a motor that doesn't chuff. Then reduce the soda in 1% increments until maximum thrust is approx. 18 lbs. This test motor took 10% soda, but I made it with a carbon black of medium purity. A motor made with a pure, electrically-conductive carbon black might take substantially more.

© COPYRIGHT 1998 DAVID G. SLEETER

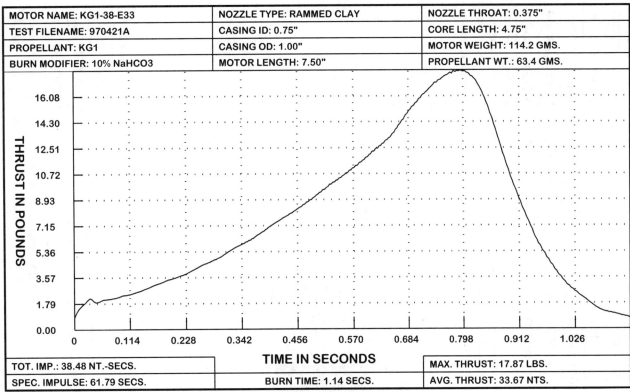

MOTOR NAME: KG1-38-E33	NOZZLE TYPE: RAMMED CLAY	NOZZLE THROAT: 0.375"
TEST FILENAME: 970421A	CASING ID: 0.75"	CORE LENGTH: 4.75"
PROPELLANT: KG1	CASING OD: 1.00"	MOTOR WEIGHT: 114.2 GMS.
BURN MODIFIER: 10% NaHCO3	MOTOR LENGTH: 7.50"	PROPELLANT WT.: 63.4 GMS.
TOT. IMP.: 38.48 NT.-SECS.		MAX. THRUST: 17.87 LBS.
SPEC. IMPULSE: 61.79 SECS.	BURN TIME: 1.14 SECS.	AVG. THRUST: 33.67 NTS.

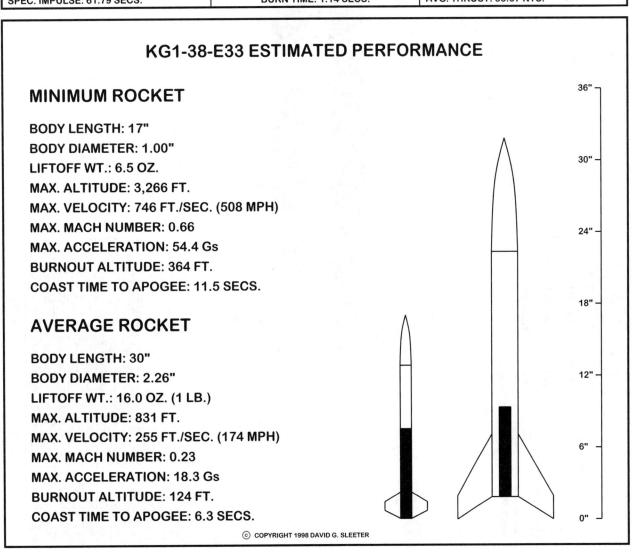

KG1-38-E33 ESTIMATED PERFORMANCE

MINIMUM ROCKET

BODY LENGTH: 17"
BODY DIAMETER: 1.00"
LIFTOFF WT.: 6.5 OZ.
MAX. ALTITUDE: 3,266 FT.
MAX. VELOCITY: 746 FT./SEC. (508 MPH)
MAX. MACH NUMBER: 0.66
MAX. ACCELERATION: 54.4 Gs
BURNOUT ALTITUDE: 364 FT.
COAST TIME TO APOGEE: 11.5 SECS.

AVERAGE ROCKET

BODY LENGTH: 30"
BODY DIAMETER: 2.26"
LIFTOFF WT.: 16.0 OZ. (1 LB.)
MAX. ALTITUDE: 831 FT.
MAX. VELOCITY: 255 FT./SEC. (174 MPH)
MAX. MACH NUMBER: 0.23
MAX. ACCELERATION: 18.3 Gs
BURNOUT ALTITUDE: 124 FT.
COAST TIME TO APOGEE: 6.3 SECS.

© COPYRIGHT 1998 DAVID G. SLEETER

TELEFLITE CORPORATION

MOTOR NAME:	KG1-44-F25
PROPELLANT TYPE:	KG1
CASING TYPE:	CONVOLUTE PAPER
NOZZLE TYPE:	RAMMED CLAY

MOTOR SHOWN WITHOUT TIME DELAY

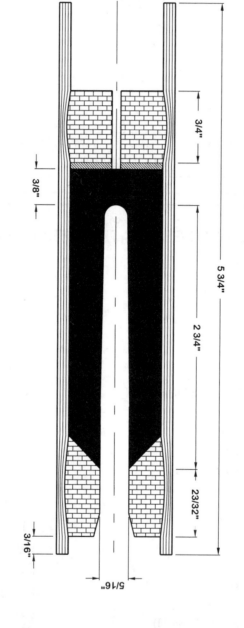

© COPYRIGHT 1998 DAVID G. SLEETER

INDEX OF REQUIRED TOOLING

CASING RETAINER	DRAWING #98-003:.7"	PAGE 92
NOZZLE MOLD	DRAWING #98-012	PAGE 101
CORE SPINDLE	DRAWING #98-054-F	PAGE 131
STARTING TAMP	DRAWING #98-092-A	PAGE 155
PROPELLANT TAMP	DRAWING #98-093-A	PAGE 156
FLAT TAMP	DRAWING #98-096-B	PAGE 159
PORTHOLE TAMP	DRAWING #98-096-A	PAGE 159
PORTHOLE WIRE	DRAWING #98-096	PAGE 159

IMPORTANT BAKING SODA NOTE: Add 30% baking soda to the propellant in your first motor. Reduce the soda in 2% increments until you get a motor that doesn't chuff. Then reduce the soda in 1% increments until maximum thrust is approx. 14 lbs. This test motor took 10% soda, but I made it with a carbon black of medium purity. A motor made with a pure, electrically-conductive carbon black might take substantially more.

TOTAL WEIGHT: 154.7 GMS.	CASING O.D.: 1.25"	TOT. IMPULSE: 44.49 NT.-SECS.	
PROPELLANT WT.: 73.9 GMS.	NOZZLE CLAY: 24.0 GMS.	AVG. THRUST: 25.72 NTS.	
BAKING SODA: See baking soda note.	BULKHEAD CLAY: 24.0 GMS.	MAX. THRUST: 13.50 LBS.	
MOTOR LENGTH: 6.0"	THROAT DIA.: 0.313"	BURN TIME: 1.73 SECS.	
CASING I.D.: 1.00"	CORE LENGTH: 2.75"	SPEC. IMPULSE: 61.29 SECS.	

MOTOR NAME: KG1-44-F25	NOZZLE TYPE: RAMMED CLAY	NOZZLE THROAT: 0.313"
TEST FILENAME: 970424A	CASING ID: 1.00	CORE LENGTH: 2.75"
PROPELLANT: KG1	CASING OD: 1.25	MOTOR WEIGHT: 156.1 GMS.
BURN MODIFIER: 10% NaHCO3	MOTOR LENGTH: 6.00	PROPELLANT WT.: 73.9 GMS.

TOT. IMP.: 44.49 NT.-SECS.		MAX. THRUST: 13.50 LBS.
SPEC. IMPULSE: 61.29 SECS.	BURN TIME: 1.73 SECS.	AVG. THRUST: 25.72 NTS.

KG1-44-F25 ESTIMATED PERFORMANCE

MINIMUM ROCKET

BODY LENGTH: 16.5"
BODY DIAMETER: 1.25"
LIFTOFF WT.: 9.0 OZ.
MAX. ALTITUDE: 2,761 FT.
MAX. VELOCITY: 580 FT./SEC. (395 MPH)
MAX. MACH NUMBER: 0.51
MAX. ACCELERATION: 27.8 Gs
BURNOUT ALTITUDE: 369 FT.
COAST TIME TO APOGEE: 10.9 SECS.

AVERAGE ROCKET

BODY LENGTH: 36"
BODY DIAMETER: 2.63"
LIFTOFF WT.: 20.5 OZ. (1.28 LBS.)
MAX. ALTITUDE: 641 FT.
MAX. VELOCITY: 208 FT./SEC. (142 MPH)
MAX. MACH NUMBER: 0.18
MAX. ACCELERATION: 10.3 Gs
BURNOUT ALTITUDE: 129 FT.
COAST TIME TO APOGEE: 5.4 SECS.

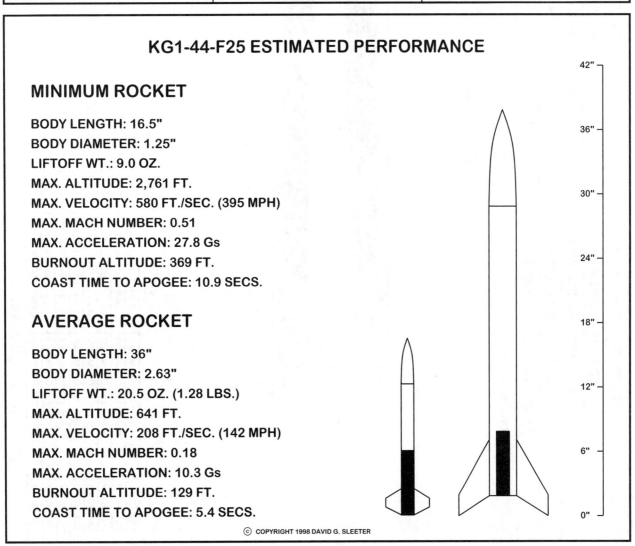

© COPYRIGHT 1998 DAVID G. SLEETER

TELEFLITE CORPORATION

MOTOR NAME:	KG1-97-G60
PROPELLANT TYPE:	KG1
CASING TYPE:	CONVOLUTE PAPER
NOZZLE TYPE:	RAMMED CLAY

MOTOR SHOWN WITHOUT TIME DELAY

INDEX OF REQUIRED TOOLING

CASING RETAINER	DRAWING #98-003: 13"	PAGE 92
NOZZLE MOLD	DRAWING #98-014	PAGE 103
CORE SPINDLE	DRAWING #98-061-I	PAGE 134
STARTING TAMP	DRAWING #98-094-A	PAGE 157
PROPELLANT TAMP	DRAWING #98-095-A	PAGE 158
FLAT TAMP	DRAWING #98-096-B	PAGE 159
PORTHOLE TAMP	DRAWING #98-096-A	PAGE 159
PORTHOLE WIRE	DRAWING #98-096	PAGE 159

Dimensions shown on drawing: 1 1/4", 1", 3/4", 11/32", 9 3/4", 6 1/2", 1", 3/16", 1/2"

IMPORTANT BAKING SODA NOTE: Add 30% baking soda to the propellant in your first motor. Reduce the soda in 2% increments until you get a motor that doesn't chuff. Then reduce the soda in 1% increments until maximum thrust is approx. 28 lbs. This test motor took 10% soda, but I made it with a carbon black of medium purity. A motor made with a pure, electrically-conductive carbon black might take substantially more.

© COPYRIGHT 1998 DAVID G. SLEETER

TOTAL WEIGHT: 256.3 GMS.	CASING O.D.: 1.25"	TOT. IMPULSE: 97.18 NT.-SECS.
PROPELLANT WT.: 157.4 GMS.	NOZZLE CLAY: 24.0 GMS.	AVG. THRUST: 60.66 NTS.
BAKING SODA: See baking soda note.	BULKHEAD CLAY: 24.0 GMS.	MAX. THRUST: 28.62 LBS.
MOTOR LENGTH: 9.75"	THROAT DIA.: 0.50"	BURN TIME: 1.60 SECS.
CASING I.D.: 1.00"	CORE LENGTH: 6.50"	SPEC. IMPULSE: 62.85 SECS.

MOTOR NAME: KG1-97-G60	NOZZLE TYPE: RAMMED CLAY	NOZZLE THROAT: 0.50"
TEST FILENAME: 970601A	CASING ID: 1.00"	CORE LENGTH: 6.50"
PROPELLANT: KG1	CASING OD: 1.25"	MOTOR WEIGHT: 256.3 GMS.
BURN MODIFIER: 10% NaHCO3	MOTOR LENGTH: 9.75"	PROPELLANT WT.: 157.4 GMS.

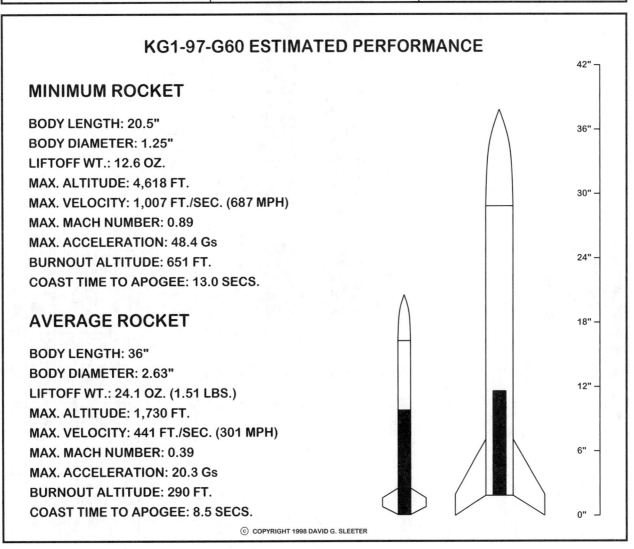

TOT. IMP.: 97.18 NT.-SECS.		MAX. THRUST: 28.62 LBS.
SPEC. IMPULSE: 62.85 SECS.	BURN TIME: 1.60 SECS.	AVG. THRUST: 60.66 NTS.

KG1-97-G60 ESTIMATED PERFORMANCE

MINIMUM ROCKET

BODY LENGTH: 20.5"
BODY DIAMETER: 1.25"
LIFTOFF WT.: 12.6 OZ.
MAX. ALTITUDE: 4,618 FT.
MAX. VELOCITY: 1,007 FT./SEC. (687 MPH)
MAX. MACH NUMBER: 0.89
MAX. ACCELERATION: 48.4 Gs
BURNOUT ALTITUDE: 651 FT.
COAST TIME TO APOGEE: 13.0 SECS.

AVERAGE ROCKET

BODY LENGTH: 36"
BODY DIAMETER: 2.63"
LIFTOFF WT.: 24.1 OZ. (1.51 LBS.)
MAX. ALTITUDE: 1,730 FT.
MAX. VELOCITY: 441 FT./SEC. (301 MPH)
MAX. MACH NUMBER: 0.39
MAX. ACCELERATION: 20.3 Gs
BURNOUT ALTITUDE: 290 FT.
COAST TIME TO APOGEE: 8.5 SECS.

© COPYRIGHT 1998 DAVID G. SLEETER

TELEFLITE CORPORATION

MOTOR NAME:	KG1-71-F40
PROPELLANT TYPE:	KG1
CASING TYPE:	CONVOLUTE PAPER
NOZZLE TYPE:	RAMMED CLAY

MOTOR SHOWN WITHOUT TIME DELAY

© COPYRIGHT 1998 DAVID G. SLEETER

INDEX OF REQUIRED TOOLING

CASING RETAINER	DRAWING #98-004; 8"	PAGE 93
NOZZLE MOLD	DRAWING #98-016	PAGE 105
CORE SPINDLE	DRAWING #98-067-G	PAGE 137
STARTING TAMP	DRAWING #98-097-A	PAGE 160
PROPELLANT TAMP	DRAWING #98-098-A	PAGE 161
FLAT TAMP	DRAWING #98-104-B	PAGE 167
PORTHOLE TAMP	DRAWING #98-104-A	PAGE 167
PORTHOLE WIRE	DRAWING #98-104	PAGE 167

CASING I.D.:	1.125"
MOTOR LENGTH:	7.0"
BAKING SODA:	See baking soda note.
PROPELLANT WT.:	112.9 GMS.
TOTAL WEIGHT:	237.3 GMS.
CASING O.D.:	1.50"
NOZZLE CLAY:	33.0 GMS.
BULKHEAD CLAY:	33.0 GMS.
THROAT DIA.:	0.375"
CORE LENGTH:	3.50"
TOT. IMPULSE:	71.65 NT.-SECS.
AVG. THRUST:	40.67 NTS.
MAX. THRUST:	21.04 LBS.
BURN TIME:	1.76 SECS.
SPEC. IMPULSE:	64.60 SECS.

IMPORTANT BAKING SODA NOTE: Add 30% baking soda to the propellant in your first motor. Reduce the baking soda in 2% increments until you get a motor that doesn't chuff. Then reduce the soda in 1% increments until maximum thrust is approx. 20 lbs. This test motor took 10% soda, but I made it with a carbon black of medium purity. A motor made with a pure, electrically-conductive carbon black might take substantially more.

MOTOR NAME: KG1-71-F40	NOZZLE TYPE: RAMMED CLAY	NOZZLE THROAT: 0.375"
TEST FILENAME: 970422C	CASING ID: 1.125"	CORE LENGTH: 3.50"
PROPELLANT: KG1	CASING OD: 1.50"	MOTOR WEIGHT: 237.3 GMS.
BURN MODIFIER: 10% NaHCO3	MOTOR LENGTH: 7.00"	PROPELLANT WT.: 112.9 GMS.

[Thrust curve: THRUST IN POUNDS vs TIME IN SECONDS, peaking at ~21 lbs near 1.408 s]

TOT. IMP.: 71.65 NT.-SECS.		MAX. THRUST: 21.04 LBS.
SPEC. IMPULSE: 64.60 SECS.	BURN TIME: 1.76 SECS.	AVG. THRUST: 40.67 NTS.

KG1-71-F40 ESTIMATED PERFORMANCE

MINIMUM ROCKET

BODY LENGTH: 20"
BODY DIAMETER: 1.50"
LIFTOFF WT.: 14.4 OZ.
MAX. ALTITUDE: 2,908 FT.
MAX. VELOCITY: 580 FT./SEC. (395 MPH)
MAX. MACH NUMBER: 0.51
MAX. ACCELERATION: 27.0 Gs
BURNOUT ALTITUDE: 378 FT.
COAST TIME TO APOGEE: 11.3 SECS.

AVERAGE ROCKET

BODY LENGTH: 48"
BODY DIAMETER: 3.10"
LIFTOFF WT.: 32.4 OZ. (2.03 LBS.)
MAX. ALTITUDE: 658 FT.
MAX. VELOCITY: 209 FT./SEC. (143 MPH)
MAX. MACH NUMBER: 0.19
MAX. ACCELERATION: 10.1 Gs
BURNOUT ALTITUDE: 132 FT.
COAST TIME TO APOGEE: 5.5 SECS.

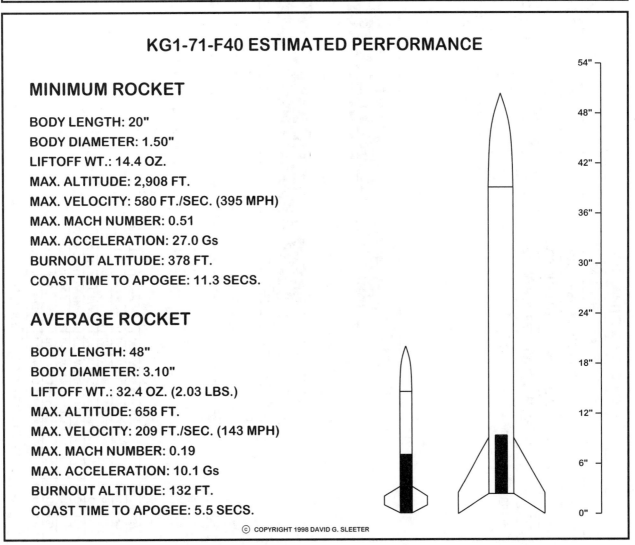

© COPYRIGHT 1998 DAVID G. SLEETER

TELEFLITE CORPORATION

MOTOR NAME:	KG1-149-G104
PROPELLANT TYPE:	KG1
CASING TYPE:	CONVOLUTE PAPER
NOZZLE TYPE:	RAMMED CLAY

MOTOR SHOWN WITHOUT TIME DELAY

Dimensions shown on diagram: 1 1/2", 1 1/8", 27/32", 3/8", 11-3/4", 8", 1 1/8", 3/16", 9/16"

IMPORTANT BAKING SODA NOTE: Add 30% baking soda to the propellant in your first motor. Reduce the soda in 2% increments until you get a motor that doesn't chuff. Then reduce the soda in 1% increments until maximum thrust is approx. 48 lbs. This test motor took 10% soda, but I made it with a carbon black of medium purity. A motor made with a pure, electrically-conductive carbon black might take substantially more.

© COPYRIGHT 1998 DAVID G. SLEETER

TOTAL WEIGHT: 385.2 GMS.	CASING O.D.: 1.50"		TOT. IMPULSE: 149.92 NT.-SECS.
PROPELLANT WT.: 219.8 GMS.	NOZZLE CLAY: 33.0 GMS.		AVG. THRUST: 104.18 NTS.
BAKING SODA: See baking soda note.	BULKHEAD CLAY: 33.0 GMS.		MAX. THRUST: 48.08 LBS.
MOTOR LENGTH: 11.75"	THROAT DIA.: 0.563"		BURN TIME: 1.44 SECS.
CASING I.D.: 1.125"	CORE LENGTH: 8.00"		SPEC. IMPULSE: 69.43 SECS.

INDEX OF REQUIRED TOOLING

CASING RETAINER	DRAWING #98-004:15"	PAGE 93
NOZZLE MOLD	DRAWING #98-018	PAGE 107
CORE SPINDLE	DRAWING #98-072-I	PAGE 140
STARTING TAMP	DRAWING #98-099-B	PAGE 162
PROPELLANT TAMP	DRAWING #98-101-B	PAGE 164
FLAT TAMP	DRAWING #98-104-B	PAGE 167
PORTHOLE TAMP	DRAWING #98-104-A	PAGE 167
PORTHOLE WIRE	DRAWING #98-104	PAGE 167

MOTOR NAME: KG1-149-G104	NOZZLE TYPE: RAMMED CLAY	NOZZLE THROAT: 0.563"
TEST FILENAME: 970426A	CASING ID: 1.125"	CORE LENGTH: 8.00"
PROPELLANT: KG1	CASING OD: 1.50"	MOTOR WEIGHT: 385.2 GMS.
BURN MODIFIER: 10% NaHCO3	MOTOR LENGTH: 11.75"	PROPELLANT WT.: 219.8 GMS.

TOT. IMP.: 149.92 NT.-SECS.		MAX. THRUST: 48.08 LBS.
SPEC. IMPULSE: 69.43 SECS.	BURN TIME: 1.44 SECS.	AVG. THRUST: 104.18 NTS.

KG1-149-G104 ESTIMATED PERFORMANCE

MINIMUM ROCKET

BODY LENGTH: 25"
BODY DIAMETER: 1.50"
LIFTOFF WT.: 19.50 OZ. (1.22 LBS.)
MAX. ALTITUDE: 4,896 FT.
MAX. VELOCITY: 994 FT./SEC. (678 MPH)
MAX. MACH NUMBER: 0.88
MAX. ACCELERATION: 50.2 Gs
BURNOUT ALTITUDE: 601 FT.
COAST TIME TO APOGEE: 13.7 SECS.

AVERAGE ROCKET

BODY LENGTH: 48"
BODY DIAMETER: 3.10"
LIFTOFF WT.: 37.50 OZ. (2.34 LBS.)
MAX. ALTITUDE: 1,768 FT.
MAX. VELOCITY: 443 FT./SEC. (302 MPH)
MAX. MACH NUMBER: 0.39
MAX. ACCELERATION: 21.8 Gs
BURNOUT ALTITUDE: 272 FT.
COAST TIME TO APOGEE: 8.7 SECS.

© COPYRIGHT 1998 DAVID G. SLEETER

TELEFLITE CORPORATION

MOTOR NAME: KG1-193-H144
PROPELLANT TYPE: KG1
CASING TYPE: CONVOLUTE PAPER
NOZZLE TYPE: STAINLESS STEEL DeLAVAL

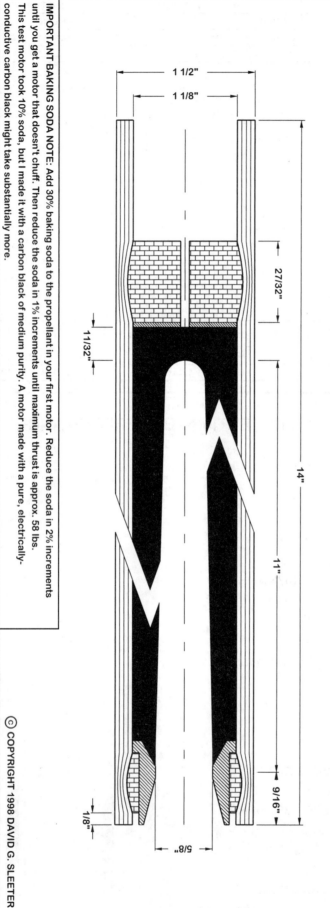

MOTOR SHOWN WITHOUT TIME DELAY

© COPYRIGHT 1998 DAVID G. SLEETER

INDEX OF REQUIRED TOOLING

NOZZLE ANVIL	DRAWING #98-109-C	PAGE 172
NOZZLE RAMMER	DRAWING #98-110-D	PAGE 173
CASING RETAINER	DRAWING #98-004; 18-1/2"	PAGE 93
NOZZLE MOLD	DRAWING #98-020	PAGE 109
CORE SPINDLE	DRAWING #98-079-F	PAGE 143
PROPELLANT TAMP	DRAWING #98-103-B	PAGE 166
FLAT TAMP	DRAWING #98-104-B	PAGE 167
PORTHOLE TAMP	DRAWING #98-104-A	PAGE 167
PORTHOLE WIRE	DRAWING #98-104	PAGE 167

CASING I.D.: 1.125"
MOTOR LENGTH: 14.00"
BAKING SODA: See baking soda note.
PROPELLANT WT.: 279.0 GMS.
TOTAL WEIGHT: 479.7 GMS.

IMPORTANT BAKING SODA NOTE: Add 30% baking soda to the propellant in your first motor. Reduce the soda in 1% increments until you get a motor that doesn't chuff. Then reduce the soda in 1% increments until maximum thrust is approx. 58 lbs. This test motor took 10% soda, but I made it with a carbon black of medium purity. A motor made with a pure, electrically-conductive carbon black might take substantially more.

CASING O.D.: 1.50"
NOZZLE: S.S. DeLAVAL
BULKHEAD CLAY: 33.0 GMS.
THROAT DIA.: 0.625"
CORE LENGTH: 11.00"

TOT. IMPULSE: 193.36 NT.-SECS.
AVG. THRUST: 144.30 NTS.
MAX. THRUST: 58.53 LBS.
BURN TIME: 1.34 SECS.
SPEC. IMPULSE: 70.55 SECS.

MOTOR NAME: KG1-193-H144	NOZZLE TYPE: S.S. DeLAVAL	NOZZLE THROAT: 0.625"
TEST FILENAME: 970626B	CASING ID: 1.125"	CORE LENGTH: 11.00"
PROPELLANT: KG1	CASING OD: 1.50"	MOTOR WEIGHT: 479.7 GMS.
BURN MODIFIER: 10% NaHCO3	MOTOR LENGTH: 14.50"	PROPELLANT WT.: 279.0 GMS.

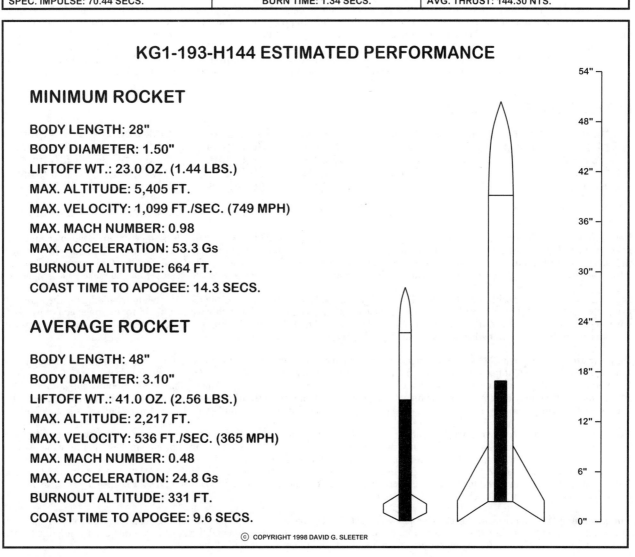

TOT. IMP.: 193.36 NT.-SECS.		MAX. THRUST: 58.53 LBS.
SPEC. IMPULSE: 70.44 SECS.	BURN TIME: 1.34 SECS.	AVG. THRUST: 144.30 NTS.

KG1-193-H144 ESTIMATED PERFORMANCE

MINIMUM ROCKET

BODY LENGTH: 28"
BODY DIAMETER: 1.50"
LIFTOFF WT.: 23.0 OZ. (1.44 LBS.)
MAX. ALTITUDE: 5,405 FT.
MAX. VELOCITY: 1,099 FT./SEC. (749 MPH)
MAX. MACH NUMBER: 0.98
MAX. ACCELERATION: 53.3 Gs
BURNOUT ALTITUDE: 664 FT.
COAST TIME TO APOGEE: 14.3 SECS.

AVERAGE ROCKET

BODY LENGTH: 48"
BODY DIAMETER: 3.10"
LIFTOFF WT.: 41.0 OZ. (2.56 LBS.)
MAX. ALTITUDE: 2,217 FT.
MAX. VELOCITY: 536 FT./SEC. (365 MPH)
MAX. MACH NUMBER: 0.48
MAX. ACCELERATION: 24.8 Gs
BURNOUT ALTITUDE: 331 FT.
COAST TIME TO APOGEE: 9.6 SECS.

© COPYRIGHT 1998 DAVID G. SLEETER

12. KG3 PROPELLANT AND MOTOR DESIGNS

● First and Most Important!

Before you proceed, if you haven't read **Chapter 1**, go back and read it now. It contains important information on the safe handling of this rocket propellant. Also pay attention to the bold warning at the beginning of **Chapter 1**. To emphasize its importance, I'll repeat that warning here.

> Do *not* add or substitute other chemicals either to-or-for the chemicals described in this book. Specifically do *not* use POTASSIUM CHLORATE, SODIUM CHLORATE, POTASSIUM PERCHLORATE, SODIUM PERCHLORATE, POTASSIUM PERMANGANATE, AMMONIUM PERCHLORATE, PHOSPHOROUS, MATCH HEADS, ALUMINUM, MAGNESIUM, or ANY METALS AT ALL! *All* of these substances will form dangerous and explosive mixtures when combined with the other ingredients described herein, *AND OTHER CHEMICALS WILL FORM DANGEROUS MIXTURES AS WELL.* When mixing and handling propellants, be sure to ground yourself and all containers and utensils according to the instructions on page 14.
>
> *If you decide to experiment on your own outside the bounds of the instructions provided in this book, consult with a professional chemist before you proceed.*

The **KG3** propellant is a milled mixture of **4 hour milled potassium nitrate**, **6 hour milled charcoal**, **carbon black**, **red gum**, and **sulfur**. If you haven't read **Chapters 5** and **6**, go back and read them now. Together they explain how to buy or make these chemicals, and how to prepare them for use in a rocket propellant.

If all other parameters are equal, then the **KG3** is the faster burning of the two potassium nitrate-based propellants. But its burn rate can vary *greatly* depending on the exact nature and purity of its chemicals. The quality of the carbon black is particularly important. A carbon black high in residual organics (i.e. oil) makes a slow burning propellant, and a pure, electrically conductive carbon black makes a fast burning propellant. The *exact* burn rate can be adjusted by substituting slower burning charcoal for some of the fast burning carbon black, or by adding or subtracting baking soda. When compressed into a solid propellant grain, the **KG3**'s density is **25** to **30 gms. per cubic inch**. To prepare the **KG3** propellant, set up a powder mill (**Chapter 3**). Then mix the propellant (**Chapter 9**), and use the figures below to determine the exact weights of the chemicals needed for the desired batch size.

◆ THE KG3 PROPELLANT

4 hr. milled potassium nitrate .72.1%	721.0 gms.	360.5 gms.	180.2 gms.	90.0 gms.
6 hr. milled charcoal 7.3%	73.0 gms.	36.5 gms.	18.3 gms.	9.2 gms.
Carbon black 4.0%	40.0 gms.	20.0 gms.	10.0 gms.	5.0 gms.
Red gum 7.0%	70.0 gms.	35.0 gms.	17.5 gms.	8.8 gms.
Sulfur 9.6%	96.0 gms.	48.0 gms.	24.0 gms.	12.0 gms.
Total batch size 100%	1000.0 gms.	500.0 gms.	250.0 gms.	125.0 gms.

Before you can use the propellant, *you must perform the pellet test* (**pages 238 - 245**). The pellet test determines the exact amount of solvent needed to form 20 grams of propellant into a dense and solid propellant grain. The correct amount of solvent is called **the solvent requirement**. *The solvent requirement is critical, and cannot be guessed at.* The pellet test is time consuming, *but you only do it twice*. Perform the *first* pellet test with the **KG3** propellant *plus 30% baking soda*, and do it *before* you make your first motor.

To build your *first* motor with the **KG3** propellant, weigh out the amount indicated in the motor drawing (rounding off the numbers makes the math easier). Then add 30% of that weight in baking soda, and mix it in thoroughly with a fork. If, for example, you've weighed out 100 grams of propellant, add 30 grams of baking soda for a total of 130 grams. Mix in the proper amount of solvent as determined by your first pellet test. Then build and dry the motor (**Chapter 10**).

Per the instructions on **page 494**, do a rudimentary test by firing the motor (with the motor pointing *down* and the exhaust aimed *upward*) from a pipe in the ground. If this first motor "chuffs" (chugs like a steam engine), build a *second* motor, and reduce the baking soda to 28%. Keep building, testing, and reducing the amount of soda in 2% increments until you get a motor that burns smoothly and *doesn't* chuff. At first the chuffing may be slow. As you approach a condition of combustion stability, the frequency of the chuffs will increase (the chuffing sound will get *faster*). When you're *very* near stability, the amplitude of the chuffs will decrease (the chuffs will sound *fainter*).

Now set up a thrust-testing stand (**page 470**), and test the motor for maximum thrust (**page 475**). In the tests that follow, reduce the soda in 1% increments until you get a motor whose maximum thrust approximates the maximum thrust printed in the motor drawing. At that point you'll have a working motor with the maximum safe chamber pressure and thrust for that particular design. As you approach a final, successful soda content, *repeat the pellet test*, and use the solvent requirement indicated by this *second* test to zero-in the *final* design. Keep accurate notes. Write down the final solvent requirement and the amount of baking soda that works the best, and *forever thereafter* use those figures to build copies of your final, successful motor.

Important note. You can also alter the chamber pressure and thrust by changing the length of the motor's core. A *longer* core will *increase* the chamber pressure and thrust, and a *shorter* core will *reduce* the chamber pressure and thrust. But, of course, changing the length of the core will also change the amount of propellant required, and the physical size of the motor.

● Improving Motor Performance

Now you have a working rocket motor, but some of the material inside the motor isn't rocket propellant. It's *baking soda*, and baking soda is *dead weight*. It doesn't contribute anything to the motor's total impulse. You added the soda to control chamber pressure and thrust, but you can *also* control those parameters by altering the balance between the carbon black and the charcoal. And unlike baking soda, the charcoal is a *fuel*. It takes part in the reaction, and it *contributes* to the motor's total impulse.

In the following hypothetical illustration, I'll call the original motor a **KG3-19-D20**. I'll stipulate that its maximum thrust is 10 lbs., and I'll stipulate that it takes 8% baking soda to make it work as intended. To *improve* the motor's performance, I'll have to get rid of some of the soda. Of course when I do that, the propellant's burn rate will increase. But I can *reverse* the increase, and *compensate* for the loss of the soda by replacing some of the carbon black with slower burning charcoal. **Important note**. *Each time you change a propellant's formula, you must repeat the pellet test*. Skipping the pellet test is *not* an option. Possible replacement formulas are:

KG3X1	KG3X2	KG3X3
4 hr. mill. potassium nitrate 72.1%	4 hr. mill. potassium nitrate 72.1%	4 hr. mill. potassium nitrate 72.1%
6 hr. mill. charcoal 8.3%	6 hr. mill. charcoal 9.3%	6 hr. mill. charcoal 10.3%
Carbon black 3.0%	Carbon black 2.0%	Carbon black 1.0%
Red gum 7.0%	Red gum 7.0%	Red gum 7.0%
Sulfur 9.6%	Sulfur 9.6%	Sulfur 9.6%

In the formulas above, though the balance between the charcoal and the carbon black changes (in 1% increments), the *total* amount of carbon fuel remains the same (11.3%). If I build a motor with the KG3X1 propellant, and add 8% baking soda, because the charcoal burns *slower* than the carbon black it replaced, the motor's maximum thrust will be *less* than 10 pounds. To restore the maximum thrust to the desired level, I'll keep building, testing, and reducing the baking soda in 1% increments until the maximum thrust is back up to 10 lbs. I'll call this second motor a **KG3X1** (named after its propellant), and I'll pretend here that it took 6% soda to make it work.

If the **KG3X1** with 6% soda burns smoothly and doesn't chuff, I'll build an identical motor with the KG3X2 propellant. Of course when I do that, the maximum thrust will drop again, so I'll keep building, testing, and *again* reducing the soda in 1% increments until the maximum thrust equals 10 lbs. I'll call this third motor a **KG3X2**, and for the purpose of this illustration, I'll pretend that it took 4% baking soda to make it work. If the **KG3X2** burns smoothly and doesn't chuff, I'll repeat the experiment with the KG3X3 propellant. If the **KG3X2** chuffs and *doesn't* burn smoothly, I'll know that the KG3X2 propellant *doesn't* work; that I need *more* than 2.0% carbon black to achieve combustion stability, and that the KG3X1 propellant with 6% baking soda works the best.

If I now build a chart recorder (**page 478**), and generate thrust-time curves for the **KG3-19-D20** and **KG3X1** motors, I'll see that the **KG3X1** has a slightly longer burn time and a greater total impulse. The KG3X1 might turn out to be a KG3-*22-E*20, and I'll adopt the KG3X1 propellant with 6% baking soda for all the motors that follow. When you perform these experiments on your own, you'll eventually reach a point where the motors begin to chuff, and the process of finding the minimum amount of carbon black needed to maintain combustion stability will be one of trial and error.

Important note. *The KG3 motors in this book were all made with propellants milled for 12 hours*. When making your own propellants, you can reduce the milling time to as little as 4 hours, and compensate by adjusting the amount of baking soda. The process of finding the minimum milling time is one of trial and error, and each time you change the milling time, *you must repeat* **the pellet test**.

The **KG3** propellant exhibits good case bonding properties with the **KG3-26-E31**, the **KG3-116-G102**, and the **KG3-140-G139** showing a small vestige of the effect discussed on **pages 248 - 249**. All the **KG3** motors were made with standard fireworks grade casings, and all but the **KG3-140-G139** were made with clay nozzles. By switching to stainless steel De Laval nozzles and the high strength casings manufactured (as of this writing) by **New England Paper Tube** (**pages 163 - 164**), you can increase the motors' maximum chamber pressure and thrust, and substantially improve their performance. Shop drawings for De Laval nozzles begin on the **page 323**.

With each motor design, I've included the thrust-time curve and motor performance data generated by the test equipment and software described on **pages 468 - 469**, plus two flight performance estimates generated by **Rogers' Aeroscience Alt4** rocket performance prediction software. The "minimum rocket" is what you might build if you were trying to break an altitude or speed record. The lift-off weight is minimal. The body diameter is equal to the o.d. of the motor, and the parachute compartment is small; approx. 5 body diameter's in length. The "average rocket" is more like what you would build for scientific research or general sport flying. The body diameter is one of the industry standards approx. twice the diameter of the motor, and there's ample space and weight allowance for a large parachute and a small scientific payload.

In each case the lift-off weight is *very* approximate, and based on an educated guess. The *extremely* accurate performance specs. *are the actual numbers generated by the Alt-4 program*. They are based on the rocket's stated length, diameter, and lift-off weight, and the *tested* motor's thrust-time data. *In the real world*, small variations in motor performance, the size and shape of the rocket's fins, and even the smoothness of its paint-job will cause the rocket's performance to vary from the stated figures.

● Important G-Force Advisory

Static testing perfects a motor's performance, but it *doesn't* predict how the hardened propellant grain will respond to high G-forces during an actual rocket flight. Baking soda is *soft*, and the carbon fuels used in these propellants vary *greatly* in hardness. So variations in the soda content and the type of carbon fuel will effect the strength of the finished propellant grain. If a soft propellant grain collapses at maximum acceleration, the motor will self-destruct, and the rocket will be lost in flight. Initial flight testing of a new motor design should *always* be conducted with this possibility in mind, particularly when testing the motor in a small, high-acceleration airframe. It is always advisable to conduct your early tests in a "throwaway" rocket, i.e. something cheap that you don't mind losing. When you're sure that you've got a successful motor-airframe combination, *then* invest in something nicer.

TELEFLITE CORPORATION

MOTOR NAME:	KG3-6-C8
PROPELLANT TYPE:	KG3
CASING TYPE:	CONVOLUTE PAPER
NOZZLE TYPE:	RAMMED CLAY

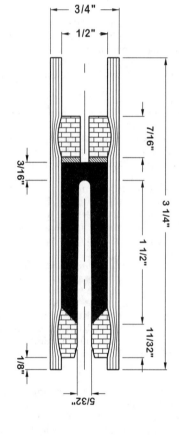

MOTOR SHOWN WITHOUT TIME DELAY

INDEX OF REQUIRED TOOLING

CASING RETAINER	DRAWING #98-001:5"	PAGE 90
NOZZLE MOLD	DRAWING #98-006	PAGE 95
CORE SPINDLE	DRAWING #98-035-C	PAGE 121
STARTING TAMP	USE PROPELLANT TAMP	
PROPELLANT TAMP	DRAWING #98-082-A	PAGE 145
FLAT TAMP	DRAWING #98-084-B	PAGE 147
PORTHOLE TAMP	DRAWING #98-084-A	PAGE 147
PORTHOLE WIRE	DRAWING #98-084	PAGE 147

TOTAL WEIGHT:	27.6 GMS.
PROPELLANT WT.:	10.7 GMS.
BAKING SODA:	See baking soda note.
MOTOR LENGTH:	3.25"
CASING I.D.:	0.50"
CASING O.D.:	0.75"
NOZZLE CLAY:	3.3 GMS.
BULKHEAD CLAY:	3.3 GMS.
THROAT DIA.:	0.1563"
CORE LENGTH:	1.50"
TOT. IMPULSE:	6.83 NT.-SECS.
AVG. THRUST:	8.38 NTS.
MAX. THRUST:	3.79 LBS.
BURN TIME:	0.81 SECS.
SPEC. IMPULSE:	64.98 SECS.

IMPORTANT BAKING SODA NOTE: Add 30% baking soda to the propellant in your first motor. Reduce the soda in 1% increments until you get a motor that doesn't chuff. Then reduce the soda in 2% increments until maximum thrust is approx. 3-1/2 lbs. This test motor took 5% soda, but I made it with a carbon black of medium purity. A motor made with a pure, electrically-conductive carbon black might take substantially more.

© COPYRIGHT 1998 DAVID G. SLEETER

MOTOR NAME: KG3-6-C8	NOZZLE TYPE: RAMMED CLAY	NOZZLE THROAT: 0.1563"
TEST FILENAME: 970419A	CASING ID: 0.50"	CORE LENGTH: 1.50"
PROPELLANT: KG3	CASING OD: 0.75"	MOTOR WEIGHT: 27.6 GMS.
BURN MODIFIER: 5% NaHCO3	MOTOR LENGTH: 3.25"	PROPELLANT WT.: 10.7 GMS.

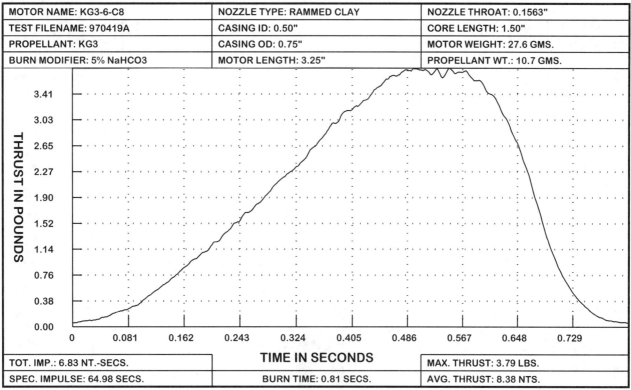

TOT. IMP.: 6.83 NT.-SECS.		MAX. THRUST: 3.79 LBS.
SPEC. IMPULSE: 64.98 SECS.	BURN TIME: 0.81 SECS.	AVG. THRUST: 8.38 NTS.

KG3-6-C8 ESTIMATED PERFORMANCE

MINIMUM ROCKET

BODY LENGTH: 11"
BODY DIAMETER: 0.75"
LIFTOFF WT.: 2.8 OZ.
MAX. ALTITUDE: 962 FT.
MAX. VELOCITY: 274 FT./SEC. (187 MPH)
MAX. MACH NUMBER: 0.24
MAX. ACCELERATION: 22.5 Gs
BURNOUT ALTITUDE: 94 FT.
COAST TIME TO APOGEE: 7.0 SECS.

AVERAGE ROCKET

BODY LENGTH: 24"
BODY DIAMETER: 1.63"
LIFTOFF WT.: 10.0 OZ.
MAX. ALTITUDE: 69 FT.
MAX. VELOCITY: 59 FT./SEC. (40 MPH)
MAX. MACH NUMBER: 0.05
MAX. ACCELERATION: 5.2 Gs
BURNOUT ALTITUDE: 20 FT.
COAST TIME TO APOGEE: 1.8 SECS.

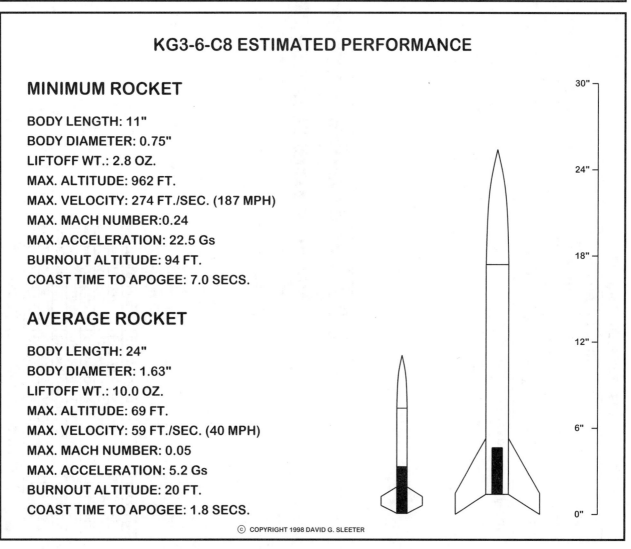

© COPYRIGHT 1998 DAVID G. SLEETER

TELEFLITE CORPORATION

MOTOR NAME: KG3-12-D17
PROPELLANT TYPE: KG3
CASING TYPE: CONVOLUTE PAPER
NOZZLE TYPE: RAMMED CLAY

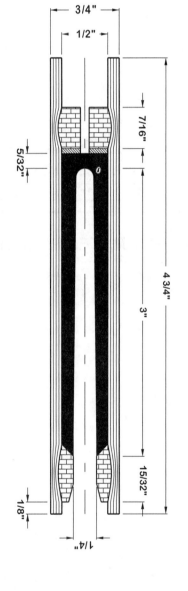

MOTOR SHOWN WITHOUT TIME DELAY

© COPYRIGHT 1998 DAVID G. SLEETER

IMPORTANT BAKING SODA NOTE: Add 30% baking soda to the propellant in your first motor. Then reduce the soda in 1% increments until you get a motor that doesn't chuff. Then reduce the soda in 2% increments until maximum thrust is approx. 9 lbs. This test motor took 5% soda, but I made it with a carbon black of medium purity. A motor made with a pure, electrically-conductive carbon black might take substantially more.

TOTAL WEIGHT: 40.7 GMS.	CASING O.D.: 0.75"
PROPELLANT WT.: 19.2 GMS.	NOZZLE CLAY: 3.3 GMS.
BAKING SODA: See baking soda note.	BULKHEAD CLAY: 3.3 GMS.
MOTOR LENGTH: 4.75"	THROAT DIA.: 0.25"
CASING I.D.: 0.50"	CORE LENGTH: 3.00"

TOT. IMPULSE: 12.00 NT.-SECS.	
AVG. THRUST: 17.60 NTS.	
MAX. THRUST: 9.10 LBS.	
BURN TIME: 0.68 SECS.	
SPEC. IMPULSE: 63.62 SECS.	

INDEX OF REQUIRED TOOLING

CASING RETAINER	DRAWING #98-001: 7"
NOZZLE MOLD	DRAWING #98-007
CORE SPINDLE	DRAWING #98-038-C
STARTING TAMP	USE PROPELLANT TAMP
PROPELLANT TAMP	DRAWING #98-083-A
FLAT TAMP	DRAWING #98-084-A
PORTHOLE TAMP	DRAWING #98-084-B
PORTHOLE WIRE	DRAWING #98-084

PAGE 90	
PAGE 96	
PAGE 123	
PAGE 146	
PAGE 147	
PAGE 147	
PAGE 147	

MOTOR NAME: KG3-12-D17	NOZZLE TYPE: RAMMED CLAY	NOZZLE THROAT: 0.25"
TEST FILENAME: 970419B	CASING ID: 0.50"	CORE LENGTH: 3.00"
PROPELLANT: KG3	CASING OD: 0.75"	MOTOR WEIGHT: 40.7 GMS.
BURN MODIFIER: 5% NaHCO3	MOTOR LENGTH: 4.75"	PROPELLANT WT.: 19.2 GMS.

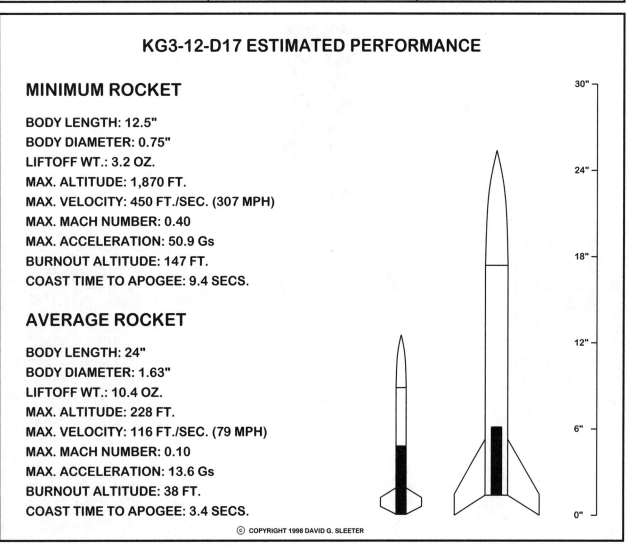

TOT. IMP.: 12.00 NT.-SECS.		MAX. THRUST: 9.10 LBS.
SPEC. IMPULSE: 63.62 SECS.	BURN TIME: 0.68 SECS.	AVG. THRUST: 17.60 NTS.

KG3-12-D17 ESTIMATED PERFORMANCE

MINIMUM ROCKET

BODY LENGTH: 12.5"
BODY DIAMETER: 0.75"
LIFTOFF WT.: 3.2 OZ.
MAX. ALTITUDE: 1,870 FT.
MAX. VELOCITY: 450 FT./SEC. (307 MPH)
MAX. MACH NUMBER: 0.40
MAX. ACCELERATION: 50.9 Gs
BURNOUT ALTITUDE: 147 FT.
COAST TIME TO APOGEE: 9.4 SECS.

AVERAGE ROCKET

BODY LENGTH: 24"
BODY DIAMETER: 1.63"
LIFTOFF WT.: 10.4 OZ.
MAX. ALTITUDE: 228 FT.
MAX. VELOCITY: 116 FT./SEC. (79 MPH)
MAX. MACH NUMBER: 0.10
MAX. ACCELERATION: 13.6 Gs
BURNOUT ALTITUDE: 38 FT.
COAST TIME TO APOGEE: 3.4 SECS.

© COPYRIGHT 1998 DAVID G. SLEETER

TELEFLITE CORPORATION

MOTOR NAME:	KG3-14-D12
PROPELLANT TYPE:	KG3
CASING TYPE:	CONVOLUTE PAPER
NOZZLE TYPE:	RAMMED CLAY

CASING I.D.: 0.75"	
MOTOR LENGTH: 4.25"	CORE LENGTH: 1.75"
BAKING SODA: See baking soda note.	THROAT DIA.: 0.25"
PROPELLANT WT.: 24.1 GMS.	BULKHEAD CLAY: 11.0 GMS.
TOTAL WEIGHT: 61.6 GMS.	NOZZLE CLAY: 11.0 GMS.
	CASING O.D.: 1.00"

IMPORTANT BAKING SODA NOTE: Add 30% baking soda to the propellant in your first motor. Reduce the soda in 2% increments until you get a motor that doesn't chuff. Then reduce the soda in 1% increments until maximum thrust is approx. 7 lbs. This test motor took 5% soda, but I made it with a carbon black of medium purity. A motor made with a pure, electrically-conductive carbon black might take substantially more.

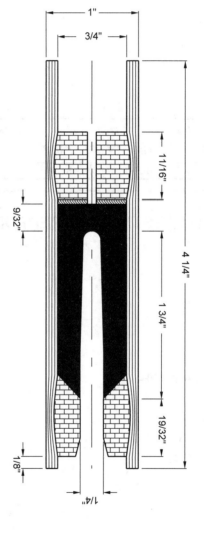

MOTOR SHOWN WITHOUT TIME DELAY

© COPYRIGHT 1998 DAVID G. SLEETER

TOT. IMPULSE: 14.61 NT.-SECS.
AVG. THRUST: 12.38 NTS.
MAX. THRUST: 7.02 LBS.
BURN TIME: 1.18 SECS.
SPEC. IMPULSE: 61.71 SECS.

INDEX OF REQUIRED TOOLING

CASING RETAINER	DRAWING #98-002: 7"	PAGE 91
NOZZLE MOLD	DRAWING #98-008	PAGE 97
CORE SPINDLE	DRAWING #98-044-D	PAGE 126
STARTING TAMP	DRAWING #98-085-A	PAGE 148
PROPELLANT TAMP	DRAWING #98-086-A	PAGE 149
FLAT TAMP	DRAWING #98-091-B	PAGE 154
PORTHOLE TAMP	DRAWING #98-091-A	PAGE 154
PORTHOLE WIRE	DRAWING #98-091	PAGE 154

MOTOR NAME: KG3-14-D12	NOZZLE TYPE: RAMMED CLAY	NOZZLE THROAT: 0.25"
TEST FILENAME: 970601D	CASING ID: 0.75"	CORE LENGTH: 1.75"
PROPELLANT: KG3	CASING OD: 1.00"	MOTOR WEIGHT: 61.6 GMS.
BURN MODIFIER: 5% NaHCO3	MOTOR LENGTH: 4.25"	PROPELLANT WT.: 24.1 GMS.

[Thrust vs. Time curve: thrust in pounds (0.00 to 6.32+) vs. time in seconds (0 to 1.062). Curve rises gradually to a peak near 0.944 s then drops sharply.]

TOT. IMP.: 14.61 NT.-SECS.		MAX. THRUST: 7.02 LBS.
SPEC. IMPULSE: 61.71 SECS.	BURN TIME: 1.18 SECS.	AVG. THRUST: 12.38 NTS.

KG3-14-D12 ESTIMATED PERFORMANCE

MINIMUM ROCKET

BODY LENGTH: 13.5"
BODY DIAMETER: 1.00"
LIFTOFF WT.: 4.7 OZ.
MAX. ALTITUDE: 1,373 FT.
MAX. VELOCITY: 350 FT./SEC. (238 MPH)
MAX. MACH NUMBER: 0.31
MAX. ACCELERATION: 25.9 Gs
BURNOUT ALTITUDE: 142 FT.
COAST TIME TO APOGEE: 8.2 SECS.

AVERAGE ROCKET

BODY LENGTH: 30"
BODY DIAMETER: 2.26"
LIFTOFF WT.: 14.2 OZ.
MAX. ALTITUDE: 149 FT.
MAX. VELOCITY: 89 FT./SEC. (61 MPH)
MAX. MACH NUMBER: 0.08
MAX. ACCELERATION: 7.2 Gs
BURNOUT ALTITUDE: 34 FT.
COAST TIME TO APOGEE: 2.6 SECS.

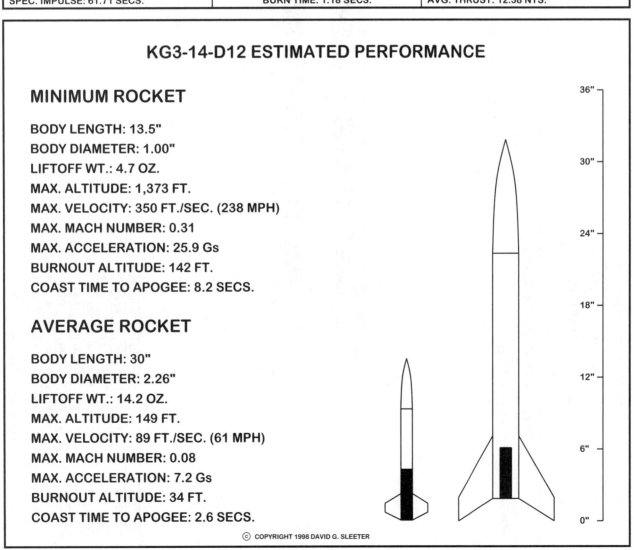

© COPYRIGHT 1998 DAVID G. SLEETER

TELEFLITE CORPORATION

MOTOR NAME:	KG3-26-E31
PROPELLANT TYPE:	KG3
CASING TYPE:	CONVOLUTE PAPER
NOZZLE TYPE:	RAMMED CLAY

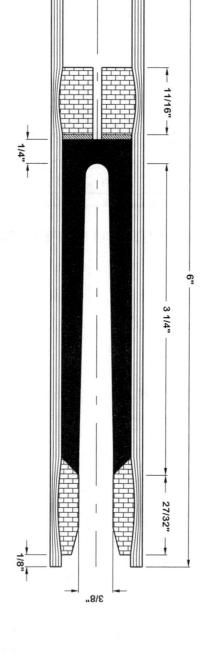

MOTOR SHOWN WITHOUT TIME DELAY

© COPYRIGHT 1998 DAVID G. SLEETER

INDEX OF REQUIRED TOOLING

CASING RETAINER	DRAWING #98-002: 7"	PAGE 91
NOZZLE MOLD	DRAWING #98-010	PAGE 99
CORE SPINDLE	DRAWING #98-049-D	PAGE 128
STARTING TAMP	DRAWING #98-087-A	PAGE 150
PROPELLANT TAMP	DRAWING #98-089-A	PAGE 152
FLAT TAMP	DRAWING #98-091-A	PAGE 154
PORTHOLE TAMP	DRAWING #98-091-B	PAGE 154
PORTHOLE WIRE	DRAWING #98-091	PAGE 154

CASING I.D.:	0.75"
MOTOR LENGTH:	6.00"
BAKING SODA:	See baking soda note.
PROPELLANT WT.:	39.9 GMS.
TOTAL WEIGHT:	84.0 GMS.
CASING O.D.:	1.00"
NOZZLE CLAY:	11.0 GMS.
BULKHEAD CLAY:	11.0 GMS.
THROAT DIA.:	0.375"
CORE LENGTH:	3.25"
TOT. IMPULSE:	26.01 NT.-SECS.
AVG. THRUST:	31.30 NTS.
MAX. THRUST:	19.83 LBS.
BURN TIME:	0.83 SECS.
SPEC. IMPULSE:	66.36 SECS.

IMPORTANT BAKING SODA NOTE: Add 30% baking soda to the propellant in your first motor. Then reduce the soda in 1% increments until you get a motor that doesn't chuff. Then reduce the soda in 2% increments until maximum thrust is approx. 20 lbs. This test motor took 5% soda, but I made it with a carbon black of medium purity. A motor made with a pure, electrically-conductive carbon black might take substantially more.

MOTOR NAME: KG3-26-E31	NOZZLE TYPE: RAMMED CLAY	NOZZLE THROAT: 0.375"
TEST FILENAME: 970421L	CASING ID: 0.75"	CORE LENGTH: 3.25"
PROPELLANT: KG3	CASING OD: 1.00"	MOTOR WEIGHT: 84.0 GMS.
BURN MODIFIER: 5% NaHCO3	MOTOR LENGTH: 6.00"	PROPELLANT WT.: 39.9 GMS.

TOT. IMP.: 26.01 NT.-SECS.		MAX. THRUST: 19.83 LBS.
SPEC. IMPULSE: 66.36 SECS.	BURN TIME: 0.83 SECS.	AVG. THRUST: 31.30 NTS.

KG3-26-E31 ESTIMATED PERFORMANCE

MINIMUM ROCKET

BODY LENGTH: 15.5"
BODY DIAMETER: 1.00"
LIFTOFF WT.: 5.5 OZ.
MAX. ALTITUDE: 2,471 FT.
MAX. VELOCITY: 581 FT./SEC. (396 MPH)
MAX. MACH NUMBER: 0.52
MAX. ACCELERATION: 67.3 Gs
BURNOUT ALTITUDE: 184 FT.
COAST TIME TO APOGEE: 10.5 SECS.

AVERAGE ROCKET

BODY LENGTH: 30"
BODY DIAMETER: 2.26"
LIFTOFF WT.: 15.0 OZ.
MAX. ALTITUDE: 473 FT.
MAX. VELOCITY: 182 FT./SEC. (124 MPH)
MAX. MACH NUMBER: 0.16
MAX. ACCELERATION: 21.3 Gs
BURNOUT ALTITUDE: 57 FT.
COAST TIME TO APOGEE: 4.9 SECS.

© COPYRIGHT 1998 DAVID G. SLEETER

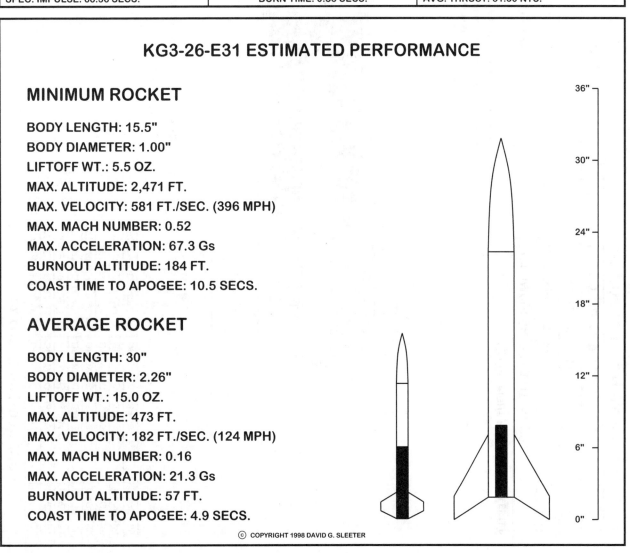

TELEFLITE CORPORATION

MOTOR NAME: KG3-37-E25
PROPELLANT TYPE: KG3
CASING TYPE: CONVOLUTE PAPER
NOZZLE TYPE: RAMMED CLAY

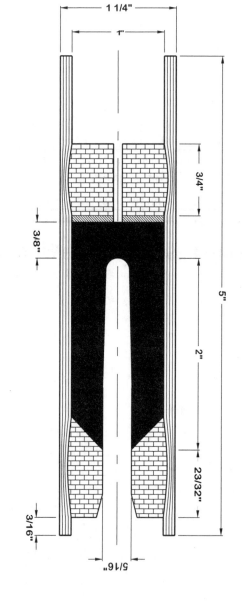

MOTOR SHOWN WITHOUT TIME DELAY

INDEX OF REQUIRED TOOLING

CASING RETAINER	DRAWING #98-003: 7"	PAGE 92
NOZZLE MOLD	DRAWING #98-012	PAGE 101
CORE SPINDLE	DRAWING #98-054-C	PAGE 131
STARTING TAMP	DRAWING #98-092-A	PAGE 155
PROPELLANT TAMP	DRAWING #98-093-A	PAGE 156
FLAT TAMP	DRAWING #98-096-A	PAGE 159
PORTHOLE TAMP	DRAWING #98-096-B	PAGE 159
PORTHOLE WIRE	DRAWING #98-096	PAGE 159

IMPORTANT BAKING SODA NOTE: Add 30% baking soda to the propellant in your first motor. Reduce the soda in 2% increments until you get a motor that doesn't chuff. Then reduce the soda in 1% increments until maximum thrust is approx. 14 lbs. This test motor took 5% soda, but I made it with a carbon black of medium purity. A motor made with a pure, electrically-conductive carbon black might take substantially more.

TOTAL WEIGHT: 131.6 GMS.
PROPELLANT WT.: 57.2 GMS.
BAKING SODA: See baking soda note.
MOTOR LENGTH: 5.00"
CASING I.D.: 1.00"

CASING O.D.: 1.25"
NOZZLE CLAY: 24.0 GMS.
BULKHEAD CLAY: 24.0 GMS.
THROAT DIA.: 0.313"
CORE LENGTH: 2.00"

TOT. IMPULSE: 37.05 NT.-SECS.
AVG. THRUST: 25.45 NTS.
MAX. THRUST: 13.24 LBS.
BURN TIME: 1.40 SECS.
SPEC. IMPULSE: 65.94 SECS.

© COPYRIGHT 1998 DAVID G. SLEETER

358

MOTOR NAME: KG3-37-E25	NOZZLE TYPE: RAMMED CLAY	NOZZLE THROAT: 0.313"
TEST FILENAME: 970601B	CASING ID: 1.00"	CORE LENGTH: 2.00"
PROPELLANT: KG3	CASING OD: 1.25"	MOTOR WEIGHT: 131.6 GMS.
BURN MODIFIER: 5% NaHCO3	MOTOR LENGTH: 5.00"	PROPELLANT WT.: 57.2 GMS.

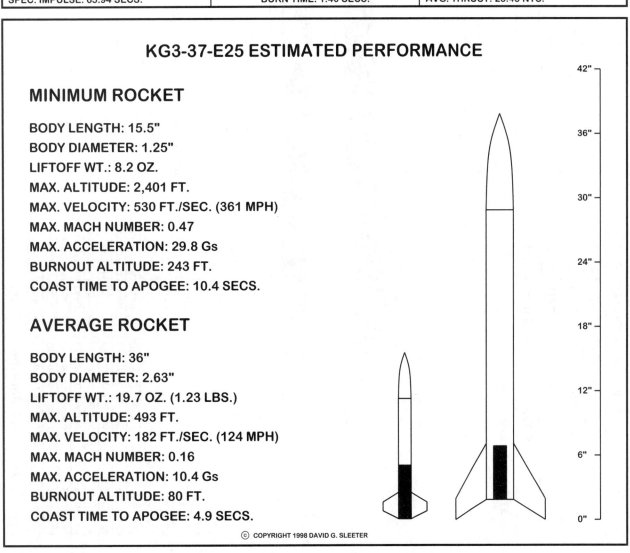

TOT. IMP.: 37.05 NT.-SECS.		MAX. THRUST: 13.24 LBS.
SPEC. IMPULSE: 65.94 SECS.	BURN TIME: 1.40 SECS.	AVG. THRUST: 25.45 NTS.

KG3-37-E25 ESTIMATED PERFORMANCE

MINIMUM ROCKET

BODY LENGTH: 15.5"
BODY DIAMETER: 1.25"
LIFTOFF WT.: 8.2 OZ.
MAX. ALTITUDE: 2,401 FT.
MAX. VELOCITY: 530 FT./SEC. (361 MPH)
MAX. MACH NUMBER: 0.47
MAX. ACCELERATION: 29.8 Gs
BURNOUT ALTITUDE: 243 FT.
COAST TIME TO APOGEE: 10.4 SECS.

AVERAGE ROCKET

BODY LENGTH: 36"
BODY DIAMETER: 2.63"
LIFTOFF WT.: 19.7 OZ. (1.23 LBS.)
MAX. ALTITUDE: 493 FT.
MAX. VELOCITY: 182 FT./SEC. (124 MPH)
MAX. MACH NUMBER: 0.16
MAX. ACCELERATION: 10.4 Gs
BURNOUT ALTITUDE: 80 FT.
COAST TIME TO APOGEE: 4.9 SECS.

© COPYRIGHT 1998 DAVID G. SLEETER

TELEFLITE CORPORATION

MOTOR NAME: KG3-78-F55
PROPELLANT TYPE: KG3
CASING TYPE: CONVOLUTE PAPER
NOZZLE TYPE: RAMMED CLAY

CASING I.D.: 1.00"
MOTOR LENGTH: 8.25"
BAKING SODA: See baking soda note.
PROPELLANT WT.: 117.9 GMS.
TOTAL WEIGHT: 213.6 GMS.

CASING O.D.: 1.25"
NOZZLE CLAY: 24.0 GMS.
BULKHEAD CLAY: 24.0 GMS.
THROAT DIA.: 0.50"
CORE LENGTH: 5.00"

TOT. IMPULSE: 78.90 NT.-SECS.
AVG. THRUST: 55.37 NTS.
MAX. THRUST: 30.07 LBS.
BURN TIME: 1.42 SECS.
SPEC. IMPULSE: 68.12 SECS.

IMPORTANT BAKING SODA NOTE: Add 30% baking soda to the propellant in your first motor. Reduce the soda in 2% increments until you get a motor that doesn't chuff. Then reduce the soda in 1% increments until maximum thrust is approx. 30 lbs. This test motor took 5% soda, but I made it with a carbon black of medium purity. A motor made with a pure, electrically-conductive carbon black might take substantially more.

© COPYRIGHT 1998 DAVID G. SLEETER

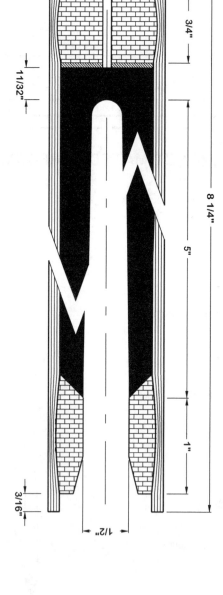

MOTOR SHOWN WITHOUT TIME DELAY

INDEX OF REQUIRED TOOLING

CASING RETAINER	DRAWING #98-003: .10"	PAGE 92
NOZZLE MOLD	DRAWING #98-014	PAGE 103
CORE SPINDLE	DRAWING #98-060-F	PAGE 134
STARTING TAMP	DRAWING #98-094-A	PAGE 157
PROPELLANT TAMP	DRAWING #98-095-A	PAGE 158
FLAT TAMP	DRAWING #98-096-B	PAGE 159
PORTHOLE TAMP	DRAWING #98-096-A	PAGE 159
PORTHOLE WIRE	DRAWING #98-096	PAGE 159

360

MOTOR NAME: KG3-78-F55	NOZZLE TYPE: RAMMED CLAY	NOZZLE THROAT: 0.50"
TEST FILENAME: 970424D	CASING ID: 1.00"	CORE LENGTH: 5.00"
PROPELLANT: KG3	CASING OD: 1.25"	MOTOR WEIGHT: 213.6 GMS.
BURN MODIFIER: 5% NaHCO3	MOTOR LENGTH: 8.25"	PROPELLANT WT.: 117.9 GMS.

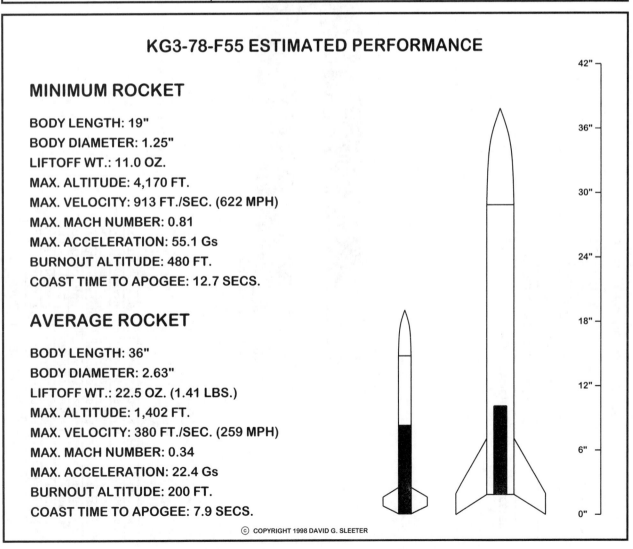

TOT. IMP.: 78.90 NT.-SECS.		MAX. THRUST: 30.07 LBS.
SPEC. IMPULSE: 68.12 SECS.	BURN TIME: 1.42 SECS.	AVG. THRUST: 55.37 NTS.

KG3-78-F55 ESTIMATED PERFORMANCE

MINIMUM ROCKET

BODY LENGTH: 19"
BODY DIAMETER: 1.25"
LIFTOFF WT.: 11.0 OZ.
MAX. ALTITUDE: 4,170 FT.
MAX. VELOCITY: 913 FT./SEC. (622 MPH)
MAX. MACH NUMBER: 0.81
MAX. ACCELERATION: 55.1 Gs
BURNOUT ALTITUDE: 480 FT.
COAST TIME TO APOGEE: 12.7 SECS.

AVERAGE ROCKET

BODY LENGTH: 36"
BODY DIAMETER: 2.63"
LIFTOFF WT.: 22.5 OZ. (1.41 LBS.)
MAX. ALTITUDE: 1,402 FT.
MAX. VELOCITY: 380 FT./SEC. (259 MPH)
MAX. MACH NUMBER: 0.34
MAX. ACCELERATION: 22.4 Gs
BURNOUT ALTITUDE: 200 FT.
COAST TIME TO APOGEE: 7.9 SECS.

© COPYRIGHT 1998 DAVID G. SLEETER

TELEFLITE CORPORATION

MOTOR NAME: KG3-63-F41
PROPELLANT TYPE: KG3
CASING TYPE: CONVOLUTE PAPER
NOZZLE TYPE: RAMMED CLAY

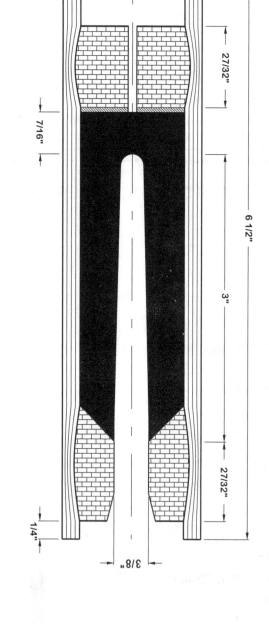

MOTOR SHOWN WITHOUT TIME DELAY

INDEX OF REQUIRED TOOLING

CASING RETAINER	DRAWING #98-004; 8"	PAGE 93
NOZZLE MOLD	DRAWING #98-016	PAGE 105
CORE SPINDLE	DRAWING #98-066-E	PAGE 137
STARTING TAMP	DRAWING #98-097-A	PAGE 160
PROPELLANT TAMP	DRAWING #98-098-A	PAGE 161
FLAT TAMP	DRAWING #98-104-B	PAGE 167
PORTHOLE TAMP	DRAWING #98-104-A	PAGE 167
PORTHOLE WIRE	DRAWING #98-104	PAGE 167

IMPORTANT BAKING SODA NOTE: Add 30% baking soda to the propellant in your first motor. Reduce the soda in 1% increments until you get a motor that doesn't chuff. Then reduce the soda in 2% increments until maximum thrust is approx. 20 lbs. This test motor took 5% soda, but I made it with a carbon black of medium purity. A motor made with a pure, electrically-conductive carbon black might take substantially more.

TOTAL WEIGHT: 220.9 GMS.
PROPELLANT WT.: 99.5 GMS.
BAKING SODA: See baking soda note.
MOTOR LENGTH: 6.5"
CASING I.D.: 1.125"

CASING O.D.: 1.50"
NOZZLE CLAY: 33.0 GMS.
BULKHEAD CLAY: 33.0 GMS.
THROAT DIA.: 0.375"
CORE LENGTH: 3.00"

TOT. IMPULSE: 61.58 NT.-SECS.
AVG. THRUST: 41.58 NTS.
MAX. THRUST: 20.42 LBS.
BURN TIME: 1.48 SECS.
SPEC. IMPULSE: 63.00 SECS.

© COPYRIGHT 1998 DAVID G. SLEETER

MOTOR NAME: KG3-63-F41	NOZZLE TYPE: RAMMED CLAY	NOZZLE THROAT: 0.375"
TEST FILENAME: 970422A	CASING ID: 1.125"	CORE LENGTH: 3.00"
PROPELLANT: KG3	CASING OD: 1.50"	MOTOR WEIGHT: 220.9 GMS.
BURN MODIFIER: 5% NaHCO3	MOTOR LENGTH: 6.50"	PROPELLANT WT.: 99.5 GMS.

TOT. IMP.: 61.58 NT.-SECS.		MAX. THRUST: 20.42 LBS.
SPEC. IMPULSE: 63.00 SECS.	BURN TIME: 1.48 SECS.	AVG. THRUST: 41.58 NTS.

KG3-63-F41 ESTIMATED PERFORMANCE

MINIMUM ROCKET

BODY LENGTH: 19.5"
BODY DIAMETER: 1.50"
LIFTOFF WT.: 13.8 OZ.
MAX. ALTITUDE: 2,536 FT.
MAX. VELOCITY: 529 FT./SEC. (361 MPH)
MAX. MACH NUMBER: 0.47
MAX. ACCELERATION: 27.4 Gs
BURNOUT ALTITUDE: 261 FT.
COAST TIME TO APOGEE: 10.8 SECS.

AVERAGE ROCKET

BODY LENGTH: 48"
BODY DIAMETER: 3.10"
LIFTOFF WT.: 31.8 OZ. (1.99 LBS.)
MAX. ALTITUDE: 532 FT.
MAX. VELOCITY: 188 FT./SEC. (128 MPH)
MAX. MACH NUMBER: 0.17
MAX. ACCELERATION: 10.0 Gs
BURNOUT ALTITUDE: 89 FT.
COAST TIME TO APOGEE: 5.1 SECS.

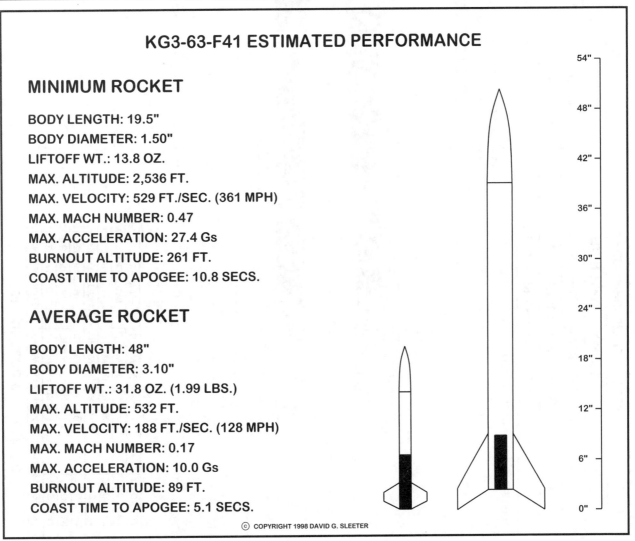

© COPYRIGHT 1998 DAVID G. SLEETER

TELEFLITE CORPORATION

MOTOR NAME:	KG3-116-G102
PROPELLANT TYPE:	KG3
CASING TYPE:	CONVOLUTE PAPER
NOZZLE TYPE:	RAMMED CLAY

MOTOR SHOWN WITHOUT TIME DELAY

Dimensions shown on diagram: 1 1/2", 1 1/8", 27/32", 3/8", 9-3/4", 6", 1 1/8", 3/16", 9/16"

© COPYRIGHT 1998 DAVID G. SLEETER

IMPORTANT BAKING SODA NOTE: Add 30% baking soda to the propellant in your first motor. Reduce the soda in 2% increments until you get a motor that doesn't chuff. Then reduce the soda in 1% increments until maximum thrust is approx. 48 lbs. This test motor took 5% soda, but I made it with a carbon black of medium purity. A motor made with a pure, electrically-conductive carbon black might take substantially more.

TOTAL WEIGHT: 305.3 GMS.	TOT. IMPULSE: 116.56 NT.-SECS.
PROPELLANT WT.: 155.4 GMS.	NOZZLE CLAY: 33.0 GMS.	AVG. THRUST: 102.61 NTS.
BAKING SODA: See baking soda note.	BULKHEAD CLAY: 33.0 GMS.	MAX. THRUST: 47.50 LBS.
MOTOR LENGTH: 9.75"	THROAT DIA.: 0.563"	BURN TIME: 1.14 SECS.
CASING I.D.: 1.125"	CORE LENGTH: 6.00"	SPEC. IMPULSE: 76.35 SECS.
CASING O.D.: 1.50"		

INDEX OF REQUIRED TOOLING

CASING RETAINER	DRAWING #98-004; 11-1/2"	PAGE 93
NOZZLE MOLD	DRAWING #98-018	PAGE 107
CORE SPINDLE	DRAWING #98-071-E	PAGE 139
STARTING TAMP	DRAWING #98-099-A	PAGE 162
PROPELLANT TAMP	DRAWING #98-101-A	PAGE 164
FLAT TAMP	DRAWING #98-104-B	PAGE 167
PORTHOLE TAMP	DRAWING #98-104-A	PAGE 167
PORTHOLE WIRE	DRAWING #98-104	PAGE 167

MOTOR NAME: KG3-116-G102	NOZZLE TYPE: RAMMED CLAY	NOZZLE THROAT: 0.563"
TEST FILENAME: 970423A	CASING ID: 1.125"	CORE LENGTH: 6.00"
PROPELLANT: KG3	CASING OD: 1.50"	MOTOR WEIGHT: 305.3 GMS.
BURN MODIFIER: 5% NaHCO3	MOTOR LENGTH: 9.75"	PROPELLANT WT.: 155.4 GMS.

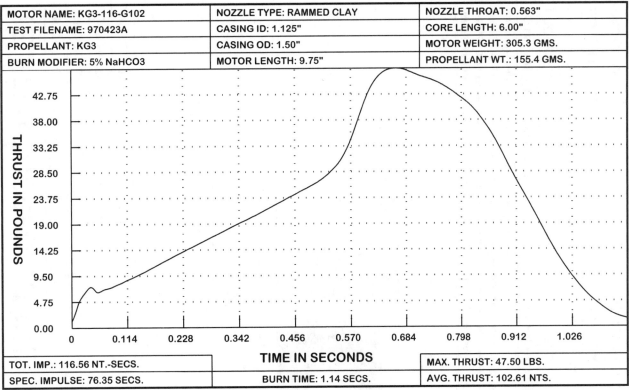

TOT. IMP.: 116.56 NT.-SECS.		MAX. THRUST: 47.50 LBS.
SPEC. IMPULSE: 76.35 SECS.	BURN TIME: 1.14 SECS.	AVG. THRUST: 102.61 NTS.

KG3-116-G102 ESTIMATED PERFORMANCE

MINIMUM ROCKET

BODY LENGTH: 23"
BODY DIAMETER: 1.50"
LIFTOFF WT.: 16.8 OZ. (1.05 LBS.)
MAX. ALTITUDE: 4,286 FT.
MAX. VELOCITY: 870 FT./SEC. (593 MPH)
MAX. MACH NUMBER: 0.77
MAX. ACCELERATION: 52.0 Gs
BURNOUT ALTITUDE: 444 FT.
COAST TIME TO APOGEE: 13.2 SECS.

AVERAGE ROCKET

BODY LENGTH: 48"
BODY DIAMETER: 3.10"
LIFTOFF WT.: 34.8 OZ. (2.18 LBS.)
MAX. ALTITUDE: 1,372 FT.
MAX. VELOCITY: 365 FT./SEC. (249 MPH)
MAX. MACH NUMBER: 0.32
MAX. ACCELERATION: 22.2 Gs
BURNOUT ALTITUDE: 189 FT.
COAST TIME TO APOGEE: 7.9 SECS.

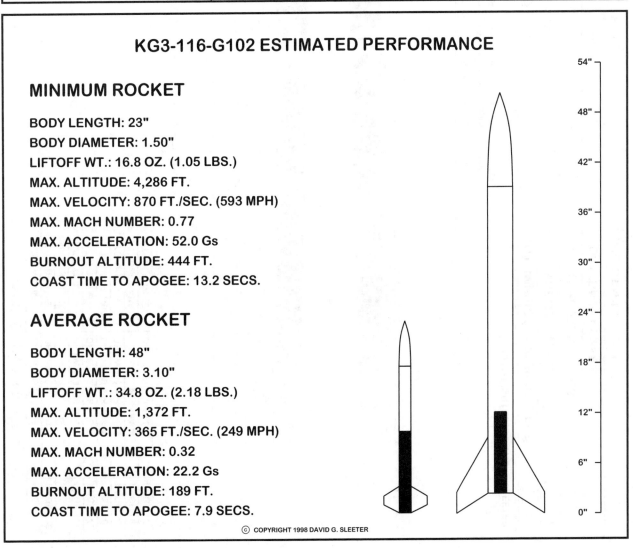

© COPYRIGHT 1998 DAVID G. SLEETER

TELEFLITE CORPORATION

MOTOR NAME: KG3-140-G139
PROPELLANT TYPE: KG3
CASING TYPE: CONVOLUTE PAPER
NOZZLE TYPE: STAINLESS STEEL DeLAVAL

MOTOR SHOWN WITHOUT TIME DELAY

Dimensions shown on diagram: 1 1/2", 1 1/8", 27/32", 11/32", 11 1/2", 8", 5/8"

IMPORTANT BAKING SODA NOTE: Add 30% baking soda to the propellant in your first motor. Reduce the soda until you get a motor that doesn't chuff. Then reduce the soda in 1% increments until maximum thrust is approx. 58 lbs. This test motor took 5% soda, but I made it with a carbon black of medium purity. A motor made with a pure, electrically-conductive carbon black might take substantially more.

TOTAL WEIGHT: 369.7 GMS.	TOT. IMPULSE: 140.10 NT.-SECS.
PROPELLANT WT.: 199.0 GMS.	CASING O.D.: 1.50"
BAKING SODA: See baking soda note.	NOZZLE: S.S. DeLAVAL
MOTOR LENGTH: 11.00"	BULKHEAD CLAY: 33.0 GMS.
CASING I.D.: 1.125"	THROAT DIA.: 0.625"
	CORE LENGTH: 8.00"

AVG. THRUST: 139.82 NTS.
MAX. THRUST: 56.98 LBS.
BURN TIME: 1.00 SECS.
SPEC. IMPULSE: 71.67 SECS.

© COPYRIGHT 1998 DAVID G. SLEETER

INDEX OF REQUIRED TOOLING

NOZZLE ANVIL	DRAWING #98-109-C	PAGE 172
NOZZLE RAMMER	DRAWING #98-110-D	PAGE 173
CASING RETAINER	DRAWING #98-004: .15"	PAGE 93
NOZZLE MOLD	DRAWING #98-020	PAGE 109
CORE SPINDLE	DRAWING #98-078-C	PAGE 143
PROPELLANT TAMP	DRAWING #98-103-A	PAGE 166
FLAT TAMP	DRAWING #98-104-B	PAGE 167
PORTHOLE TAMP	DRAWING #98-104-A	PAGE 167
PORTHOLE WIRE	DRAWING #98-104	PAGE 167

366

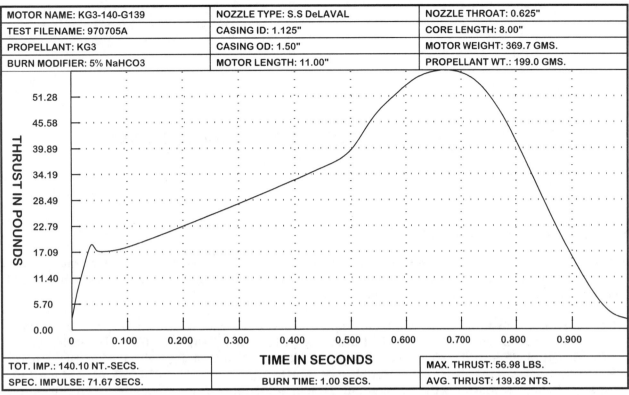

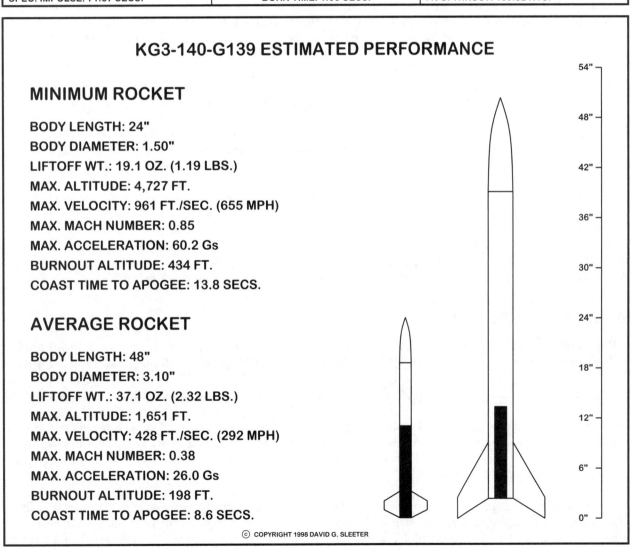

13. KS PROPELLANT AND MOTOR DESIGNS

● First and Most Important!

Before you proceed, if you haven't read **Chapter 1**, go back and read it now. It contains important information on the safe handling of this rocket propellant. Also pay attention to the bold warning at the beginning of **Chapter 1**. To emphasize its importance, I'll repeat that warning here.

> Do *not* add or substitute other chemicals either to-or-for the chemicals described in this book. Specifically do *not* use POTASSIUM CHLORATE, SODIUM CHLORATE, POTASSIUM PERCHLORATE, SODIUM PERCHLORATE, POTASSIUM PERMANGANATE, AMMONIUM PERCHLORATE, PHOSPHOROUS, MATCH HEADS, ALUMINUM, MAGNESIUM, or ANY METALS AT ALL! *All* of these substances will form dangerous and explosive mixtures when combined with the other ingredients described herein, *AND OTHER CHEMICALS WILL FORM DANGEROUS MIXTURES AS WELL.* When mixing and handling propellants, be sure to ground yourself and all containers and utensils according to the instructions on page 14.
>
> If you decide to experiment on your own outside the bounds of the instructions provided in this book, consult with a professional chemist before you proceed.

The **KS** propellant is a dry-packed mixture of **4 hour milled potassium nitrate** or **precipitated potassium nitrate**, and **powdered confectioner's sugar**. If you haven't read **Chapters 5** and **6**, go back and read them now. Together they explain where to buy these chemicals, and how to prepare them for use in a rocket propellant.

Because it is a powder, the **KS** propellant *cannot* be compressed into a truly solid propellant grain. It has no structural strength of its own, and relies on the strength of the motor casing to keep it intact. You will therefore notice that in all but the 1/2" ID motors, I've kept the maximum recommended thrust for each of the **KS** motors *low*. When compressed into a motor casing, its density is **20** to **25 gms. per cubic inch**. As with the other propellants, its burn rate can be adjusted by adding or subtracting baking soda. To prepare the **KS** propellant, follow the instructions on **page 252**, and use the figures below to determine the exact weights of the chemicals needed for the desired batch size.

◆ THE KS PROPELLANT

Powdered potassium nitrate .. 60.0%	600.0 gms.	300.0 gms.	150.0 gms.	75.0 gms.
Powdered sugar 40%%	400.0 gms.	200.0 gms.	100.0 gms.	50.0 gms.
Total batch size 100%	1000.0 gms.	500.0 gms.	250.0 gms.	125.0 gms.

To build your *first* motor with the **KS** propellant, weigh out the amount indicated in the motor drawing (rounding off the numbers makes the math easier). Then add 20% of that weight in baking soda, and mix it in thoroughly with a fork. If, for example, you've weighed out 50 grams of propellant, add 10 grams of baking soda for a total of 60 grams. Then build the motor per the instructions in **Chapter 10**.

Per the instructions on **page 494**, do a rudimentary test by firing the motor (with the motor pointing *down* and the exhaust aimed *upward*) from a pipe in the ground. If this first motor "chuffs" (chugs like a steam engine), build a *second* motor, and reduce the baking soda to 18%. Keep building, testing, and reducing the amount of soda in 2% increments until you get a motor that burns smoothly and *doesn't* chuff. At first the chuffing may be slow. As you approach a condition of combustion stability, the frequency of the chuffs will increase (the chuffing sound will get *faster*). When you are *very* near stability, the amplitude of the chuffs will decrease (the chuffs will sound *fainter*).

Now set up a thrust-testing stand (**page 470**), and test the motor for maximum thrust (**page 475**). In the tests that follow, reduce the soda in 1% increments until you get a motor whose maximum thrust approximates the maximum thrust printed in the motor drawing. At that point you'll have a working motor with the maximum safe chamber pressure and thrust for that particular design. Keep accurate notes. Write down the amount of baking soda that works the best, and *forever thereafter* use that amount of soda to build copies of your final, successful motor.

Important note. You can also alter the chamber pressure and thrust by changing the length of the motor's core. A *longer* core will *increase* the chamber pressure and thrust, and a *shorter* core will *reduce* the chamber pressure and thrust. But, of course, changing the length of the core will also change the amount of propellant required, and the physical size of the motor.

Because the **KS** propellant is packed dry with no binder, it uses the packing pressure generated during loading to insure the seal against the casing wall. The case bonding issues associated with the other propellants are not relevant. All the **KS** motors were made with standard fireworks grade casings and clay nozzles. By switching to stainless steel De Laval nozzles and the high strength casings manufactured (as of this writing) by **New England Paper Tube** (**pages 163** and **164**), you can substantially improve their performance. Shop drawings for De Laval nozzles begin on **page 323**.

Important note. *For the reasons discussed on* **page 192**, *the KS propellant continues to self-mix, both in storage, and after it has been loaded into a rocket motor.* As it does so, it shrinks, and its burn rate increases, eventually exceeding the motor's design limits. For this reason, **both the KS propellant and all KS motors must be used within 48 hours of when they are made**.

With each motor design, I've included the thrust-time curve and motor performance data generated by the test equipment and software described on **pages 468 - 469**, plus two flight performance estimates generated by **Rogers' Aeroscience Alt4** rocket performance prediction software. The "minimum rocket" is what you might build if you were trying to break an altitude or speed record. The lift-off weight is minimal. The body diameter is equal to the o.d. of the motor, and the parachute compartment is small; approx. 5 body diameter's in length. The "average rocket" is more like what you would build for scientific research or general sport flying. The body diameter is one of the industry standards approx. twice the diameter of the motor, and there's ample space and weight allowance for a large parachute and a small scientific payload.

In each case the lift-off weight is *very* approximate, and based on an educated guess. The *extremely* accurate performance specs. *are the actual numbers generated by the Alt-4 program.* They are based on the rocket's stated length, diameter, and lift-off weight, and the *tested* motor's thrust-time data. *In the real world*, small variations in motor performance, the size and shape of the rocket's fins, and even the smoothness of its paint-job will cause the rocket's performance to vary from the stated figures.

TELEFLITE CORPORATION

MOTOR NAME: KS-6-C7
PROPELLANT TYPE: KS
CASING TYPE: CONVOLUTE PAPER
NOZZLE TYPE: RAMMED CLAY

IMPORTANT BAKING SODA NOTE: Add 20% baking soda to the propellant in your first motor. Reduce the soda until you get a motor that doesn't chuff. Then reduce the soda in 1% increments until maximum thrust is approx. 3-1/2 lbs. This test motor took 6% soda, but variations in the purity and particle size of the chemicals might make your motor take more or less.

CASING I.D.: 0.50"	
MOTOR LENGTH: 3.50"	
BAKING SODA: See baking soda note.	
PROPELLANT WT.: 9.9 GMS.	
TOTAL WEIGHT: 27.8 GMS.	
CASING O.D.: 0.75"	TOT. IMPULSE: 6.97 NT.-SECS.
NOZZLE CLAY: 3.3 GMS.	AVG. THRUST: 7.70 NTS.
BULKHEAD CLAY: 3.3 GMS.	MAX. THRUST: 3.88 LBS.
THROAT DIA.: 0.1563"	BURN TIME: 0.90 SECS.
CORE LENGTH: 1.75"	SPEC. IMPULSE: 71.67 SECS.

© COPYRIGHT 1998 DAVID G. SLEETER

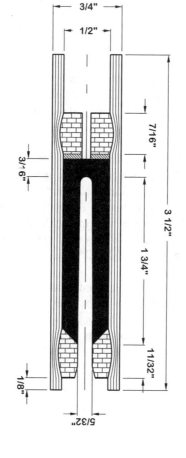

MOTOR SHOWN WITHOUT TIME DELAY

INDEX OF REQUIRED TOOLING

CASING RETAINER	DRAWING #98-001: .5"	PAGE 90
NOZZLE MOLD	DRAWING #98-006	PAGE 95
CORE SPINDLE	DRAWING #98-035-D	PAGE 121
STARTING TAMP	USE PROPELLANT TAMP	
PROPELLANT TAMP	DRAWING #98-082-A	PAGE 145
FLAT TAMP	DRAWING #98-084-B	PAGE 147
PORTHOLE TAMP	DRAWING #98-084-A	PAGE 147
PORTHOLE WIRE	DRAWING #98-084	PAGE 147

MOTOR NAME: KS-6-C7	NOZZLE TYPE: RAMMED CLAY	NOZZLE THROAT: 0.1563"
TEST FILENAME: 960618A	CASING ID: 0.50"	CORE LENGTH: 1.75"
PROPELLANT: KG3	CASING OD: 0.75"	MOTOR WEIGHT: 27.8 GMS.
BURN MODIFIER: 6% NaHCO3	MOTOR LENGTH: 3.50"	PROPELLANT WT.: 9.9 GMS.

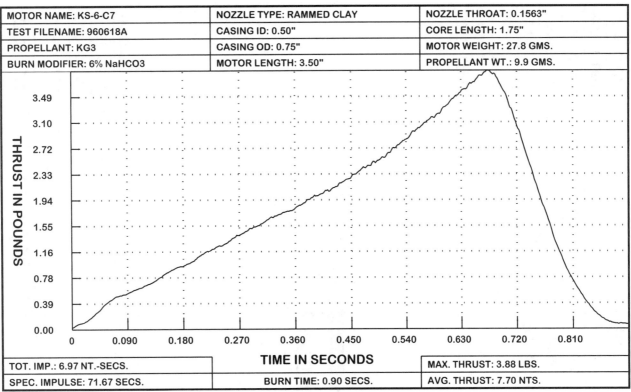

TOT. IMP.: 6.97 NT.-SECS.	TIME IN SECONDS	MAX. THRUST: 3.88 LBS.
SPEC. IMPULSE: 71.67 SECS.	BURN TIME: 0.90 SECS.	AVG. THRUST: 7.70 NTS.

KS-6-C7 ESTIMATED PERFORMANCE

MINIMUM ROCKET

BODY LENGTH: 11"
BODY DIAMETER: 0.75"
LIFTOFF WT.: 2.75 OZ.
MAX. ALTITUDE: 1,002 FT.
MAX. VELOCITY: 281 FT./SEC. (191 MPH)
MAX. MACH NUMBER: 0.25
MAX. ACCELERATION: 23.5 Gs
BURNOUT ALTITUDE: 105 FT.
COAST TIME TO APOGEE: 7.1 SECS.

AVERAGE ROCKET

BODY LENGTH: 24"
BODY DIAMETER: 1.63"
LIFTOFF WT.: 10.00 OZ.
MAX. ALTITUDE: 65 FT.
MAX. VELOCITY: 56 FT./SEC. (38 MPH)
MAX. MACH NUMBER: 0.05
MAX. ACCELERATION: 5.4 Gs
BURNOUT ALTITUDE: 20 FT.
COAST TIME TO APOGEE: 1.7 SECS.

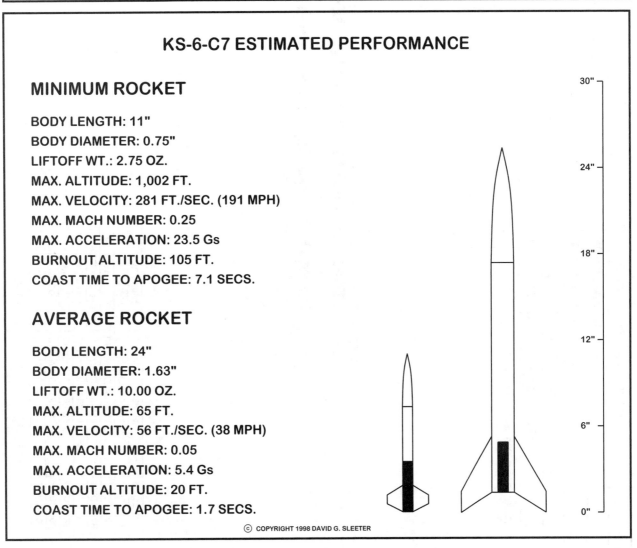

© COPYRIGHT 1998 DAVID G. SLEETER

TELEFLITE CORPORATION

MOTOR NAME: KS-14-D20
PROPELLANT TYPE: KS
CASING TYPE: CONVOLUTE PAPER
NOZZLE TYPE: RAMMED CLAY

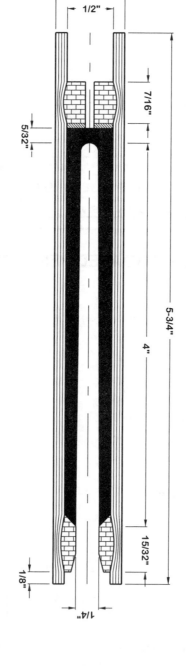

MOTOR SHOWN WITHOUT TIME DELAY

INDEX OF REQUIRED TOOLING

CASING RETAINER	DRAWING #98-001: 7"	PAGE 90
NOZZLE MOLD	DRAWING #98-007	PAGE 96
CORE SPINDLE	DRAWING #98-039-G	PAGE 123
STARTING TAMP	USE PROPELLANT TAMP	
PROPELLANT TAMP	DRAWING #98-083-B	PAGE 146
FLAT TAMP	DRAWING #98-084-B	PAGE 147
PORTHOLE TAMP	DRAWING #98-084-A	PAGE 147
PORTHOLE WIRE	DRAWING #98-084	PAGE 147

IMPORTANT BAKING SODA NOTE: Add 20% baking soda to the propellant in your first motor. Reduce the soda in 2% increments until you get a motor that doesn't chuff. Then reduce the soda in 1% increments until maximum thrust is approx. 8 lbs. This test motor took 6% soda, but variations in the purity and particle size of the chemicals might make your motor take more or less.

TOTAL WEIGHT: 44.3 GMS.	CASING O.D.: 0.75"	TOT. IMPULSE: 14.19 NT.-SECS.
PROPELLANT WT.: 19.2 GMS.	NOZZLE CLAY: 3.3 GMS.	AVG. THRUST: 20.07 NTS.
BAKING SODA: See baking soda note.	BULKHEAD CLAY: 3.3 GMS.	MAX. THRUST: 8.56 LBS.
MOTOR LENGTH: 5.75"	THROAT DIA.: 0.25"	BURN TIME: 0.71 SECS.
CASING I.D.: 0.50"	CORE LENGTH: 4.00"	SPEC. IMPULSE: 75.23 SECS.

© COPYRIGHT 1998 DAVID G. SLEETER

MOTOR NAME: KS-14-D20	NOZZLE TYPE: RAMMED CLAY	NOZZLE THROAT: 0.25"
TEST FILENAME: 960618B	CASING ID: 0.50"	CORE LENGTH: 4.00"
PROPELLANT: KG3	CASING OD: 0.75"	MOTOR WEIGHT: 44.3 GMS.
BURN MODIFIER: 6% NaHCO3	MOTOR LENGTH: 5.75"	PROPELLANT WT.: 19.2 GMS.

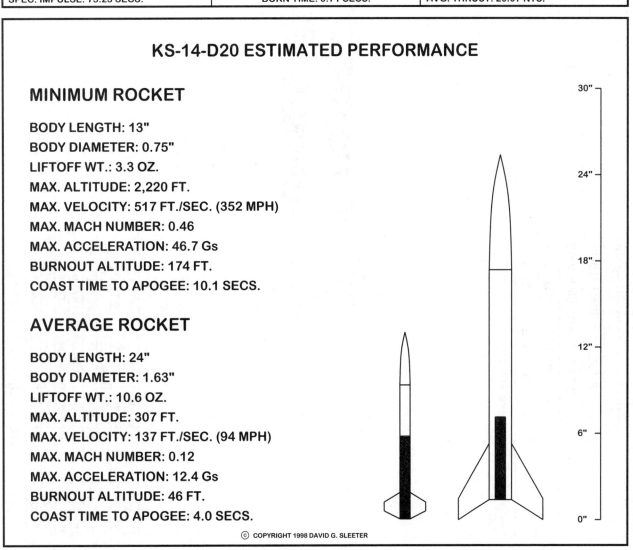

TOT. IMP.: 14.19 NT.-SECS.		MAX. THRUST: 8.56 LBS.
SPEC. IMPULSE: 75.23 SECS.	BURN TIME: 0.71 SECS.	AVG. THRUST: 20.07 NTS.

KS-14-D20 ESTIMATED PERFORMANCE

MINIMUM ROCKET

BODY LENGTH: 13"
BODY DIAMETER: 0.75"
LIFTOFF WT.: 3.3 OZ.
MAX. ALTITUDE: 2,220 FT.
MAX. VELOCITY: 517 FT./SEC. (352 MPH)
MAX. MACH NUMBER: 0.46
MAX. ACCELERATION: 46.7 Gs
BURNOUT ALTITUDE: 174 FT.
COAST TIME TO APOGEE: 10.1 SECS.

AVERAGE ROCKET

BODY LENGTH: 24"
BODY DIAMETER: 1.63"
LIFTOFF WT.: 10.6 OZ.
MAX. ALTITUDE: 307 FT.
MAX. VELOCITY: 137 FT./SEC. (94 MPH)
MAX. MACH NUMBER: 0.12
MAX. ACCELERATION: 12.4 Gs
BURNOUT ALTITUDE: 46 FT.
COAST TIME TO APOGEE: 4.0 SECS.

© COPYRIGHT 1998 DAVID G. SLEETER

TELEFLITE CORPORATION

MOTOR NAME: KS-14-D10
PROPELLANT TYPE: KS
CASING TYPE: CONVOLUTE PAPER
NOZZLE TYPE: RAMMED CLAY

MOTOR SHOWN WITHOUT TIME DELAY

© COPYRIGHT 1998 DAVID G. SLEETER

IMPORTANT BAKING SODA NOTE: Add 20% baking soda to the propellant in your first motor. Reduce the soda in 2% increments until you get a motor that doesn't chuff. Then reduce the soda in 1% increments until maximum thrust is approx. 5 lbs. This test motor took 6% soda, but variations in the purity and particle size of the chemicals might make your motor take more or less.

TOTAL WEIGHT: 61.6 GMS.
PROPELLANT WT.: 23.2 GMS.
BAKING SODA: See baking soda note.
MOTOR LENGTH: 4.50"
CASING I.D.: 0.75"

CASING O.D.: 1.00"
NOZZLE CLAY: 11.0 GMS.
BULKHEAD CLAY: 11.0 GMS.
THROAT DIA.: 0.25"
CORE LENGTH: 2.00"

TOT. IMPULSE: 14.63 NT.-SECS.
AVG. THRUST: 10.64 NTS.
MAX. THRUST: 5.29 LBS.
BURN TIME: 1.38 SECS.
SPEC. IMPULSE: 64.19 SECS.

INDEX OF REQUIRED TOOLING

CASING RETAINER	DRAWING #98-002: 7"	PAGE 91
NOZZLE MOLD	DRAWING #98-008	PAGE 97
CORE SPINDLE	DRAWING #98-044-E	PAGE 126
STARTING TAMP	DRAWING #98-085-A	PAGE 148
PROPELLANT TAMP	DRAWING #98-086-A	PAGE 149
FLAT TAMP	DRAWING #98-091-B	PAGE 154
PORTHOLE TAMP	DRAWING #98-091-A	PAGE 154
PORTHOLE WIRE	DRAWING #98-091	PAGE 154

MOTOR NAME: KS-14-D10	NOZZLE TYPE: RAMMED CLAY	NOZZLE THROAT: 0.25"
TEST FILENAME: 960618C	CASING ID: 0.75"	CORE LENGTH: 2.00"
PROPELLANT: KG3	CASING OD: 1.00"	MOTOR WEIGHT: 61.6 GMS.
BURN MODIFIER: 6% NaHCO3	MOTOR LENGTH: 4.50"	PROPELLANT WT.: 23.2 GMS.

TOT. IMP.: 14.63 NT.-SECS.		MAX. THRUST: 5.29 LBS.
SPEC. IMPULSE: 64.19 SECS.	BURN TIME: 1.38 SECS.	AVG. THRUST: 10.64 NTS.

KS-14-D10 ESTIMATED PERFORMANCE

MINIMUM ROCKET

BODY LENGTH: 14"
BODY DIAMETER: 1.00"
LIFTOFF WT.: 4.7 OZ.
MAX. ALTITUDE: 1,347 FT.
MAX. VELOCITY: 340 FT./SEC. (232 MPH)
MAX. MACH NUMBER: 0.30
MAX. ACCELERATION: 19.1 Gs
BURNOUT ALTITUDE: 185 FT.
COAST TIME TO APOGEE: 8.0 SECS.

AVERAGE ROCKET

BODY LENGTH: 30"
BODY DIAMETER: 2.26"
LIFTOFF WT.: 14.2 OZ.
MAX. ALTITUDE: 137 FT.
MAX. VELOCITY: 83 FT./SEC. (57 MPH)
MAX. MACH NUMBER: 0.07
MAX. ACCELERATION: 5.2 Gs
BURNOUT ALTITUDE: 42 FT.
COAST TIME TO APOGEE: 2.4 SECS.

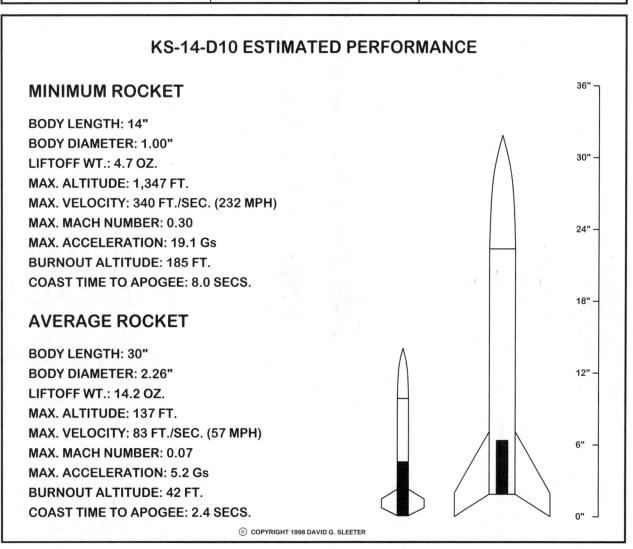

© COPYRIGHT 1998 DAVID G. SLEETER

TELEFLITE CORPORATION

MOTOR NAME:	KS-30-E30
PROPELLANT TYPE:	KS
CASING TYPE:	CONVOLUTE PAPER
NOZZLE TYPE:	RAMMED CLAY

IMPORTANT BAKING SODA NOTE: Add 20% baking soda to the propellant in your first motor. Reduce the soda in 2% increments until you get a motor that doesn't chuff. Then reduce the soda in 1% increments until maximum thrust is approx. 14 lbs. This test motor took 6% soda, but variations in the purity and particle size of the chemicals might make your motor take more or less.

TOTAL WEIGHT:	92.9 GMS.
PROPELLANT WT.:	44.9 GMS.
BAKING SODA:	See baking soda note.
MOTOR LENGTH:	7.25"
CASING I.D.:	0.75"

CASING O.D.:	1.00"
NOZZLE CLAY:	11.0 GMS.
BULKHEAD CLAY:	11.0 GMS.
THROAT DIA.:	0.375"
CORE LENGTH:	4.50"

TOT. IMPULSE:	30.87 NT.-SECS.
AVG. THRUST:	30.47 NTS.
MAX. THRUST:	13.42 LBS.
BURN TIME:	1.01 SECS.
SPEC. IMPULSE:	69.99 SECS.

MOTOR SHOWN WITHOUT TIME DELAY

© COPYRIGHT 1998 DAVID G. SLEETER

INDEX OF REQUIRED TOOLING

CASING RETAINER	DRAWING #98-002; 9"	PAGE 91
NOZZLE MOLD	DRAWING #98-010	PAGE 99
CORE SPINDLE	DRAWING #98-050-I	PAGE 129
STARTING TAMP	DRAWING #98-087-A	PAGE 150
PROPELLANT TAMP	DRAWING #98-089-A	PAGE 152
FLAT TAMP	DRAWING #98-091-A	PAGE 154
PORTHOLE TAMP	DRAWING #98-091-B	PAGE 154
PORTHOLE WIRE	DRAWING #98-091	PAGE 154

MOTOR NAME: KS-30-E30	NOZZLE TYPE: RAMMED CLAY	NOZZLE THROAT: 0.375"
TEST FILENAME: 960618D	CASING ID: 0.75"	CORE LENGTH: 4.50"
PROPELLANT: KG3	CASING OD: 1.00"	MOTOR WEIGHT: 92.9 GMS.
BURN MODIFIER: 6% NaHCO3	MOTOR LENGTH: 7.25"	PROPELLANT WT.: 44.9 GMS.

TOT. IMP.: 30.87 NT.-SECS.		MAX. THRUST: 13.43 LBS.
SPEC. IMPULSE: 69.99 SECS.	BURN TIME: 1.01 SECS.	AVG. THRUST: 30.47 NTS.

KS-30-E30 ESTIMATED PERFORMANCE

MINIMUM ROCKET

BODY LENGTH: 16.5"
BODY DIAMETER: 1.00"
LIFTOFF WT.: 5.8 OZ.
MAX. ALTITUDE: 2,826 FT.
MAX. VELOCITY: 649 FT./SEC. (443 MPH)
MAX. MACH NUMBER: 0.58
MAX. ACCELERATION: 43.3 Gs
BURNOUT ALTITUDE: 289 FT.
COAST TIME TO APOGEE: 11.0 SECS.

AVERAGE ROCKET

BODY LENGTH: 30"
BODY DIAMETER: 2.26"
LIFTOFF WT.: 15.3 OZ.
MAX. ALTITUDE: 617 FT.
MAX. VELOCITY: 211 FT./SEC. (144 MPH)
MAX. MACH NUMBER: 0.19
MAX. ACCELERATION: 13.9 Gs
BURNOUT ALTITUDE: 94 FT.
COAST TIME TO APOGEE: 5.5 SECS.

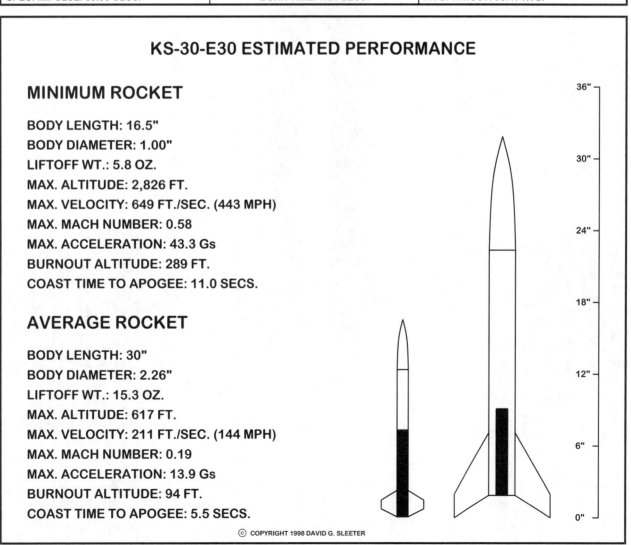

© COPYRIGHT 1998 DAVID G. SLEETER

TELEFLITE CORPORATION

MOTOR NAME: KS-38-E22
PROPELLANT TYPE: KS
CASING TYPE: CONVOLUTE PAPER
NOZZLE TYPE: RAMMED CLAY

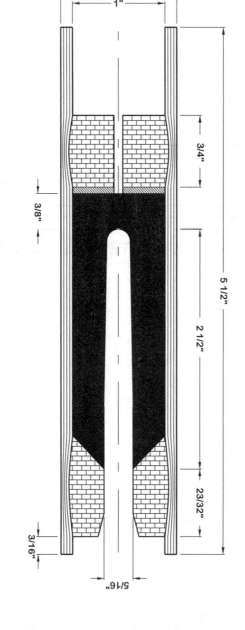

MOTOR SHOWN WITHOUT TIME DELAY

© COPYRIGHT 1998 DAVID G. SLEETER

INDEX OF REQUIRED TOOLING

CASING RETAINER	DRAWING #98-003: 7"	PAGE 92
NOZZLE MOLD	DRAWING #98-012	PAGE 101
CORE SPINDLE	DRAWING #98-054-E	PAGE 131
STARTING TAMP	DRAWING #98-092-A	PAGE 155
PROPELLANT TAMP	DRAWING #98-093-A	PAGE 156
FLAT TAMP	DRAWING #98-096-B	PAGE 159
PORTHOLE TAMP	DRAWING #98-096-A	PAGE 159
PORTHOLE WIRE	DRAWING #98-096	PAGE 159

CASING I.D.: 1.00"
MOTOR LENGTH: 5.75"
BAKING SODA: See baking soda note.
PROPELLANT WT.: 56.5 GMS.
TOTAL WEIGHT: 134.4 GMS.

IMPORTANT BAKING SODA NOTE: Add 20% baking soda to the propellant in your first motor. Reduce the soda in 2% increments until you get a motor that doesn't chuff. Then reduce the soda in 1% increments until maximum thrust is approx. 10 lbs. This test motor took 6% soda, but variations in the purity and particle size of the chemicals might make your motor take more or less.

CASING O.D.: 1.25"	TOT. IMPULSE: 38.45 NT.-SECS.
NOZZLE CLAY: 24.0 GMS.	AVG. THRUST: 22.63 NTS.
BULKHEAD CLAY: 24.0 GMS.	MAX. THRUST: 10.97 LBS.
THROAT DIA.: 0.313	BURN TIME: 1.70 SECS.
CORE LENGTH: 2.50"	SPEC. IMPULSE: 69.28 SECS.

378

MOTOR NAME: KS-38-E22	NOZZLE TYPE: RAMMED CLAY	NOZZLE THROAT: 0.313"
TEST FILENAME: 960618E	CASING ID: 1.00"	CORE LENGTH: 2.50"
PROPELLANT: KG3	CASING OD: 1.25"	MOTOR WEIGHT: 134.4 GMS.
BURN MODIFIER: 6% NaHCO3	MOTOR LENGTH: 5.50"	PROPELLANT WT.: 56.5 GMS.

[Thrust vs. Time curve: Thrust in Pounds (0.00 to 9.87) vs. Time in Seconds (0 to 1.530)]

TOT. IMP.: 38.45 NT.-SECS.		MAX. THRUST: 10.97 LBS.
SPEC. IMPULSE: 69.28 SECS.	BURN TIME: 1.70 SECS.	AVG. THRUST: 22.63 NTS.

KS-38-E22 ESTIMATED PERFORMANCE

MINIMUM ROCKET

BODY LENGTH: 16"
BODY DIAMETER: 1.25"
LIFTOFF WT.: 8.2 OZ.
MAX. ALTITUDE: 2,484 FT.
MAX. VELOCITY: 535 FT./SEC. (365 MPH)
MAX. MACH NUMBER: 0.48
MAX. ACCELERATION: 24.1 Gs
BURNOUT ALTITUDE: 317 FT.
COAST TIME TO APOGEE: 10.4 SECS.

AVERAGE ROCKET

BODY LENGTH: 36"
BODY DIAMETER: 2.63"
LIFTOFF WT.: 19.7 OZ. (1.23 LBS.)
MAX. ALTITUDE: 512 FT.
MAX. VELOCITY: 181 FT./SEC. (124 MPH)
MAX. MACH NUMBER: 0.16
MAX. ACCELERATION: 8.4 Gs
BURNOUT ALTITUDE: 104 FT.
COAST TIME TO APOGEE: 4.9 SECS.

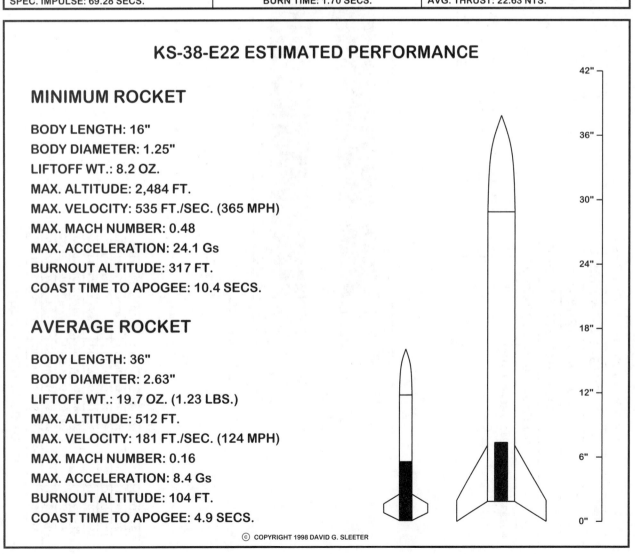

© COPYRIGHT 1998 DAVID G. SLEETER

TELEFLITE CORPORATION

MOTOR NAME: KS-50-F25
PROPELLANT TYPE: KS
CASING TYPE: CONVOLUTE PAPER
NOZZLE TYPE: RAMMED CLAY

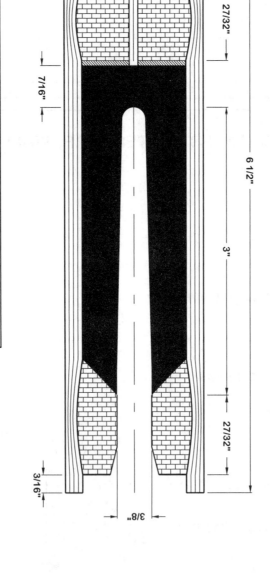

MOTOR SHOWN WITHOUT TIME DELAY

INDEX OF REQUIRED TOOLING

CASING RETAINER	DRAWING #98-004; 8"	PAGE 93
NOZZLE MOLD	DRAWING #98-016	PAGE 105
CORE SPINDLE	DRAWING #98-066-E	PAGE 137
STARTING TAMP	DRAWING #98-097-A	PAGE 160
PROPELLANT TAMP	DRAWING #98-098-A	PAGE 161
FLAT TAMP	DRAWING #98-104-B	PAGE 167
PORTHOLE TAMP	DRAWING #98-104-A	PAGE 167
PORTHOLE WIRE	DRAWING #98-104	PAGE 167

IMPORTANT BAKING SODA NOTE: Add 20% baking soda to the propellant in your first motor. Reduce the soda until you get a motor that doesn't chuff. Then reduce the soda in 1% increments until maximum thrust is approx. 12 lbs. This test motor took 6% soda, but variations in the purity and particle size of the chemicals might make your motor take more or less.

TOTAL WEIGHT: 199.8 GMS.
PROPELLANT WT.: 80.1 GMS.
BAKING SODA: See baking soda note.
MOTOR LENGTH: 6.75"
CASING I.D.: 1.125"

CASING O.D.: 1.50"
NOZZLE CLAY: 33.0 GMS.
BULKHEAD CLAY: 33.0 GMS.
THROAT DIA.: 0.375"
CORE LENGTH: 3.00"

TOT. IMPULSE: 50.84 NT.-SECS.
AVG. THRUST: 25.03 NTS.
MAX. THRUST: 12.38 LBS.
BURN TIME: 2.03 SECS.
SPEC. IMPULSE: 64.61 SECS.

© COPYRIGHT 1998 DAVID G. SLEETER

MOTOR NAME: KS-50-F25	NOZZLE TYPE: RAMMED CLAY	NOZZLE THROAT: 0.375"
TEST FILENAME: 960618F	CASING ID: 1.125"	CORE LENGTH: 3.00"
PROPELLANT: KG3	CASING OD: 1.50"	MOTOR WEIGHT: 199.8 GMS.
BURN MODIFIER: 6% NaHCO3	MOTOR LENGTH: 6.50"	PROPELLANT WT.: 80.1 GMS.

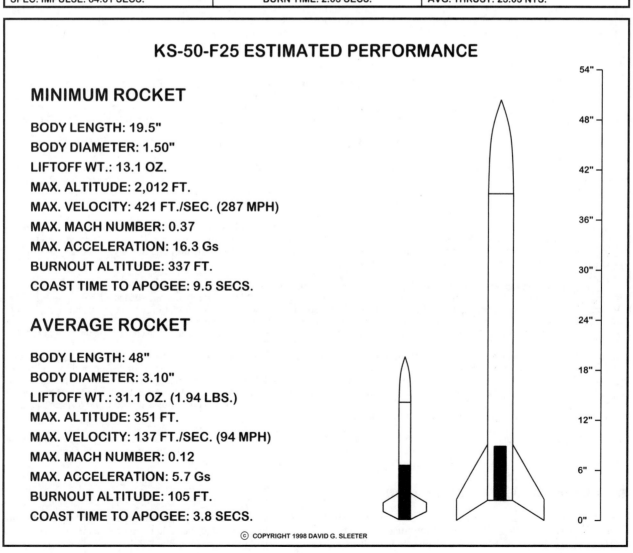

TOT. IMP.: 50.84 NT.-SECS.		MAX. THRUST: 12.38 LBS.
SPEC. IMPULSE: 64.61 SECS.	BURN TIME: 2.03 SECS.	AVG. THRUST: 25.03 NTS.

KS-50-F25 ESTIMATED PERFORMANCE

MINIMUM ROCKET

BODY LENGTH: 19.5"
BODY DIAMETER: 1.50"
LIFTOFF WT.: 13.1 OZ.
MAX. ALTITUDE: 2,012 FT.
MAX. VELOCITY: 421 FT./SEC. (287 MPH)
MAX. MACH NUMBER: 0.37
MAX. ACCELERATION: 16.3 Gs
BURNOUT ALTITUDE: 337 FT.
COAST TIME TO APOGEE: 9.5 SECS.

AVERAGE ROCKET

BODY LENGTH: 48"
BODY DIAMETER: 3.10"
LIFTOFF WT.: 31.1 OZ. (1.94 LBS.)
MAX. ALTITUDE: 351 FT.
MAX. VELOCITY: 137 FT./SEC. (94 MPH)
MAX. MACH NUMBER: 0.12
MAX. ACCELERATION: 5.7 Gs
BURNOUT ALTITUDE: 105 FT.
COAST TIME TO APOGEE: 3.8 SECS.

© COPYRIGHT 1998 DAVID G. SLEETER

14. NG6 PROPELLANT AND MOTOR DESIGNS

● First and Most Important!

Before you proceed, if you haven't read **Chapter 1**, go back and read it now. It contains important information on the safe handling of this rocket propellant. Also pay attention to the bold warning at the beginning of **Chapter 1**. To emphasize its importance, I'll repeat that warning here.

> Do *not* add or substitute other chemicals either to-or-for the chemicals described in this book. Specifically do *not* use POTASSIUM CHLORATE, SODIUM CHLORATE, POTASSIUM PERCHLORATE, SODIUM PERCHLORATE, POTASSIUM PERMANGANATE, AMMONIUM PERCHLORATE, PHOSPHOROUS, MATCH HEADS, ALUMINUM, MAGNESIUM, or ANY METALS AT ALL! *All* of these substances will form dangerous and explosive mixtures when combined with the other ingredients described herein, *AND OTHER CHEMICALS WILL FORM DANGEROUS MIXTURES AS WELL.* When mixing and handling propellants, be sure to ground yourself and all containers and utensils according to the instructions on page 14.
>
> *If you decide to experiment on your own outside the bounds of the instructions provided in this book, consult with a professional chemist before you proceed.*

The **NG6** propellant is a milled mixture of **4 hour milled sodium nitrate**, **carbon black**, **red gum**, and **sulfur**. If you haven't read **Chapters 5** and **6**, go back and read them now. Together they explain how to buy or make these chemicals, and how to prepare them for use in a rocket propellant.

If all other parameters are equal, then the **NG6** is the faster burning of the two sodium nitrate-based propellants. But its burn rate can vary *greatly* depending on the exact nature and purity of its chemicals. The quality of the carbon black is particularly important. A carbon black high in residual organics (i.e. oil) makes a slow burning propellant, and a pure, electrically conductive carbon black makes a fast burning propellant. The *exact* burn rate can be adjusted by substituting slower burning anthracite or charcoal for some of the fast burning carbon black, or by adding or subtracting baking soda. When compressed into a solid propellant grain, the **NG6**'s density is **25** to **30 gms. per cubic inch**. To prepare the **NG6** propellant, set up a powder mill (**Chapter 3**). Then mix the propellant (**Chapter 9**), and use the figures below to determine the exact weights of the chemicals needed for the desired batch size.

◆ THE NG6 PROPELLANT

4 hr. milled sodium nitrate	68.9%	689.0 gms.	344.5 gms.	172.3 gms.	86.1 gms.
Carbon black	13.2%	132.0 gms.	66.0 gms.	33.0 gms.	16.5 gms.
Red gum	7.0%	70.0 gms.	35.0 gms.	17.5 gms.	8.8 gms.
Sulfur	10.9%	109.0 gms.	54.5 gms.	27.2 gms.	13.6 gms.
Total	100%	1000.0 gms.	500.0 gms.	250.0 gms.	125.0 gms.

Before you can use the propellant, *you must perform the pellet test* (**pages 238 - 245**). The pellet test determines the exact amount of solvent needed to form 20 grams of propellant into a dense and solid propellant grain. The correct amount of solvent is called **the solvent requirement**. *The solvent requirement is critical, and cannot be guessed at*. The pellet test is time consuming, *but you only do it twice*. Perform the *first* pellet test with the **NG6** propellant *plus 30% baking soda*, and do it *before* you make your first motor.

To build your *first* motor with the **NG6** propellant, weigh out the amount indicated in the motor drawing (rounding off the numbers makes the math easier). Then add 30% of that weight in baking soda, and mix it in thoroughly with a fork. If, for example, you've weighed out 100 grams of propellant, add 30 grams of baking soda for a total of 130 grams. Mix in the proper amount of solvent as determined by your first pellet test. Then build and dry the motor (**Chapter 10**).

Per the instructions on **page 494**, do a rudimentary test by firing the motor (with the motor pointing *down* and the exhaust aimed *upward*) from a pipe in the ground. If this first motor "chuffs" (chugs like a steam engine), build a *second* motor, and reduce the baking soda to 28%. Keep building, testing, and reducing the amount of soda in 2% increments until you get a motor that burns smoothly and *doesn't* chuff. At first the chuffing may be slow. As you approach a condition of combustion stability, the frequency of the chuffs will increase (the chuffing sound will get *faster*). When you are *very* near stability, the amplitude of the chuffs will decrease (the chuffs will sound *fainter*).

Now set up a thrust-testing stand (**page 470**), and test the motor for maximum thrust (**page 475**). In the tests that follow, reduce the soda in 1% increments until you get a motor whose maximum thrust approximates the maximum thrust printed in the motor drawing. At that point you'll have a working motor with the maximum safe chamber pressure and thrust for that particular design. As you approach a final, successful soda content, *repeat the pellet test*, and use the solvent requirement indicated by this *second* test to zero-in the *final* design. Keep accurate notes. Write down the final solvent requirement and the amount of baking soda that works the best, and *forever thereafter* use those figures to build copies of your final, successful motor.

Important note. You can also alter the chamber pressure and thrust by changing the length of the motor's core. A *longer* core will *increase* the chamber pressure and thrust, and a *shorter* core will *reduce* the chamber pressure and thrust. But, of course, changing the length of the core will also change the amount of propellant required, and the physical size of the motor.

● Improving Motor Performance

Now you have a working rocket motor, but some of the material inside the motor isn't rocket propellant. It's *baking soda*, and baking soda is *dead weight*. It doesn't contribute anything to the motor's total impulse. You added the soda to control chamber pressure and thrust, but you can also control those parameters by replacing some of the fast burning carbon black with slower burning charcoal or anthracite (which I call "other carbon" in the text below). And unlike the baking soda, this "other carbon" is a *fuel*. It takes part in the reaction, and *contributes* to the motor's total impulse.

In the following hypothetical illustration, I'll call the original motor an **NG6-21-E23**. I'll stipulate that its maximum thrust is 10 lbs., and I'll stipulate that it takes 15% baking soda to make it work as intended. To *improve* the motor's performance, I'll have to get rid of some of the soda. Of course when I do that, the propellant's burn rate will increase. But I can *reverse* the increase, and *compensate* for the loss of the soda by replacing some of the carbon black with slower burning "other carbon". The amounts of sodium nitrate, red gum, and sulfur will remain at 68.9%, 7.0%, and 10.9% respectively, but the carbon fuel ratios will change. **Important note**. *Each time you change a propellant's carbon fuel ratio, you must repeat* **the pellet test**. Possible carbon replacement ratios are:

NG6X1	NG6X2	NG6X3	NG6X4
Carbon black 12.2%	Carbon black 11.2%	Carbon black 10.2%	Carbon black 9.2%
Other carbon 1.0%	Other carbon 2.0%	Other carbon 3.0%	Other carbon 4.0%
NG6X5	NG6X6	NG6X7	NG6X8
Carbon black 8.2%	Carbon black 7.2%	Carbon black 6.2%	Carbon black 5.2%
Other carbon 5.0%	Other carbon 6.0%	Other carbon 7.0%	Other carbon 8.0%
NG6X9	NG6X10	NG6X11	NG6X12
Carbon black 4.2%	Carbon black 3.2%	Carbon black 2.2%	Carbon black 1.2%
Other carbon 9.0%	Other carbon 10.0%	Other carbon 11.0%	Other carbon 12.0%

In the ratios above, though the balance between the carbon black and the "other carbon" changes (in 1% increments), the total amount of carbon fuel remains the same (13.2%). If I build a motor with the NG6X3 ratio, and add 15% baking soda, because the "other carbon" burns *slower* than the carbon black it replaced, the motor's maximum thrust will be *less* than 10 pounds. To restore the maximum thrust to the desired level, I'll keep building, testing, and reducing the baking soda in 1% increments until the maximum thrust is back up to 10 lbs. I'll call this second motor an **NG6X3** (named after its propellant), and I'll pretend here that it took 12% soda to make it work.

If the **NG6X3** with 12% soda burns smoothly and doesn't chuff, I'll build an identical motor with the NG6X4 propellant. Of course, when I do that, the maximum thrust will drop again, so I'll keep building, testing, and *again* reducing the soda in 1% increments until the maximum thrust equals 10 lbs. I'll call this third motor an **NG6X4**, and for the purpose of this illustration, I'll pretend that it took 10% baking soda to make it work. If the **NG6X4** burns smoothly and doesn't chuff, I might repeat the experiment with the NG6X5 propellant. If the **NG6X4** chuffs and *doesn't* burn smoothly, I'll know that the NG6X4 propellant *doesn't* work; that I need *more* than 9.2% carbon black to achieve combustion stability, and that the NG6X3 propellant with 12% baking soda works the best.

If I now build a chart recorder (**page 478**), and generate thrust-time curves for the **NG6-21-E23** and **NG6X3** motors, I'll see that the **NG6X3** has a slightly longer burn time and a greater total impulse. The **NG6X3** might turn out to be an **NG6-*25*-E23**, and I'll adopt the NG6X3 propellant with 12% baking soda for all the motors that follow. When you perform these experiments on your own, you'll eventually reach a point where the motors begin to chuff, and the process of finding the minimum amount of carbon black needed to maintain combustion stability will be one of trial and error.

Important note. *The NG6 motors in this book were all made with propellants milled for 12 hours.* When making your own propellants, you can reduce the milling time to as little as 4 hours, and compensate by adjusting the amount of baking soda. The process of finding the minimum milling time is one of trial and error, and each time you change the milling time, *you must repeat* **the pellet test**.

The **NG6** propellant exhibits excellent case bonding properties with no evidence of the effect discussed on **pages 248 - 249**. All the NG6 motors were made with standard fireworks grade casings, and all but the **NG6-227-H137** were made with clay nozzles. By switching to stainless steel De Laval nozzles and the high strength casings manufactured (as of this writing) by **New England Paper Tube** (**pages 163 - 164**), you can increase the motors' maximum chamber pressure and thrust, and substantially improve their performance. Shop drawings for De Laval nozzles begin on the **page 323**.

With each motor design, I've included the thrust-time curve and motor performance data generated by the test equipment and software described on **pages 468 - 469**, plus two flight performance estimates generated by **Rogers' Aeroscience Alt4** rocket performance prediction software. The "minimum rocket" is what you might build if you were trying to break an altitude or speed record. The lift-off weight is minimal. The body diameter is equal to the o.d. of the motor, and the parachute compartment is small; approx. 5 body diameter's in length. The "average rocket" is more like what you would build for scientific research or general sport flying. The body diameter is one of the industry standards approx. twice the diameter of the motor, and there's ample space and weight allowance for a large parachute and a small scientific payload.

In each case the lift-off weight is *very* approximate, and based on an educated guess. The *extremely* accurate performance specs. *are the actual numbers generated by the Alt-4 program*. They are based on the rocket's stated length, diameter, and lift-off weight, and the *tested* motor's thrust-time data. *In the real world*, small variations in motor performance, the size and shape of the rocket's fins, and even the smoothness of its paint-job will cause the rocket's performance to vary from the stated figures.

● Important G-Force Advisory

Static testing perfects a motor's performance, but it *doesn't* predict how the hardened propellant grain will respond to high G-forces during an actual rocket flight. Baking soda is *soft*, and the carbon fuels used in these propellants vary *greatly* in hardness. So variations in the soda content and the type of carbon fuel will effect the strength of the finished propellant grain. If a soft propellant grain collapses at maximum acceleration, the motor will self-destruct, and the rocket will be lost in flight. Initial flight testing of a new motor design should *always* be conducted with this possibility in mind, particularly when testing the motor in a small, high-acceleration airframe. It is always advisable to conduct your early tests in a "throwaway" rocket, i.e. something cheap that you don't mind losing. When you're sure that you've got a successful motor-airframe combination, *then* invest in something nicer.

TELEFLITE CORPORATION

MOTOR NAME: NG6-9-C8
PROPELLANT TYPE: NG6
CASING TYPE: CONVOLUTE PAPER
NOZZLE TYPE: RAMMED CLAY

MOTOR SHOWN WITHOUT TIME DELAY

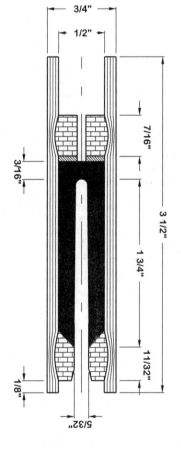

INDEX OF REQUIRED TOOLING

CASING RETAINER	DRAWING #98-001.5"
NOZZLE MOLD	DRAWING #98-006
CORE SPINDLE	DRAWING #98-035-D
STARTING TAMP	USE PROPELLANT TAMP
PROPELLANT TAMP	DRAWING #98-082-A
FLAT TAMP	DRAWING #98-084-B
PORTHOLE TAMP	DRAWING #98-084-A
PORTHOLE WIRE	DRAWING #98-084

	PAGE 90
	PAGE 95
	PAGE 121
	PAGE 145
	PAGE 147
	PAGE 147
	PAGE 147

CASING I.D.: 0.50"
MOTOR LENGTH: 3.50"
BAKING SODA: See baking soda note.
PROPELLANT WT.: 13.9 GMS.
TOTAL WEIGHT: 31.9 GMS.

CASING O.D.: 0.75"	TOT. IMPULSE: 9.00 NT.-SECS.
NOZZLE CLAY: 3.3 GMS.	AVG. THRUST: 8.03 NTS.
BULKHEAD CLAY: 3.3 GMS.	MAX. THRUST: 4.23 LBS.
THROAT DIA.: 0.1563"	BURN TIME: 1.12 SECS.
CORE LENGTH: 1.75"	SPEC. IMPULSE: 65.91 SECS.

IMPORTANT BAKING SODA NOTE: Add 30% baking soda to the propellant in your first motor. Reduce the soda in 2% increments until you get a motor that doesn't chuff. Then reduce the soda in 1% increments until maximum thrust is approx. 4 lbs. This test motor took 7% soda, but I made it with a carbon black of medium purity. A motor made with a pure, electrically-conductive carbon black might take substantially more.

© COPYRIGHT 1998 DAVID G. SLEETER

MOTOR NAME: NG6-9-C8	NOZZLE TYPE: RAMMED CLAY	NOZZLE THROAT: 0.1563"
TEST FILENAME: 970608D	CASING ID: 0.50"	CORE LENGTH: 1.75"
PROPELLANT: NG6	CASING OD: 0.75"	MOTOR WEIGHT: 31.9 GMS.
BURN MODIFIER: 7% NaHCO3	MOTOR LENGTH: 3.50"	PROPELLANT WT.: 13.9 GMS.

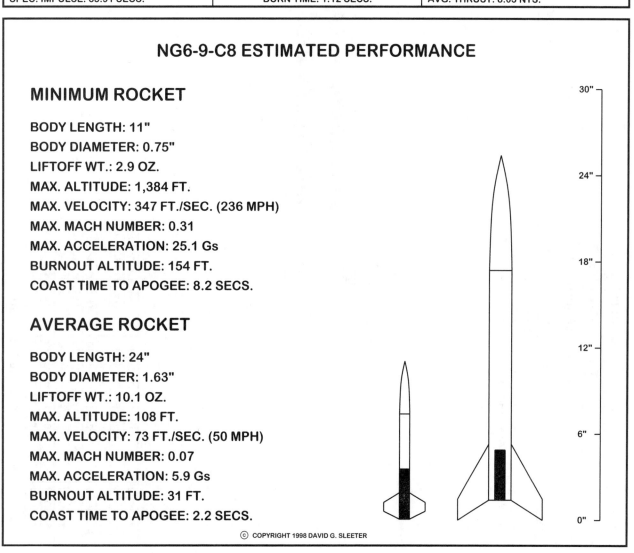

TOT. IMP.: 9.00 NT.-SECS.		MAX. THRUST: 4.23 LBS.
SPEC. IMPULSE: 65.91 SECS.	BURN TIME: 1.12 SECS.	AVG. THRUST: 8.03 NTS.

NG6-9-C8 ESTIMATED PERFORMANCE

MINIMUM ROCKET

BODY LENGTH: 11"
BODY DIAMETER: 0.75"
LIFTOFF WT.: 2.9 OZ.
MAX. ALTITUDE: 1,384 FT.
MAX. VELOCITY: 347 FT./SEC. (236 MPH)
MAX. MACH NUMBER: 0.31
MAX. ACCELERATION: 25.1 Gs
BURNOUT ALTITUDE: 154 FT.
COAST TIME TO APOGEE: 8.2 SECS.

AVERAGE ROCKET

BODY LENGTH: 24"
BODY DIAMETER: 1.63"
LIFTOFF WT.: 10.1 OZ.
MAX. ALTITUDE: 108 FT.
MAX. VELOCITY: 73 FT./SEC. (50 MPH)
MAX. MACH NUMBER: 0.07
MAX. ACCELERATION: 5.9 Gs
BURNOUT ALTITUDE: 31 FT.
COAST TIME TO APOGEE: 2.2 SECS.

© COPYRIGHT 1998 DAVID G. SLEETER

TELEFLITE CORPORATION

MOTOR NAME:	NG6-10-D7
PROPELLANT TYPE:	NG6
CASING TYPE:	CONVOLUTE PAPER
NOZZLE TYPE:	RAMMED CLAY

INDEX OF REQUIRED TOOLING

CASING RETAINER	DRAWING #98-001: 5" PAGE 90
NOZZLE MOLD	DRAWING #98-006 PAGE 95
CORE SPINDLE	DRAWING #98-035-F PAGE 121
STARTING TAMP	USE PROPELLANT TAMP
PROPELLANT TAMP	DRAWING #98-082-A PAGE 145
FLAT TAMP	DRAWING #98-084-B PAGE 147
PORTHOLE TAMP	DRAWING #98-084-A PAGE 147
PORTHOLE WIRE	DRAWING #98-084 PAGE 147

MOTOR SHOWN WITHOUT TIME DELAY

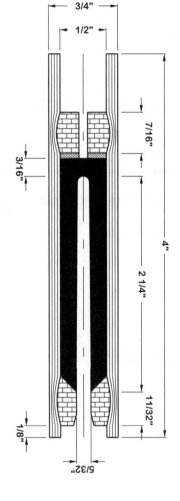

CASING I.D.:	0.50"
MOTOR LENGTH:	4.00"
BAKING SODA:	See baking soda note.
PROPELLANT WT.:	18.2 GMS.
TOTAL WEIGHT:	37.9 GMS.
CASING O.D.:	0.75"
NOZZLE CLAY:	3.3 GMS.
BULKHEAD CLAY:	3.3 GMS.
THROAT DIA.:	0.1563"
CORE LENGTH:	2.25"
TOT. IMPULSE:	10.14 NT.-SECS.
AVG. THRUST:	7.70 NTS.
MAX. THRUST:	3.42 LBS.
BURN TIME:	1.32 SECS.
SPEC. IMPULSE:	56.72 SECS.

IMPORTANT BAKING SODA NOTE: Add 30% baking soda to the propellant in your first motor. Reduce the soda in 1% increments until you get a motor that doesn't chuff. Then reduce the soda in 2% increments until maximum thrust is approx. 3-1/2 lbs. This test motor took 15% soda, but I made it with a carbon black of medium purity. A motor made with a pure, electrically-conductive carbon black might take substantially more.

© COPYRIGHT 1998 DAVID G. SLEETER

MOTOR NAME: NG6-10-D7	NOZZLE TYPE: RAMMED CLAY	NOZZLE THROAT: 0.1563"
TEST FILENAME: 970602A	CASING ID: 0.50"	CORE LENGTH: 2.25"
PROPELLANT: NG6	CASING OD: 0.75"	MOTOR WEIGHT: 37.9 GMS.
BURN MODIFIER: 15% NaHCO3	MOTOR LENGTH: 4.00"	PROPELLANT WT.: 18.2 GMS.

TOT. IMP.: 10.14 NT.-SECS.		MAX. THRUST: 3.42 LBS.
SPEC. IMPULSE: 56.72 SECS.	BURN TIME: 1.32 SECS.	AVG. THRUST: 7.70 NTS.

NG6-10-D7 ESTIMATED PERFORMANCE

MINIMUM ROCKET

BODY LENGTH: 11.5"
BODY DIAMETER: 0.75"
LIFTOFF WT.: 3.1 OZ.
MAX. ALTITUDE: 1,546 FT.
MAX. VELOCITY: 367 FT./SEC. (250 MPH)
MAX. MACH NUMBER: 0.33
MAX. ACCELERATION: 18.9 Gs
BURNOUT ALTITUDE: 211 FT.
COAST TIME TO APOGEE: 8.5 SECS.

AVERAGE ROCKET

BODY LENGTH: 24"
BODY DIAMETER: 1.63"
LIFTOFF WT.: 10.3 OZ.
MAX. ALTITUDE: 128 FT.
MAX. VELOCITY: 78 FT./SEC. (53 MPH)
MAX. MACH NUMBER: 0.07
MAX. ACCELERATION: 4.5 Gs
BURNOUT ALTITUDE: 42 FT.
COAST TIME TO APOGEE: 2.3 SECS.

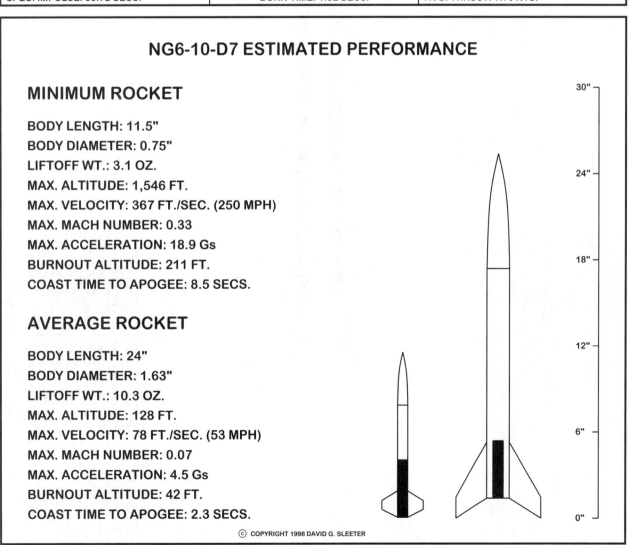

© COPYRIGHT 1998 DAVID G. SLEETER

TELEFLITE CORPORATION

MOTOR NAME:	NG6-18-D23
PROPELLANT TYPE:	NG6
CASING TYPE:	CONVOLUTE PAPER
NOZZLE TYPE:	RAMMED CLAY

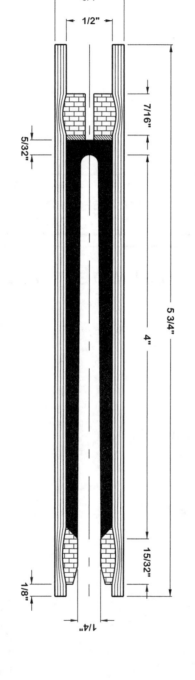

MOTOR SHOWN WITHOUT TIME DELAY

© COPYRIGHT 1998 DAVID G. SLEETER

INDEX OF REQUIRED TOOLING

CASING RETAINER	DRAWING #98-001: 7"	PAGE 90
NOZZLE MOLD	DRAWING #98-007	PAGE 96
CORE SPINDLE	DRAWING #98-039-G	PAGE 123
STARTING TAMP	USE PROPELLANT TAMP	
PROPELLANT TAMP	DRAWING #98-082-B	PAGE 145
FLAT TAMP	DRAWING #98-084-B	PAGE 147
PORTHOLE TAMP	DRAWING #98-084-A	PAGE 147
PORTHOLE WIRE	DRAWING #98-084	PAGE 147

IMPORTANT BAKING SODA NOTE: Add 30% baking soda to the propellant in your first motor. Then reduce the soda in 2% increments until you get a motor that doesn't chuff. Then reduce the soda in 1% increments until maximum thrust is approx. 10 lbs. This test motor took 7% soda, but I made it with a carbon black of medium purity. A motor made with a pure, electrically-conductive carbon black might take substantially more.

TOTAL WEIGHT: 51.2 GMS.	CASING O.D.: 0.75"
PROPELLANT WT.: 25.6 GMS.	NOZZLE CLAY: 3.3 GMS.
BAKING SODA: See baking soda note.	BULKHEAD CLAY: 3.3 GMS.
MOTOR LENGTH: 5.75"	THROAT DIA.: 0.25"
CASING I.D.: 0.50"	CORE LENGTH: 4.00"

TOT. IMPULSE: 18.20 NT.-SECS.
AVG. THRUST: 23.79 NTS.
MAX. THRUST: 9.66 LBS.
BURN TIME: 0.76 SECS.
SPEC. IMPULSE: 72.37 SECS.

388

MOTOR NAME: NG6-18-D23	NOZZLE TYPE: RAMMED CLAY	NOZZLE THROAT: 0.25"
TEST FILENAME: 970606A	CASING ID: 0.50"	CORE LENGTH: 4.00"
PROPELLANT: NG6	CASING OD: 0.75"	MOTOR WEIGHT: 51.2 GMS.
BURN MODIFIER: 7% NaHCO3	MOTOR LENGTH: 5.75"	PROPELLANT WT.: 25.6 GMS.

[Thrust vs. Time curve: thrust in pounds (0.00 to 8.69) vs. time in seconds (0 to 0.684)]

TOT. IMP.: 18.20 NT.-SECS.		MAX. THRUST: 9.66 LBS.
SPEC. IMPULSE: 72.37 SECS.	BURN TIME: 0.76 SECS.	AVG. THRUST: 23.79 NTS.

NG6-18-D23 ESTIMATED PERFORMANCE

MINIMUM ROCKET

BODY LENGTH: 13"
BODY DIAMETER: 0.75"
LIFTOFF WT.: 3.6 OZ.
MAX. ALTITUDE: 2,795 FT.
MAX. VELOCITY: 626 FT./SEC. (427 MPH)
MAX. MACH NUMBER: 0.56
MAX. ACCELERATION: 49.2 Gs
BURNOUT ALTITUDE: 219 FT.
COAST TIME TO APOGEE: 11.1 SECS.

AVERAGE ROCKET

BODY LENGTH: 24"
BODY DIAMETER: 1.63"
LIFTOFF WT.: 10.8 OZ.
MAX. ALTITUDE: 484 FT.
MAX. VELOCITY: 179 FT./SEC. (122 MPH)
MAX. MACH NUMBER: 0.16
MAX. ACCELERATION: 14.0 Gs
BURNOUT ALTITUDE: 63 FT.
COAST TIME TO APOGEE: 5.0 SECS.

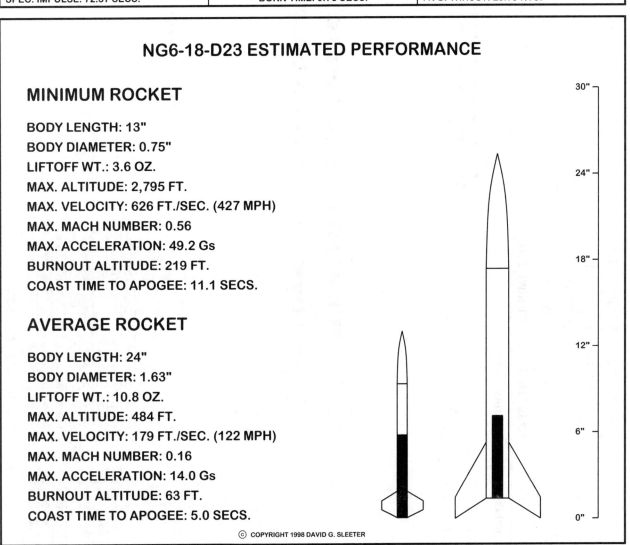

© COPYRIGHT 1998 DAVID G. SLEETER

TELEFLITE CORPORATION

MOTOR NAME: NG6-21-E23
PROPELLANT TYPE: NG6
CASING TYPE: CONVOLUTE PAPER
NOZZLE TYPE: RAMMED CLAY

MOTOR SHOWN WITHOUT TIME DELAY

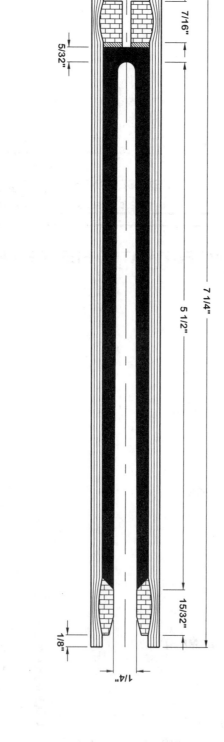

INDEX OF REQUIRED TOOLING

CASING RETAINER	DRAWING #98-001: 9"	PAGE 90
NOZZLE MOLD	DRAWING #98-007	PAGE 96
CORE SPINDLE	DRAWING #98-041-M	PAGE 124
STARTING TAMP	USE PROPELLANT TAMP	
PROPELLANT TAMP	DRAWING #98-083-C	PAGE 146
FLAT TAMP	DRAWING #98-084-B	PAGE 147
PORTHOLE TAMP	DRAWING #98-084-A	PAGE 147
PORTHOLE WIRE	DRAWING #98-084	PAGE 147

IMPORTANT BAKING SODA NOTE: Add 30% baking soda to the propellant in your first motor. Reduce the soda in 1% increments until you get a motor that doesn't chuff. Then reduce the soda in 2% increments. This test motor took 15% soda, but I made it with a carbon black of medium purity. A motor made with a pure, electrically-conductive carbon black might take substantially more.

CASING I.D.: 0.50"	CORE LENGTH: 5.50"	SPEC. IMPULSE: 62.53 SECS.
MOTOR LENGTH: 7.25"	THROAT DIA.: 0.25"	BURN TIME: 0.88 SECS.
BAKING SODA: See baking soda note.	BULKHEAD CLAY: 3.3 GMS.	MAX. THRUST: 10.12 LBS.
PROPELLANT WT.: 34.4 GMS.	NOZZLE CLAY: 3.3 GMS.	AVG. THRUST: 23.96 NTS.
TOTAL WEIGHT: 64.1 GMS.	CASING O.D.: 0.75"	TOT. IMPULSE: 21.13 NT.-SECS.

© COPYRIGHT 1998 DAVID G. SLEETER

MOTOR NAME: NG6-21-E23	NOZZLE TYPE: RAMMED CLAY	NOZZLE THROAT: 0.25"
TEST FILENAME: 970608E	CASING ID: 0.50"	CORE LENGTH: 5.50"
PROPELLANT: NG6	CASING OD: 0.75"	MOTOR WEIGHT: 64.1 GMS.
BURN MODIFIER: 15% NaHCO3	MOTOR LENGTH: 7.25"	PROPELLANT WT.: 34.4 GMS.

[Thrust vs. Time curve]

TOT. IMP.: 21.13 NT.-SECS.		MAX. THRUST: 10.12 LBS.
SPEC. IMPULSE: 62.53 SECS.	BURN TIME: 0.88 SECS.	AVG. THRUST: 23.96 NTS.

NG6-21-E23 ESTIMATED PERFORMANCE

MINIMUM ROCKET

BODY LENGTH: 14.5"
BODY DIAMETER: 0.75"
LIFTOFF WT.: 4.0 OZ.
MAX. ALTITUDE: 2,938 FT.
MAX. VELOCITY: 652 FT./SEC. (445 MPH)
MAX. MACH NUMBER: 0.58
MAX. ACCELERATION: 47.6 Gs
BURNOUT ALTITUDE: 314 FT.
COAST TIME TO APOGEE: 11.2 SECS.

AVERAGE ROCKET

BODY LENGTH: 24"
BODY DIAMETER: 1.63"
LIFTOFF WT.: 11.3 OZ.
MAX. ALTITUDE: 587 FT.
MAX. VELOCITY: 196 FT./SEC. (134 MPH)
MAX. MACH NUMBER: 0.17
MAX. ACCELERATION: 14.2 Gs
BURNOUT ALTITUDE: 96 FT.
COAST TIME TO APOGEE: 5.4 SECS.

© COPYRIGHT 1998 DAVID G. SLEETER

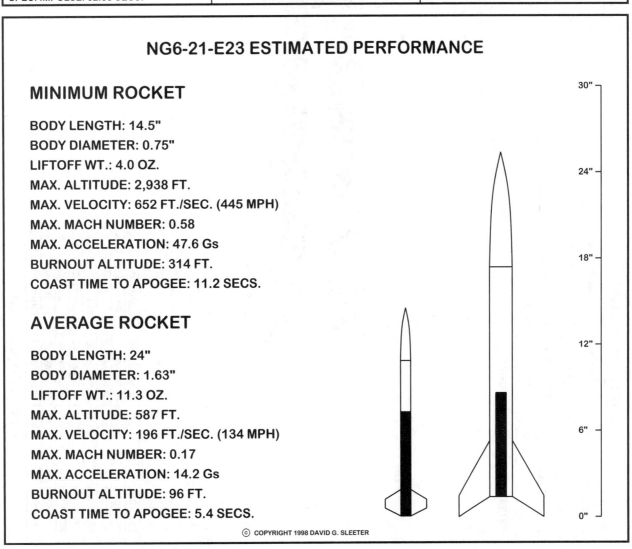

TELEFLITE CORPORATION

MOTOR NAME:	NG6-29-E17
PROPELLANT TYPE:	NG6
CASING TYPE:	CONVOLUTE PAPER
NOZZLE TYPE:	RAMMED CLAY

MOTOR SHOWN WITHOUT TIME DELAY

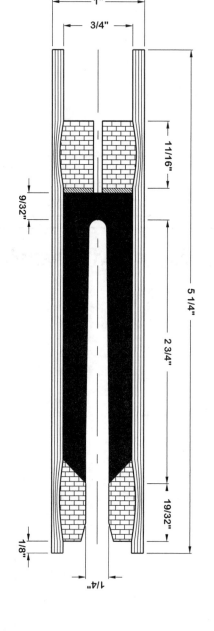

INDEX OF REQUIRED TOOLING

CASING RETAINER	DRAWING #98-002: 7"	PAGE 91
NOZZLE MOLD	DRAWING #98-008	PAGE 97
CORE SPINDLE	DRAWING #98-045-H	PAGE 126
STARTING TAMP	DRAWING #98-085-A	PAGE 148
PROPELLANT TAMP	DRAWING #98-086-A	PAGE 149
FLAT TAMP	DRAWING #98-091-A	PAGE 154
PORTHOLE TAMP	DRAWING #98-091-B	PAGE 154
PORTHOLE WIRE	DRAWING #98-091	PAGE 154

© COPYRIGHT 1998 DAVID G. SLEETER

TOTAL WEIGHT:	83.6 GMS.
PROPELLANT WT.:	41.4 GMS.
BAKING SODA:	See baking soda note.
MOTOR LENGTH:	5.25"
CASING I.D.:	0.75"

CASING O.D.:	1.00"
NOZZLE CLAY:	11.0 GMS.
BULKHEAD CLAY:	11.0 GMS.
THROAT DIA.:	0.25"
CORE LENGTH:	2.75"

TOT. IMPULSE:	29.52 NT.-SECS.
AVG. THRUST:	17.55 NTS.
MAX. THRUST:	8.34 LBS.
BURN TIME:	1.68 SECS.
SPEC. IMPULSE:	72.59 SECS.

IMPORTANT BAKING SODA NOTE: Add 30% baking soda to the propellant in your first motor. Reduce the soda in 2% increments until you get a motor that doesn't chuff. Then reduce the soda in 1% increments until maximum thrust is approx. 9 lbs. This test motor took 7% soda, but I made it with a carbon black of medium purity. A motor made with a pure, electrically-conductive carbon black might take substantially more.

392

MOTOR NAME: NG6-29-E17	NOZZLE TYPE: RAMMED CLAY	NOZZLE THROAT: 0.25"
TEST FILENAME: 970602B	CASING ID: 0.75"	CORE LENGTH: 2.75"
PROPELLANT: NG6	CASING OD: 1.00"	MOTOR WEIGHT: 83.6 GMS.
BURN MODIFIER: 7% NaHCO3	MOTOR LENGTH: 5.25"	PROPELLANT WT.: 41.4 GMS.

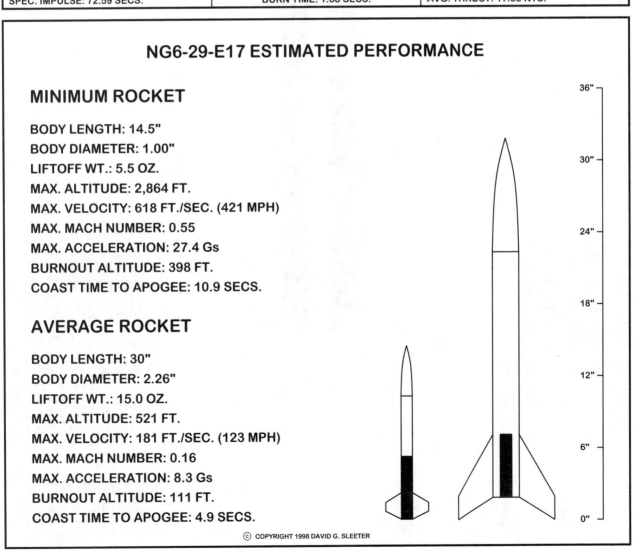

TOT. IMP.: 29.52 NT.-SECS.		MAX. THRUST: 8.34 LBS.
SPEC. IMPULSE: 72.59 SECS.	BURN TIME: 1.68 SECS.	AVG. THRUST: 17.55 NTS.

NG6-29-E17 ESTIMATED PERFORMANCE

MINIMUM ROCKET

BODY LENGTH: 14.5"
BODY DIAMETER: 1.00"
LIFTOFF WT.: 5.5 OZ.
MAX. ALTITUDE: 2,864 FT.
MAX. VELOCITY: 618 FT./SEC. (421 MPH)
MAX. MACH NUMBER: 0.55
MAX. ACCELERATION: 27.4 Gs
BURNOUT ALTITUDE: 398 FT.
COAST TIME TO APOGEE: 10.9 SECS.

AVERAGE ROCKET

BODY LENGTH: 30"
BODY DIAMETER: 2.26"
LIFTOFF WT.: 15.0 OZ.
MAX. ALTITUDE: 521 FT.
MAX. VELOCITY: 181 FT./SEC. (123 MPH)
MAX. MACH NUMBER: 0.16
MAX. ACCELERATION: 8.3 Gs
BURNOUT ALTITUDE: 111 FT.
COAST TIME TO APOGEE: 4.9 SECS.

© COPYRIGHT 1998 DAVID G. SLEETER

TELEFLITE CORPORATION

MOTOR NAME: NG6-41-F21
PROPELLANT TYPE: NG6
CASING TYPE: CONVOLUTE PAPER
NOZZLE TYPE: RAMMED CLAY

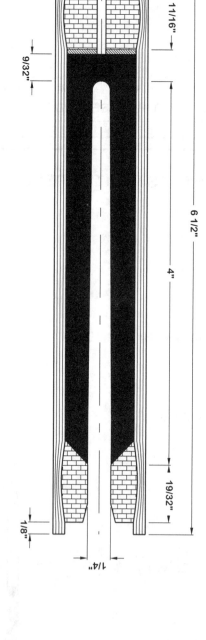

MOTOR SHOWN WITHOUT TIME DELAY

IMPORTANT BAKING SODA NOTE: Add 30% baking soda to the propellant in your first motor. Reduce the soda in 2% increments until you get a motor that doesn't chuff. Then reduce the soda in 1% increments until maximum thrust is approx. 10 lbs. This test motor took 15% soda, but I made it with a carbon black of medium purity. A motor made with a pure, electrically-conductive carbon black might take substantially more.

TOTAL WEIGHT: 108.2 GMS.	
PROPELLANT WT.: 61.6 GMS.	
BAKING SODA: See baking soda note.	
MOTOR LENGTH: 6.50"	
CASING I.D.: 0.75"	

CASING O.D.: 1.00"	TOT. IMPULSE: 41.31 NT.-SECS.
NOZZLE CLAY: 11.0 GMS.	AVG. THRUST: 21.30 NTS.
BULKHEAD CLAY: 11.0 GMS.	MAX. THRUST: 9.78 LBS.
THROAT DIA.: 0.25"	BURN TIME: 1.94 SECS.
CORE LENGTH: 4.00"	SPEC. IMPULSE: 68.27 SECS.

© COPYRIGHT 1998 DAVID G. SLEETER

INDEX OF REQUIRED TOOLING

CASING RETAINER	DRAWING #98-002: 9"	PAGE 91
NOZZLE MOLD	DRAWING #98-008	PAGE 97
CORE SPINDLE	DRAWING #98-046-M	PAGE 127
STARTING TAMP	DRAWING #98-085-B	PAGE 148
PROPELLANT TAMP	DRAWING #98-086-B	PAGE 149
FLAT TAMP	DRAWING #98-091-B	PAGE 154
PORTHOLE TAMP	DRAWING #98-091-A	PAGE 154
PORTHOLE WIRE	DRAWING #98-091	PAGE 154

MOTOR NAME: NG6-41-F21	NOZZLE TYPE: RAMMED CLAY	NOZZLE THROAT: 0.25"
TEST FILENAME: 970423G	CASING ID: 0.75"	CORE LENGTH: 4.00"
PROPELLANT: NG6	CASING OD: 1.00"	MOTOR WEIGHT: 108.2 GMS.
BURN MODIFIER: 15% NaHCO3	MOTOR LENGTH: 6.50"	PROPELLANT WT.: 61.6 GMS.

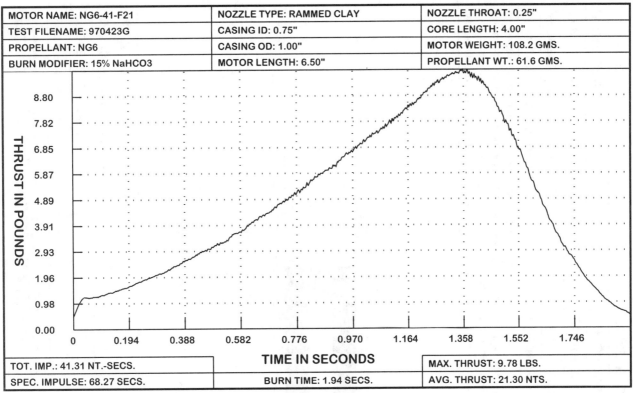

TOT. IMP.: 41.31 NT.-SECS.		MAX. THRUST: 9.78 LBS.
SPEC. IMPULSE: 68.27 SECS.	BURN TIME: 1.94 SECS.	AVG. THRUST: 21.30 NTS.

NG6-41-F21 ESTIMATED PERFORMANCE

MINIMUM ROCKET

BODY LENGTH: 16"
BODY DIAMETER: 1.00"
LIFTOFF WT.: 6.3 OZ.
MAX. ALTITUDE: 3,634 FT.
MAX. VELOCITY: 769 FT./SEC. (524 MPH)
MAX. MACH NUMBER: 0.68
MAX. ACCELERATION: 29.0 Gs
BURNOUT ALTITUDE: 643 FT.
COAST TIME TO APOGEE: 11.6 SECS.

AVERAGE ROCKET

BODY LENGTH: 30"
BODY DIAMETER: 2.26"
LIFTOFF WT.: 15.8 OZ.
MAX. ALTITUDE: 884 FT.
MAX. VELOCITY: 250 FT./SEC. (170 MPH)
MAX. MACH NUMBER: 0.22
MAX. ACCELERATION: 9.5 Gs
BURNOUT ALTITUDE: 204 FT.
COAST TIME TO APOGEE: 6.2 SECS.

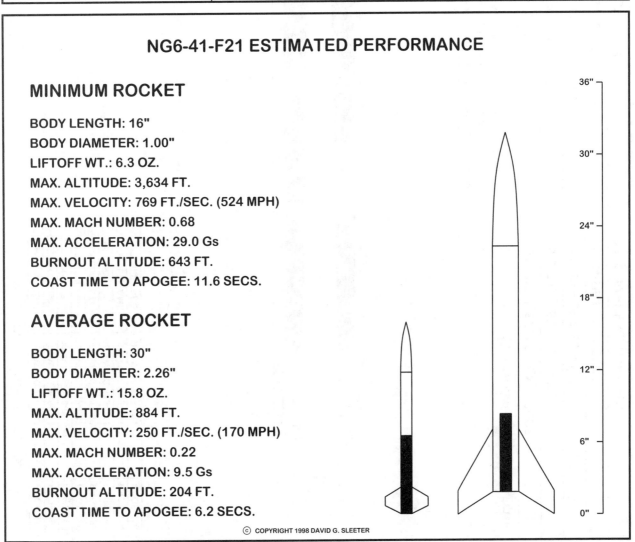

© COPYRIGHT 1998 DAVID G. SLEETER

TELEFLITE CORPORATION

MOTOR NAME: NG6-56-F41
PROPELLANT TYPE: NG6
CASING TYPE: CONVOLUTE PAPER
NOZZLE TYPE: RAMMED CLAY

MOTOR SHOWN WITHOUT TIME DELAY

INDEX OF REQUIRED TOOLING

CASING RETAINER	DRAWING #98-002; 11"	PAGE 91
NOZZLE MOLD	DRAWING #98-010	PAGE 99
CORE SPINDLE	DRAWING #98-051-M	PAGE 129
STARTING TAMP	DRAWING #98-087-B	PAGE 150
PROPELLANT TAMP	DRAWING #98-089-B	PAGE 152
FLAT TAMP	DRAWING #98-091-B	PAGE 154
PORTHOLE TAMP	DRAWING #98-091-A	PAGE 154
PORTHOLE WIRE	DRAWING #98-091	PAGE 154

IMPORTANT BAKING SODA NOTE: Add 30% baking soda to the propellant in your first motor. Reduce the soda in 2% increments until you get a motor that doesn't chuff. Then reduce the soda in 1% increments until maximum thrust is approx. 18 lbs. This test motor took 7% soda, but I made it with a carbon black of medium purity. A motor made with a pure, electrically-conductive carbon black might take substantially more.

TOTAL WEIGHT: 126.9 GMS.
PROPELLANT WT.: 73.3 GMS.
BAKING SODA: See baking soda note.
MOTOR LENGTH: 8.25"
CASING I.D.: 0.75"
CASING O.D.: 1.00"
NOZZLE CLAY: 11.0 GMS.
BULKHEAD CLAY: 11.0 GMS.
THROAT DIA.: 0.375"
CORE LENGTH: 5.50"

TOT. IMPULSE: 56.51 NT.-SECS.
AVG. THRUST: 41.25 NTS.
MAX. THRUST: 18.37 LBS.
BURN TIME: 1.37 SECS.
SPEC. IMPULSE: 78.48 SECS.

© COPYRIGHT 1998 DAVID G. SLEETER

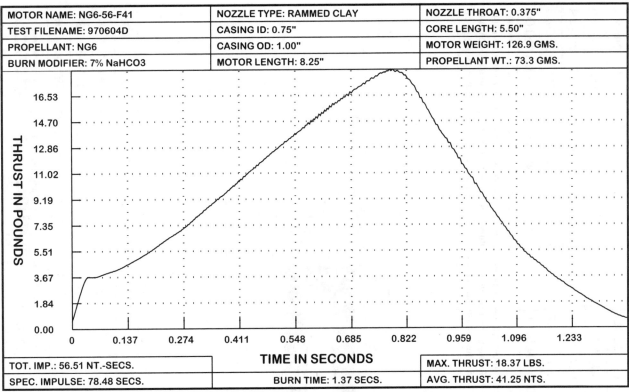

MOTOR NAME: NG6-56-F41	NOZZLE TYPE: RAMMED CLAY	NOZZLE THROAT: 0.375"
TEST FILENAME: 970604D	CASING ID: 0.75"	CORE LENGTH: 5.50"
PROPELLANT: NG6	CASING OD: 1.00"	MOTOR WEIGHT: 126.9 GMS.
BURN MODIFIER: 7% NaHCO3	MOTOR LENGTH: 8.25"	PROPELLANT WT.: 73.3 GMS.

TOT. IMP.: 56.51 NT.-SECS.		MAX. THRUST: 18.37 LBS.
SPEC. IMPULSE: 78.48 SECS.	BURN TIME: 1.37 SECS.	AVG. THRUST: 41.25 NTS.

NG6-56-F41 ESTIMATED PERFORMANCE

MINIMUM ROCKET

BODY LENGTH: 17.5"
BODY DIAMETER: 1.00"
LIFTOFF WT.: 7.0 OZ.
MAX. ALTITUDE: 4,398 FT.
MAX. VELOCITY: 979 FT./SEC. (667 MPH)
MAX. MACH NUMBER: 0.87
MAX. ACCELERATION: 50.1 Gs
BURNOUT ALTITUDE: 689 FT.
COAST TIME TO APOGEE: 12.6 SECS.

AVERAGE ROCKET

BODY LENGTH: 30"
BODY DIAMETER: 2.26"
LIFTOFF WT.: 16.5 OZ. (1.03 LBS.)
MAX. ALTITUDE: 1,394 FT.
MAX. VELOCITY: 360 FT./SEC. (246 MPH)
MAX. MACH NUMBER: 0.32
MAX. ACCELERATION: 18.0 Gs
BURNOUT ALTITUDE: 255 FT.
COAST TIME TO APOGEE: 7.7 SECS.

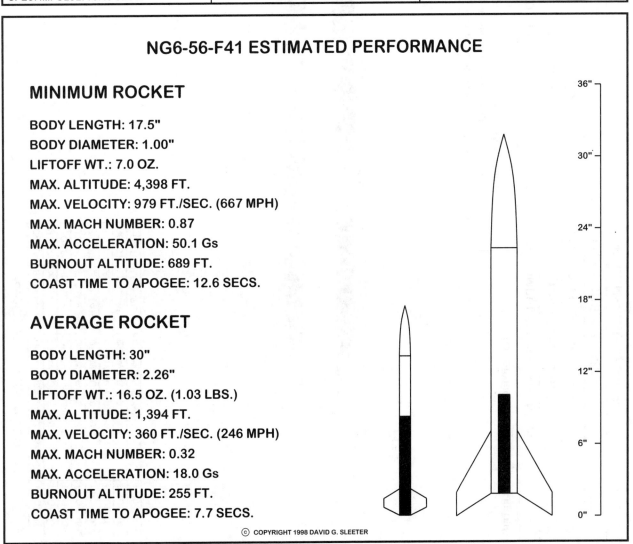

© COPYRIGHT 1998 DAVID G. SLEETER

TELEFLITE CORPORATION

MOTOR NAME: NG6-71-F42
PROPELLANT TYPE: NG6
CASING TYPE: CONVOLUTE PAPER
NOZZLE TYPE: RAMMED CLAY

MOTOR SHOWN WITHOUT TIME DELAY

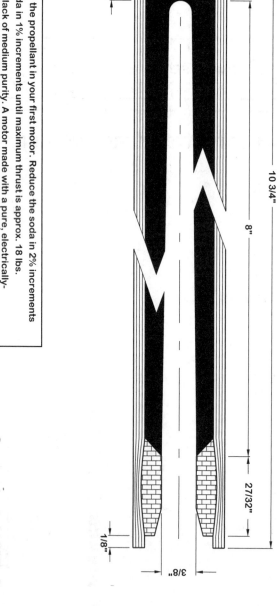

INDEX OF REQUIRED TOOLING

CASING RETAINER	DRAWING #98-002, 13"	PAGE 91
NOZZLE MOLD	DRAWING #98-010	PAGE 99
CORE SPINDLE	DRAWING #98-053-S	PAGE 130
STARTING TAMP	DRAWING #98-088-C	PAGE 151
PROPELLANT TAMP	DRAWING #98-090-C	PAGE 153
FLAT TAMP	DRAWING #98-091-B	PAGE 154
PORTHOLE TAMP	DRAWING #98-091-A	PAGE 154
PORTHOLE WIRE	DRAWING #98-091	PAGE 154

IMPORTANT BAKING SODA NOTE: Add 30% baking soda to the propellant in your first motor. Reduce the soda in 2% increments until you get a motor that doesn't chuff. Then reduce the soda in 1% increments until maximum thrust is approx. 18 lbs. This test motor took 15% soda, but I made it with a carbon black of medium purity. A motor made with a pure, electrically-conductive carbon black might take substantially more.

TOTAL WEIGHT: 170.5 GMS.
PROPELLANT WT.: 107.5 GMS.
BAKING SODA: See baking soda note.
MOTOR LENGTH: 10.75"
CASING I.D.: 0.75"

CASING O.D.: 1.00"
NOZZLE CLAY: 11.0 GMS.
BULKHEAD CLAY: 11.0 GMS.
THROAT DIA.: 0.375"
CORE LENGTH: 8.00"

TOT. IMPULSE: 71.32 NT.-SECS.
AVG. THRUST: 42.03 NTS.
MAX. THRUST: 18.35 LBS.
BURN TIME: 1.70 SECS.
SPEC. IMPULSE: 67.54 SECS.

© COPYRIGHT 1998 DAVID G. SLEETER

MOTOR NAME: NG6-71-F42	NOZZLE TYPE: RAMMED CLAY	NOZZLE THROAT: 0.375"
TEST FILENAME: 970606D	CASING ID: 0.75"	CORE LENGTH: 8.00"
PROPELLANT: NG6	CASING OD: 1.00"	MOTOR WEIGHT: 170.5 GMS.
BURN MODIFIER: 15% NaHCO3	MOTOR LENGTH: 10.75"	PROPELLANT WT.: 107.5 GMS.

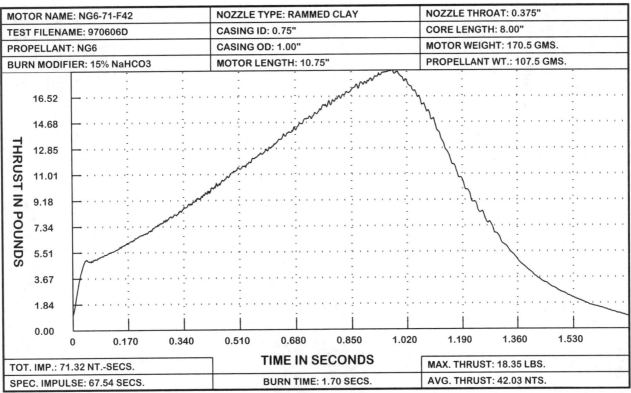

TOT. IMP.: 71.32 NT.-SECS.		MAX. THRUST: 18.35 LBS.
SPEC. IMPULSE: 67.54 SECS.	BURN TIME: 1.70 SECS.	AVG. THRUST: 42.03 NTS.

NG6-71-F42 ESTIMATED PERFORMANCE

MINIMUM ROCKET

BODY LENGTH: 20"
BODY DIAMETER: 1.00"
LIFTOFF WT.: 8.5 OZ.
MAX. ALTITUDE: 4,703 FT.
MAX. VELOCITY: 1,036 FT./SEC. (706 MPH)
MAX. MACH NUMBER: 0.92
MAX. ACCELERATION: 43.8 Gs
BURNOUT ALTITUDE: 952 FT.
COAST TIME TO APOGEE: 12.6 SECS.

AVERAGE ROCKET

BODY LENGTH: 30"
BODY DIAMETER: 2.26"
LIFTOFF WT.: 18.0 OZ. (1.13 LBS.)
MAX. ALTITUDE: 1,736 FT.
MAX. VELOCITY: 415 FT./SEC. (283 MPH)
MAX. MACH NUMBER: 0.37
MAX. ACCELERATION: 16.8 Gs
BURNOUT ALTITUDE: 385 FT.
COAST TIME TO APOGEE: 8.3 SECS.

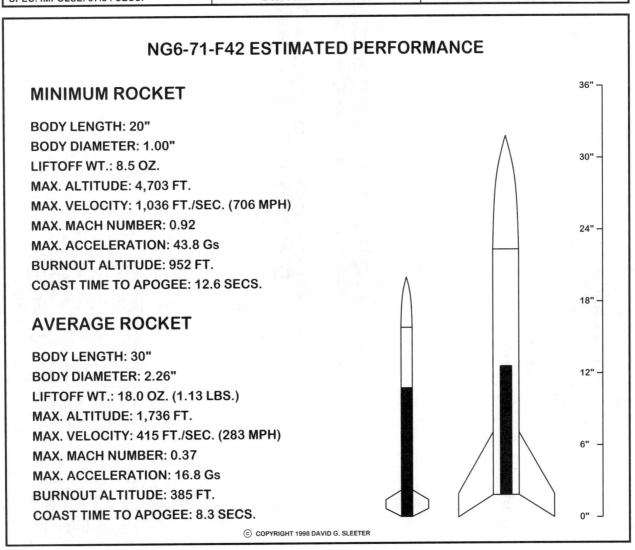

© COPYRIGHT 1998 DAVID G. SLEETER

TELEFLITE CORPORATION

MOTOR NAME: NG6-71-F32
PROPELLANT TYPE: NG6
CASING TYPE: CONVOLUTE PAPER
NOZZLE TYPE: RAMMED CLAY

MOTOR SHOWN WITHOUT TIME DELAY

Dimensions shown on drawing:
- 1 1/4" (casing O.D.)
- 1" (casing I.D.)
- 3/8"
- 3/4"
- 6 1/4"
- 3 1/4"
- 23/32"
- 3/16"
- 5/16"

© COPYRIGHT 1998 DAVID G. SLEETER

TOTAL WEIGHT: 179.4 GMS.	CASING O.D.: 1.25"	TOT. IMPULSE: 71.97 NT.-SECS.
PROPELLANT WT.: 93.5 GMS.	NOZZLE CLAY: 24.0 GMS.	AVG. THRUST: 32.39 NTS.
BAKING SODA: See baking soda note.	BULKHEAD CLAY: 24.0 GMS.	MAX. THRUST: 16.86 LBS.
MOTOR LENGTH: 6.25"	THROAT DIA.: 0.313"	BURN TIME: 2.22 SECS.
CASING I.D.: 1.00"	CORE LENGTH: 3.25"	SPEC. IMPULSE: 78.36 SECS.

IMPORTANT BAKING SODA NOTE: Add 30% baking soda to the propellant in your first motor. Reduce the soda in 2% increments until you get a motor that doesn't chuff. Then reduce the soda in 1% increments until maximum thrust is approx. 16 lbs. This test motor took 7% soda, but I made it with a carbon black of medium purity. A motor made with a pure, electrically-conductive carbon black might take substantially more.

INDEX OF REQUIRED TOOLING

CASING RETAINER	DRAWING #98-003: 10"	PAGE 92
NOZZLE MOLD	DRAWING #98-012	PAGE 101
CORE SPINDLE	DRAWING #98-055-H	PAGE 131
STARTING TAMP	DRAWING #98-092-A	PAGE 155
PROPELLANT TAMP	DRAWING #98-093-A	PAGE 156
FLAT TAMP	DRAWING #98-096-A	PAGE 159
PORTHOLE TAMP	DRAWING #98-096-B	PAGE 159
PORTHOLE WIRE	DRAWING #98-096	PAGE 159

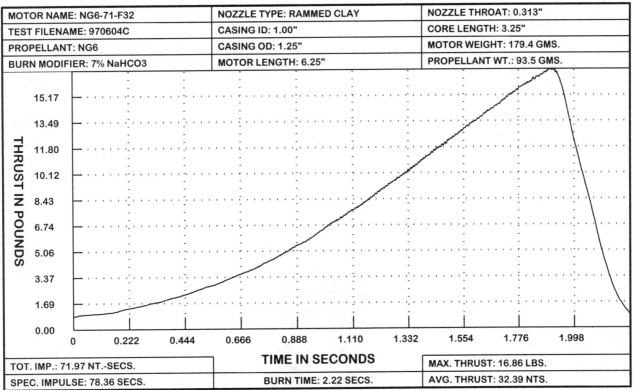

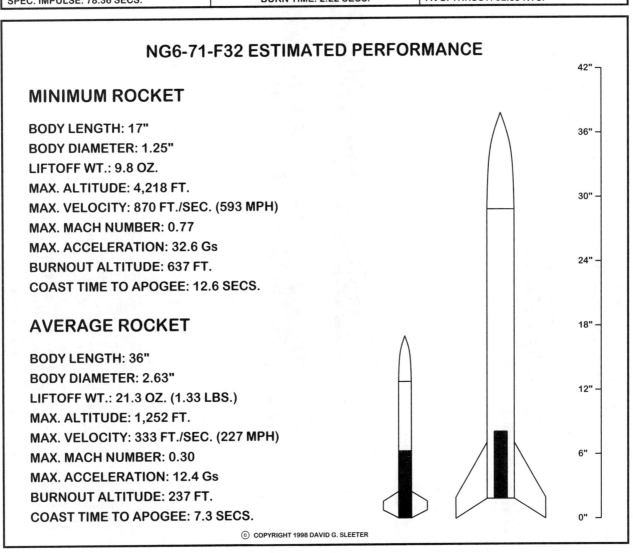

TELEFLITE CORPORATION

MOTOR NAME:	NG6-87-G30
PROPELLANT TYPE:	NG6
CASING TYPE:	CONVOLUTE PAPER
NOZZLE TYPE:	RAMMED CLAY

IMPORTANT BAKING SODA NOTE: Add 30% baking soda to the propellant in your first motor. Reduce the soda in 2% increments until you get a motor that doesn't chuff. Then reduce the soda in 1% increments until maximum thrust is approx. 15 lbs. This test motor took 15% soda, but I made it with a carbon black of medium purity. A motor made with a pure, electrically-conductive carbon black might take substantially more.

CASING I.D.:	1.00"
MOTOR LENGTH:	7.50"
BAKING SODA:	See baking soda note.
PROPELLANT WT.:	125.7 GMS.
TOTAL WEIGHT:	219.8 GMS.

CASING O.D.:	1.25"
THROAT DIA.:	0.313
BULKHEAD CLAY:	24.0 GMS.
NOZZLE CLAY:	24.0 GMS.

CORE LENGTH:	4.50"
BURN TIME:	2.90 SECS.
MAX. THRUST:	15.76 LBS.
AVG. THRUST:	30.13 NTS.
TOT. IMPULSE:	87.42 NT.-SECS.
SPEC. IMPULSE:	70.80 SECS.

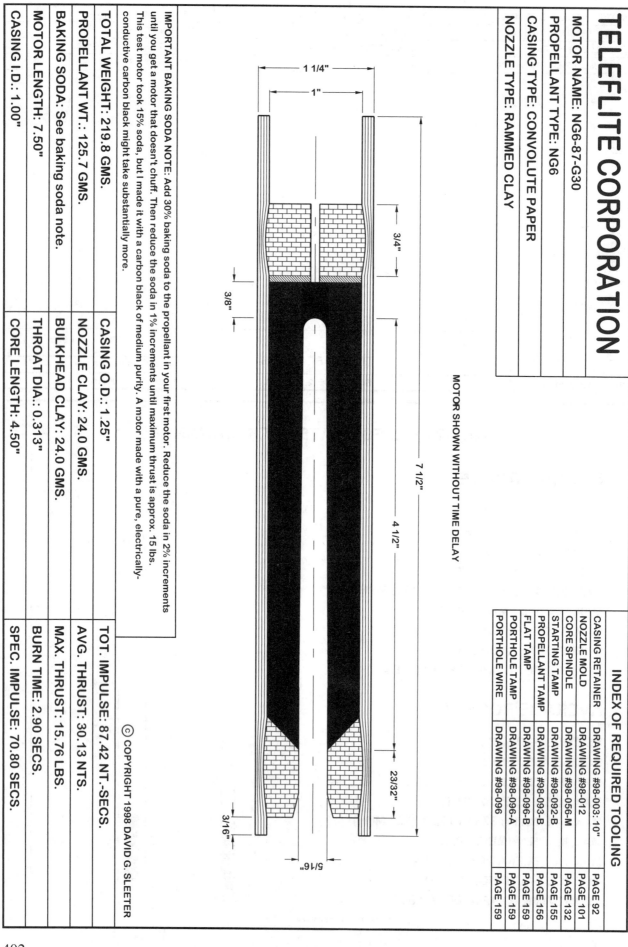

MOTOR SHOWN WITHOUT TIME DELAY

© COPYRIGHT 1998 DAVID G. SLEETER

INDEX OF REQUIRED TOOLING

CASING RETAINER	DRAWING #98-003: 10"	PAGE 92
NOZZLE MOLD	DRAWING #98-012	PAGE 101
CORE SPINDLE	DRAWING #98-056-M	PAGE 132
STARTING TAMP	DRAWING #98-092-B	PAGE 155
PROPELLANT TAMP	DRAWING #98-093-B	PAGE 156
FLAT TAMP	DRAWING #98-096-B	PAGE 159
PORTHOLE TAMP	DRAWING #98-096-A	PAGE 159
PORTHOLE WIRE	DRAWING #98-096	PAGE 159

MOTOR NAME: NG6-87-G30	NOZZLE TYPE: RAMMED CLAY	NOZZLE THROAT: 0.313"
TEST FILENAME: 970601I	CASING ID: 1.00"	CORE LENGTH: 4.50"
PROPELLANT: NG6	CASING OD: 1.25"	MOTOR WEIGHT: 219.8 GMS.
BURN MODIFIER: 15% NaHCO3	MOTOR LENGTH: 7.50"	PROPELLANT WT.: 125.7 GMS.

TOT. IMP.: 87.42 NT.-SECS.	TIME IN SECONDS	MAX. THRUST: 15.76 LBS.
SPEC. IMPULSE: 70.80 SECS.	BURN TIME: 2.90 SECS.	AVG. THRUST: 30.13 NTS.

NG6-87-G30 ESTIMATED PERFORMANCE

MINIMUM ROCKET

BODY LENGTH: 18"
BODY DIAMETER: 1.25"
LIFTOFF WT.: 11.3 OZ.
MAX. ALTITUDE: 4,682 FT.
MAX. VELOCITY: 929 FT./SEC. (633 MPH)
MAX. MACH NUMBER: 0.83
MAX. ACCELERATION: 27.5 Gs
BURNOUT ALTITUDE: 817 FT.
COAST TIME TO APOGEE: 13.0 SECS.

AVERAGE ROCKET

BODY LENGTH: 36"
BODY DIAMETER: 2.63"
LIFTOFF WT.: 22.8 OZ. (1.43 LBS.)
MAX. ALTITUDE: 1,484 FT.
MAX. VELOCITY: 369 FT./SEC. (252 MPH)
MAX. MACH NUMBER: 0.33
MAX. ACCELERATION: 10.8 Gs
BURNOUT ALTITUDE: 312 FT.
COAST TIME TO APOGEE: 7.8 SECS.

© COPYRIGHT 1998 DAVID G. SLEETER

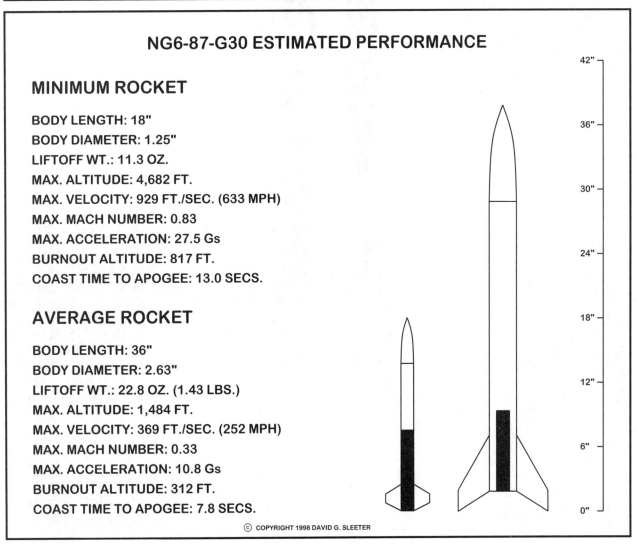

TELEFLITE CORPORATION

MOTOR NAME:	NG6-134-G78
PROPELLANT TYPE:	NG6
CASING TYPE:	CONVOLUTE PAPER
NOZZLE TYPE:	RAMMED CLAY

MOTOR SHOWN WITHOUT TIME DELAY

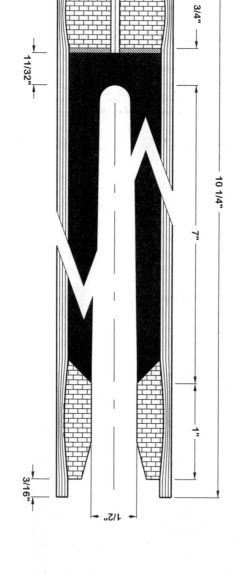

© COPYRIGHT 1998 DAVID G. SLEETER

IMPORTANT BAKING SODA NOTE: Add 30% baking soda to the propellant in your first motor. Reduce the soda in 1% increments until you get a motor that doesn't chuff. Then reduce the soda in 2% increments until maximum thrust is approx. 32 lbs. This test motor took 7% soda, but I made it with a carbon black of medium purity. A motor made with a pure, electrically-conductive carbon black might take substantially more.

TOTAL WEIGHT:	280.5 GMS.
PROPELLANT WT.:	177.0 GMS.
BAKING SODA:	See baking soda note.
MOTOR LENGTH:	10.25"
CASING I.D.:	1.00"

CASING O.D.:	1.25"
NOZZLE CLAY:	24.0 GMS.
BULKHEAD CLAY:	24.0 GMS.
THROAT DIA.:	0.50"
CORE LENGTH:	7.00"

TOT. IMPULSE:	134.19 NT.-SECS.
AVG. THRUST:	78.85 NTS.
MAX. THRUST:	32.56 LBS.
BURN TIME:	1.70 SECS.
SPEC. IMPULSE:	77.18 SECS.

INDEX OF REQUIRED TOOLING

CASING RETAINER	DRAWING #98-003: 13"	PAGE 92
NOZZLE MOLD	DRAWING #98-014	PAGE 103
CORE SPINDLE	DRAWING #98-061-J	PAGE 134
STARTING TAMP	DRAWING #98-094-A	PAGE 157
PROPELLANT TAMP	DRAWING #98-095-A	PAGE 158
FLAT TAMP	DRAWING #98-096-B	PAGE 159
PORTHOLE TAMP	DRAWING #98-096-A	PAGE 159
PORTHOLE WIRE	DRAWING #98-096	PAGE 159

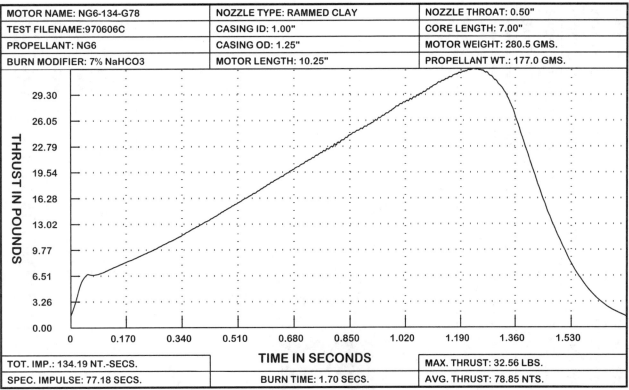

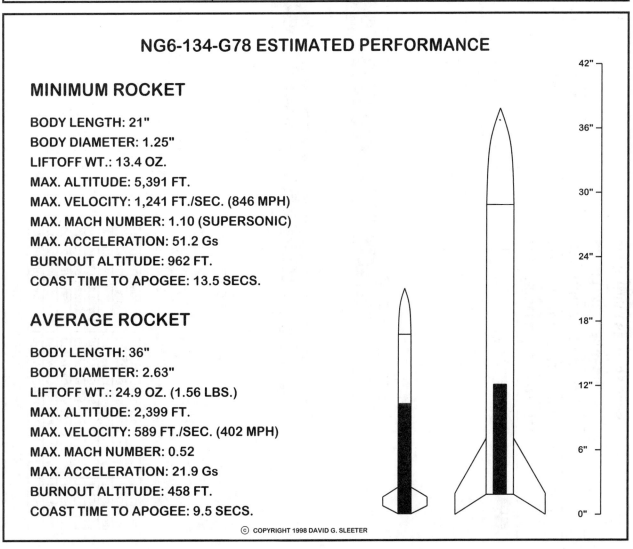

TELEFLITE CORPORATION

MOTOR NAME: NG6-155-G88
PROPELLANT TYPE: NG6
CASING TYPE: CONVOLUTE PAPER
NOZZLE TYPE: RAMMED CLAY

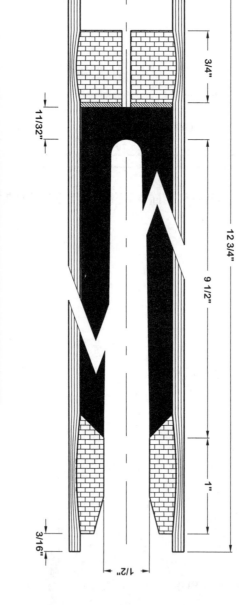

MOTOR SHOWN WITHOUT TIME DELAY

IMPORTANT BAKING SODA NOTE: Add 30% baking soda to the propellant in your first motor. Reduce the soda in 2% increments until you get a motor that doesn't chuff. Then reduce the soda in 1% increments until maximum thrust is approx. 36 lbs. This test motor took 15% soda, but I made it with a carbon black of medium purity. A motor made with a pure, electrically-conductive carbon black might take substantially more.

CASING I.D.: 1.00"	CORE LENGTH: 9.50"
MOTOR LENGTH: 12.75"	THROAT DIA.: 0.50"
BAKING SODA: See baking soda note.	BULKHEAD CLAY: 24.0 GMS.
PROPELLANT WT.: 215.7 GMS.	NOZZLE CLAY: 24.0 GMS.
TOTAL WEIGHT: 338.2 GMS.	CASING O.D.: 1.25"

TOT. IMPULSE: 155.77 NT.-SECS.	
AVG. THRUST: 88.81 NTS.	
MAX. THRUST: 36.13 LBS.	
BURN TIME: 1.75 SECS.	
SPEC. IMPULSE: 73.51 SECS.	

© COPYRIGHT 1998 DAVID G. SLEETER

INDEX OF REQUIRED TOOLING

CASING RETAINER	DRAWING #98-003:-16"	PAGE 92
NOZZLE MOLD	DRAWING #98-014	PAGE 103
CORE SPINDLE	DRAWING #98-063-O	PAGE 135
STARTING TAMP	DRAWING #98-094-B	PAGE 157
PROPELLANT TAMP	DRAWING #98-095-B	PAGE 158
FLAT TAMP	DRAWING #98-096-B	PAGE 159
PORTHOLE TAMP	DRAWING #98-096-A	PAGE 159
PORTHOLE WIRE	DRAWING #98-096	PAGE 159

MOTOR NAME: NG6-155-G88	NOZZLE TYPE: RAMMED CLAY	NOZZLE THROAT: 0.50"
TEST FILENAME: 970608B	CASING ID: 1.00"	CORE LENGTH: 9.50"
PROPELLANT: NG6	CASING OD: 1.25"	MOTOR WEIGHT: 338.2 GMS.
BURN MODIFIER: 15% NaHCO3	MOTOR LENGTH: 12.75"	PROPELLANT WT.: 215.7 GMS.

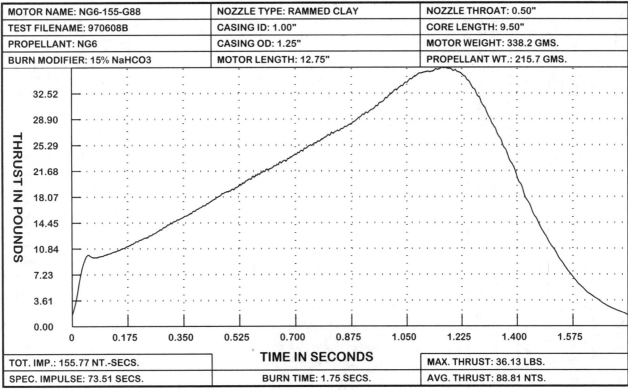

TOT. IMP.: 155.77 NT.-SECS.		MAX. THRUST: 36.13 LBS.
SPEC. IMPULSE: 73.51 SECS.	BURN TIME: 1.75 SECS.	AVG. THRUST: 88.81 NTS.

NG6-155-G88 ESTIMATED PERFORMANCE

MINIMUM ROCKET

BODY LENGTH: 23.5"
BODY DIAMETER: 1.25"
LIFTOFF WT.: 15.5 OZ.
MAX. ALTITUDE: 5,557 FT.
MAX. VELOCITY: 1,245 FT./SEC. (849 MPH)
MAX. MACH NUMBER: 1.11 (SUPERSONIC)
MAX. ACCELERATION: 49.6 Gs
BURNOUT ALTITUDE: 1,067 FT.
COAST TIME TO APOGEE: 13.6 SECS.

AVERAGE ROCKET

BODY LENGTH: 36"
BODY DIAMETER: 2.63"
LIFTOFF WT.: 27.0 OZ. (1.69 LBS.)
MAX. ALTITUDE: 2,667 FT.
MAX. VELOCITY: 635 FT./SEC. (433 MPH)
MAX. MACH NUMBER: 0.56
MAX. ACCELERATION: 23.0 Gs
BURNOUT ALTITUDE: 547 FT.
COAST TIME TO APOGEE: 9.9 SECS.

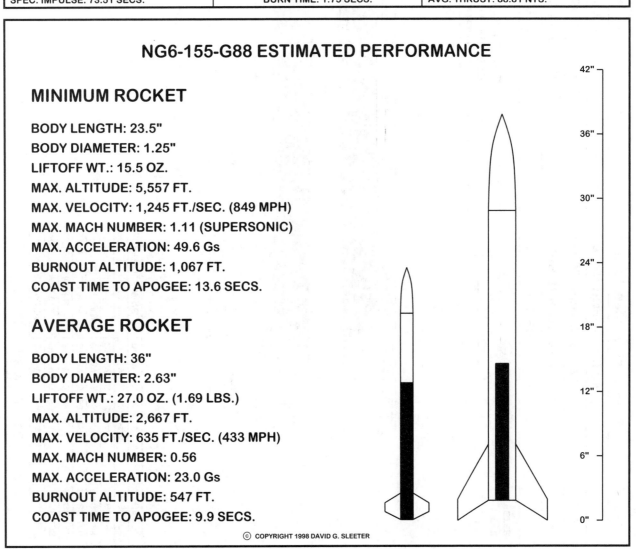

© COPYRIGHT 1998 DAVID G. SLEETER

TELEFLITE CORPORATION

MOTOR NAME:	NG6-94-G36
PROPELLANT TYPE:	NG6
CASING TYPE:	CONVOLUTE PAPER
NOZZLE TYPE:	RAMMED CLAY

MOTOR SHOWN WITHOUT TIME DELAY

Dimensions shown on diagram: 1 1/2", 1 1/8", 27/32", 7/16", 7 1/4", 3 3/4", 27/32", 3/16", 3/8"

© COPYRIGHT 1998 DAVID G. SLEETER

IMPORTANT BAKING SODA NOTE: Add 30% baking soda to the propellant in your first motor. Reduce the soda to the propellant in 1% increments until you get a motor that doesn't chuff. Then reduce the soda in 2% increments until maximum thrust is approx. 19 lbs. This test motor took 7% soda, but I made it with a carbon black of medium purity. A motor made with a pure, electrically-conductive carbon black might take substantially more.

TOTAL WEIGHT: 259.2 GMS.	CASING O.D.: 1.50"	TOT. IMPULSE: 94.24 NT.-SECS.	
PROPELLANT WT.: 128.4 GMS.	NOZZLE CLAY: 33.0 GMS.	AVG. THRUST: 36.27 NTS.	
BAKING SODA: See baking soda note.	BULKHEAD CLAY: 33.0 GMS.	MAX. THRUST: 19.27 LBS.	
MOTOR LENGTH: 7.25"	THROAT DIA.: 0.375"	BURN TIME: 2.60 SECS.	
CASING I.D.: 1.125"	CORE LENGTH: 3.75"	SPEC. IMPULSE: 74.72 SECS.	

INDEX OF REQUIRED TOOLING

CASING RETAINER	DRAWING #98-004; 11-1/2"	PAGE 93
NOZZLE MOLD	DRAWING #98-016	PAGE 105
CORE SPINDLE	DRAWING #98-067-H	PAGE 137
STARTING TAMP	DRAWING #98-097-A	PAGE 160
PROPELLANT TAMP	DRAWING #98-098-A	PAGE 161
FLAT TAMP	DRAWING #98-104-B	PAGE 167
PORTHOLE TAMP	DRAWING #98-104-A	PAGE 167
PORTHOLE WIRE	DRAWING #98-104	PAGE 167

MOTOR NAME: NG6-94-G36	NOZZLE TYPE: RAMMED CLAY	NOZZLE THROAT: 0.375"
TEST FILENAME: 970602E	CASING ID: 1.125"	CORE LENGTH: 3.75"
PROPELLANT: NG6	CASING OD: 1.50"	MOTOR WEIGHT: 259.2 GMS.
BURN MODIFIER: 7% NaHCO3	MOTOR LENGTH: 7.25"	PROPELLANT WT.: 128.4 GMS.

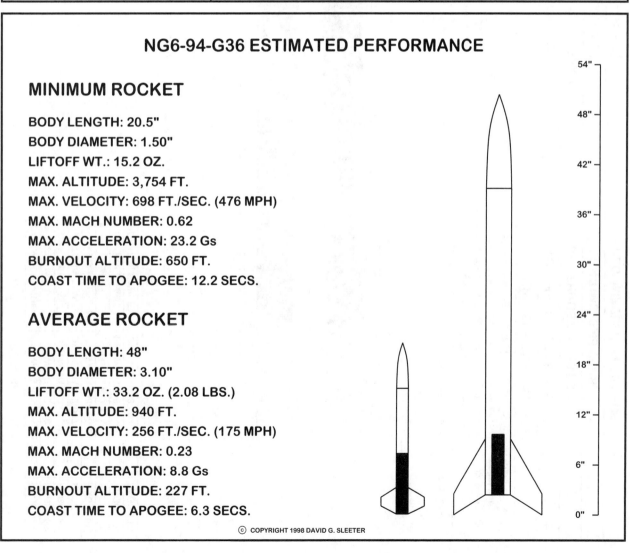

TOT. IMP.: 94.24 NT.-SECS.		MAX. THRUST: 19.27 LBS.
SPEC. IMPULSE: 74.72 SECS.	BURN TIME: 2.60 SECS.	AVG. THRUST: 36.27 NTS.

NG6-94-G36 ESTIMATED PERFORMANCE

MINIMUM ROCKET

BODY LENGTH: 20.5"
BODY DIAMETER: 1.50"
LIFTOFF WT.: 15.2 OZ.
MAX. ALTITUDE: 3,754 FT.
MAX. VELOCITY: 698 FT./SEC. (476 MPH)
MAX. MACH NUMBER: 0.62
MAX. ACCELERATION: 23.2 Gs
BURNOUT ALTITUDE: 650 FT.
COAST TIME TO APOGEE: 12.2 SECS.

AVERAGE ROCKET

BODY LENGTH: 48"
BODY DIAMETER: 3.10"
LIFTOFF WT.: 33.2 OZ. (2.08 LBS.)
MAX. ALTITUDE: 940 FT.
MAX. VELOCITY: 256 FT./SEC. (175 MPH)
MAX. MACH NUMBER: 0.23
MAX. ACCELERATION: 8.8 Gs
BURNOUT ALTITUDE: 227 FT.
COAST TIME TO APOGEE: 6.3 SECS.

© COPYRIGHT 1998 DAVID G. SLEETER

TELEFLITE CORPORATION

MOTOR NAME:	NG6-129-G37
PROPELLANT TYPE:	NG6
CASING TYPE:	CONVOLUTE PAPER
NOZZLE TYPE:	RAMMED CLAY

INDEX OF REQUIRED TOOLING

CASING RETAINER	DRAWING #98-004; 11-1/2"	PAGE 93
NOZZLE MOLD	DRAWING #98-016	PAGE 105
CORE SPINDLE	DRAWING #98-068-M	PAGE 138
STARTING TAMP	DRAWING #98-097-B	PAGE 160
PROPELLANT TAMP	DRAWING #98-098-B	PAGE 161
FLAT TAMP	DRAWING #98-104-B	PAGE 167
PORTHOLE TAMP	DRAWING #98-104-A	PAGE 167
PORTHOLE WIRE	DRAWING #98-104	PAGE 167

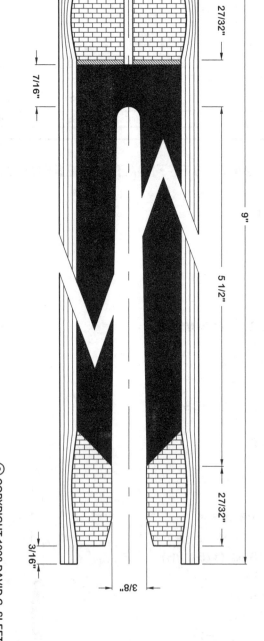

MOTOR SHOWN WITHOUT TIME DELAY

© COPYRIGHT 1998 DAVID G. SLEETER

IMPORTANT BAKING SODA NOTE: Add 30% baking soda to the propellant in your first motor. Reduce the soda in 2% increments until you get a motor that doesn't chuff. Then reduce the soda in 1% increments until maximum thrust is approx. 20 lbs. This test motor took 15% soda, but I made it with a carbon black of medium purity. A motor made with pure, electrically conductive carbon black might take substantially more.

EXTENDED BURN TIME WARNING! This motor's burn time is longer than 3 seconds. Do not fly this motor on a stick. Fly it only in a properly balanced, aerodynamic rocket body. Use a tall launch tower, and fly the rocket only in calm weather.

TOTAL WEIGHT: 331.1 GMS.	CASING O.D.: 1.50"	TOT. IMPULSE: 129.59 NT.-SECS.
PROPELLANT WT.: 184.0 GMS.	NOZZLE CLAY: 33.0 GMS.	AVG. THRUST: 37.56 NTS.
BAKING SODA: See baking soda note.	BULKHEAD CLAY: 33.0 GMS.	MAX. THRUST: 19.30 LBS.
MOTOR LENGTH: 9.00"	THROAT DIA.: 0.375"	BURN TIME: 3.45 SECS.
CASING I.D.: 1.125"	CORE LENGTH: 5.50"	SPEC. IMPULSE: 71.70 SECS.

MOTOR NAME: NG6-129-G37	NOZZLE TYPE: RAMMED CLAY	NOZZLE THROAT: 0.375"
TEST FILENAME: 970424E	CASING ID: 1.125"	CORE LENGTH: 5.50"
PROPELLANT: NG6	CASING OD: 1.50"	MOTOR WEIGHT: 331.1 GMS.
BURN MODIFIER: 15% NaHCO3	MOTOR LENGTH: 9.00"	PROPELLANT WT.: 184.0 GMS.

TOT. IMP.: 129.59 NT.-SECS.		MAX. THRUST: 19.30 LBS.
SPEC. IMPULSE: 71.70 SECS.	BURN TIME: 3.45 SECS.	AVG. THRUST: 37.56 NTS.

NG6-129-G37 ESTIMATED PERFORMANCE

MINIMUM ROCKET

BODY LENGTH: 22"
BODY DIAMETER: 1.50"
LIFTOFF WT.: 17.7 OZ. (1.11 LBS.)
MAX. ALTITUDE: 4,653 FT.
MAX. VELOCITY: 828 FT./SEC. (565 MPH)
MAX. MACH NUMBER: 0.74
MAX. ACCELERATION: 20.5 Gs
BURNOUT ALTITUDE: 918 FT.
COAST TIME TO APOGEE: 13.1 SECS.

AVERAGE ROCKET

BODY LENGTH: 48"
BODY DIAMETER: 3.10"
LIFTOFF WT.: 35.7 OZ. (2.23 LBS.)
MAX. ALTITUDE: 1,358 FT.
MAX. VELOCITY: 326 FT./SEC. (222 MPH)
MAX. MACH NUMBER: 0.29
MAX. ACCELERATION: 8.2 Gs
BURNOUT ALTITUDE: 342 FT.
COAST TIME TO APOGEE: 7.4 SECS.

© COPYRIGHT 1998 DAVID G. SLEETER

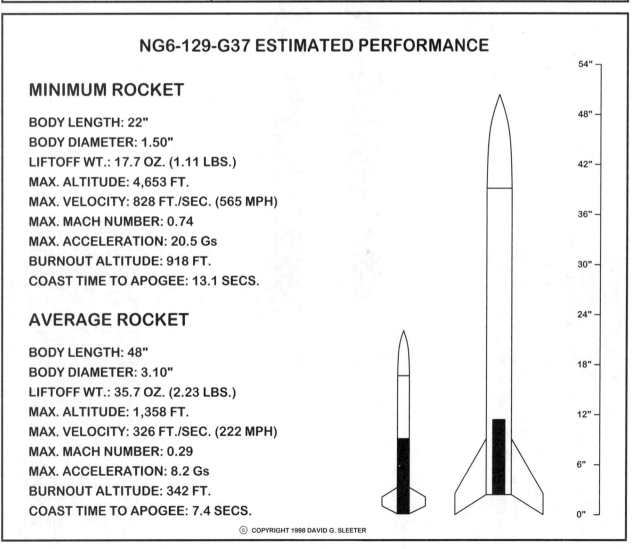

TELEFLITE CORPORATION

MOTOR NAME: NG6-191-H100
PROPELLANT TYPE: NG6
CASING TYPE: CONVOLUTE PAPER
NOZZLE TYPE: RAMMED CLAY

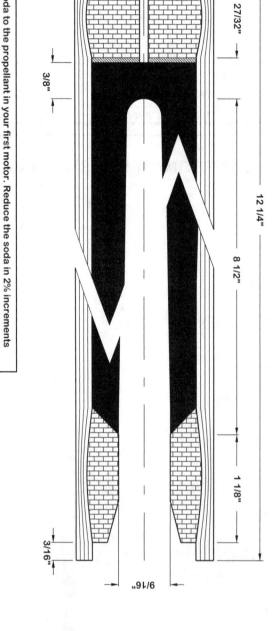

MOTOR SHOWN WITHOUT TIME DELAY

© COPYRIGHT 1998 DAVID G. SLEETER

IMPORTANT BAKING SODA NOTE: Add 30% baking soda to the propellant in your first motor. Reduce the soda in 1% increments until you get a motor that doesn't chuff. Then reduce the soda in 2% increments until maximum thrust is approx. 42 lbs. This test motor took 7% soda, but I made it with a carbon black of medium purity. A motor made with a pure, electrically-conductive carbon black might take substantially more.

TOTAL WEIGHT: 425.2 GMS.	CASING O.D.: 1.50"
PROPELLANT WT.: 251.1 GMS.	NOZZLE CLAY: 33.0 GMS.
BAKING SODA: See baking soda note.	BULKHEAD CLAY: 33.0 GMS.
MOTOR LENGTH: 12.25"	THROAT DIA.: 0.563"
CASING I.D.: 1.125"	CORE LENGTH: 8.50"
TOT. IMPULSE: 191.57 NT.-SECS.	
AVG. THRUST: 99.98 NTS.	
MAX. THRUST: 42.65 LBS.	
BURN TIME: 1.92 SECS.	
SPEC. IMPULSE: 77.66 SECS.	

INDEX OF REQUIRED TOOLING

CASING RETAINER	DRAWING #98-004: 15"	PAGE 93
NOZZLE MOLD	DRAWING #98-018	PAGE 107
CORE SPINDLE	DRAWING #98-072-J	PAGE 140
STARTING TAMP	DRAWING #98-099-B	PAGE 162
PROPELLANT TAMP	DRAWING #98-101-B	PAGE 164
FLAT TAMP	DRAWING #98-104-B	PAGE 167
PORTHOLE TAMP	DRAWING #98-104-A	PAGE 167
PORTHOLE WIRE	DRAWING #98-104	PAGE 167

MOTOR NAME: NG6-191-H100	NOZZLE TYPE: RAMMED CLAY	NOZZLE THROAT: 0.563"
TEST FILENAME: 970604F	CASING ID: 1.125"	CORE LENGTH: 8.50"
PROPELLANT: NG6	CASING OD: 1.50"	MOTOR WEIGHT: 425.2 GMS.
BURN MODIFIER: 7% NaHCO3	MOTOR LENGTH: 12.25"	PROPELLANT WT.: 251.1 GMS.

TOT. IMP.: 191.57 NT.-SECS.		MAX. THRUST: 42.65 LBS.
SPEC. IMPULSE: 77.66 SECS.	BURN TIME: 1.92 SECS.	AVG. THRUST: 99.98 NTS.

NG6-191-H100 ESTIMATED PERFORMANCE

MINIMUM ROCKET

BODY LENGTH: 25.5"
BODY DIAMETER: 1.50"
LIFTOFF WT.: 21.0 OZ. (1.31 LBS.)
MAX. ALTITUDE: 5,688 FT.
MAX. VELOCITY: 1,112 FT./SEC. (758 MPH)
MAX. MACH NUMBER: 0.99
MAX. ACCELERATION: 40.7 Gs
BURNOUT ALTITUDE: 1,026 FT.
COAST TIME TO APOGEE: 14.2 SECS.

AVERAGE ROCKET

BODY LENGTH: 48"
BODY DIAMETER: 3.10"
LIFTOFF WT.: 39.0 OZ. (2.44 LBS.)
MAX. ALTITUDE: 2,257 FT.
MAX. VELOCITY: 518 FT./SEC. (353 MPH)
MAX. MACH NUMBER: 0.46
MAX. ACCELERATION: 18.0 Gs
BURNOUT ALTITUDE: 482 FT.
COAST TIME TO APOGEE: 9.3 SECS.

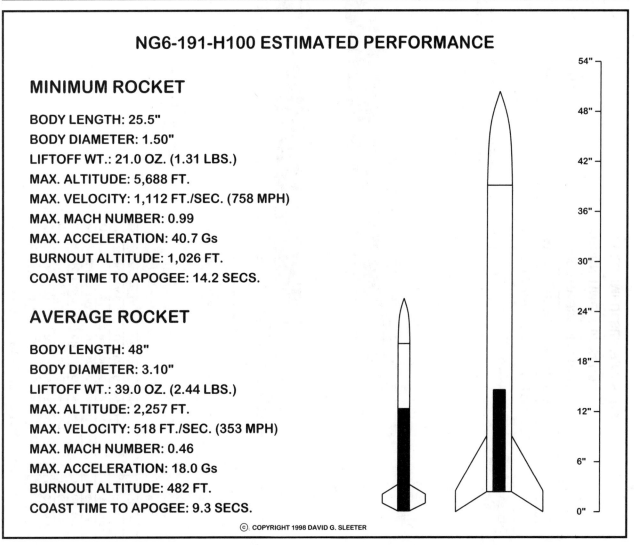

© COPYRIGHT 1998 DAVID G. SLEETER

TELEFLITE CORPORATION

MOTOR NAME:	NG6-313-H119
PROPELLANT TYPE:	NG6
CASING TYPE:	CONVOLUTE PAPER
NOZZLE TYPE:	RAMMED CLAY

MOTOR SHOWN WITHOUT TIME DELAY

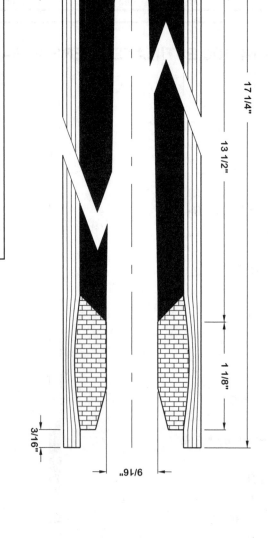

IMPORTANT BAKING SODA NOTE: Add 30% baking soda to the propellant in your first motor. Reduce the soda in 1% increments until you get a motor that doesn't chuff. Then reduce the soda in 2% increments until maximum thrust is approx. 48 lbs. This test motor took 15% soda, but I made it with a carbon black of medium purity. A motor made with a pure, electrically-conductive carbon black might take substantially more.

© COPYRIGHT 1998 DAVID G. SLEETER

TOTAL WEIGHT: 622.0 GMS.	CASING O.D.: 1.50"		TOT. IMPULSE: 313.91 NT.-SECS.
PROPELLANT WT.: 402.5 GMS.	NOZZLE CLAY: 33.0 GMS.		AVG. THRUST: 119.86 NTS.
BAKING SODA: See baking soda note.	BULKHEAD CLAY: 33.0 GMS.		MAX. THRUST: 47.36 LBS.
MOTOR LENGTH: 17.25"	THROAT DIA.: 0.563"		BURN TIME: 2.62 SECS.
CASING I.D.: 1.125"	CORE LENGTH: 13.50"		SPEC. IMPULSE: 79.39 SECS.

INDEX OF REQUIRED TOOLING

CASING RETAINER	DRAWING #98-004: 22"	PAGE 93
NOZZLE MOLD	DRAWING #98-018	PAGE 107
CORE SPINDLE	DRAWING #98-075-T	PAGE 141
STARTING TAMP	DRAWING #98-100-C	PAGE 163
PROPELLANT TAMP	DRAWING #98-102-C	PAGE 165
FLAT TAMP	DRAWING #98-104-B	PAGE 167
PORTHOLE TAMP	DRAWING #98-104-A	PAGE 167
PORTHOLE WIRE	DRAWING #98-104	PAGE 167

MOTOR NAME: NG6-313-H119	NOZZLE TYPE: RAMMED CLAY	NOZZLE THROAT: 0.563"
TEST FILENAME: 970606E	CASING ID: 1.125"	CORE LENGTH: 13.50"
PROPELLANT: NG6	CASING OD: 1.50"	MOTOR WEIGHT: 622.0 GMS.
BURN MODIFIER: 15% NaHCO3	MOTOR LENGTH: 17.25"	PROPELLANT WT.: 402.5 GMS.

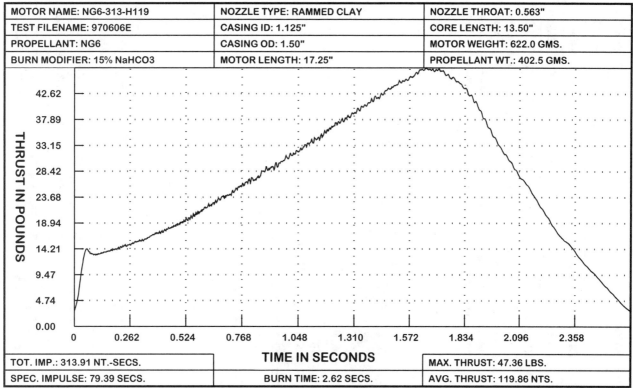

TOT. IMP.: 313.91 NT.-SECS.		MAX. THRUST: 47.36 LBS.
SPEC. IMPULSE: 79.39 SECS.	BURN TIME: 2.62 SECS.	AVG. THRUST: 119.86 NTS.

NG6-313-H119 ESTIMATED PERFORMANCE

MINIMUM ROCKET

BODY LENGTH: 30.5"
BODY DIAMETER: 1.50"
LIFTOFF WT.: 28.0 OZ. (1.75 LBS.)
MAX. ALTITUDE: 6,728 FT.
MAX. VELOCITY: 1,281 FT./SEC. (873 MPH)
MAX. MACH NUMBER: 1.14 (SUPERSONIC)
MAX. ACCELERATION: 34.3 Gs
BURNOUT ALTITUDE: 1,692 FT.
COAST TIME TO APOGEE: 14.6 SECS.

AVERAGE ROCKET

BODY LENGTH: 48"
BODY DIAMETER: 3.10"
LIFTOFF WT.: 46.0 OZ. (2.88 LBS.)
MAX. ALTITUDE: 3,381 FT.
MAX. VELOCITY: 701 FT./SEC. (478 MPH)
MAX. MACH NUMBER: 0.62
MAX. ACCELERATION: 17.0 Gs
BURNOUT ALTITUDE: 920 FT.
COAST TIME TO APOGEE: 10.6 SECS.

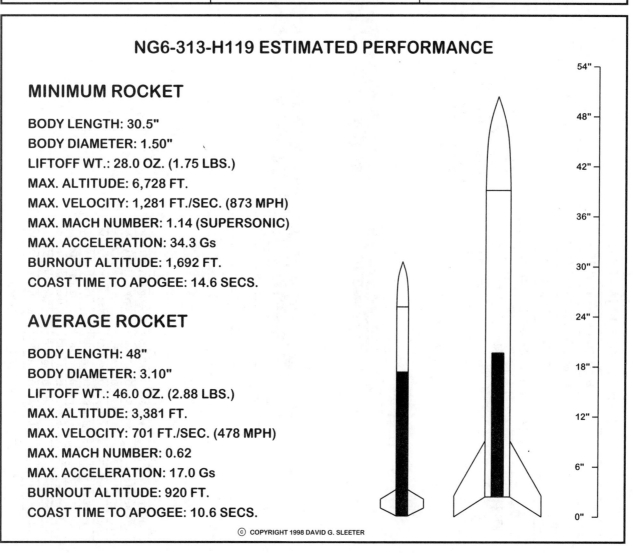

© COPYRIGHT 1998 DAVID G. SLEETER

TELEFLITE CORPORATION

MOTOR NAME:	NG6-227-H137
PROPELLANT TYPE:	NG6
CASING TYPE:	CONVOLUTE PAPER
NOZZLE TYPE:	RAMMED CLAY

MOTOR SHOWN WITHOUT TIME DELAY

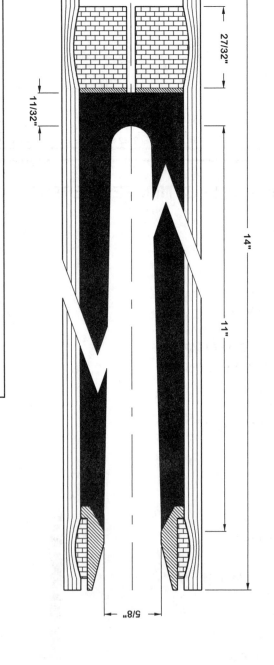

© COPYRIGHT 1998 DAVID G. SLEETER

INDEX OF REQUIRED TOOLING

NOZZLE ANVIL	DRAWING #98-109-C	PAGE 172
NOZZLE RAMMER	DRAWING #98-110-D	PAGE 173
CASING RETAINER	DRAWING #98-004: 18-1/2"	PAGE 93
NOZZLE MOLD	DRAWING #98-020	PAGE 109
CORE SPINDLE	DRAWING #98-079-F	PAGE 143
PROPELLANT TAMP	DRAWING #98-103-B	PAGE 166
FLAT TAMP	DRAWING #98-104-B	PAGE 167
PORTHOLE TAMP	DRAWING #98-104-A	PAGE 167
PORTHOLE WIRE	DRAWING #98-104	PAGE 167

IMPORTANT BAKING SODA NOTE: Add 30% baking soda to the propellant in your first motor. Then reduce the soda in 1% increments until you get a motor that doesn't chuff. Then reduce the soda in 2% increments until maximum thrust is approx. 55 lbs. This test motor took 7% soda, but I made it with a carbon black of medium purity. A motor made with a pure, electrically-conductive carbon black might take substantially more.

TOTAL WEIGHT:	498.3 GMS.
PROPELLANT WT.:	293.6 GMS.
BAKING SODA:	See baking soda note.
MOTOR LENGTH:	14.00"
CASING I.D.:	1.125"
CASING O.D.:	1.50"
NOZZLE:	S.S. DeLAVAL
BULKHEAD CLAY:	33.0 GMS.
THROAT DIA.:	0.625"
CORE LENGTH:	11.00"
TOT. IMPULSE:	227.97 NT.-SECS.
AVG. THRUST:	137.00 NTS.
MAX. THRUST:	55.49 LBS.
BURN TIME:	1.66 SECS.
SPEC. IMPULSE:	79.04 SECS.

416

MOTOR NAME: NG6-227-H137	NOZZLE TYPE: RAMMED CLAY	NOZZLE THROAT: 0.625"
TEST FILENAME: 970606E	CASING ID: 1.125"	CORE LENGTH: 11.00"
PROPELLANT: NG6	CASING OD: 1.50"	MOTOR WEIGHT: 498.3 GMS.
BURN MODIFIER: 7% NaHCO3	MOTOR LENGTH: 14.50"	PROPELLANT WT.: 293.6 GMS.

[Thrust curve: THRUST IN POUNDS vs TIME IN SECONDS, peaks around 55 lbs near 1.1 sec]

TOT. IMP.: 227.97 NT.-SECS.		MAX. THRUST: 55.49 LBS.
SPEC. IMPULSE: 79.04 SECS.	BURN TIME: 1.66 SECS.	AVG. THRUST: 137.00 NTS.

NG6-227-H137 ESTIMATED PERFORMANCE

MINIMUM ROCKET

BODY LENGTH: 28"
BODY DIAMETER: 1.50"
LIFTOFF WT.: 23.6 OZ. (1.48 LBS.)
MAX. ALTITUDE: 5,909 FT.
MAX. VELOCITY: 1,189 FT./SEC. (811 MPH)
MAX. MACH NUMBER: 1.06 (SUPERSONIC)
MAX. ACCELERATION: 48.3 Gs
BURNOUT ALTITUDE: 988 FT.
COAST TIME TO APOGEE: 14.5 SECS.

AVERAGE ROCKET

BODY LENGTH: 48"
BODY DIAMETER: 3.10"
LIFTOFF WT.: 41.6 OZ. (2.60 LBS.)
MAX. ALTITUDE: 2,610 FT.
MAX. VELOCITY: 602 FT./SEC. (410 MPH)
MAX. MACH NUMBER: 0.53
MAX. ACCELERATION: 22.6 Gs
BURNOUT ALTITUDE: 504 FT.
COAST TIME TO APOGEE: 10.0 SECS.

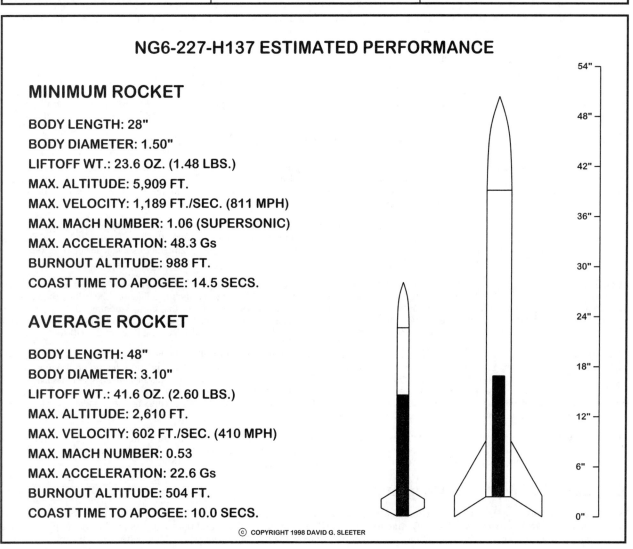

© COPYRIGHT 1998 DAVID G. SLEETER

15. NV6 PROPELLANT AND MOTOR DESIGNS

● First and Most Important!

Before you proceed, if you haven't read **Chapter 1**, go back and read it now. It contains important information on the safe handling of this rocket propellant. Also pay attention to the bold warning at the beginning of **Chapter 1**. To emphasize its importance, I'll repeat that warning here.

> Do *not* add or substitute other chemicals either to-or-for the chemicals described in this book. Specifically do *not* use POTASSIUM CHLORATE, SODIUM CHLORATE, POTASSIUM PERCHLORATE, SODIUM PERCHLORATE, POTASSIUM PERMANGANATE, AMMONIUM PERCHLORATE, PHOSPHOROUS, MATCH HEADS, ALUMINUM, MAGNESIUM, or ANY METALS AT ALL! *All* of these substances will form dangerous and explosive mixtures when combined with the other ingredients described herein, *AND OTHER CHEMICALS WILL FORM DANGEROUS MIXTURES AS WELL.* When mixing and handling propellants, be sure to ground yourself and all containers and utensils according to the instructions on page 14.
>
> *If you decide to experiment on your own outside the bounds of the instructions provided in this book, consult with a professional chemist before you proceed.*

The **NV6** propellant is a milled mixture of **4 hour milled sodium nitrate**, **carbon black**, **vinsol resin**, and **sulfur**. If you haven't read **Chapters 5** and **6**, go back and read them now. Together they explain how to buy or make these chemicals, and how to prepare them for use in a rocket propellant.

If all other parameters are equal, then the **NV6** is the slower burning of the two sodium nitrate-based propellants. But its burn rate can vary *greatly* depending on the exact nature and purity of its chemicals. The quality of the carbon black is particularly important. A carbon black high in residual organics (i.e. oil) makes a slow burning propellant, and a pure, electrically conductive carbon black makes a fast burning propellant. The *exact* burn rate can be adjusted by substituting slower burning anthracite or charcoal for some of the fast burning carbon black, or by adding or subtracting baking soda. When compressed into a solid propellant grain, the **NV6**'s density is **25 to 30 gms. per cubic inch**. To prepare the **NV6** propellant, set up a powder mill (**Chapter 3**). Then mix the propellant (**Chapter 9**), and use the figures below to determine the exact weights of the chemicals needed for the desired batch size.

◆ THE NV6 PROPELLANT

4 hr. milled sodium nitrate	68.9%	689.0 gms.	344.5 gms.	172.3 gms.	86.1 gms.
Carbon black	13.2%	132.0 gms.	66.0 gms.	33.0 gms.	16.5 gms.
Vinsol resin	7.0%	70.0 gms.	35.0 gms.	17.5 gms.	8.8 gms.
Sulfur	10.9%	109.0 gms.	54.5 gms.	27.2 gms.	13.6 gms.
Total	100%	1000.0 gms.	500.0 gms.	250.0 gms.	125.0 gms.

Before you can use the propellant, *you must perform the pellet test* (**pages 238 - 245**). The pellet test determines the exact amount of solvent needed to form 20 grams of propellant into a dense and solid propellant grain. The correct amount of solvent is called **the solvent requirement**. *The solvent requirement is critical, and cannot be guessed at.* The pellet test is time consuming, *but you only do it twice.* Perform the *first* pellet test with the **NV6** propellant *plus 30% baking soda*, and do it *before* you make your first motor.

To build your *first* motor with the **NV6** propellant, weigh out the amount indicated in the motor drawing (rounding off the numbers makes the math easier). Then add 30% of that weight in baking soda, and mix it in thoroughly with a fork. If, for example, you've weighed out 100 grams of propellant, add 30 grams of baking soda for a total of 130 grams. Mix in the proper amount of solvent as determined by your first pellet test. Then build and dry the motor (**Chapter 10**).

Per the instructions on **page 494**, do a rudimentary test by firing the motor (with the motor pointing *down* and the exhaust aimed *upward*) from a pipe in the ground. If this first motor "chuffs" (chugs like a steam engine), build a *second* motor, and reduce the baking soda to 28%. Keep building, testing, and reducing the amount of soda in 2% increments until you get a motor that burns smoothly and *doesn't* chuff. At first the chuffing may be slow. As you approach a condition of combustion stability, the frequency of the chuffs will increase (the chuffing sound will get *faster*). When you are *very* near stability, the amplitude of the chuffs will decrease (the chuffs will sound *fainter*).

Now set up a thrust-testing stand (**page 470**), and test the motor for maximum thrust (**page 475**). In the tests that follow, reduce the soda in 1% increments until you get a motor whose maximum thrust approximates the maximum thrust printed in the motor drawing. At that point you'll have a working motor with the maximum safe chamber pressure and thrust for that particular design. As you approach a final, successful soda content, *repeat the pellet test*, and use the solvent requirement indicated by this *second* test to zero-in the *final* design. Keep accurate notes. Write down the final solvent requirement and the amount of baking soda that works the best, and *forever thereafter* use those figures to build copies of your final, successful motor.

Important note. You can also alter the chamber pressure and thrust by changing the length of the motor's core. A *longer* core will *increase* the chamber pressure and thrust, and a *shorter* core will *reduce* the chamber pressure and thrust. But, of course, changing the length of the core will also change the amount of propellant required, and the physical size of the motor.

● Improving Motor Performance

Now you have a working rocket motor, but some of the material inside the motor isn't rocket propellant. It's *baking soda*, and baking soda is *dead weight*. It doesn't contribute anything to the motor's total impulse. You added the soda to control chamber pressure and thrust, but you can also control those parameters by replacing some of the fast burning carbon black with slower burning charcoal or anthracite (which I call "other carbon" in the text below). And unlike the baking soda, this "other carbon" is a *fuel*. It takes part in the reaction, and *contributes* to the motor's total impulse.

In the following hypothetical illustration, I'll call the original motor an **NV6-61-F20**. I'll stipulate that its maximum thrust is 10 lbs., and I'll stipulate that it takes 12% baking soda to make it work as intended. To *improve* the motor's performance, I'll have to get rid of some of the soda. Of course when I do that, the propellant's burn rate will increase. But I can *reverse* the increase, and *compensate* for the loss of the soda by replacing some of the carbon black with slower burning "other carbon". The amounts of sodium nitrate, vinsol resin, and sulfur will remain at 68.9%, 7.0%, and 10.9% respectively, but the carbon fuel ratios will change. **Important note**. *Each time you change a propellant's carbon fuel ratio, you must repeat* **the pellet test**. Possible carbon replacement ratios are:

NV6X1	NV6X2	NV6X3	NV6X4
Carbon black 12.2%	Carbon black 11.2%	Carbon black 10.2%	Carbon black 9.2%
Other carbon 1.0%	Other carbon 2.0%	Other carbon 3.0%	Other carbon 4.0%
NV6X5	NV6X6	NV6X7	NV6X8
Carbon black 8.2%	Carbon black 7.2%	Carbon black 6.2%	Carbon black 5.2%
Other carbon 5.0%	Other carbon 6.0%	Other carbon 7.0%	Other carbon 8.0%
NV6X9	NV6X10	NV6X11	NV6X12
Carbon black 4.2%	Carbon black 3.2%	Carbon black 2.2%	Carbon black 1.2%
Other carbon 9.0%	Other carbon 10.0%	Other carbon 11.0%	Other carbon 12.0%

In the ratios above, though the balance between the carbon black and the "other carbon" changes (in 1% increments), the total amount of carbon fuel remains the same (13.2%). If I build a motor with the NV6X3 ratio, and add 12% baking soda, because the "other carbon" burns *slower* than the carbon black it replaced, the motor's maximum thrust will be *less* than 10 pounds. To restore the maximum thrust to the desired level, I'll keep building, testing, and reducing the baking soda in 1% increments until the maximum thrust is back up to 10 lbs. I'll call this second motor an **NV6X3** (named after its carbon fuel ratio), and I'll pretend here that it took 10% soda to make it work.

If the **NV6X3** with 10% soda burns smoothly and doesn't chuff, I'll build an identical motor with the NV6X4 ratio. Of course when I do that, the maximum thrust will drop again, so I'll keep building, testing, and *again* reducing the soda in 1% increments until the maximum thrust equals 10 lbs. I'll call this third motor an **NV6X4**, and for the purpose of this illustration, I'll pretend that it took 8% baking soda to make it work. If the **NV6X4** burns smoothly and doesn't chuff, I might repeat the experiment with the NV6X5 ratio. If the **NV6X4** chuffs and *doesn't* burn smoothly, I'll know that the NV6X4 ratio *doesn't* work; that I need *more* than 9.2% carbon black to achieve combustion stability, and that the NV6X3 ratio with 10% baking soda works the best.

If I now build a chart recorder (**page 478**), and generate thrust-time curves for the **NV6-61-F20** and **NV6X3** motors, I'll see that the **NV6X3** has a slightly longer burn time, and a greater total impulse. The **NV6X3** might turn out to be an **NV6-*70*-F20**, and I'll adopt the NV6X3 ratio with 10% baking soda for all the motors that follow. When you perform these experiments on your own, you'll eventually reach a point where the motors begin to chuff, and the process of finding the minimum amount of carbon black needed to maintain combustion stability will be one of trial and error.

Important note. *The NV6 motors in this book were all made with propellants milled for 12 hours*. When making your own propellants, you can reduce the milling time to as little as 4 hours, and compensate by adjusting the amount of baking soda. The process of finding the minimum milling time is one of trial and error, and each time you change the milling time, *you must repeat* **the pellet test**.

The **NV6** propellant exhibits excellent case bonding properties with no evidence of the effect discussed on **pages 248 - 249**. All the **NV6** motors were made with standard fireworks grade casings and clay nozzles. By switching to stainless steel De Laval nozzles and the high strength casings manufactured (as of this writing) by **New England Paper Tube** (**pages 163 - 164**), you can increase the motors' maximum chamber pressure and thrust, and substantially improve their performance. Shop drawings for De Laval nozzles begin on the **page 323**.

With each motor design, I've included the thrust-time curve and motor data generated by the test equipment and software described on **pages 468 - 469**, plus two flight performance estimates generated by **Rogers' Aeroscience Alt4** rocket performance prediction software. The "minimum rocket" is what you might build if you were trying to break an altitude or speed record. The lift-off weight is minimal. The body diameter is equal to the o.d. of the motor, and the parachute compartment is small; approx. 5 body diameter's in length. The "average rocket" is more like what you would build for scientific research or general sport flying. The body diameter is one of the industry standards approx. twice the diameter of the motor, and there's ample space and weight allowance for a large parachute and a small scientific payload.

In each case the lift-off weight is *very* approximate, and based on an educated guess. The *extremely* accurate performance specs. *are the actual numbers generated by the Alt-4 program*. They are based on the rocket's stated length, diameter, and lift-off weight, and the *tested* motor's thrust-time data. *In the real world*, small variations in motor performance, the size and shape of the rocket's fins, and even the smoothness of its paint-job will cause the rocket's performance to vary from the stated figures.

● Important G-Force Advisory

Static testing perfects a motor's performance, but it *doesn't* predict how the hardened propellant grain will respond to high G-forces during an actual rocket flight. Baking soda is *soft*, and the carbon fuels used in these propellants vary *greatly* in hardness. So variations in the soda content and the type of carbon fuel will effect the strength of the finished propellant grain. If a soft propellant grain collapses at maximum acceleration, the motor will self-destruct, and the rocket will be lost in flight. Initial flight testing of a new motor design should *always* be conducted with this possibility in mind, particularly when testing the motor in a small, high-acceleration airframe. It is always advisable to conduct your early tests in a "throwaway" rocket, i.e. something cheap that you don't mind losing. When you're sure that you've got a successful motor-airframe combination, *then* invest in something nicer.

TELEFLITE CORPORATION

MOTOR NAME:	NV6-15-D8
PROPELLANT TYPE:	NV6
CASING TYPE:	CONVOLUTE PAPER
NOZZLE TYPE:	RAMMED CLAY

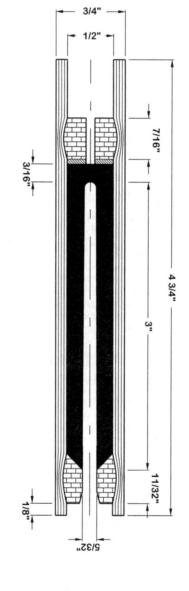

MOTOR SHOWN WITHOUT TIME DELAY

INDEX OF REQUIRED TOOLING

CASING RETAINER	DRAWING #98-001: 7"	PAGE 90
NOZZLE MOLD	DRAWING #98-006	PAGE 95
CORE SPINDLE	DRAWING #98-036-I	PAGE 122
STARTING TAMP	USE PROPELLANT TAMP	
PROPELLANT TAMP	DRAWING #98-082-A	PAGE 145
FLAT TAMP	DRAWING #98-084-A	PAGE 147
PORTHOLE TAMP	DRAWING #98-084-A	PAGE 147
PORTHOLE WIRE	DRAWING #98-084	PAGE 147

IMPORTANT BAKING SODA NOTE: Add 30% baking soda to the propellant in your first motor. Reduce the soda in 2% increments until you get a motor that doesn't chuff. Then reduce the soda in 1% increments until maximum thrust is approx. 3-1/2 lbs. This test motor took 5% soda, but I made it with a carbon black of medium purity. A motor made with a pure, electrically-conductive carbon black might take substantially more.

TOTAL WEIGHT: 45.6 GMS.	CASING O.D.: 0.75"
PROPELLANT WT.: 22.7 GMS.	NOZZLE CLAY: 3.3 GMS.
BAKING SODA: See baking soda note.	BULKHEAD CLAY: 3.3 GMS.
MOTOR LENGTH: 4.75"	THROAT DIA.: 0.1563"
CASING I.D.: 0.50"	CORE LENGTH: 3.00"

TOT. IMPULSE: 15.94 NT.-SECS.	
AVG. THRUST: 8.00 NTS.	
MAX. THRUST: 3.89 LBS.	
BURN TIME: 1.99 SECS.	
SPEC. IMPULSE: 71.48 SECS.	

© COPYRIGHT 1998 DAVID G. SLEETER

MOTOR NAME: NV6-15-D8	NOZZLE TYPE: RAMMED CLAY	NOZZLE THROAT: 0.1563"
TEST FILENAME: 970614B	CASING ID: 0.50"	CORE LENGTH: 3.00"
PROPELLANT: NG6	CASING OD: 0.75"	MOTOR WEIGHT: 45.6 GMS.
BURN MODIFIER: 5% NaHCO3	MOTOR LENGTH: 4.75"	PROPELLANT WT.: 22.7 GMS.

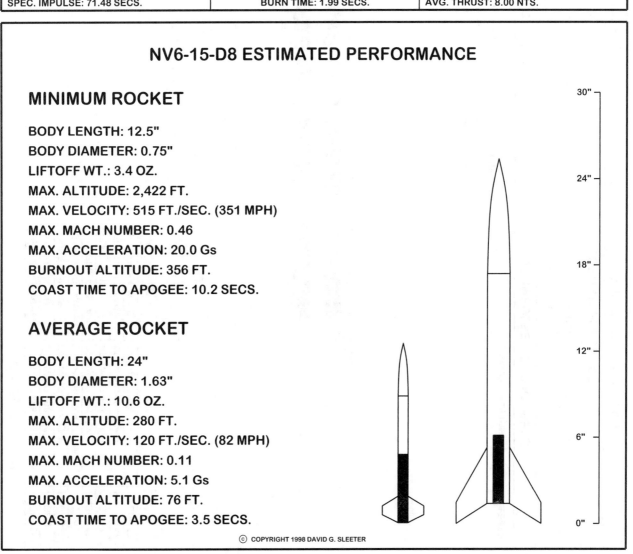

TOT. IMP.: 15.94 NT.-SECS.		MAX. THRUST: 3.89 LBS.
SPEC. IMPULSE: 71.48 SECS.	BURN TIME: 1.99 SECS.	AVG. THRUST: 8.00 NTS.

NV6-15-D8 ESTIMATED PERFORMANCE

MINIMUM ROCKET

BODY LENGTH: 12.5"
BODY DIAMETER: 0.75"
LIFTOFF WT.: 3.4 OZ.
MAX. ALTITUDE: 2,422 FT.
MAX. VELOCITY: 515 FT./SEC. (351 MPH)
MAX. MACH NUMBER: 0.46
MAX. ACCELERATION: 20.0 Gs
BURNOUT ALTITUDE: 356 FT.
COAST TIME TO APOGEE: 10.2 SECS.

AVERAGE ROCKET

BODY LENGTH: 24"
BODY DIAMETER: 1.63"
LIFTOFF WT.: 10.6 OZ.
MAX. ALTITUDE: 280 FT.
MAX. VELOCITY: 120 FT./SEC. (82 MPH)
MAX. MACH NUMBER: 0.11
MAX. ACCELERATION: 5.1 Gs
BURNOUT ALTITUDE: 76 FT.
COAST TIME TO APOGEE: 3.5 SECS.

© COPYRIGHT 1998 DAVID G. SLEETER

TELEFLITE CORPORATION

MOTOR NAME:	NV6-17-D8
PROPELLANT TYPE:	NV6
CASING TYPE:	CONVOLUTE PAPER
NOZZLE TYPE:	RAMMED CLAY

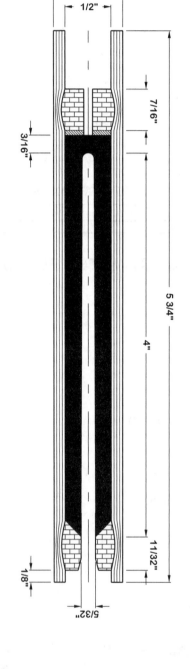

MOTOR SHOWN WITHOUT TIME DELAY

INDEX OF REQUIRED TOOLING

CASING RETAINER	DRAWING #98-001: 9"
NOZZLE MOLD	DRAWING #98-006
CORE SPINDLE	DRAWING #98-037-M
STARTING TAMP	USE PROPELLANT TAMP
PROPELLANT TAMP	DRAWING #98-082-B
FLAT TAMP	DRAWING #98-084-B
PORTHOLE TAMP	DRAWING #98-084-A
PORTHOLE WIRE	DRAWING #98-084
	PAGE 90
	PAGE 95
	PAGE 122
	PAGE 145
	PAGE 147
	PAGE 147
	PAGE 147

IMPORTANT BAKING SODA NOTE: Add 30% baking soda to the propellant in your first motor. Reduce the soda in 2% increments until you get a motor that doesn't chuff. Then reduce the soda in 1% increments until maximum thrust is approx. 3-1/2 lbs. This test motor took 12% soda, but I made it with a carbon black of medium purity. A motor made with a pure, electrically-conductive carbon black might take substantially more.

© COPYRIGHT 1998 DAVID G. SLEETER

TOTAL WEIGHT:	53.3 GMS.
PROPELLANT WT.:	27.8 GMS.
BAKING SODA:	See baking soda note.
MOTOR LENGTH:	5.75"
CASING I.D.:	0.50"

CASING O.D.:	0.75"
NOZZLE CLAY:	3.3 GMS.
BULKHEAD CLAY:	3.3 GMS.
THROAT DIA.:	0.1563"
CORE LENGTH:	4.00"

TOT. IMPULSE:	17.55 NT.-SECS.
AVG. THRUST:	8.40 NTS.
MAX. THRUST:	3.50 LBS.
BURN TIME:	2.09 SECS.
SPEC. IMPULSE:	64.26 SECS.

MOTOR NAME: NV6-17-D8	NOZZLE TYPE: RAMMED CLAY	NOZZLE THROAT: 0.1563"
TEST FILENAME: 970614C	CASING ID: 0.50"	CORE LENGTH: 4.00"
PROPELLANT: NV6	CASING OD: 0.75"	MOTOR WEIGHT: 53.3 GMS.
BURN MODIFIER: 12% NaHCO3	MOTOR LENGTH: 5.75"	PROPELLANT WT.: 27.8 GMS.

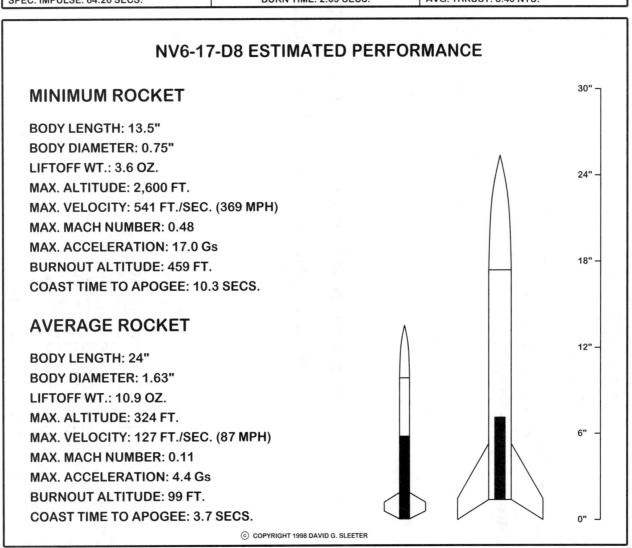

TOT. IMP.: 17.55 NT.-SECS.		MAX. THRUST: 3.50 LBS.
SPEC. IMPULSE: 64.26 SECS.	BURN TIME: 2.09 SECS.	AVG. THRUST: 8.40 NTS.

NV6-17-D8 ESTIMATED PERFORMANCE

MINIMUM ROCKET

BODY LENGTH: 13.5"
BODY DIAMETER: 0.75"
LIFTOFF WT.: 3.6 OZ.
MAX. ALTITUDE: 2,600 FT.
MAX. VELOCITY: 541 FT./SEC. (369 MPH)
MAX. MACH NUMBER: 0.48
MAX. ACCELERATION: 17.0 Gs
BURNOUT ALTITUDE: 459 FT.
COAST TIME TO APOGEE: 10.3 SECS.

AVERAGE ROCKET

BODY LENGTH: 24"
BODY DIAMETER: 1.63"
LIFTOFF WT.: 10.9 OZ.
MAX. ALTITUDE: 324 FT.
MAX. VELOCITY: 127 FT./SEC. (87 MPH)
MAX. MACH NUMBER: 0.11
MAX. ACCELERATION: 4.4 Gs
BURNOUT ALTITUDE: 99 FT.
COAST TIME TO APOGEE: 3.7 SECS.

© COPYRIGHT 1998 DAVID G. SLEETER

TELEFLITE CORPORATION

MOTOR NAME:	NV6-27-E21
PROPELLANT TYPE:	NV6
CASING TYPE:	CONVOLUTE PAPER
NOZZLE TYPE:	RAMMED CLAY

MOTOR SHOWN WITHOUT TIME DELAY

Dimensions shown on drawing: 3/4", 1/2", 5/32", 7/16", 8 1/2", 6 3/4", 15/32", 1/8", 1/4"

INDEX OF REQUIRED TOOLING

CASING RETAINER	DRAWING #98-001: 11"	PAGE 90
NOZZLE MOLD	DRAWING #98-007	PAGE 96
CORE SPINDLE	DRAWING #98-042-R	PAGE 125
STARTING TAMP	USE PROPELLANT TAMP	
PROPELLANT TAMP	DRAWING #98-083-C	PAGE 146
FLAT TAMP	DRAWING #98-084-A	PAGE 147
PORTHOLE TAMP	DRAWING #98-084-B	PAGE 147
PORTHOLE WIRE	DRAWING #98-084	PAGE 147

IMPORTANT BAKING SODA NOTE: Add 30% baking soda to the propellant in your first motor. Reduce the soda in 2% increments until you get a motor that doesn't chuff. Then reduce the soda in 1% increments until maximum thrust is approx. 10 lbs. This test motor took 5% soda, but I made it with a carbon black of medium purity. A motor made with a pure, electrically-conductive carbon black might take substantially more.

TOTAL WEIGHT: 72.9 GMS.	CASING O.D.: 0.75"	TOT. IMPULSE: 27.85 NT.-SECS.
PROPELLANT WT.: 38.7 GMS.	NOZZLE CLAY: 3.3 GMS.	AVG. THRUST: 21.39 NTS.
BAKING SODA: See baking soda note.	BULKHEAD CLAY: 3.3 GMS.	MAX. THRUST: 9.68 LBS.
MOTOR LENGTH: 8.50"	THROAT DIA.: 0.25"	BURN TIME: 1.30 SECS.
CASING I.D.: 0.50"	CORE LENGTH: 6.75"	SPEC. IMPULSE: 73.26 SECS.

© COPYRIGHT 1998 DAVID G. SLEETER

MOTOR NAME: NV6-27-E21	NOZZLE TYPE: RAMMED CLAY	NOZZLE THROAT: 0.25"
TEST FILENAME: 970619B	CASING ID: 0.50"	CORE LENGTH: 6.75"
PROPELLANT: NV6	CASING OD: 0.75"	MOTOR WEIGHT: 72.9 GMS.
BURN MODIFIER: 5% NaHCO3	MOTOR LENGTH: 8.50"	PROPELLANT WT.: 38.7 GMS.

TOT. IMP.: 27.85 NT.-SECS.		MAX. THRUST: 9.68 LBS.
SPEC. IMPULSE: 73.26 SECS.	BURN TIME: 1.30 SECS.	AVG. THRUST: 21.39 NTS.

NV6-27-E21 ESTIMATED PERFORMANCE

MINIMUM ROCKET

BODY LENGTH: 16"
BODY DIAMETER: 0.75"
LIFTOFF WT.: 4.3 OZ.
MAX. ALTITUDE: 3,578 FT.
MAX. VELOCITY: 772 FT./SEC. (527 MPH)
MAX. MACH NUMBER: 0.69
MAX. ACCELERATION: 41.5 Gs
BURNOUT ALTITUDE: 513 FT.
COAST TIME TO APOGEE: 11.8 SECS.

AVERAGE ROCKET

BODY LENGTH: 24"
BODY DIAMETER: 1.63"
LIFTOFF WT.: 11.6 OZ.
MAX. ALTITUDE: 871 FT.
MAX. VELOCITY: 244 FT./SEC. (167 MPH)
MAX. MACH NUMBER: 0.22
MAX. ACCELERATION: 13.1 Gs
BURNOUT ALTITUDE: 162 FT.
COAST TIME TO APOGEE: 6.4 SECS.

© COPYRIGHT 1998 DAVID G. SLEETER

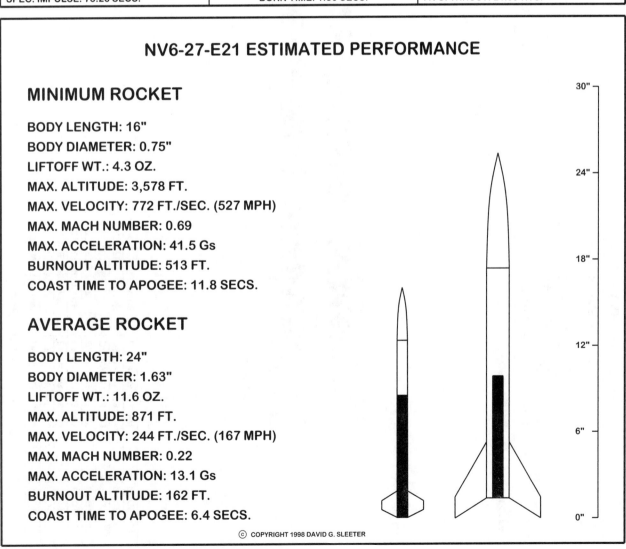

TELEFLITE CORPORATION

MOTOR NAME:	NV6-49-F20
PROPELLANT TYPE:	NV6
CASING TYPE:	CONVOLUTE PAPER
NOZZLE TYPE:	RAMMED CLAY

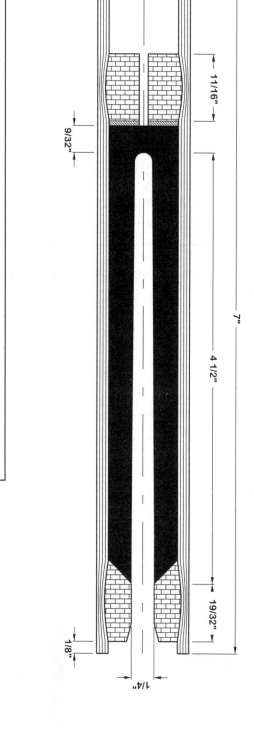

MOTOR SHOWN WITHOUT TIME DELAY

INDEX OF REQUIRED TOOLING

CASING RETAINER	DRAWING #98-002: 9"	PAGE 91
NOZZLE MOLD	DRAWING #98-008	PAGE 97
CORE SPINDLE	DRAWING #98-047-O	PAGE 127
STARTING TAMP	DRAWING #98-085-B	PAGE 148
PROPELLANT TAMP	DRAWING #98-086-B	PAGE 149
FLAT TAMP	DRAWING #98-091-B	PAGE 154
PORTHOLE TAMP	DRAWING #98-091-A	PAGE 154
PORTHOLE WIRE	DRAWING #98-091	PAGE 154

IMPORTANT BAKING SODA NOTE: Add 30% baking soda to the propellant in your first motor. Reduce the soda in 1% increments until you get a motor that doesn't chuff. Then reduce the soda in 2% increments until maximum thrust is approx. 10 lbs. This test motor took 5% soda, but I made it with a carbon black of medium purity. A motor made with a pure, electrically-conductive carbon black might take substantially more.

© COPYRIGHT 1998 DAVID G. SLEETER

TOTAL WEIGHT: 117.8 GMS.	CASING O.D.: 1.00"	TOT. IMPULSE: 49.97 NT.-SECS.	
PROPELLANT WT.: 67.9 GMS.	NOZZLE CLAY: 11.0 GMS.	AVG. THRUST: 20.93 NTS.	
BAKING SODA: See baking soda note.	BULKHEAD CLAY: 11.0 GMS.	MAX. THRUST: 9.74 LBS.	
MOTOR LENGTH: 7.00"	THROAT DIA.: 0.25"	BURN TIME: 2.39 SECS.	
CASING I.D.: 0.75"	CORE LENGTH: 4.50"	SPEC. IMPULSE: 74.92 SECS.	

MOTOR NAME: NV6-49-F20	NOZZLE TYPE: RAMMED CLAY	NOZZLE THROAT: 0.25"
TEST FILENAME: 970611B	CASING ID: 0.75"	CORE LENGTH: 4.50"
PROPELLANT: NV6	CASING OD: 1.00"	MOTOR WEIGHT: 117.8 GMS.
BURN MODIFIER: 5% NaHCO3	MOTOR LENGTH: 7.00"	PROPELLANT WT.: 67.9 GMS.

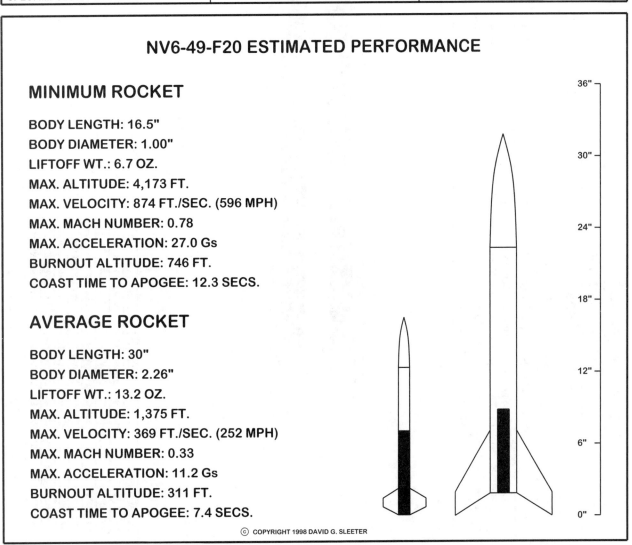

TOT. IMP.: 49.97 NT.-SECS.		MAX. THRUST: 9.74 LBS.
SPEC. IMPULSE: 74.92 SECS.	BURN TIME: 2.39 SECS.	AVG. THRUST: 20.93 NTS.

NV6-49-F20 ESTIMATED PERFORMANCE

MINIMUM ROCKET

BODY LENGTH: 16.5"
BODY DIAMETER: 1.00"
LIFTOFF WT.: 6.7 OZ.
MAX. ALTITUDE: 4,173 FT.
MAX. VELOCITY: 874 FT./SEC. (596 MPH)
MAX. MACH NUMBER: 0.78
MAX. ACCELERATION: 27.0 Gs
BURNOUT ALTITUDE: 746 FT.
COAST TIME TO APOGEE: 12.3 SECS.

AVERAGE ROCKET

BODY LENGTH: 30"
BODY DIAMETER: 2.26"
LIFTOFF WT.: 13.2 OZ.
MAX. ALTITUDE: 1,375 FT.
MAX. VELOCITY: 369 FT./SEC. (252 MPH)
MAX. MACH NUMBER: 0.33
MAX. ACCELERATION: 11.2 Gs
BURNOUT ALTITUDE: 311 FT.
COAST TIME TO APOGEE: 7.4 SECS.

© COPYRIGHT 1998 DAVID G. SLEETER

TELEFLITE CORPORATION

MOTOR NAME:	NV6-61-F20
PROPELLANT TYPE:	NV6
CASING TYPE:	CONVOLUTE PAPER
NOZZLE TYPE:	RAMMED CLAY

INDEX OF REQUIRED TOOLING

CASING RETAINER	DRAWING #98-002: 11"	PAGE 91
NOZZLE MOLD	DRAWING #98-008	PAGE 97
CORE SPINDLE	DRAWING #98-048-U	PAGE 128
STARTING TAMP	DRAWING #98-085-C	PAGE 148
PROPELLANT TAMP	DRAWING #98-086-C	PAGE 149
FLAT TAMP	DRAWING #98-091-B	PAGE 154
PORTHOLE TAMP	DRAWING #98-091-A	PAGE 154
PORTHOLE WIRE	DRAWING #98-091	PAGE 154

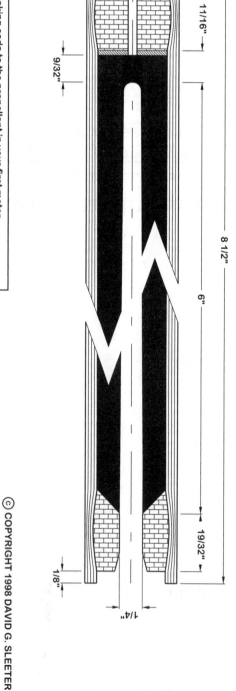

MOTOR SHOWN WITHOUT TIME DELAY

© COPYRIGHT 1998 DAVID G. SLEETER

EXTENDED BURN TIME WARNING!
This motor's burn time is longer than 3 seconds. Do not fly this motor on a stick. Fly it only in a properly balanced, aerodynamic rocket body. Use a tall launch tower, and fly the rocket only in calm weather.

IMPORTANT BAKING SODA NOTE: Add 30% baking soda to the propellant in your first motor. Reduce the soda in 2% increments until you get a motor that doesn't chuff. Then reduce the soda in 1% increments until maximum thrust is approx. 10 lbs. This test motor took 12% soda, but I made it with a carbon black of medium purity. A motor made with pure, electrically conductive carbon black might take substantially more.

TOTAL WEIGHT:	142.7 GMS.
PROPELLANT WT.:	86.2 GMS.
BAKING SODA:	See baking soda note.
MOTOR LENGTH:	8.50 "
CASING I.D.:	0.75 "

CASING O.D.:	1.00 "
NOZZLE CLAY:	11.0 GMS.
BULKHEAD CLAY:	11.0 GMS.
THROAT DIA.:	0.25 "
CORE LENGTH:	6.00 "

TOT. IMPULSE:	61.57 NT.-SECS.
AVG. THRUST:	20.08 NTS.
MAX. THRUST:	9.64 LBS.
BURN TIME:	3.07 SECS.
SPEC. IMPULSE:	72.71 SECS.

MOTOR NAME: NV6-61-F20	NOZZLE TYPE: RAMMED CLAY	NOZZLE THROAT: 0.25"
TEST FILENAME: 970612A	CASING ID: 0.75"	CORE LENGTH: 6.00"
PROPELLANT: NV6	CASING OD: 1.00"	MOTOR WEIGHT: 142.7 GMS.
BURN MODIFIER: 12% NaHCO3	MOTOR LENGTH: 8.50"	PROPELLANT WT.: 86.2 GMS.

TOT. IMP.: 61.57 NT.-SECS.		MAX. THRUST: 9.64 LBS.
SPEC. IMPULSE: 72.71 SECS.	BURN TIME: 3.07 SECS.	AVG. THRUST: 20.08 NTS.

NV6-61-F20 ESTIMATED PERFORMANCE

MINIMUM ROCKET

BODY LENGTH: 18"
BODY DIAMETER: 1.00"
LIFTOFF WT.: 7.5 OZ.
MAX. ALTITUDE: 4,630 FT.
MAX. VELOCITY: 942 FT./SEC. (642 MPH)
MAX. MACH NUMBER: 0.84
MAX. ACCELERATION: 24.2 Gs
BURNOUT ALTITUDE: 1,042 FT.
COAST TIME TO APOGEE: 12.4 SECS.

AVERAGE ROCKET

BODY LENGTH: 30"
BODY DIAMETER: 2.26"
LIFTOFF WT.: 14.0 OZ.
MAX. ALTITUDE: 1,656 FT.
MAX. VELOCITY: 413 FT./SEC. (281 MPH)
MAX. MACH NUMBER: 0.37
MAX. ACCELERATION: 10.3 Gs
BURNOUT ALTITUDE: 451 FT.
COAST TIME TO APOGEE: 7.7 SECS.

© COPYRIGHT 1998 DAVID G. SLEETER

TELEFLITE CORPORATION

MOTOR NAME:	NV6-83-G44
PROPELLANT TYPE:	NV6
CASING TYPE:	CONVOLUTE PAPER
NOZZLE TYPE:	RAMMED CLAY

MOTOR SHOWN WITHOUT TIME DELAY

IMPORTANT BAKING SODA NOTE: Add 30% baking soda to the propellant in your first motor. Reduce the soda until you get a motor that doesn't chuff. Then reduce the soda in 1% increments until maximum thrust is approx. 20 lbs. This test motor took 5% soda, but I made it with a carbon black of medium purity. A motor made with a pure, electrically-conductive carbon black might take substantially more.

TOTAL WEIGHT:	179.7 GMS.
PROPELLANT WT.:	112.3 GMS.
BAKING SODA:	See baking soda note.
MOTOR LENGTH:	11.75"
CASING I.D.:	0.75"
CASING O.D.:	1.00"
NOZZLE CLAY:	11.0 GMS.
BULKHEAD CLAY:	11.0 GMS.
THROAT DIA.:	0.375"
CORE LENGTH:	9.00"
TOT. IMPULSE:	83.76 NT.-SECS.
AVG. THRUST:	44.01 NTS.
MAX. THRUST:	19.63 LBS.
BURN TIME:	1.90 SECS.
SPEC. IMPULSE:	75.93 SECS.

© COPYRIGHT 1998 DAVID G. SLEETER

INDEX OF REQUIRED TOOLING

CASING RETAINER	DRAWING #98-002: 13"
NOZZLE MOLD	DRAWING #98-010
CORE SPINDLE	DRAWING #98-053-U
STARTING TAMP	DRAWING #98-088-D
PROPELLANT TAMP	DRAWING #98-090-D
FLAT TAMP	DRAWING #98-091-B
PORTHOLE TAMP	DRAWING #98-091-A
PORTHOLE WIRE	DRAWING #98-091

	PAGE 91
	PAGE 99
	PAGE 130
	PAGE 151
	PAGE 153
	PAGE 154
	PAGE 154
	PAGE 154

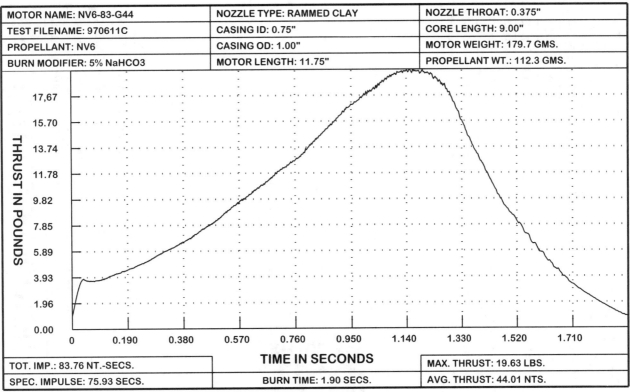

MOTOR NAME: NV6-83-G44	NOZZLE TYPE: RAMMED CLAY	NOZZLE THROAT: 0.375"
TEST FILENAME: 970611C	CASING ID: 0.75"	CORE LENGTH: 9.00"
PROPELLANT: NV6	CASING OD: 1.00"	MOTOR WEIGHT: 179.7 GMS.
BURN MODIFIER: 5% NaHCO3	MOTOR LENGTH: 11.75"	PROPELLANT WT.: 112.3 GMS.
TOT. IMP.: 83.76 NT.-SECS.		MAX. THRUST: 19.63 LBS.
SPEC. IMPULSE: 75.93 SECS.	BURN TIME: 1.90 SECS.	AVG. THRUST: 44.01 NTS.

NV6-83-G44 ESTIMATED PERFORMANCE

MINIMUM ROCKET

BODY LENGTH: 21"
BODY DIAMETER: 1.00"
LIFTOFF WT.: 8.8 OZ.
MAX. ALTITUDE: 4,984 FT.
MAX. VELOCITY: 1,126 FT./SEC. (768 MPH)
MAX. MACH NUMBER: 1.0 (SONIC)
MAX. ACCELERATION: 44.2 Gs
BURNOUT ALTITUDE: 1,061 FT.
COAST TIME TO APOGEE: 12.7 SECS.

AVERAGE ROCKET

BODY LENGTH: 30"
BODY DIAMETER: 2.26"
LIFTOFF WT.: 18.3 OZ. (1.14 LBS.)
MAX. ALTITUDE: 2,048 FT.
MAX. VELOCITY: 479 FT./SEC. (327 MPH)
MAX. MACH NUMBER: 0.43
MAX. ACCELERATION: 17.7 Gs
BURNOUT ALTITUDE: 448 FT.
COAST TIME TO APOGEE: 8.9 SECS.

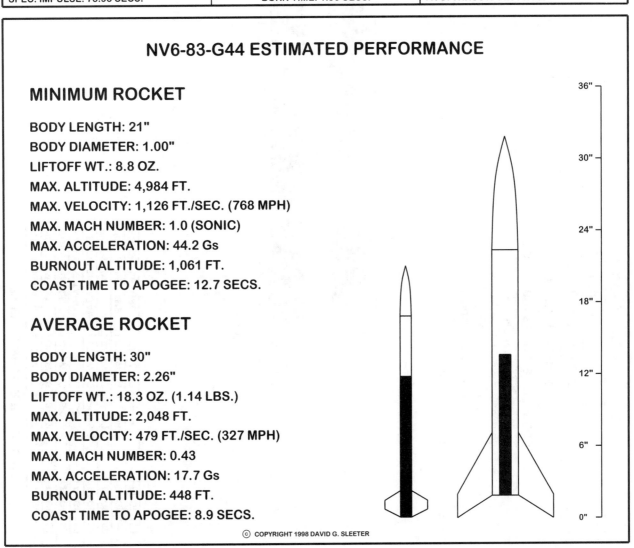

© COPYRIGHT 1998 DAVID G. SLEETER

TELEFLITE CORPORATION

MOTOR NAME: NV6-102-G32
PROPELLANT TYPE: NV6
CASING TYPE: CONVOLUTE PAPER
NOZZLE TYPE: RAMMED CLAY

MOTOR SHOWN WITHOUT TIME DELAY

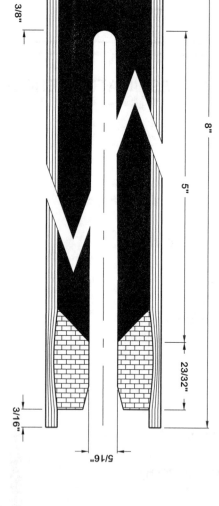

IMPORTANT BAKING SODA NOTE: Add 30% baking soda to the propellant in your first motor. Reduce the soda in 2% increments until you get a motor that doesn't chuff. Then reduce the soda in 1% increments until maximum thrust is approx. 15 lbs. This test motor took 5% soda, but I made it with a carbon black of medium purity. A motor made with pure, electrically conductive carbon black might take substantially more.

TOTAL WEIGHT: 234.7 GMS.
PROPELLANT WT.: 139.2 GMS.
BAKING SODA: See baking soda note.
MOTOR LENGTH: 8.00"
CASING I.D.: 1.00"

CASING O.D.: 1.25"
NOZZLE CLAY: 24.0 GMS.
BULKHEAD CLAY: 24.0 GMS.
THROAT DIA.: 0.313"
CORE LENGTH: 5.00"

TOT. IMPULSE: 102.99 NT.-SECS.
AVG. THRUST: 32.84 NTS.
MAX. THRUST: 14.62 LBS.
BURN TIME: 3.14 SECS.
SPEC. IMPULSE: 75.32 SECS.

EXTENDED BURN TIME WARNING! This motor's burn time is longer than 3 seconds. Do not fly this motor on a stick. Fly it only in a properly balanced, aerodynamic rocket body. Use a tall launch tower, and fly the rocket only in calm weather.

© COPYRIGHT 1998 DAVID G. SLEETER

INDEX OF REQUIRED TOOLING

CASING RETAINER	DRAWING #98-003:-10"	PAGE 92
NOZZLE MOLD	DRAWING #98-012	PAGE 101
CORE SPINDLE	DRAWING #98-057-O	PAGE 132
STARTING TAMP	DRAWING #98-092-B	PAGE 155
PROPELLANT TAMP	DRAWING #98-093-B	PAGE 156
FLAT TAMP	DRAWING #98-096-B	PAGE 159
PORTHOLE TAMP	DRAWING #98-096-A	PAGE 159
PORTHOLE WIRE	DRAWING #98-096	PAGE 159

MOTOR NAME: NV6-102-G32	NOZZLE TYPE: RAMMED CLAY	NOZZLE THROAT: 0.313"
TEST FILENAME: 970613C	CASING ID: 1.00"	CORE LENGTH: 5.00"
PROPELLANT: NV6	CASING OD: 1.25"	MOTOR WEIGHT: 234.7 GMS.
BURN MODIFIER: 5% NaHCO3	MOTOR LENGTH: 8.00"	PROPELLANT WT.: 139.2 GMS.

TOT. IMP.: 102.99 NT.-SECS.		MAX. THRUST: 14.62 LBS.
SPEC. IMPULSE: 75.32 SECS.	BURN TIME: 3.14 SECS.	AVG. THRUST: 32.84 NTS.

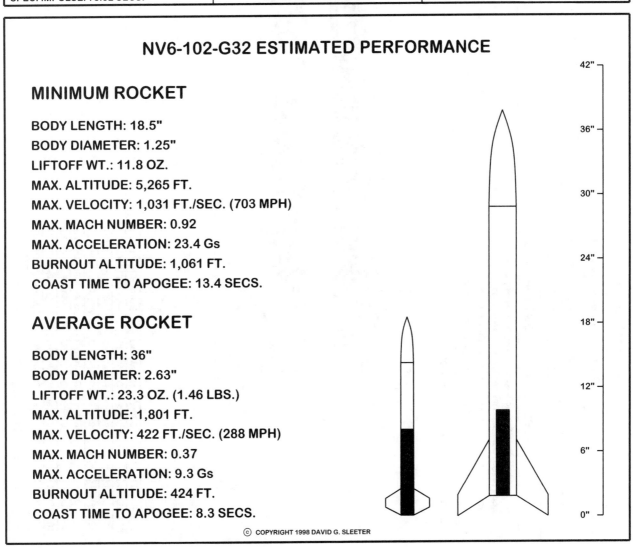

NV6-102-G32 ESTIMATED PERFORMANCE

MINIMUM ROCKET

BODY LENGTH: 18.5"
BODY DIAMETER: 1.25"
LIFTOFF WT.: 11.8 OZ.
MAX. ALTITUDE: 5,265 FT.
MAX. VELOCITY: 1,031 FT./SEC. (703 MPH)
MAX. MACH NUMBER: 0.92
MAX. ACCELERATION: 23.4 Gs
BURNOUT ALTITUDE: 1,061 FT.
COAST TIME TO APOGEE: 13.4 SECS.

AVERAGE ROCKET

BODY LENGTH: 36"
BODY DIAMETER: 2.63"
LIFTOFF WT.: 23.3 OZ. (1.46 LBS.)
MAX. ALTITUDE: 1,801 FT.
MAX. VELOCITY: 422 FT./SEC. (288 MPH)
MAX. MACH NUMBER: 0.37
MAX. ACCELERATION: 9.3 Gs
BURNOUT ALTITUDE: 424 FT.
COAST TIME TO APOGEE: 8.3 SECS.

© COPYRIGHT 1998 DAVID G. SLEETER

TELEFLITE CORPORATION

MOTOR NAME: NV6-142-G32
PROPELLANT TYPE: NV6
CASING TYPE: CONVOLUTE PAPER
NOZZLE TYPE: RAMMED CLAY

CASING I.D.: 1.00"	
MOTOR LENGTH: 10.25"	
BAKING SODA: See baking soda note.	
PROPELLANT WT.: 194.4 GMS.	
TOTAL WEIGHT: 306.1 GMS.	

IMPORTANT BAKING SODA NOTE: Add 30% baking soda to the propellant in your first motor. Reduce the soda in 2% increments until you get a motor that doesn't chuff. Then reduce the soda in 1% increments until maximum thrust is approx. 16 lbs. This test motor took 12% soda, but I made it with a carbon black of medium purity. A motor made with pure, electrically conductive carbon black might take substantially more.

CASING O.D.: 1.25"
NOZZLE CLAY: 24.0 GMS.
BULKHEAD CLAY: 24.0 GMS.
THROAT DIA.: 0.313"
CORE LENGTH: 7.25"

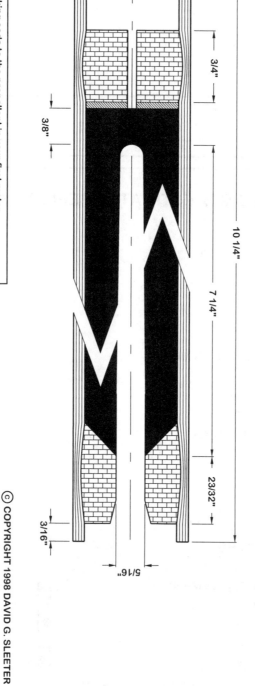

MOTOR SHOWN WITHOUT TIME DELAY

EXTENDED BURN TIME WARNING!

This motor's burn time is longer than 3 seconds. Do not fly this motor on a stick. Fly it only in a properly balanced, aerodynamic rocket body. Use a tall launch tower, and fly the rocket only in calm weather.

© COPYRIGHT 1998 DAVID G. SLEETER

TOT. IMPULSE: 142.04 NT.-SECS.
AVG. THRUST: 32.99 NTS.
MAX. THRUST: 15.77 LBS.
BURN TIME: 4.31 SECS.
SPEC. IMPULSE: 74.38 SECS.

INDEX OF REQUIRED TOOLING

CASING RETAINER	DRAWING #98-003: 13"	PAGE 92
NOZZLE MOLD	DRAWING #98-012	PAGE 101
CORE SPINDLE	DRAWING #98-059-X	PAGE 133
STARTING TAMP	DRAWING #98-092-C	PAGE 155
PROPELLANT TAMP	DRAWING #98-093-C	PAGE 156
FLAT TAMP	DRAWING #98-096-B	PAGE 159
PORTHOLE TAMP	DRAWING #98-096-A	PAGE 159
PORTHOLE WIRE	DRAWING #98-096	PAGE 159

MOTOR NAME: NV6-142-G32	NOZZLE TYPE: RAMMED CLAY	NOZZLE THROAT: 0.313"
TEST FILENAME: 970615B	CASING ID: 1.00"	CORE LENGTH: 7.25"
PROPELLANT: NV6	CASING OD: 1.25"	MOTOR WEIGHT: 306.1 GMS.
BURN MODIFIER: 12% NaHCO3	MOTOR LENGTH: 10.25"	PROPELLANT WT.: 194.4 GMS.

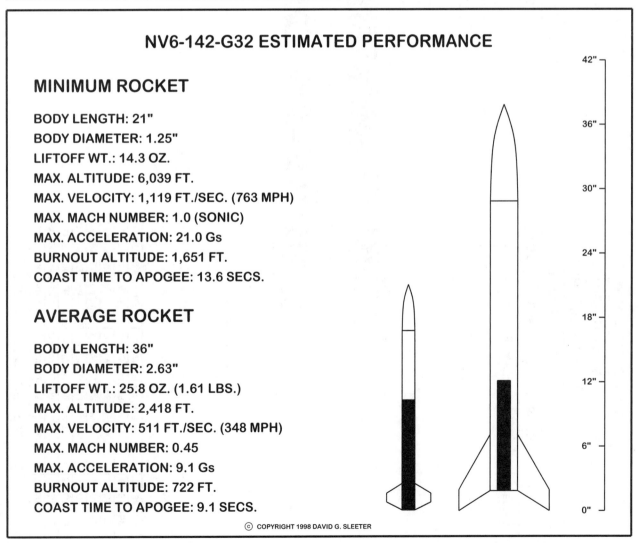

TOT. IMP.: 142.04 NT.-SECS.	TIME IN SECONDS	MAX. THRUST: 15.77 LBS.
SPEC. IMPULSE: 74.38 SECS.	BURN TIME: 4.31 SECS.	AVG. THRUST: 32.99 NTS.

NV6-142-G32 ESTIMATED PERFORMANCE

MINIMUM ROCKET

BODY LENGTH: 21"
BODY DIAMETER: 1.25"
LIFTOFF WT.: 14.3 OZ.
MAX. ALTITUDE: 6,039 FT.
MAX. VELOCITY: 1,119 FT./SEC. (763 MPH)
MAX. MACH NUMBER: 1.0 (SONIC)
MAX. ACCELERATION: 21.0 Gs
BURNOUT ALTITUDE: 1,651 FT.
COAST TIME TO APOGEE: 13.6 SECS.

AVERAGE ROCKET

BODY LENGTH: 36"
BODY DIAMETER: 2.63"
LIFTOFF WT.: 25.8 OZ. (1.61 LBS.)
MAX. ALTITUDE: 2,418 FT.
MAX. VELOCITY: 511 FT./SEC. (348 MPH)
MAX. MACH NUMBER: 0.45
MAX. ACCELERATION: 9.1 Gs
BURNOUT ALTITUDE: 722 FT.
COAST TIME TO APOGEE: 9.1 SECS.

© COPYRIGHT 1998 DAVID G. SLEETER

TELEFLITE CORPORATION

MOTOR NAME: NV6-217-H75
PROPELLANT TYPE: NV6
CASING TYPE: CONVOLUTE PAPER
NOZZLE TYPE: RAMMED CLAY

CASING I.D.: 1.00"
MOTOR LENGTH: 15.50"
BAKING SODA: See baking soda note.
PROPELLANT WT.: 280.4 GMS.
TOTAL WEIGHT: 420.8 GMS.

IMPORTANT BAKING SODA NOTE: Add 30% baking soda to the propellant in your first motor. Reduce the soda in 2% increments until you get a motor that doesn't chuff. Then reduce the soda in 1% increments until maximum thrust is approx. 30 lbs. This test motor took 5% soda, but I made it with a carbon black of medium purity. A motor made with a pure, electrically-conductive carbon black might take substantially more.

MOTOR SHOWN WITHOUT TIME DELAY

Dimensions shown on diagram: 1 1/4", 1", 3/4", 11/32", 15 1/2", 12", 1", 3/16", 1/2"

CORE LENGTH: 12.00"
THROAT DIA.: 0.50"
BULKHEAD CLAY: 24.0 GMS.
NOZZLE CLAY: 24.0 GMS.
CASING O.D.: 1.25"

SPEC. IMPULSE: 79.07 SECS.
BURN TIME: 2.89 SECS.
MAX. THRUST: 30.52 LBS.
AVG. THRUST: 75.29 NTS.
TOT. IMPULSE: 217.81 NT.-SECS.

© COPYRIGHT 1998 DAVID G. SLEETER

INDEX OF REQUIRED TOOLING

CASING RETAINER	DRAWING #98-003: 19"	PAGE 92
NOZZLE MOLD	DRAWING #98-014	PAGE 103
CORE SPINDLE	DRAWING #98-064-T	PAGE 136
STARTING TAMP	DRAWING #98-094-C	PAGE 157
PROPELLANT TAMP	DRAWING #98-095-C	PAGE 158
FLAT TAMP	DRAWING #98-096-C	PAGE 159
PORTHOLE TAMP	DRAWING #98-096-A	PAGE 159
PORTHOLE WIRE	DRAWING #98-096-B	PAGE 159

MOTOR NAME: NV6-217-H75	NOZZLE TYPE: RAMMED CLAY	NOZZLE THROAT: 0.50"
TEST FILENAME: 970619C	CASING ID: 1.00"	CORE LENGTH: 12.00"
PROPELLANT: NV6	CASING OD: 1.25"	MOTOR WEIGHT: 420.8 GMS.
BURN MODIFIER: 5% NaHCO3	MOTOR LENGTH: 15.50"	PROPELLANT WT.: 280.4 GMS.

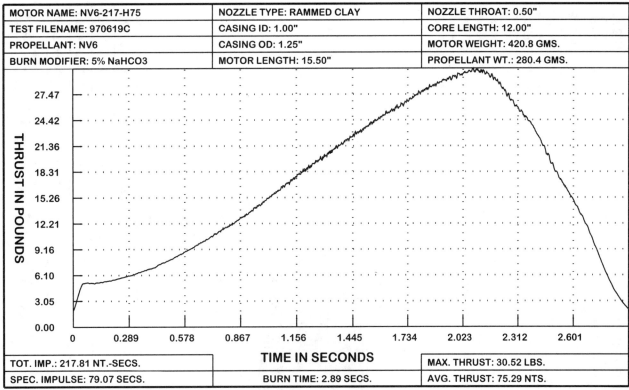

TOT. IMP.: 217.81 NT.-SECS.		MAX. THRUST: 30.52 LBS.
SPEC. IMPULSE: 79.07 SECS.	BURN TIME: 2.89 SECS.	AVG. THRUST: 75.29 NTS.

NV6-217-H75 ESTIMATED PERFORMANCE

MINIMUM ROCKET

BODY LENGTH: 26"
BODY DIAMETER: 1.25"
LIFTOFF WT.: 18.4 OZ. (1.15 LBS.)
MAX. ALTITUDE: 6,462 FT.
MAX. VELOCITY: 1,349 FT./SEC. (920 MPH)
MAX. MACH NUMBER: 1.2 (SUPERSONIC)
MAX. ACCELERATION: 33.5 Gs
BURNOUT ALTITUDE: 1,721 FT.
COAST TIME TO APOGEE: 13.9 SECS.

AVERAGE ROCKET

BODY LENGTH: 36"
BODY DIAMETER: 2.63"
LIFTOFF WT.: 29.9 OZ. (1.87 LBS.)
MAX. ALTITUDE: 3,471 FT.
MAX. VELOCITY: 752 FT./SEC. (513 MPH)
MAX. MACH NUMBER: 0.67
MAX. ACCELERATION: 16.7 Gs
BURNOUT ALTITUDE: 941 FT.
COAST TIME TO APOGEE: 10.6 SECS.

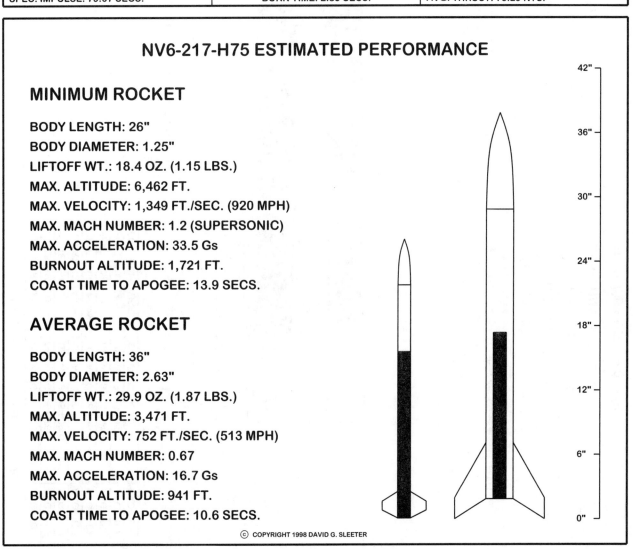

© COPYRIGHT 1998 DAVID G. SLEETER

TELEFLITE CORPORATION

MOTOR NAME:	NV6-158-G43
PROPELLANT TYPE:	NV6
CASING TYPE:	CONVOLUTE PAPER
NOZZLE TYPE:	RAMMED CLAY

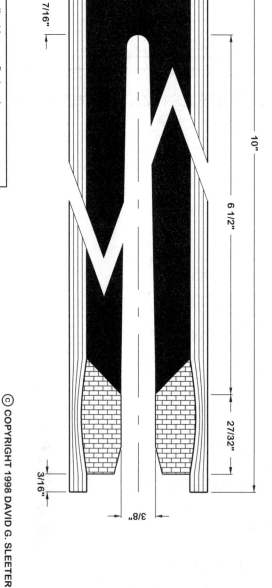

MOTOR SHOWN WITHOUT TIME DELAY

© COPYRIGHT 1998 DAVID G. SLEETER

IMPORTANT BAKING SODA NOTE: Add 30% baking soda to the propellant in your first motor. Reduce the soda in 2% increments until you get a motor that doesn't chuff. Then reduce the soda in 1% increments until maximum thrust is approx. 20 lbs. This test motor took 5% soda, but I made it with a carbon black of medium purity. A motor made with pure, electrically conductive carbon black might take substantially more.

EXTENDED BURN TIME WARNING!
This motor's burn time is longer than 3 seconds. Do not fly this motor on a stick. Fly it only in a properly balanced, aerodynamic rocket body. Use a tall launch tower, and fly the rocket only in calm weather.

TOTAL WEIGHT:	372.8 GMS.
PROPELLANT WT.:	214.6 GMS.
BAKING SODA:	See baking soda note.
MOTOR LENGTH:	10.00"
CASING I.D.:	1.125"
CASING O.D.:	1.50"
NOZZLE CLAY:	33.0 GMS.
BULKHEAD CLAY:	33.0 GMS.
THROAT DIA.:	0.375"
CORE LENGTH:	6.50"
TOT. IMPULSE:	158.99 NT.-SECS.
AVG. THRUST:	43.97 NTS.
MAX. THRUST:	20.43 LBS.
BURN TIME:	3.62 SECS.
SPEC. IMPULSE:	75.42 SECS.

INDEX OF REQUIRED TOOLING

CASING RETAINER	DRAWING #98-004; 11-1/2" PAGE 93
NOZZLE MOLD	DRAWING #98-016 PAGE 105
CORE SPINDLE	DRAWING #98-069-O PAGE 138
STARTING TAMP	DRAWING #98-097-C PAGE 160
PROPELLANT TAMP	DRAWING #98-098-C PAGE 161
FLAT TAMP	DRAWING #98-104-B PAGE 167
PORTHOLE TAMP	DRAWING #98-104-A PAGE 167
PORTHOLE WIRE	DRAWING #98-104 PAGE 167

MOTOR NAME: NV6-158-G43	NOZZLE TYPE: RAMMED CLAY	NOZZLE THROAT: 0.375"
TEST FILENAME: 970611E	CASING ID: 1.125"	CORE LENGTH: 6.50"
PROPELLANT: NV6	CASING OD: 1.50"	MOTOR WEIGHT: 372.8 GMS.
BURN MODIFIER: 5% NaHCO3	MOTOR LENGTH: 10.00"	PROPELLANT WT.: 214.6 GMS.

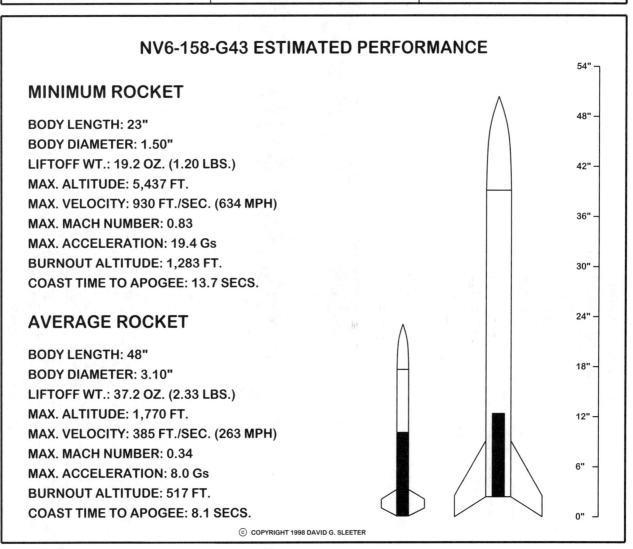

TOT. IMP.: 158.99 NT.-SECS.		MAX. THRUST: 20.10 LBS.
SPEC. IMPULSE: 75.42 SECS.	BURN TIME: 3.62 SECS.	AVG. THRUST: 43.97 NTS.

NV6-158-G43 ESTIMATED PERFORMANCE

MINIMUM ROCKET

BODY LENGTH: 23"
BODY DIAMETER: 1.50"
LIFTOFF WT.: 19.2 OZ. (1.20 LBS.)
MAX. ALTITUDE: 5,437 FT.
MAX. VELOCITY: 930 FT./SEC. (634 MPH)
MAX. MACH NUMBER: 0.83
MAX. ACCELERATION: 19.4 Gs
BURNOUT ALTITUDE: 1,283 FT.
COAST TIME TO APOGEE: 13.7 SECS.

AVERAGE ROCKET

BODY LENGTH: 48"
BODY DIAMETER: 3.10"
LIFTOFF WT.: 37.2 OZ. (2.33 LBS.)
MAX. ALTITUDE: 1,770 FT.
MAX. VELOCITY: 385 FT./SEC. (263 MPH)
MAX. MACH NUMBER: 0.34
MAX. ACCELERATION: 8.0 Gs
BURNOUT ALTITUDE: 517 FT.
COAST TIME TO APOGEE: 8.1 SECS.

© COPYRIGHT 1998 DAVID G. SLEETER

TELEFLITE CORPORATION

MOTOR NAME:	NV6-193-H47
PROPELLANT TYPE:	NV6
CASING TYPE:	CONVOLUTE PAPER
NOZZLE TYPE:	RAMMED CLAY

INDEX OF REQUIRED TOOLING

CASING RETAINER	DRAWING #98-004: 15"
NOZZLE MOLD	DRAWING #98-016
CORE SPINDLE	DRAWING #98-070-S
STARTING TAMP	DRAWING #98-097-C
PROPELLANT TAMP	DRAWING #98-098-C
FLAT TAMP	DRAWING #98-104-B
PORTHOLE TAMP	DRAWING #98-104-A
PORTHOLE WIRE	DRAWING #98-104
	PAGE 93
	PAGE 105
	PAGE 139
	PAGE 160
	PAGE 161
	PAGE 167
	PAGE 167
	PAGE 167

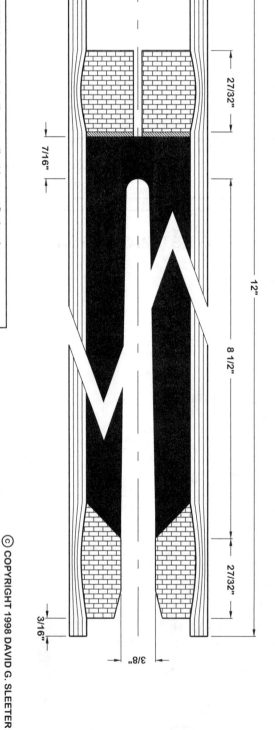

MOTOR SHOWN WITHOUT TIME DELAY

© COPYRIGHT 1998 DAVID G. SLEETER

IMPORTANT BAKING SODA NOTE: Add 30% baking soda to the propellant in your first motor. Reduce the soda in 2% increments until you get a motor that doesn't chuff. Then reduce the soda in 1% increments until maximum thrust is approx. 20 lbs. This test motor took 12% soda, but I made it with a carbon black of medium purity. A motor made with pure, electrically conductive carbon black might take substantially more.

EXTENDED BURN TIME WARNING!
This motor's burn time is longer than 3 seconds. Do not fly this motor on a stick. Fly it only in a properly balanced, aerodynamic rocket body. Use a tall launch tower, and fly the rocket only in calm weather.

TOTAL WEIGHT:	455.5 GMS.
PROPELLANT WT.:	282.9 GMS.
BAKING SODA:	See baking soda note.
MOTOR LENGTH:	12.00"
CASING I.D.:	1.125"

CASING O.D.:	1.50"
NOZZLE CLAY:	33.0 GMS.
BULKHEAD CLAY:	33.0 GMS.
THROAT DIA.:	0.375"
CORE LENGTH:	8.50"

TOT. IMPULSE:	193.39 NT.-SECS.
AVG. THRUST:	47.30 NTS.
MAX. THRUST:	20.43 LBS.
BURN TIME:	4.09 SECS.
SPEC. IMPULSE:	69.59 SECS.

MOTOR NAME: NV6-193-H47	NOZZLE TYPE: RAMMED CLAY	NOZZLE THROAT: 0.375"
TEST FILENAME: 970613A	CASING ID: 1.125"	CORE LENGTH: 8.50"
PROPELLANT: NV6	CASING OD: 1.50"	MOTOR WEIGHT: 455.5 GMS.
BURN MODIFIER: 12% NaHCO3	MOTOR LENGTH: 12.00"	PROPELLANT WT.: 282.9 GMS.

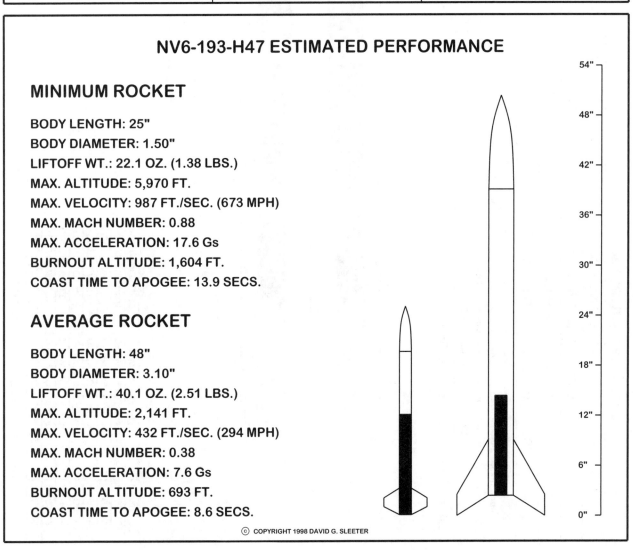

TOT. IMP.: 193.39 NT.-SECS.		MAX. THRUST: 20.43 LBS.
SPEC. IMPULSE: 69.59 SECS.	BURN TIME: 4.09 SECS.	AVG. THRUST: 47.30 NTS.

NV6-193-H47 ESTIMATED PERFORMANCE

MINIMUM ROCKET

BODY LENGTH: 25"
BODY DIAMETER: 1.50"
LIFTOFF WT.: 22.1 OZ. (1.38 LBS.)
MAX. ALTITUDE: 5,970 FT.
MAX. VELOCITY: 987 FT./SEC. (673 MPH)
MAX. MACH NUMBER: 0.88
MAX. ACCELERATION: 17.6 Gs
BURNOUT ALTITUDE: 1,604 FT.
COAST TIME TO APOGEE: 13.9 SECS.

AVERAGE ROCKET

BODY LENGTH: 48"
BODY DIAMETER: 3.10"
LIFTOFF WT.: 40.1 OZ. (2.51 LBS.)
MAX. ALTITUDE: 2,141 FT.
MAX. VELOCITY: 432 FT./SEC. (294 MPH)
MAX. MACH NUMBER: 0.38
MAX. ACCELERATION: 7.6 Gs
BURNOUT ALTITUDE: 693 FT.
COAST TIME TO APOGEE: 8.6 SECS.

© COPYRIGHT 1998 DAVID G. SLEETER

TELEFLITE CORPORATION

MOTOR NAME:	NV6-340-1100
PROPELLANT TYPE:	NV6
CASING TYPE:	CONVOLUTE PAPER
NOZZLE TYPE:	RAMMED CLAY

CASING I.D.:	1.125"
MOTOR LENGTH:	18.75"
BAKING SODA:	See baking soda note.
PROPELLANT WT.:	426.6 GMS.
TOTAL WEIGHT:	658.9 GMS.

IMPORTANT BAKING SODA NOTE: Add 30% baking soda to the propellant in your first motor. Reduce the soda in 2% increments until you get a motor that doesn't chuff. Then reduce the soda in 1% increments until maximum thrust is approx. 40 lbs. This test motor took 5% soda, but I made it with a carbon black of medium purity. A motor made with pure, electrically conductive carbon black might take substantially more.

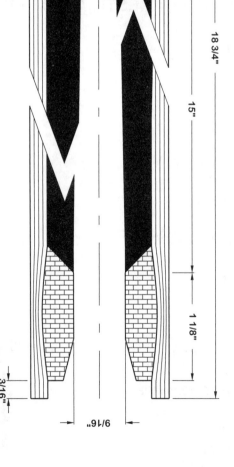

MOTOR SHOWN WITHOUT TIME DELAY

EXTENDED BURN TIME WARNING!
This motor's burn time is longer than 3 seconds. Do not fly this motor on a stick. Fly it only in a properly balanced, aerodynamic rocket body. Use a tall launch tower, and fly the rocket only in calm weather.

© COPYRIGHT 1998 DAVID G. SLEETER

CASING O.D.:	1.50"
NOZZLE CLAY:	33.0 GMS.
BULKHEAD CLAY:	33.0 GMS.
THROAT DIA.:	0.563"
CORE LENGTH:	15.00"

TOT. IMPULSE:	340.91 NT.-SECS.
AVG. THRUST:	100.12 NTS.
MAX. THRUST:	40.38 LBS.
BURN TIME:	3.40 SECS.
SPEC. IMPULSE:	81.35 SECS.

INDEX OF REQUIRED TOOLING

CASING RETAINER	DRAWING #98-004; 22" PAGE 93
NOZZLE MOLD	DRAWING #98-018 PAGE 107
CORE SPINDLE	DRAWING #98-076-W PAGE 142
STARTING TAMP	DRAWING #98-100-D PAGE 163
PROPELLANT TAMP	DRAWING #98-102-D PAGE 165
FLAT TAMP	DRAWING #98-104-B PAGE 167
PORTHOLE TAMP	DRAWING #98-104-A PAGE 167
PORTHOLE WIRE	DRAWING #98-104 PAGE 167

MOTOR NAME: NV6-340-I100	NOZZLE TYPE: RAMMED CLAY	NOZZLE THROAT: 0.563"
TEST FILENAME: 970615A	CASING ID: 1.125"	CORE LENGTH: 15.00"
PROPELLANT: NV6	CASING OD: 1.50"	MOTOR WEIGHT: 658.9 GMS.
BURN MODIFIER: 5% NaHCO3	MOTOR LENGTH: 18.75"	PROPELLANT WT.: 426.6 GMS.

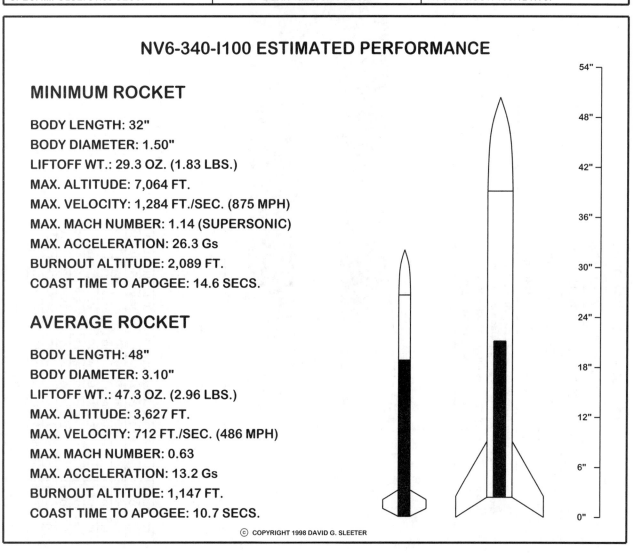

TOT. IMP.: 340.91 NT.-SECS.		MAX. THRUST: 40.38 LBS.
SPEC. IMPULSE: 81.35 SECS.	BURN TIME: 3.40 SECS.	AVG. THRUST: 100.12 NTS.

NV6-340-I100 ESTIMATED PERFORMANCE

MINIMUM ROCKET

BODY LENGTH: 32"
BODY DIAMETER: 1.50"
LIFTOFF WT.: 29.3 OZ. (1.83 LBS.)
MAX. ALTITUDE: 7,064 FT.
MAX. VELOCITY: 1,284 FT./SEC. (875 MPH)
MAX. MACH NUMBER: 1.14 (SUPERSONIC)
MAX. ACCELERATION: 26.3 Gs
BURNOUT ALTITUDE: 2,089 FT.
COAST TIME TO APOGEE: 14.6 SECS.

AVERAGE ROCKET

BODY LENGTH: 48"
BODY DIAMETER: 3.10"
LIFTOFF WT.: 47.3 OZ. (2.96 LBS.)
MAX. ALTITUDE: 3,627 FT.
MAX. VELOCITY: 712 FT./SEC. (486 MPH)
MAX. MACH NUMBER: 0.63
MAX. ACCELERATION: 13.2 Gs
BURNOUT ALTITUDE: 1,147 FT.
COAST TIME TO APOGEE: 10.7 SECS.

© COPYRIGHT 1998 DAVID G. SLEETER

TELEFLITE CORPORATION

MOTOR NAME:	NV6-417-I65
PROPELLANT TYPE:	NV6
CASING TYPE:	CONVOLUTE PAPER
NOZZLE TYPE:	RAMMED CLAY

MOTOR SHOWN WITHOUT TIME DELAY

© COPYRIGHT 1998 DAVID G. SLEETER

IMPORTANT BAKING SODA NOTE: Add 30% baking soda to the propellant in your first motor. Reduce the soda in 2% increments until you get a motor that doesn't chuff. Then reduce the soda in 1% increments until maximum thrust is approx. 32 lbs. This test motor took 12% soda, but I made it with a carbon black of medium purity. A motor made with pure, electrically conductive carbon black might take substantially more.

TOTAL WEIGHT: 903.6 GMS.	CASING O.D.: 2.00"	TOT. IMPULSE: 417.23 NT.-SECS.
PROPELLANT WT.: 573.2 GMS.	NOZZLE CLAY: 80.0 GMS.	AVG. THRUST: 65.92 NTS.
BAKING SODA: See baking soda note.	BULKHEAD CLAY: 80.0 GMS.	MAX. THRUST: 32.57 LBS.
MOTOR LENGTH: 14.5"	THROAT DIA.: 0.50"	BURN TIME: 6.33 SECS.
CASING I.D.: 1.50"	CORE LENGTH: 10.00"	SPEC. IMPULSE: 74.10 SECS.

EXTENDED BURN TIME WARNING!
This motor's burn time is longer than 3 seconds. Do not fly this motor on a stick. Fly it only in a properly balanced, aerodynamic rocket body. Use a tall launch tower, and fly the rocket only in calm weather.

INDEX OF REQUIRED TOOLING

CASING RETAINER	DRAWING #98-005: 18-1/2"	PAGE 94
NOZ. MOLD BASE	DRAWING #98-021	PAGE 110
NOZ. MOLD INSERT	DRAWING #98-022	PAGE 111
CORE SPINDLE	DRAWING #98-081-E	PAGE 144
STARTING TAMP	DRAWING #98-105-A	PAGE 168
PROPELLANT TAMP	DRAWING #98-105-B	PAGE 168
FLAT TAMP	DRAWING #98-106-B	PAGE 169
PORTHOLE TAMP	DRAWING #98-106-A	PAGE 169
PORTHOLE WIRE	DRAWING #98-106	PAGE 169

MOTOR NAME: NV6-417-I65	NOZZLE TYPE: RAMMED CLAY	NOZZLE THROAT: 0.50"
TEST FILENAME: 970623A	CASING ID: 1.50"	CORE LENGTH: 10.00"
PROPELLANT: NV6	CASING OD: 2.00"	MOTOR WEIGHT: 903.6 GMS.
BURN MODIFIER: 12% NaHCO3	MOTOR LENGTH: 14.50"	PROPELLANT WT.: 573.2 GMS.

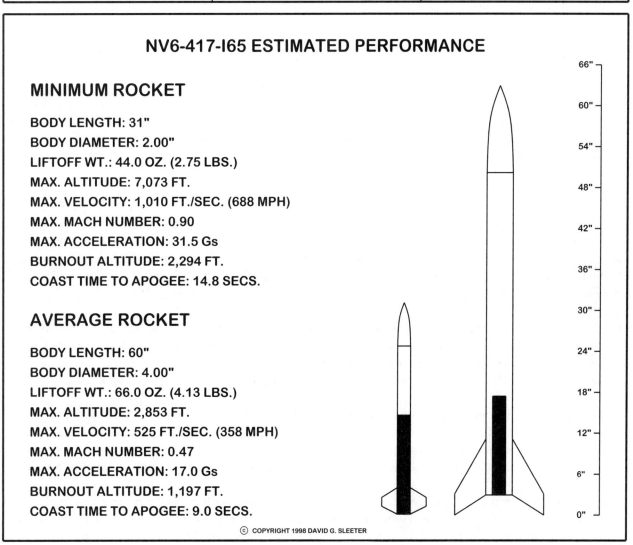

TOT. IMP.: 417.23 NT.-SECS.		MAX. THRUST: 32.57 LBS.
SPEC. IMPULSE: 74.10 SECS.	BURN TIME: 6.33 SECS.	AVG. THRUST: 65.92 NTS.

NV6-417-I65 ESTIMATED PERFORMANCE

MINIMUM ROCKET

BODY LENGTH: 31"
BODY DIAMETER: 2.00"
LIFTOFF WT.: 44.0 OZ. (2.75 LBS.)
MAX. ALTITUDE: 7,073 FT.
MAX. VELOCITY: 1,010 FT./SEC. (688 MPH)
MAX. MACH NUMBER: 0.90
MAX. ACCELERATION: 31.5 Gs
BURNOUT ALTITUDE: 2,294 FT.
COAST TIME TO APOGEE: 14.8 SECS.

AVERAGE ROCKET

BODY LENGTH: 60"
BODY DIAMETER: 4.00"
LIFTOFF WT.: 66.0 OZ. (4.13 LBS.)
MAX. ALTITUDE: 2,853 FT.
MAX. VELOCITY: 525 FT./SEC. (358 MPH)
MAX. MACH NUMBER: 0.47
MAX. ACCELERATION: 17.0 Gs
BURNOUT ALTITUDE: 1,197 FT.
COAST TIME TO APOGEE: 9.0 SECS.

© COPYRIGHT 1998 DAVID G. SLEETER

16. THE ORIGINAL 1983 MOTOR BUILDING PROCESS

A solid rocket propellant needs a binder to hold it together, and my original circa-1983 propellant used the propellant's own potassium nitrate. A small amount of water, added before loading, dissolved some of the nitrate. When the nitrate dried, it recrystallized, and formed the loose propellant into a solid propellant grain.

As the water evaporated, the dampened propellant shrank. To compensate, I dampened the casings ahead of time, and thereby caused them to shrink right along with the propellant. The dampening process combined with the *very* long drying time made this motor building method inconvenient. Nevertheless, it worked, and those of you unable to find either red gum or vinsol resin may wish to experiment with it. What I publish here is the original process with a few improvements that I've made in the years since.

Propellant Formulas and Propellant Preparation

In 1983 I used garden charcoal as the sole carbon fuel, and for the purpose of making the propellant's burn rate adjustable, I experimented with *two* stoichiometric versions of black powder. The fast burning mixture contained the classic proportions of 75% potassium nitrate, 15% charcoal, and 10% sulfur, and the slow burning mixture contained 70% potassium nitrate, 30% charcoal, and no sulfur. But I quickly learned that the slow-burning version didn't burn slow *enough*, and its high charcoal content made the dried propellant grain porous and fragile. To slow the 70/30 mixture even further, and make it structurally stronger at the same time, I added extra potassium nitrate.

Through a process of trial and error, I reduced the charcoal to 16.7%, and increased the potassium nitrate to 83.3%. Then I generated the table of 21 formulas shown in ***Figure 16-1***. They range from the classic mixture with 10% sulfur and 15% charcoal, to the slow-burning version with no sulfur and 16.7% charcoal. As you can see, I reduced the sulfur in 1/2% increments along the way. Your initial reaction might be that the formulas with excess nitrate should exhibit combustion instability, but the fact is that they *don't*, because without the chemically-complex molecules of red gum or vinsol resin, combustion instability isn't a problem.

Over the years I've gotten feedback from many readers, and those using garden charcoal say that their motors work best with the formulas in the 1-1/2% to 4-1/2% sulfur range. Those using coal or hardwood charcoal have reported success with sulfur contents as high as 8%. When making these propellants yourself, begin with the prepared ingredients described in **Chapter 7**, and use the same milling times recommended for the other propellants in this book.

You can start with the 10% sulfur formula, and experiment with baking soda as a burn rate modifier, but the results might be unpredictable. In the presence of water, some of the sulfur oxidizes to sulfur dioxide, which dissolves in the water, creating weak sulfurous acid. There *might* be enough acid to neutralize some of the baking soda, and in the presence of water, other acids and contaminants from the coal or charcoal might react with the soda in unpredictable ways.

Because the water evaporates so slowly, you can dampen an entire bowl of propellant, and work from it for several hours *if you keep it covered and out of the sun*. For my original work, I added 40 cc. of water per pound, but I *strongly* advise you to perform **the pellet test** (**page 240**) *with water*, and determine the correct amount of water for yourself.

NO SULFUR KNO_3 = 416.7 gms. Charcoal = 83.3 gms.	1/2% SULFUR KNO_3 = 414.6 gms. Charcoal = 82.9 gms. Sulfur = 2.5 gms.	1% SULFUR KNO_3 = 412.5 gms. Charcoal = 82.5 gms. Sulfur = 5.0 gms.	1-1/2% SULFUR KNO_3 = 410.4 gms. Charcoal = 82.1 gms. Sulfur = 7.5 gms.
2% SULFUR KNO_3 = 408.3 gms. Charcoal = 81.7 gms. Sulfur = 10.0 gms.	2-1/2% SULFUR KNO_3 = 406.2 gms. Charcoal = 81.3 gms. Sulfur = 12.5 gms.	3% SULFUR KNO_3 = 404.2 gms. Charcoal = 80.8 gms. Sulfur = 15.0 gms.	3-1/2% SULFUR KNO_3 = 402.1 gms. Charcoal = 80.4 gms. Sulfur = 17.5 gms.
4% SULFUR KNO_3 = 400.0 gms. Charcoal = 80.0 gms. Sulfur = 20.0 gms.	4-1/2% SULFUR KNO_3 = 397.9 gms. Charcoal = 79.6 gms. Sulfur = 22.5 gms.	5% SULFUR KNO_3 = 395.8 gms. Charcoal = 79.2 gms. Sulfur = 25.0 gms.	5-1/2% SULFUR KNO_3 = 393.7 gms. Charcoal = 78.8 gms. Sulfur = 27.5 gms.
6% SULFUR KNO_3 = 391.7 gms. Charcoal = 78.3 gms. Sulfur = 30.0 gms.	6-1/2% SULFUR KNO_3 = 389.6 gms. Charcoal = 77.9 gms. Sulfur = 32.5 gms.	7% SULFUR KNO_3 = 387.5 gms. Charcoal = 77.5 gms. Sulfur = 35.0 gms.	7-1/2% SULFUR KNO_3 = 385.4 gms. Charcoal = 77.1 gms. Sulfur = 37.5 gms.
8% SULFUR KNO_3 = 383.3 gms. Charcoal = 76.7 gms. Sulfur = 40.0 gms.	8-1/2% SULFUR KNO_3 = 381.2 gms. Charcoal = 76.3 gms. Sulfur = 42.5 gms.	9% SULFUR KNO_3 = 379.2 gms. Charcoal = 75.8 gms. Sulfur = 45.0 gms.	9-1/2% SULFUR KNO_3 = 377.1 gms. Charcoal = 75.4 gms. Sulfur = 47.5 gms.
10% SULFUR KNO_3 = 375.0 gms. Charcoal = 75.0 gms. Sulfur = 50.0 gms.			

Figure 16-1. *The 21 propellant formulas from my first book. They range from 10% sulfur down to no sulfur. As the sulfur is reduced, the formulas become increasingly rich in potassium nitrate. Potassium nitrate over and above the amount needed for a stoichiometric reaction acts as an inert material, and thereby helps slow the propellant's burn rate. The excess nitrate also strengthens the dried propellant grain. The chemical weights shown produce a 500 gram batch.*

Preparing the Casings

● The Original Method

To dampen the casings, I cut them to length, sanded their ends, and placed them in a covered, plastic trash can on a bed of rags, surrounding a juice can full or water. I placed a large, cloth wick in the water to help it evaporate. If I needed more moisture, I stretched a wet towel over the top of the trash can, and pressed the lid down on top of it.

Figure 16-2 is the original illustration from my original book. The tubes sat in this device for up to 4 days before they were ready, and it took a lot of practice to learn how to use it properly. The guesswork involved caused a lot of frustration, and a general dissatisfaction with the entire motor building system.

● A Better Method

To remedy the situation, I developed a simpler and more accurate way of dampening the tubes that I'm publishing here for the first time. In this *new* method, the moisture is carried into the tubes by denatured alcohol, and the mixture of alcohol and water can be adjusted to impart the *exact* amount of moisture desired. The process relies upon the fact that paper is made of wood fiber, and that water causes the fibers to expand, *but alcohol does not*. The results are reliable and repeatable, and the tubes are ready to use in 24 hours. **Important note.** This method *cannot* be used for dampening *homemade* casings, because the alcohol will dissolve the white glue used to make them.

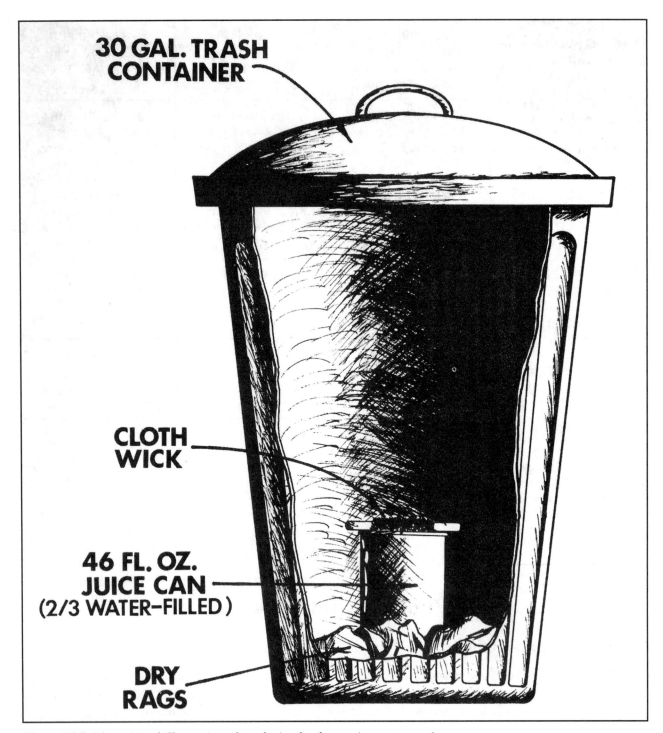

Figure 16-2. The original illustration of my device for dampening motor casings.

● **Hydrometers**

The percentages of alcohol and water must be correct. *They **cannot** be guessed at.* Also, as you use the solution, or let it sit in storage, some of the alcohol evaporates. This throws the percentages out of balance, and you need an accurate way to measure what those percentages are. The tool that does this is called a **hydrometer**.

A hydrometer is a sealed glass tube with a weight in the bottom, filled with air so that it floats. The *level* at which it floats is determined by the density of the liquid in which it is floating, and you can read that density by lining up the surface of the liquid with a printed paper scale inside the glass. Water is heavy, and alcohol is light, so by floating a properly designed hydrometer in a water-alcohol solution, you can tell exactly how much water and alcohol the solution contains.

There are many kinds of hydrometers, and the one that you need is called an **alcohol hydrometer**. Alcohol hydrometers are used by the liquor and wine industries, and you can buy one from the laboratory equipment dealers listed in *The Yellow Pages*.

In the United States, at the time of this writing, the main manufacturer of hydrometers is the **Eveready Thermometer Company** (aka **ERTCO**), and most of the places that sell hydrometers have an **ERTCO** section in their catalogues. *Figure 16-3* is a photo of the ERTCO hydrometer that I use. I bought it from the same place where I buy my chemicals. I paid $12 in 1991, but a short time later I learned that I could have gotten one for *just $7* at a beer and winemaking supply shop.

The liquor and wine industries use two scales for measuring the alcohol content of their products. One is the **Tralle scale**, and one is the **proof scale**. Most hydrometers have only one scale. My **ERTCO** hydrometer has *both*.

The Tralle scale is a direct reading of the percentage of alcohol in water, and for rocket makers, it's the simplest scale to use. The proof scale is the Tralle scale *doubled*. A reading of 90 on the proof scale equals a reading of 45 on the Tralle scale. To clarify, when the label on a bottle of vodka says "90 proof", it means that the vodka contains 45% alcohol.

The solution for soaking casings with 1/8" thick walls should be 90% alcohol, so the Tralle reading should be 90, and the proof reading should be 180. **The solution for soaking casings with 3/16" thick walls should be 85% alcohol**, so the Tralle reading should be 85, and the proof reading should be 170.

Hydrometers are *very* sensitive to temperature, *and the above readings are only good at 60° Fahrenheit*. At 70° F., a Tralle hydrometer measuring 90% alcohol would actually read 91-1/2. At 90° the Tralle reading would be about 94. To see *exactly* what the reading should be for a given temperature, look at the chart in *Figure 16-4*. It graphs an **ERTCO** hydrometer's Tralle reading as it changes with temperature.

To use the chart, look on the horizontal axis for the temperature of the solution you are measuring. Then imagine a vertical line extending straight up from that point, and see where it crosses the dotted diagonal line that begins at the percentage you want. From that crossing point, look straight to the vertical scale at the left to find the correct hydrometer reading. *Figure 16-5* shows how to do it for an 88% solution of alcohol in water at a temperature of 85° Fahrenheit. As you can see, the Tralle reading is actually 91-1/2.

● **Thermometers**

To accurately measure the temperature, you need a good darkroom thermometer. Darkroom thermometers are used to measure the temperatures of photo processing solutions, and you can buy one at a camera store. *Figure 16-6* is a photo of the type that I use. It works with a stainless steel probe and a dial, and it's faster and more accurate than the simple glass kind.

The only "caveat" is that *some* darkroom thermometers (like the one in the photo) are adjustable, and you adjust them by twisting the circular scale with respect to the needle's position. Before you can use one of these adjustable thermometers, you have to verify that it's been properly calibrated. The shops that sell these thermometers can advise you.

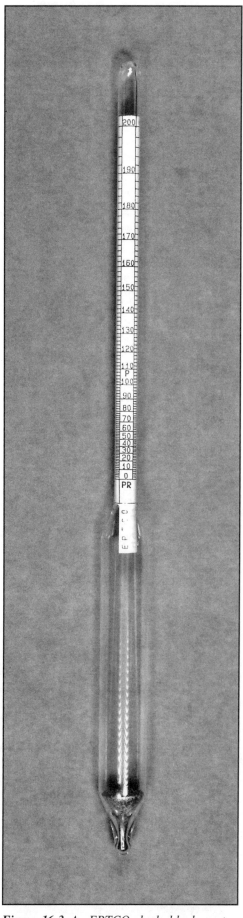

Figure 16-3. *An ERTCO alcohol hydrometer.*

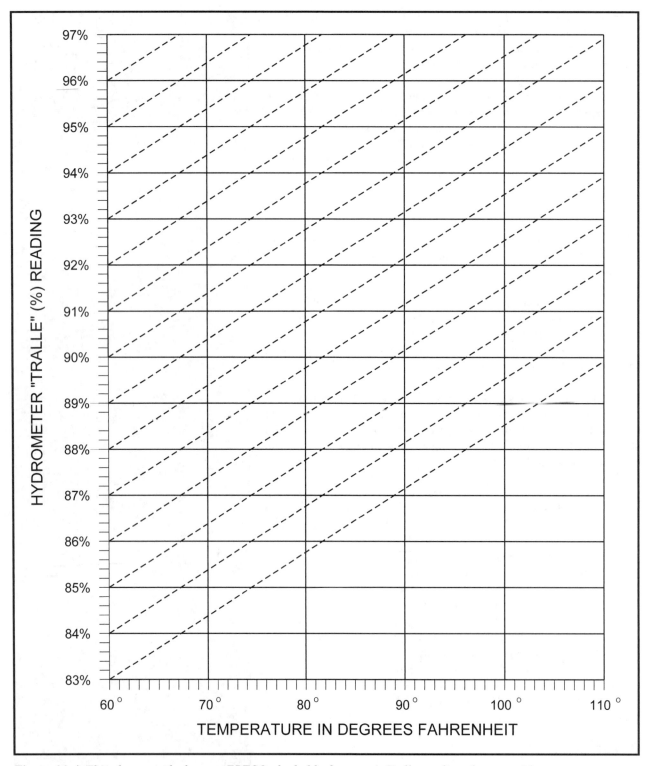

Figure 16-4. This chart graphs how an ERTCO alcohol hydrometer's Tralle reading changes with temperature.

● Hydrometer Cylinders

Hydrometers are long and thin, so you need something *tall* and thin in which to float them. The same places that sell hydrometers also sell a container called a **hydrometer cylinder** (*Figure 16-7*), specifically designed for this purpose. Some cylinders are made of glass, and some are made of an unbreakable plastic.

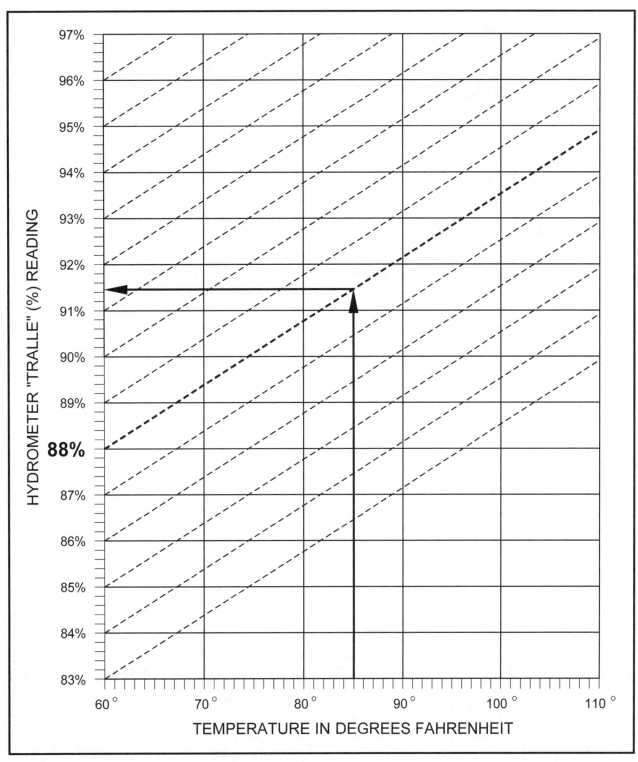

Figure 16-5. This example shows how to find the Tralle reading for an 88% alcohol-water solution at a temperature of 85° Fahrenheit.

● The Dampening Process

To dampen the casings, start with pure, denatured alcohol (sold at paint stores), and prepare the correct solution by slowly adding water until a test sample, poured into the hydrometer cylinder, reads the *temperature-corrected* Tralle scale percentage that you want. Then cut the casings to length, sand their ends, and immerse them in the solution (***Figure 16-8***). For the first 12 to 24 hours, as they absorb the water, they will soften and expand. *After* 24 hours, having reached equilibrium, their moisture content will remain constant, and you can safely store them in the solution for up to a week.

Figure 16-6. A dial-type darkroom thermometer. Quicker and more accurate than the standard glass type.

The Motor Building Process

To build a motor with one of these casings, take the casing out of the solution, and wipe it inside and out with a dry rag. Then clamp it into the casing retainer, and build the motor per to the instructions in **Chapter 10**. Because the moisture has made the tube expand, you will probably *not* need the extra wraps of paper discussed on **page 274**. Also, as you ram in the propellant, you may see the alcohol solution oozing out around the nozzle mold and the top of the casing. This is *normal*, and though it is messy, it doesn't harm the motor in any way.

When you are finished, pull out the nozzle former and the core spindle. Remove the motor from the casing retainer, *and give the outside of the motor casing a close and careful inspection*. A slight, lengthwise wrinkle where the casing retainer pinched together is normal. If the outside is otherwise smooth and free of defects, place the motor on a rack to dry. If you see *any* of the ring-shaped wrinkles shown in *Figure 16-9* (another drawing from my original book), the propellant grain will crack as it dries. **Because a cracked propellant grain causes the motor to explode shortly after ignition, this motor is defective, and you should discard it by soaking it in water.**

The ring-shaped wrinkles develop when a casing that is *too wet* and *too soft* is compressed downward in the casing retainer during loading. To fix the problem, add enough alcohol to the solution to reduce the water by 1%. Let the remaining tubes to soak for another 24 hours; then build another motor. Continue raising the alcohol (and reducing the water) until the finished motors come out wrinkle-free.

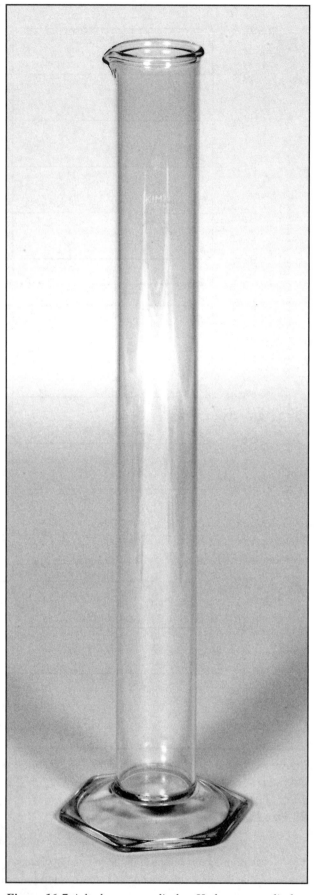

Figure 16-7. A hydrometer cylinder. Hydrometer cylinders are used to float the hydrometer during a test.

452

Figure 16-8. Prior to use, the motor casings are soaked in a solution of alcohol and water for 24 hours.

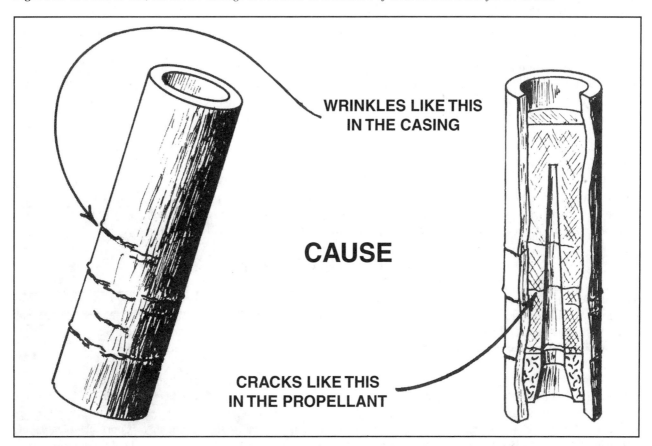

Figure 16-9. Ring-shaped wrinkles in the finished motor casing cause cracks in the propellant grain. If you see any wrinkles like these, the motor is defective, and you should discard it by soaking it in water.

453

Figure 16-10. A plastic beer cooler with 5 lbs. of silica gel and some homemade racks makes a good motor dryer.

If a motor looks smooth when finished, but explodes shortly after ignition, the problem may be casing-propellant separation during drying. Casing-propellant separation is a sign that the casings are *too dry*, and you'll have to *decrease* the percentage of alcohol (and thereby *increase* the amount of water). Keep a record of the alcohol percentage that works, and *forever thereafter* use that percentage to build *all* of the motors made with that particular casing.

The glue in most commercial paper casings is soluble in water, but not in alcohol, and the small percentage of water in the soaking solution will *not* make the tubes unravel. It *will*, however, soften the glue. If you soak the casings for too long, the glue will gradually leach out into the solution. As this happens, the solution turns brown and leaves a sticky residue when it dries. When the residue becomes obnoxious, discard the solution, and make up a fresh batch. As the glue leaches out of the casings, the casings weaken, and the solution becomes contaminated with glue. I've built successful motors using casings soaked for up to a month, but for the *best* result, limit the total soaking time to 72 hours.

Drying the Motors

The charcoal in the propellant is a natural absorbent, and once it is damp, it *stays* damp for a *long time*. These motors therefore take a long time to dry. In normal weather even the smallest take 10 days, and the G and H motors take a month or more. To find out exactly how long a particular motor takes, weigh it right after you've finished making it, and weigh it again every 2 days thereafter. When it stops losing weight, it (and all the motors like it) will be dry.

In my original book I suggested that people speed the drying process with a food dehydrator, but I now advise against this practice. Food dehydrators work by blowing a breeze of warm air across the items to be dried. If the air is *too* warm, it can damage the motors in ways that can't be seen by a visual inspection, and in damp weather the dehydrators are ineffective. I now advise my readers to use the apparatus shown in *Figure 16-10*. It amounts to a portable, plastic beer cooler with 5 lbs. of **silica gel** in the bottom, and a stack of wood-and-hardware cloth racks to hold the motors. **Silica gel** is a reusable, synthetic desiccant made from a specially processed form of sodium silicate. The little capsules that say "Do Not Eat" found in bottles of vitamins usually contain silica gel. They are placed there to keep the vitamins dry.

MOTOR	CASING ID	NOZZLE THROAT	CORE LENGTH	APPROX. PROP. WT.	MAXIMUM THRUST	TOTAL IMPULSE
A3	1/2"	5/32" dia.	3/4"	1/4 oz.	3-1/2 lbs.	2.4 Nt.-secs.
B7	1/2"	3/16" dia.	1-1/4"	3/8 oz.	5 lbs.	4.3 Nt.-secs.
C9	1/2"	1/4" dia.	2-1/4"	1/2 oz.	9 lbs.	5.7 Nt.-secs.
D18	3/4"	1/4" dia.	1-1/2"	1 oz.	12-1/4 lbs.	11.7 Nt.-secs.
D25	3/4"	5/16" dia.	2-3/8"	1-1/2 oz.	18 lbs.	18.5 Nt.-secs.
E30	3/4"	3/8" dia.	3-3/8"	2 oz.	27 lbs.	26 Nt.-secs.
F50	1"	7/16" dia.	3-1/2"	3 oz.	36 lbs.	53 Nt.-secs.
G75	1"	9/16" dia.	5-1/2"	4 oz.	45 lbs.	80 Nt.-secs.
G100	1-1/8"	9/16" dia.	5-1/2"	5 oz.	60-80 lbs.	100 Nt.-secs.

Figure 16-11. *The basic design parameters for the 9 rocket motors in my original book.*

Silica gel is sold in the packet form by laboratory and chemical dealers, but you can also buy it by the pound. It absorbs up to 10% of its own weight in water, and when saturated, can be dried out in a kitchen oven, and used again and again. There are two forms of silica gel, and for drying these motors you will need both. To fill the beer cooler in *Figure 16-10* you'll need 5 lbs. of the "regular" gel, and a 4 oz. bottle of the "indicating" gel. The regular gel granules are clear, and the indicating granules are dyed blue. When the dyed granules are dry, they remain blue. When wet, they turn yellow or pink. By mixing 4 ounces of the indicator gel with 5 lbs. of the regular gel, you can tell when the mixture is wet by watching the color of the indicator granules.

To use the device in *Figure 16-10*, place it in a warm room. Place the wet motors on the racks, and keep the lid on until the motors are dry. When the indicating granules turn yellow or pink, you can dry the gel by placing it in an oven-safe container, and heating it at 180° to 200° Fahrenheit until the crystals turn blue again. Let the gel cool, place it back in the beer cooler, and you're ready to dry another batch of motors. This device will cut the normal drying time in half, causing the largest motors to dry in two to three weeks. Because the drying occurs at room temperature, there is no danger of damage, and if you keep the silica gel clean, it can be used *many* times before it has to be replaced. At the time of this writing you, can expect to pay about $5.00 per lb. for the regular gel, and maybe twice that amount for the indicating gel.

Important note. You *cannot* vacuum dry motors made with water, because water spontaneously freezes in a vacuum. This phenomenon is, in fact, the basis for the process known as "freeze drying". If you place motors and/or casings wet with water in a vacuum, the water will freeze. Then, as the water-ice slowly sublimes into the vacuum, the casings will *not* shrink properly, and the propellant's oxidizer will *not* grow the large crystals needed to form it into a solid propellant grain.

General Advice

My original book contained the 9 motor designs listed in *Figure 16-11*. Unlike the *new* motors, I made them with steeply tapered cores, and I advised my readers to place their igniters *no deeper than halfway to the front of the core*. The original propellants burned quickly, and even the G100 had a burn time of less than one second. The tapered core and the partially inserted igniter provided a slower ignition time, and a few seconds of fire and smoke before lift-off. This was my attempt to provide a more realistic rocket flight; a feature considered important to people flying scale models, or rockets with fragile payloads. Unfortunately, the idea backfired, because people would often place the igniters *too deep*. When they did this, the motors would explode shortly after lift-off, generating another point of dissatisfaction with this motor building system.

To those of you planning to experiment with this method, I now advise you to start fresh. Forget the steeply tapered cores and the partially inserted igniters, and use the same tooling used to build the other motors in this book. Slow the propellant's burn rate by replacing the charcoal with 12 hour anthracite, and do your early experiments with core spindles shorter than what you think you will need to make a working motor. Then lengthen the cores a little at a time until you achieve the performance you want.

17. ELECTRIC IGNITERS and a HOMEMADE IGNITION SYSTEM

To *safely* start a rocket motor, you should always use an **electric igniter**, also called a **squib**, or an **electric match**. Electric igniters come in many forms and sizes, but the basic operating principle is always the same. A thin, metal fuse wire, or sometimes a strip of foil, is embedded in a pyrotechnic composition. When an electric current passes through the wire, the wire gets hot and lights the composition, which then ignites the rocket motor. There are many ways to make an igniter, and this is how I make them.

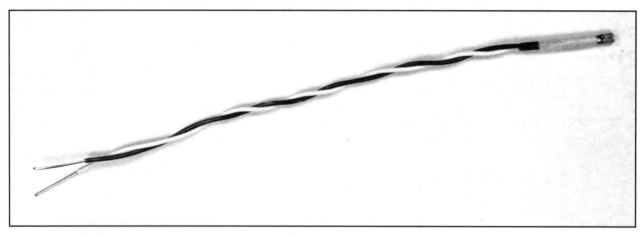

Figure 17-1. A finished paper-match igniter.

A Paper-Match Igniter

You can make a **paper-match igniter** (*Figure 17-1*) right at the launch site. You can use it immediately, and you can bring the materials with you.

MATERIALS LIST

1. A hank of thermostat-doorbell wire
2. A short length of stereo speaker wire
3. A book of paper matches
4. A roll of 3/4" masking tape
5. A single edge razor blade
6. A pair of sturdy pliers

Thermostat-doorbell wire and **stereo speaker wire** (*Figure 17-2*) are sold by the foot and by the roll at hardware stores. The doorbell wire is made from a twisted pair of stiff, single strand wires with red and white plastic coatings. The stereo speaker wire looks like regular lamp cord, but it's thinner. It is typically called "18-2". It is also coated with plastic, and the wire inside is fabricated from multiple strands of very thin wire, which you can unravel, and use for the fuse wires in these igniters.

1. Separate an 8" length of doorbell wire into a red and a white strand. Then strip the ends of each strand back about 1/2". I call these strands "lead wires" (pronounced "leed"). Strip the plastic from the end of a length of stereo speaker wire, and cut the multistrand wire inside into a pile of single strands about 2" long. Hereafter I'll call these thin, *bare* wires, "fuse wires". Cut or tear off a single paper match, and cut a 3/4" square of masking tape (*Figure 17-3*).

456

Figure 17-2. Hanks of doorbell wire (left) and stereo speaker wire (right). You can buy them at a hardware store.

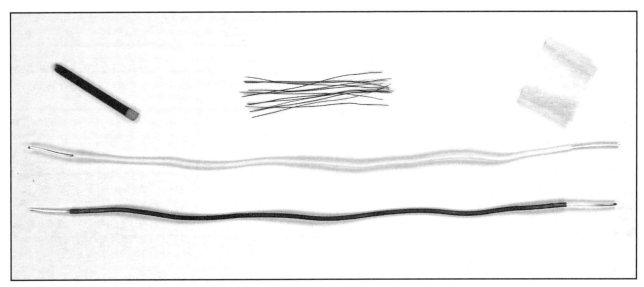

Figure 17-3. Two lead wires, a pile of fuse wires, a paper match, and a piece of masking tape. These are the things you'll need to make a paper-match igniter.

2. With the pliers, pinch a little flat spot on one end of each lead wire. Pick up a lead wire in one hand, and a fuse wire in the other. Then twist one end of the fuse wire around the flattened end of the lead wire with 3 to 4 wraps. The flat spot keeps the fuse wire from slipping off. Switch the assembly you've just made to your other hand. Pick up the *second* lead wire, and wrap the free end of the fuse wire around the flattened end of this second lead wire. The result should be a red and a white plastic-coated lead wire connected by a thin, bare fuse wire (*Figure 17-4*).

3. With a single edge razor blade, cut a slit in the tip of the paper match (*Figure 17-5*). If the match tip crumbles when you cut it, throw away the crumbly match. Hold a *new* match up to your mouth, and breathe on it for 30 seconds. The moisture in your breath will soften the tip, and the composition on the tip will then cut cleanly and smoothly. Fold the wire assembly that you made in **step 2** down over the match so that the fuse wire passes through the slit in the match's tip, and the two lead wires are separated by the match's paper stem (*Figure 17-6*).

4. Wrap the whole thing together with the square of masking tape. Then twist the lead wires together, and the igniter is finished (*Figure 17-1*).

These single paper-match igniters fit perfectly into the cores of rocket motors with 1/4" and 5/16" dia. nozzle throats. For larger motors, tape on extra matches (*Figure 17-7*). A paper-match igniter should have enough matches to insure a certain ignition, yet fit loosely into the core of the motor for which it was made.

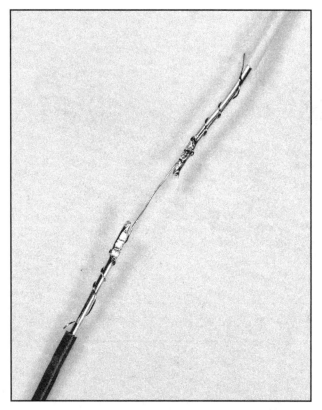

Figure 17-4. The three-wire assembly is made from two thick lead wires and a thin fuse wire. The flattened ends of the lead wires keep the fuse wire from slipping off.

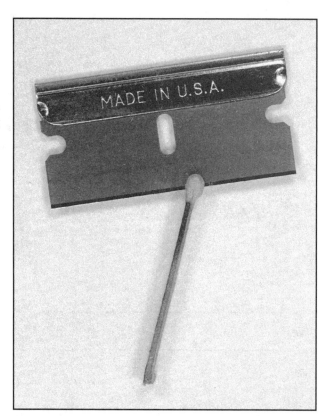

Figure 17-5. A single edge razor blade cuts a slit in the tip of the paper match.

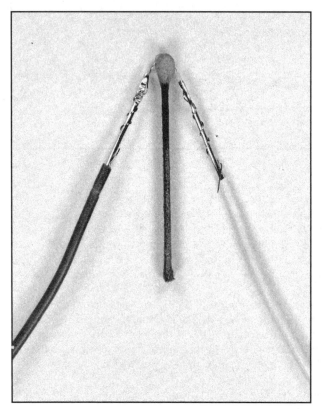

Figure 17-6. The three-wire assembly is folded down over the paper match, so that the fuse wire passes through the slit in the match's tip, and the lead wires are separated by the match's paper stem.

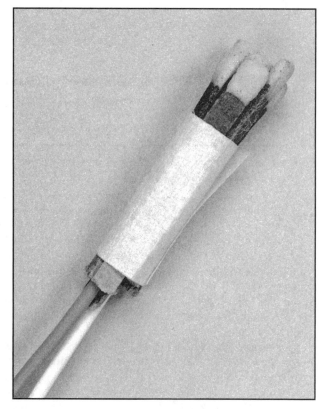

Figure 17-7. For motors with larger cores, extra matches are taped around a single match igniter.

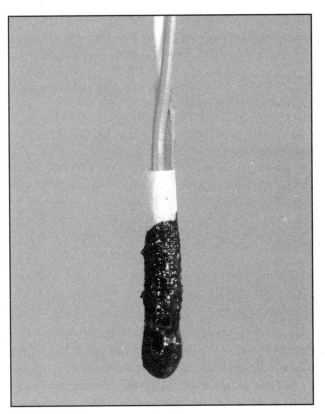

Figure 17-8. The black powder syrup should form a smooth and uniform coating.

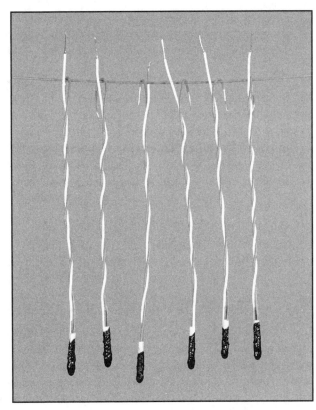

Figure 17-9. Half a dozen high-heat igniters hung up to dry. Eight hours in a warm, dry room is usually sufficient.

A High-Heat Igniter

A paper-match igniter works well with an easy-to-light propellant, but some of the propellants in this book are *hard* to light. That is, they require a hotter blast of flame to get them going. If you plan a day ahead, you can make a *much* hotter igniter that will light *anything*, and here's how to do it.

MATERIALS LIST

1. Some finished paper-match igniters
2. A few ounces of commercial black powder
3. A few grams of dextrin
4. A cup and something to stir with
5. A small amount of water

You can buy **commercial black powder** at a gun shop or a sporting goods store, but most states require a special "magazine" for commercial black powder storage, and they limit the number of cans that a dealer can display. For this reason, many dealers don't want to fool with black powder, and it might take a few phone calls to find someone who sells it. As of this writing, you can expect to pay $10 to $15 for a 1 lb. can, and any grade will work. For this application the grain size doesn't matter.

If you can't find commercial black powder, you can make a good substitute in your powder mill. Set up the mill per the instructions in **Chapter 3**, and follow the batch size recommendations in *Figure 3-10*. Then mill a 12 hour batch made from **75% potassium nitrate**, **15% carbon black**, and **10% powdered sulfur**. To make a smaller amount, cut the batch to 1/4 of the recommended size, and reduce the milling time to 3 hours.

Though I included **dextrin** in the materials list, I *didn't* include it in the formula above, because dextrin is not an absolute necessity. I don't use it myself. Dextrin is a water-soluble glue that is often used in pyrotechnics, and you can buy it from a chemical dealer. These high heat igniters work fine without it, but the dried black powder composition that makes them work can be fragile and crumbly. Adding 2% dextrin makes a sturdier finished product.

1. Pour one to two ounces of commercial black powder (or the homemade powder described above) into a cup, and add water in small amounts, stirring as you go. When the powder becomes wet, stop and let it soak for a few minutes. Then add a little more water, and stir again. Keep adding water until the mixture is the consistency of a thick syrup.

2. Dip a paper-match igniter's tip into this black powder syrup, so that the syrup covers all of the tip and some of the masking tape as well. Stir it around gently, and lift it out. If the syrup is the right consistency, the match will come out with a smooth and uniform coating (*Figure 17-8*). If the syrup is too stiff, it will form a thick glob, and you'll have to thin it by adding more water. If the syrup is too thin, it will drip back into the cup, and you'll have to thicken it by adding more powder. Some trial and error will be necessary.

3. Bend the end of one lead wire into a hook, and hang the igniter up to dry. If you stretch a long wire between two supports, you can hang the igniters side by side, all in a row (*Figure 17-9*).

If you make these igniters in the evening, and hang them in a warm room, they'll be dry and ready to use by morning. Igniters made in this way generate an *incredible* amount of flame. I used them for all the motor tests done for this book (more than 700 tests), and I didn't experience a single failure.

● A Fast-Drying Igniter Composition

If you have a powder mill, and you're building the motors in this book, you can make an *excellent* igniter syrup from the propellant itself by mixing it with acetone. You mix the syrup, and dip the igniters in the same way that you do when using water. But the red gum or the vinsol resin is the binder, and, because the acetone evaporates so fast, the igniters are dry and ready to use in 15 minutes. Because the acetone evaporates so fast, the syrup rapidly thickens, and you'll have to periodically thin it as you work. When you're finished making igniters, whether the syrup is made with water or acetone, you can dry it out, store it in a safe place, and reuse it later by dissolving it again in the appropriate solvent.

A Mini-Igniter

The paper match igniters and the high-heat igniters work fine for motors with nozzle throats down to 1/4" dia. For motors with smaller throats, you'll need these **mini-igniters**.

MATERIALS LIST

1. The thinnest, single strand, plastic coated wire you can find
2. Some bare, uncoated wire that's even thinner
3. A piece of stiff paper or an old business card
4. Some 3/32" or 1/16" dia. heat shrink tubing
5. A cup of the same black powder syrup used to make the high-heat igniters
6. A sharp pair of scissors
7. A match, a cigarette lighter, or a heat gun

You can buy **single strand, plastic coated hookup wire** down to 26 gauge at most electronics stores, and as of this writing, Radio Shack sells plastic coated **magnet wire** that is even thinner (about .008" diameter). I use it for the lead wires in these igniters. As of this writing, my company, **The Teleflite Corporation**, sells .003" dia. nichrome wire, and because I have it on hand, I use it for the fuse wires in these igniters. Nichrome is ideal, but copper works too. It just takes more electric current to make it work.

Electronics stores sell the **heat shrink tubing**. Heat shrink tubing is made of a special plastic that has a molecular memory. During the manufacturing process, it is heated and stretched. When you warm it with a match or a heat gun, it shrinks back to its original size, which is about half of its stated diameter. You can buy it in a wide range of sizes, and you can cut it with a pair of scissors. You can buy an inexpensive **heat gun** at a hobby shop. Heat guns look and work like hair dryers, *but they're hotter*, and hobbyists use them to shrink-fit the plastic covering on model airplanes.

1. Cut two 6" lengths of plastic-coated magnet wire, and strip their ends back 3/8". Cut a piece of thin fuse wire (in this example the .003" nichrome) about 1-1/2" long. With the scissors, cut a 1/16"- wide strip of paper from the business card, and cut a little notch in one end. Then cut a piece of heat shrink tubing about 3/4" long (*Figure 17-10*).

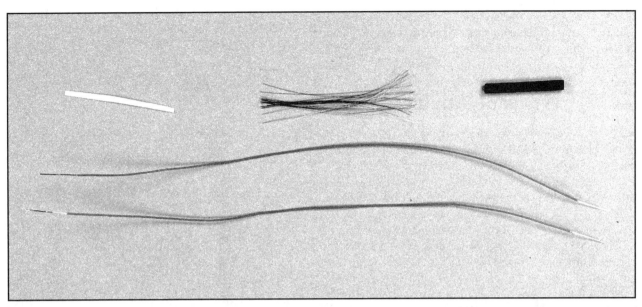

Figure 17-10. Two 6" lengths of magnet wire, a 1-1/2" length of fuse wire, a strip of paper cut from a business card, and a short length of heat shrink tubing. These are the things you will need to make a mini-igniter.

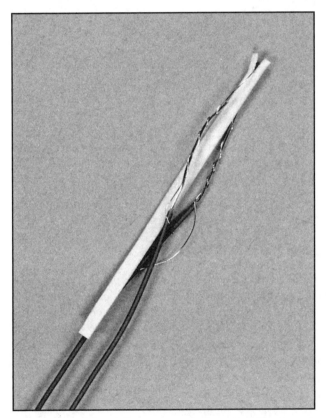

Figure 17-11. The sharp bend in the fuse wire fits through the notch in the paper strip.

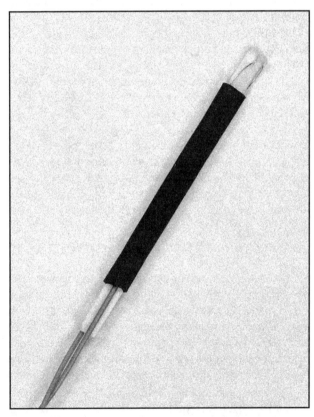

Figure 17-12. A short piece of heat shrink tubing locks everything together.

2. Wrap the two lead wires together with the fuse wire, and put a sharp bend in the middle of the fuse wire. Then slide the strip of business card paper between the two lead wires, so that the bend in the fuse wire fits through the notch in the end of the strip. (*Figure 17-11*).

3. Holding the entire assembly in one hand. Slip the heat-shrink tubing over the end, and leave 1/4" of the notched end of the paper strip exposed. Then warm the heat-shrink tubing with a match or a heat gun. As the tubing gets hot, it will shrink down over the wire-paper assembly, cinch it up tight, and lock everything together (*Figure 17-12*). Finally, dip the tip of the igniter into the black powder syrup, stir it around, lift it out, and hang it up to dry.

Figure 17-13 shows a finished mini-igniter curled up around a dime. Because these igniters are so small, it takes manual dexterity and good eyesight to make them. You'll also find patience and practice to be of considerable help.

General Advice

Always make an igniter's lead wires *extra long*, so that the tip of the igniter reaches all the way to the front of the motor's core *with several inches to spare*. The core of an **NV6-340-I100** is 15" long, so the wires should be *at least* 21" long. Also make sure that the two lead wires, and the tail ends of the fuse wire, are *completely* insulated from one another by the paper strip or the stem of the match. When an igniter fails, it is usually because a short in this area has prevented the electric current from reaching the fuse wire at the igniter's tip.

Important note. *These igniters will work with commercially made rocket motors, but they will **not** work with the commercial rocket ignition systems sold in hobby shops.* These homemade igniters require more electric current than a commercial system can supply. Commercial igniters are made on expensive, high tech equipment. Estes igniters are actually *welded* together with an amazing device that *could* be the world's tiniest spot welder. These homemade igniters are merely *wrapped* together, and the electrical contact between the wires isn't good enough to pass the tiny amounts of current that make a commercial igniter work.

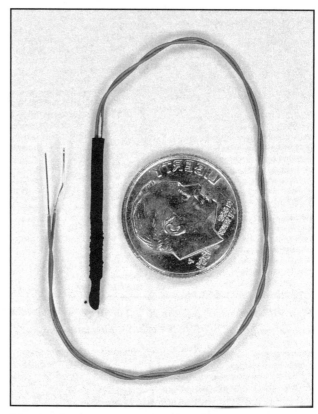

Figure 17-13. A finished mini-igniter. I've curled it around a dime to illustrate its tiny size.

You'll therefore need a bigger battery, and a car battery provides plenty of power. If, at first, such a big battery seems inconvenient, remember that, when flying rockets, you *drive* to the launch site. The battery is right there under the hood of your car, so you might as well use it. When you also consider the fact that, as of this writing, *some* commercial igniters cost more than a dollar each, a good look at what you are spending might convince you to make your *own* igniters and the homemade ignition system described below.

A Homemade Ignition System

You can assemble this ignition system from common electronic and hardware items, and, as of this writing, you can buy most of them (*Figure 17-14*) at Pep Boys, Walmart, and Radio Shack. The system consists of a heavy power cable with a pair of battery clamps, a keyed rotary switch for arming the system, a red light to indicate when the system is armed, and a push-button switch that sends power to the igniter, and launches the rocket. The part numbers below are valid as of the year, 2003. As time passes, the numbers will probably change, but items that are similar will work. **Important note.** When buying the jumper cables, chose one of the cheaper sets made with copper wire, but *thinner* cable. If the cable is too thick, it won't fit through the cable clamps and the other components used in the system's construction.

MATERIALS LIST

1. One plastic Radio Shack 7" x 5" x 3" Project Enclosure, cat. #270-1807 (or similar)
2. One Radio Shack "Jumbo" 12 volt lamp assembly (with bulb), cat. #272-336 (or similar)
3. One "300 amp" "Conduct-Tite" #85988 keyed auto battery isolator switch (sold at Pep Boys) (or similar)
4. One sealed 60 amp "Conduct-Tite" #85984 auto push-button starter switch (sold at Pep Boys) (or similar)
5. One set of automotive battery "jumper cables" (Walmart's were the cheapest)
6. 5 feet of "audio"-type stereo speaker wire (looks like lamp cord, but thinner)
7. One 10 amp, 125 volt, 2-prong plug, and one 20 amp, 125 volt, grounded 3-prong socket
8. One package of Radio Shack "Mini" alligator clips cat. #270-380A (or similar)
9. Two 3/4" cable clamps, plus the appropriate nuts, bolts, wire nut, and solderable wire-mounting lugs

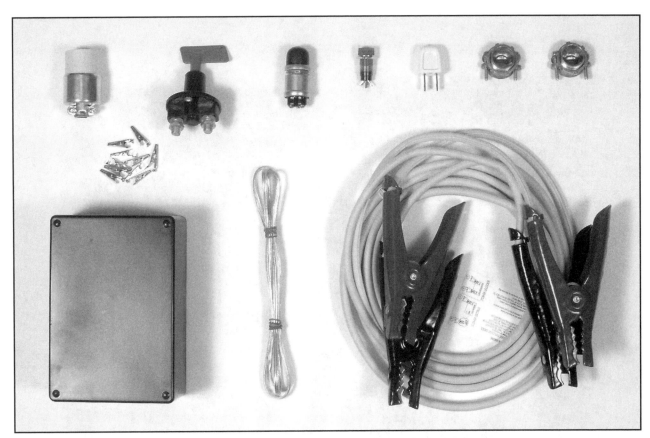

Figure 17-14. The parts for assembling a homemade ignition system. I bought most of them at Radio Shack, Pep Boys, and Walmart. I bought the wire, the plugs, and the cable clamps at a hardware store.

Figure 17-15. The plastic Radio Shack "project enclosure" with the keyed, auto battery isolator switch, the push-button ignition switch, and the 12 volt lamp assembly. For clarity, I've removed the red plastic key, and laid it next to the box.

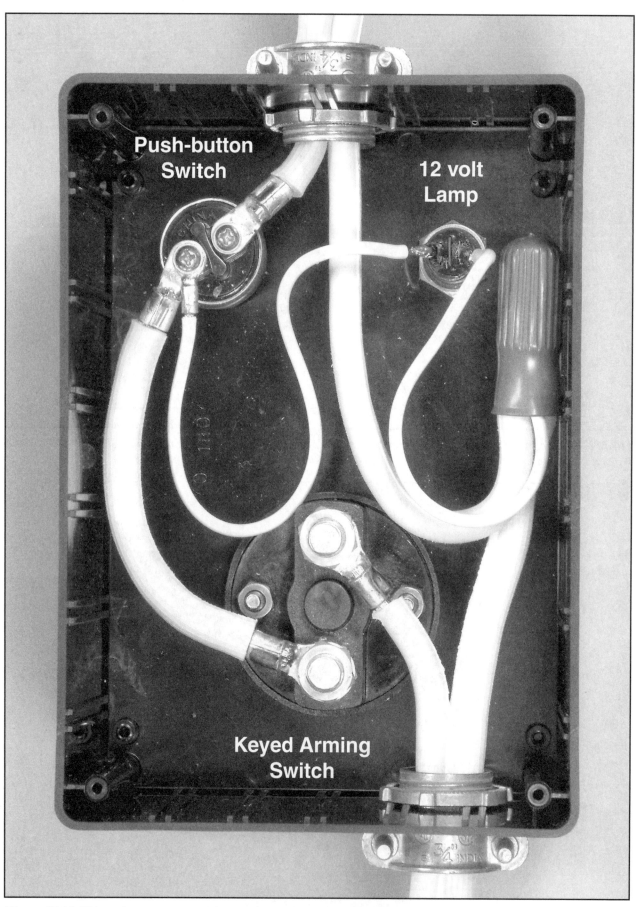

Figure 17-16. *The wiring arrangement for the homemade ignition system.* **Up** *is toward the 3-prong socket, and* **down** *is toward the battery clamps. Use a good grade of electronic or PC solder for the soldered connections.*

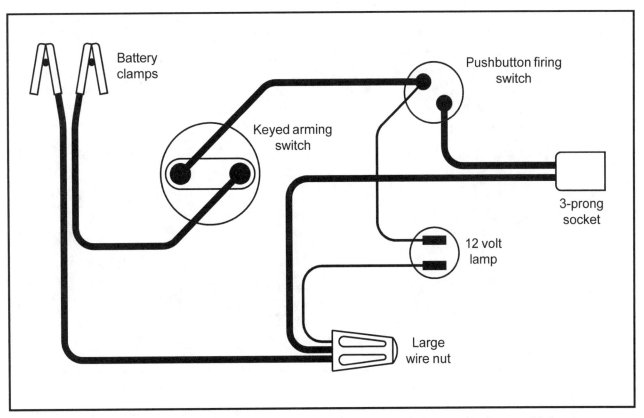

Figure 17-17. The wiring diagram for the homemade ignition system. **Note.** *If you turn this diagram 90 degrees* **counter-clockwise**, *it will look more like the photo in* ***Figure 17-16***.

1. In the plastic "project enclosure" (the plastic box), drill the appropriate-size holes for the lamp housing, the switches, and the cable clamps, and mount the various components as shown in ***Figure 17-15***.

2. Cut the clamps from one end of the jumper cable set. Then from this cut end, take the pieces of cable needed for the box's internal wiring, and the double wire leading out to the grounded 3-prong socket. Strip the ends of 2 pieces of this cable, and connect them to the 2 hot prongs of the 3-prong socket. Note that the socket's round ground pin is *not* connected. You're using a 3-prong socket, *only* because that's what you'll need to connect to the standard electrical extension cord that leads to the launch pad.

3. With a good grade of PC solder, crimp and solder on the necessary connector lugs, and make the connections shown in ***Figure 17-16***, and the wiring diagram in ***Figure 17-17***. Use thinner wire (i.e. 18 gauge) for the 12 volt lamp, and use a large wire nut to make the final connections shown at the right side of the photo.

4. Strip one end of the 5 foot length of speaker wire, crimp on a pair of alligator clips, and install the male, two-prong plug on the opposite end (***Figure 17-18***). Replace the cover on the "project enclosure" (the plastic box), and the ignition system is finished (***Figure 17-19***).

● How to Use the Ignition System

1. Turn the arming switch *off*, and remove the red key from the switch. Then lift the car's hood, and connect the battery clamps to the battery (***Figure 17-20***). Note that polarity is *not* important. Plug a standard 25-to-50 foot outdoor electrical extension cord into the system's 3-prong socket, and walk the opposite end out to the launch pad. Then plug the 5-foot wire (with the alligator clips) into the extension cord.

2. Insert the igniter into the motor until it touches the front of the core (***Figure 17-21***). Make a short, L-shaped bend in the igniter leads. Tape them to the end of the motor with a single strip of masking tape (***Figure 17-22***), and set up the rocket on the launch tower. Then with the two alligator clips, connect the system to the igniter's lead wires (***Figure 17-23***).

3. Walk back to the car, and insert the red key into the arming switch. Turn it clockwise until it snaps, and the red light comes on, indicating that the system is armed. Then push the black push-button to launch the rocket.

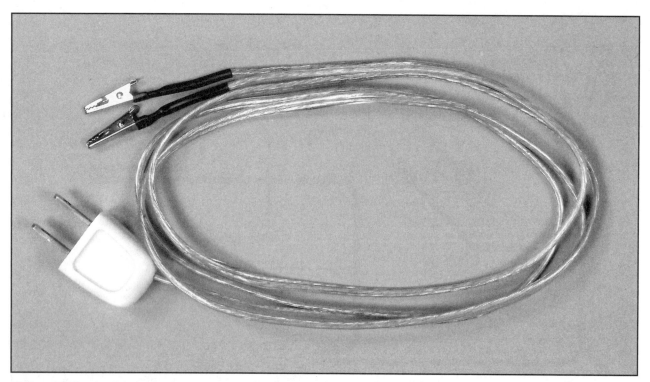

Figure 17-18. The 5 foot length of speaker wire with alligator clips and the male, two prong plug.

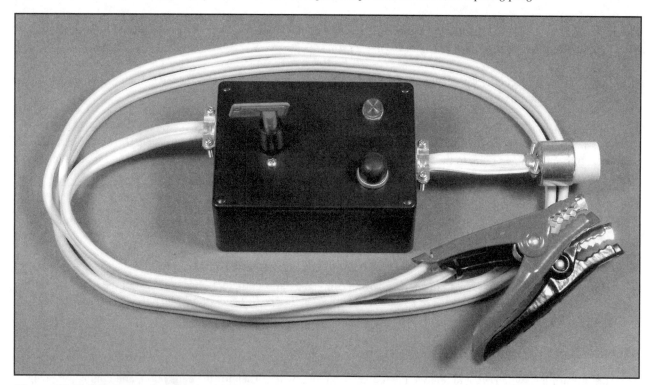

Figure 17-19. The finished rocket ignition system.

In *Figure 17-23* you'll notice something covering the connection between the alligator clips and the wire. When a rocket motor fires, it violently ejects the igniter, and in the process, the clips can be bent and broken off. To protect them, and make the joints between the wire and the clips last longer, I've covered them with two, short pieces of heat shrink tubing.

After a dozen or two firings, the 5-foot wire with the alligator clips will become burned and unusable, and you can replace it with a new one. I've tested this system with extension cords up to 75 feet long, and in every case it delivers a full 12 volts to the launch pad, and all the power needed for lighting the homemade igniters.

Figure 17-20. The ignition system is connected to the car's battery. Note that polarity is **not** important.

Figure 17-21. The igniter is inserted until its tip **touches** the front of the motor's core.

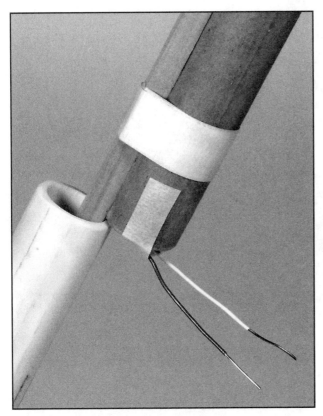

Figure 17-22. The igniter leads are secured to the end of the motor with a strip of masking tape.

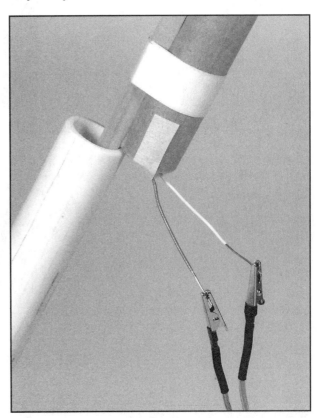

Figure 17-23. The ignition system is connected to the igniter with two alligator clips.

18. TEST EQUIPMENT

My Own Test Equipment

Important note. *At the beginning of this chapter, I've provided a description of my own test equipment so that you understand how my data was gathered. But you do **not** need equipment like this to test your own motors.* In the pages ahead, I'll show you how to make a homemade test stand and a homemade chart recorder with things you can buy at a hardware store.

To accurately predict a rocket motor's flight performance, you need an accurate chart of how its thrust varies with time. In the rocket industry, the accepted way of creating the chart is to have the motor push against a device called a **load cell**. The load cell outputs to a computer running software that graphs and interprets the data. *Figure 18-1* is a photo of the device that I made to gather the data for this book.

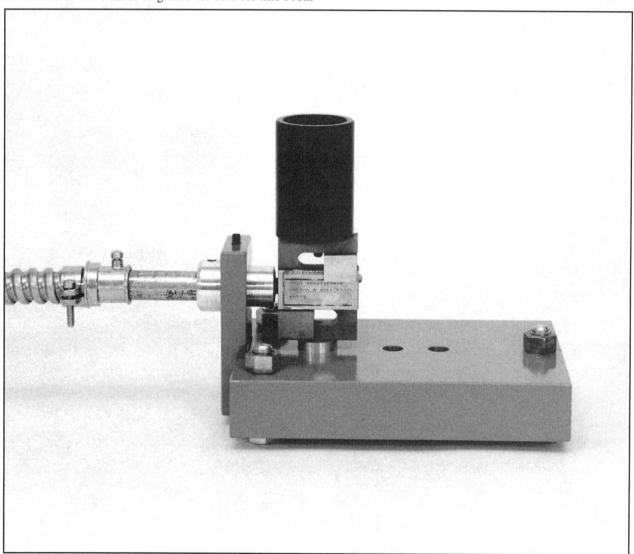

Figure 18-1. *The thrust testing stand used to gather the data for this book. It amounts to motor holder bolted to an S-bar load cell, which is bolted to a heavy block of steel. Data from the load cell feeds through a cable on the left to a computer in my shop.*

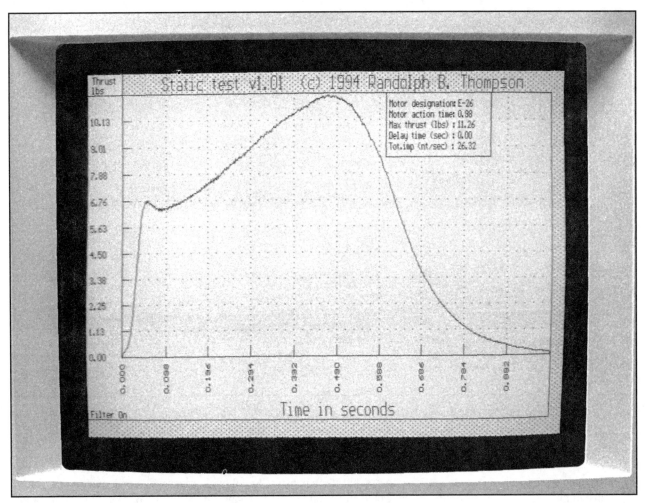

*Figure 18-2. Data from the load cell is processed and displayed by a 486 computer set up with an analog-digital board and testing software written by **Randy Thompson**.*

It consists of an S-bar type load cell bolted to a heavy block of steel (painted gray), supported by the heads of three, short bolts that form a sturdy tripod. On top of the load cell is an adapter that holds the motor being tested. The rocket exhaust fires *up*, the motor pushes *down*, and the load cell measures the force of the push. A load cell is an electronic device. To protect its cable during a test, I've added a bracket on the left that supports a 10 foot length of flexible, steel conduit. The cable passes through the conduit; then out to the amplifier and power supply that make the load cell work.

A load cell is an interesting device. The one in the photo is made of hardened steel, and the central bar of the **S** has a cavity containing a diamond-shaped printed circuit called a **strain gauge**. A 10 volt potential is applied between the left and right points of the diamond. When the **S** is compressed, the diamond is warped. The distortion is only a few thousandths of an inch, but the stress causes electrical resistances in the *sides* of the diamond to fall out of balance, and a *secondary* voltage develops between the top and the bottom of the diamond. The size of the secondary voltage is proportional to the amount of compression, so by measuring this voltage, you can determine the force that caused it.

Load cells are very sensitive, and when properly made, they are accurate to 1/4% of their full scale. They are also very expensive. The one in the photo cost $300. If it were accidentally overstressed, as might happen if a motor exploded, it would be damaged beyond repair. To prevent such damage, I've machined two, small aluminum blocks, and placed them between the bars of the **S**. They act as stops to keep the load cell from being compressed beyond its design limits.

A d.c. electric current is provided by an accurate and steady power supply. The tiny, secondary voltage is boosted by an amplifier, then sent to an analog-digital board in a 486 computer (***Figure 18-2***). The force vs. time data is graphed and printed; then saved in both ASCII data and XY file formats. The testing software, written for me by **Randy Thompson**, can be zeroed for a motor of any weight, and calibrated for any load cell. It can even compensate for a damaged load cell that doesn't give accurate readings.

Figure 18-3. A homemade thrust-testing stand. The one in this photo is suitable for testing motors up through F.

A Homemade Rocket Motor Test stand

The electronic equipment is accurate and easy to use, but it's expensive. And though prices may come down in the future, at the time of this writing, to buy everything new would cost $500 to $1,000. Thanks to the physics of rocket motor operation, a motor's performance parameters are mathematically interrelated. Therefore, if you're testing a motor who's design and propellant formula are the same as one of the motors in this book, you can adjust its performance to the *approximate* specs. quoted in this book by simply insuring that it generates the same maximum thrust. When testing motors up through **F**, you can measure the maximum thrust with the device shown in *Figure 18-3*.

It consists of a long wooden arm that hinges on a plywood platform, mounted on a flat plywood base. The free end of the arm pulls down on a spring supported by an arrangement of pipe fittings. On top of the arm, a few inches away from the spring, is a rocket motor holder made from a plastic water pipe cap. By switching pipe cap sizes, you can accommodate various motor diameters, and by switching springs, you can test motors with various thrusts.

To one side of the arm is a vertical piece of sheet metal that holds a piece of note paper. An ultra-fine point marking pen with water-based ink is mounted on the underside of the arm with a rubber band, so that is scribes a vertical line on the paper when the arm moves up and down. At the time of this writing, a Pilot (brand name) "Razor Point liquid ink marker" with black ink works well. A sheet metal disk moving up and down in a 3 lb. soft margarine tub filled with heavyweight gear oil acts as an oscillation damper. Note that in the photo above, I've left-out the margarine tub to give you a clear view of the damper's metal disk. For a photo of the stand with the tub in place, see *Figure 18-24* on **page 483**.

You can buy the materials at a hardware store. **Important note**. *Pick a hinge that is loose and exhibits some side-to-side wobble*. The one I bought is a **Stanley** cabinet hinge, but any brand will work, as long as it swings *loosely*. The springs in this example are made by the **Gardner Spring Co**. of **Tulsa, Oklahoma**. They come in a variety of sizes, but you *don't* have to use these exact ones. The Gardner springs are 6" long (including the end loops), but you can use longer or shorter springs if you adjust the height of the other parts accordingly. I bought the gear oil at an auto parts store.

When setting up the stand, try to pick springs that stretch no more than 2 to 2-1/2 inches under a full-thrust load. This will keep the maximum travel of the pen within the limits required by the chart recorder described in the pages ahead. The following instructions explain how to make the stand shown in the photos. *Figure 18-4* on the **page 471** is an exploded view of the stand's components, and the capital letters in the materials list refer to the letters in *Figure 18-4*.

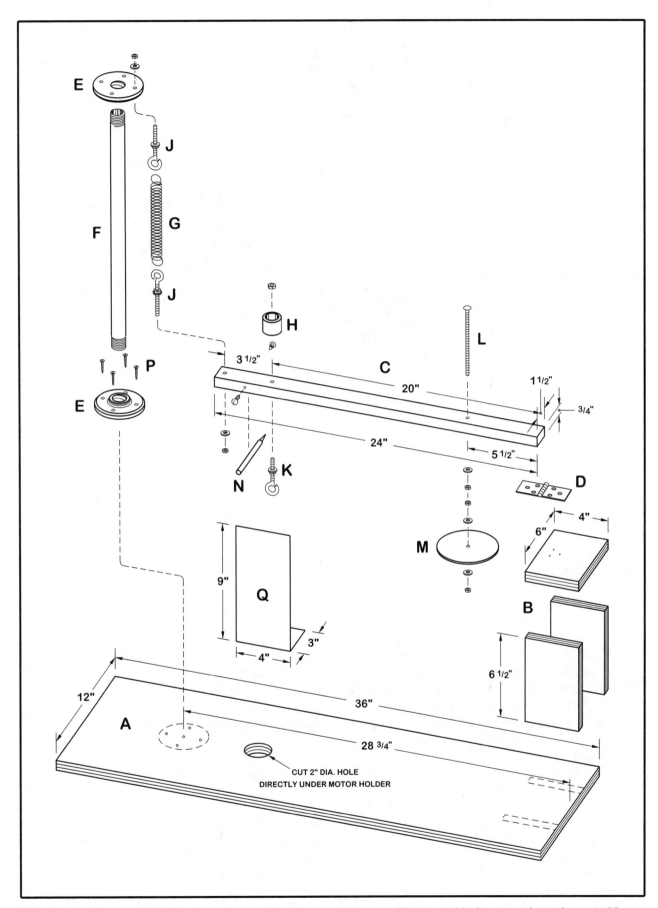

Figure 18-4. An exploded view of the test stand's components. The capital letters in this drawing refer to the capital letters in the materials list on page 472.

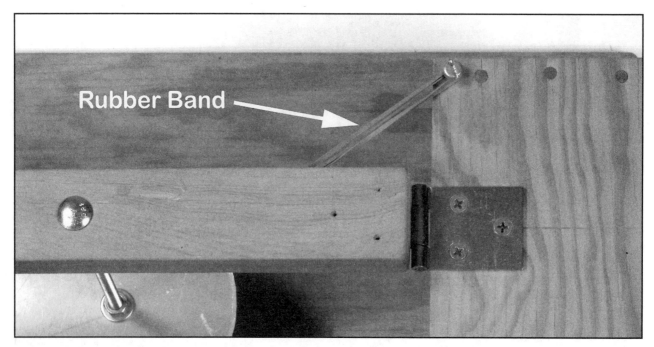

Figure 18-5. A rubber band is stretched between a wood screw on top of the hinge platform and another screw (not visible in this photo) on the underside of the thrust arm. The rubber band pulls the arm slightly to the side, and holds the pen against the paper.

MATERIALS LIST

- **A.** Test stand base: One piece 3/4" CDX or better grade plywood, 12" wide x 36" long.
- **B.** Hinge platform: Two pieces 3/4" CDX or better grade plywood, 4" wide x 6-1/2" long.
- **B.** Hinge platform: One piece 3/4" CDX or better grade plywood 4" wide x 6" long.
- **C.** Thrust arm: One piece clear pine or hardwood "one-by-two", 3/4" thick x 1-1/2" wide x 24" long.
- **D.** Thrust arm hinge: One 1-1/2" wide Stanley cabinet hinge (or similar) with mounting screws.
- **E.** Spring support assembly: Two 3/4" galvanized floor flanges.
- **F.** Spring support assembly: One 3/4" x 18" galvanized pipe nipple, threaded on both ends.
- **G.** A selection of 6" extension springs in various widths & wire sizes.
- **H.** Motor holders: One each of 1/2", 3/4", 1", and 1-1/2" PVC plastic, "slip fit" type pipe caps.
- **J.** Spring support assembly: Two 1/4" x 3" eyebolts with 2 nuts & 2 washers for each bolt.
- **K.** Motor holder assembly: One 1/4" x 2" eye bolt with 2 nuts & one washer.
- **L.** Support rod for oscillation damper: One 1/4" x 6" length of threaded rod, or a same-size carriage bolt threaded its entire length, with 3 nuts and 3 washers.
- **M.** Oscillation damper: One stiff, sheet metal disk, 4-1/2" dia., with a 1/4" dia. hole drilled in its center.
- **N.** One ultra-fine point marking pen with water-based ink.
- **P.** Four #12 x 1" F.H. wood screws.
- **Q.** Sheet metal paper holder: One piece of stiff aluminum or galvanized sheet metal, 4" X 12".
  Four #8 x 3/4" R.H. wood screws.
  Assorted nails and woodworking glue.
  Oil container for oscillation damper: One plastic 3 lb. soft margarine tub with lid.
  1-1/2 quarts of SAE 140-weight gear oil.

1. Cut the 12" x 36" plywood base (*Figure 18-4-A*) to size. Mark the positions of the hinge platform and the pipe fitting spring support, and cut a 2" dia. hole directly below the future location of the PVC motor holder. Cut the plywood pieces for the hinge platform (*Figure 18-4-B*). Assemble them as shown, and mount them at the indicated location on the plywood base. Use small nails and *glue* to insure a rigid, sturdy, and permanent structure.

2. With a pipe wrench and a vise, screw the 3/4" floor flanges (*Figure 18-4-E*) onto the ends of the 3/4" x 18" pipe nipple (*Figure 18-4-F*), *and make sure that they are tight*. It is important that the spring support stand vertical, so when buying the floor flanges, examine them carefully, and try to avoid flanges that are warped. If necessary,

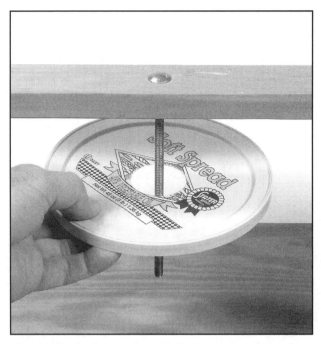

Figure 18-6. The carriage bolt, extending down from the thrust arm, passes through a hole in the lid of a 3 lb. soft margarine tube

Figure 18-7. The sheet metal disk is mounted on the lower end of the bolt with two nuts and two washers.

you can true up the face of the base flange on a metal lathe. To do so, screw it firmly onto one end of the pipe nipple. Mount the nipple, with the flange attached, in a metal lathe, and clean up the bottom of the flange with a face cut. Then mount the finished pipe assembly at the indicated location on the base with the four #12 x 1" F.H. wood screws (*Figure 18-4-P*).

3. Cut the "one-by-two" pine (or hardwood) thrust arm to length (*Figure 18-4-C*), and predrill the mounting holes for the spring support's eyebolt (1/4"), the motor holder's eyebolt (1/4"), and the oscillation damper's carriage bolt (1/4"). Locate and mount the two #8 x 3/4" R.H. wood screws that hold the pen's rubber band, and the two #8 x 3/4" R.H. wood screws for the rubber band that keeps tension on the thrust arm (*Figure 18-5*). Then connect the opposite end of the arm to the platform with the hinge (*Figure 18-4-D*).

4. Drill a 1/4" hole through the center of each plastic pipe cap (*Figure 18-4-H*), and mount the selected pipe cap on the thrust arm (*Figure 18-4-C*) with the 1/4" x 2" eye bolt (*Figure 18-4-K*). Mount the 1/4" x 6" carriage bolt (*Figure 18-4-L*) in the indicated position. Cut a 1-1/2" dia. hole in the center of the margarine tub's lid, and slip the lid over the bolt (*Figure 18-6*). Then mount the 4-1/2" dia. sheet metal disk (*Figure 18-4-M*) on the bolt's lower end with two nuts and two washers (*Figure 18-7*).

5. Fill the margarine tub with approx. 1-1/2 quarts of 140 weight gear oil. Place it on the base of the test stand. Lower the disk into the oil, and adjust the position of the tub, so that the disk moves freely up and down without touching the sides of the tub. Snap on the lid, and firmly affix the tub & lid to the test stand's base with two half-wide strips of duct tape.

6. **Important note.** *Extension springs are manufactured with their coils prestressed, so they remain tightly closed when at rest. It takes a significant amount of force just to **begin** to pull an extension spring open, and testing a motor with a spring in this condition will generate false data.* ***You must, therefore, pre-stretch each spring before you can use it.***

To relieve the stress in a *small* spring, slip one of the end loops over a punch or a screwdriver clamped ***tightly*** in a vise. Insert the shank of another screwdriver through the opposite loop, and pull gently (*Figure 18-8*). Allow the spring to relax. Then look *closely* at the coils. If the coils are closed, perform the stretch again, pulling harder, and pulling the coils farther apart. Then repeat the stretch with increasing force until *all* the coils remain slightly open when the spring is at rest (*Figure 18-9*). When *all* the coils remain open, the stress has been relieved, and the spring is ready to use.

Important safety note. *With larger springs, use a lever to perform the actual stretch, and be careful to keep your body, your arms, and your hands **well out-of-the-way** of the spring to prevent injury in the event that the spring breaks loose from the vise, and flies back in your direction.*

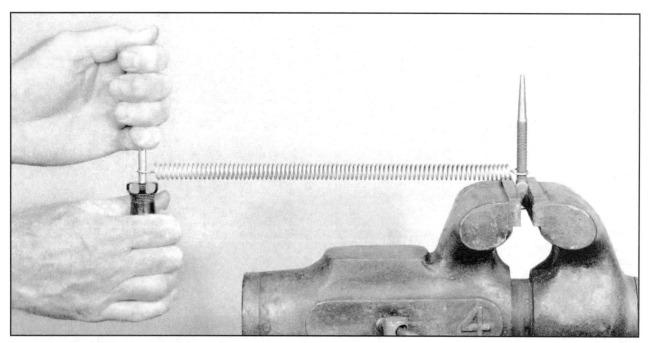

Figure 18-8. Springs used for motor testing must be pre-stretched to remove the stresses created during the manufacturing process.

Figure 18-9. The spring in this photo has been stretched to a point where its coils are slightly open, and is now suitable for testing rocket motors

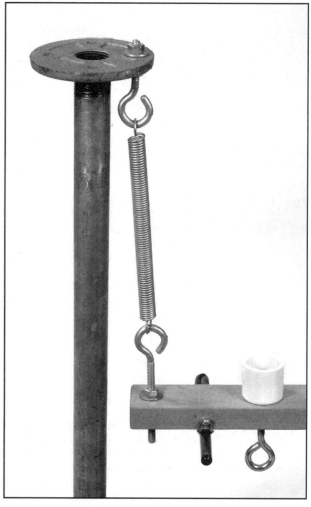

Figure 18-10. The spring is connected to the thrust arm and the spring support with two eye bolts.

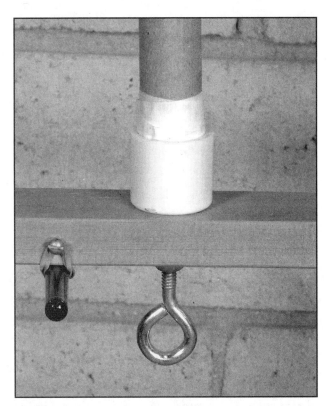 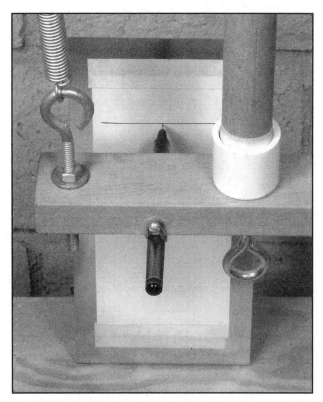

Figure 18-11. A few wraps of masking tape fit the motor snugly into the test stand's motor holder.

Figure 18-12. A short horizontal line, drawn where the pen touches the paper, represents a thrust of 0 lbs.

6. Select the appropriate spring, and connect it to the thrust arm and the spring support with the two eye bolts (*Figure 18-10*). Use the nuts and the washers on the eye bolts to adjust the position of the thrust arm *until the arm is level*. Eye bolts are manufactured with their end loops tightly closed. Before you can use them, you'll have to clamp them in a vise, and pry them open just a bit.

8. With a vise as a helper, bend the sheet metal paper holder to shape (*Figure 18-4-Q*). Set it on the test stand base, and hold it in place with a spring clamp or a heavy weight. It should *not* be mounted permanently. Mount the paper in place with two strips of masking tape. Strap the marking pen *tightly* in position with a rubber band, and back near the thrust arm hinge, mount the rubber band that keeps tension on the arm, and holds the pen against the paper (*Figure 18-5*).

Important note. *This equipment is suitable for testing motors up through F, but the "one by two" thrust arm (Figure 18-4-C), and the Stanley cabinet hinge (Figure 18-4-D), are not strong enough to safely handle anything larger.* Before you test a **G** or an **H** motor, replace the "one by two" arm with an arm made from a "one by four" (a piece of wood 3/4" thick x 3-1/2" wide), and replace the Stanley cabinet hinge with a larger hinge accordingly.

● The Test for Maximum Thrust

Before you can use the stand, you have to calibrate it. That is, you have to find out exactly how far down the pen moves when subjected to the force of a known weight. The weight should be an even number close to the maximum expected thrust of the motor-to-be-tested. In the following demonstration I'll be testing an experimental NV6-powered motor with a 3/4" i.d. casing, a 1/4" dia. nozzle throat, and a propellant mass (m_p) of 61.0 grams. I'm adjusting the propellant's baking soda content to achieve a maximum thrust of 10 lbs., so I'll use a 10 lb. calibration weight. To calibrate the stand, and perform the test, proceed as follows.

1. Place the stand *firmly* on a pair of saw horses or sturdy chairs.

2. Wrap the front of the motor with masking tape until it fits snugly into the motor holder (*Figure 18-11*).

3. Tape a piece of paper to the sheet metal paper holder. Slide the paper up against pen's tip, and mark the point where it touches the paper with a short, horizontal line (*Figure 18-12*). I call this the **zero line**, and it represents a thrust of **0 lbs.**

*Figure 18-13. The bucket is filled with sand until the total weight, **including the S-hooks and the chain**, equals **10.0 lbs**.*

Figure 18-14. The sand-filled bucket hangs (through the hole) from the eyebolt under the motor holder, and the "calibration line" is scribed on the paper.

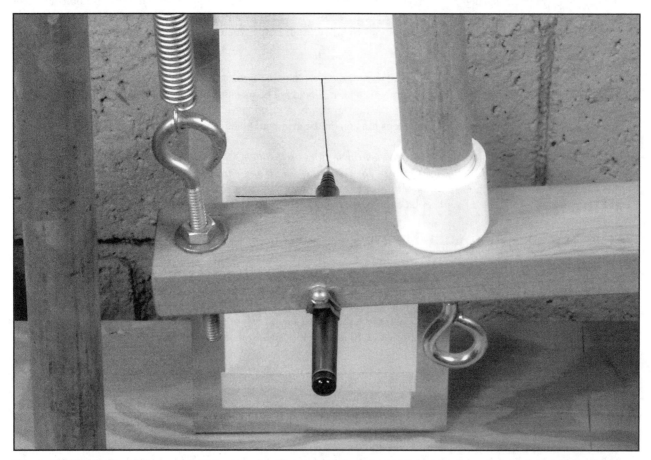

Figure 18-15. As the rocket motor fires, it pushes down on the arm, and the pen makes a vertical streak on the paper.

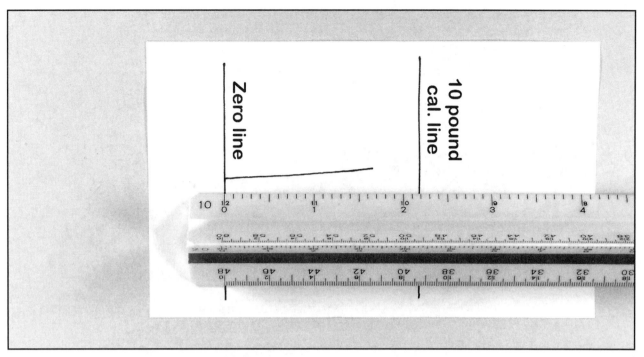

*Figure 18-16. An engineering scale, graduated in **tenths of an inch**, measures the length of the pen streak, and the distance between the **zero line** and the **calibration line**.*

4. Buy a small bucket (with a handle), a pair of S-hooks, and a short length of lightweight chain. Place the bucket on an accurate scale, toss in the S-hooks and the chain, and fill the bucket with sand until the *total* weight of the bucket *plus* the sand *plus* the S-hooks *plus* the chain equals 10.0 lbs. (*Figure 18-13*).

5. With the hooks and the chain, hang the sand-filled bucket from the eyebolt directly under the motor holder, so that the chain passes through the hole in the test stand's base. Once again, slide the paper up against the point of the pen, and draw a second line where it touches the paper (*Figure 18-14*). I call this line the **calibration line**, and for *this* test, it represents a thrust of **10.0 lbs.**

6. Slide the paper away from the pen, and remove the bucket, the hooks, and the chain. Slide the paper against the pen one more time. Then retire to a safe distance, and fire the rocket motor with an electric igniter. As the motor fires, it will push down on the thrust arm, and cause the pen to make a vertical streak on the paper (*Figure 18-15*). When the test is finished, remove the paper from the sheet metal holder, and take it to your desk for analysis.

● **Interpreting the Results**

With a ruler or an engineering scale divided in tenths of an inch, measure the length of the pen streak, and the distance between the zero line and the calibration line (*Figure 18-16*).

To calculate the maximum thrust (F_{max}), divide the length of the pen streak (L_{pen}) by the distance between the zero line and the calibration line (L_{cal}), and multiply the result by the calibration weight (F_{cal}).

EXAMPLE 18-1:

In this case the pen streak is 1.65" long. The distance between the zero line and the calibration line is 2.18", and the calibration weight is 10.0 lbs. The motor's maximum thrust (F_{max}) is calculated as follows:

$$F_{max} = (L_{pen} / L_{cal}) \times F_{cal}$$

$$F_{max} = (1.65" / 2.18") \times 10.0 \text{ lbs.} = 7.57 \text{ lbs.}$$

The motor's maximum thrust is therefore 7.57 lbs. The ideal thrust for this motor is 10 lbs., so this motor is underpowered. In this case, I'll keep building motors, testing them, and decreasing the propellant's baking soda content in 1% increments until I achieve a maximum thrust of approx. 10 lbs.

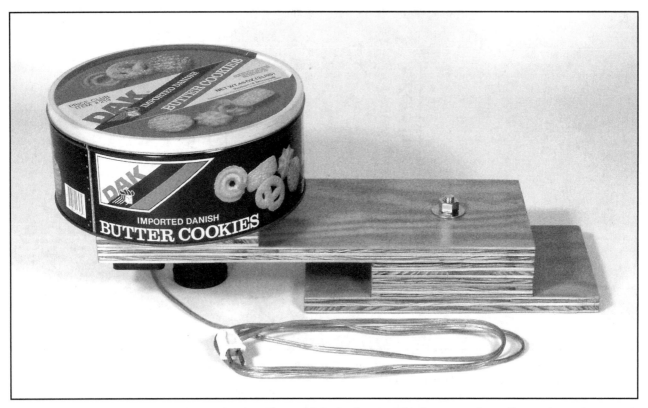

Figure 18-17. When mounted on the test stand in *Figure 18-3*, this homemade chart recorder will record a rocket motor's thrust-time curve. It is amazingly accurate, and it provides very useful data.

A Homemade Chart Recorder

I've personally designed, built, and tested all of the motors in this book. If you've copied one of these designs, used the same propellant, and adjusted your motor's maximum thrust to match the published value, *then the other performance parameters will be at least "within the ball park" of the stated values*. If you've adjusted the carbon fuel ratios so that the *baking soda* content is the same, the match will be even closer, and the thrust-time curve should be a pretty good match too. If you want to gather your *own* data, and generate your *own* thrust-time curves, you can do it with a homemade chart recorder (*Figure 18-17*).

The chart drum in the photos is made from a sheet metal, **3 pound Dak Imported Danish Butter Cookie box**. You'll find these cookies (or similar cookies) in all the markets around Christmas Time, and at other times of the year you'll see the empty boxes at swap meets and yard sales. I bought the one in the photo at The Price Club. The Dak box is 10-1/4" dia. by 4-1/4" tall. Other cookie and fruit cake boxes will work as well, but try to avoid anything smaller than 8" dia.

The cookie box revolves on a 6" dia. "Lazy Susan" bearing. The bearing that I used for this prototype is made by **Faultless**, but any brand will do. You'll find them at the big hardware warehouses like Home Depot and Lowe's. The cookie box and bearing are driven by a **Hurst model PA, part #3202-014**, 6 rpm synchronous motor. The motor is the major expense; about $60 at the time of this writing. But with a little ingenuity you can substitute another motor, *as long as it maintains a constant speed of at least 4 rpm, and not more than 10 rpm*.

Low speed synchronous and gearhead motors are common surplus items, and most cities have dealers who specialize in electronic and industrial surplus. For example, at the time of this writing, there is a well known place in Pasadena, Calif., called **C&H Sales**. They sell things like power supplies, meters, transformers, test equipment, small electronic parts, used optics, and assorted other industrial "stuff". Their stock is constantly changing, but in one of their recent catalogues, they offered a 5 RPM, 120 volt, gearhead motor for $7.95. The exact motor dimensions, the mounting hole pattern, and the shaft size are different than the ones on the Hurst motor, but with a minor modification of the chart recorder design, this motor would work just fine. To find a surplus dealer, look in *The Yellow Pages* under the **SURPLUS** headings, or search the Internet.

If you can't find a surplus motor, and you *must* buy something new, try to find the same motor described in these instructions. Hurst synchronous motors are made by **Hurst Manufacturing, 1551 E. Broadway, Princeton, IN 47670**. They are

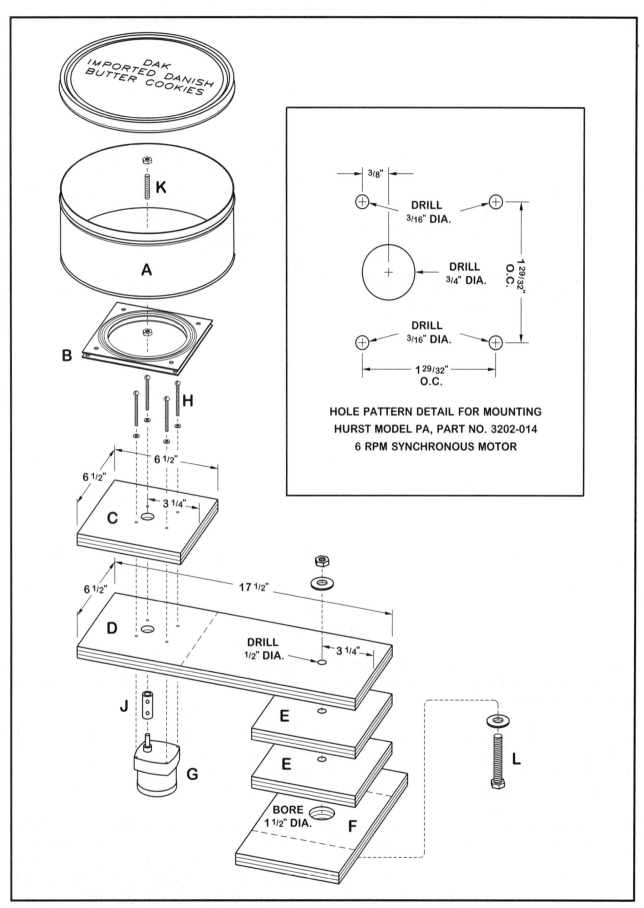

Figure 18-18. An exploded view of the chart recorder assembly. The capital letters in this drawing refer to the capital letters in the materials list on **page 481**.

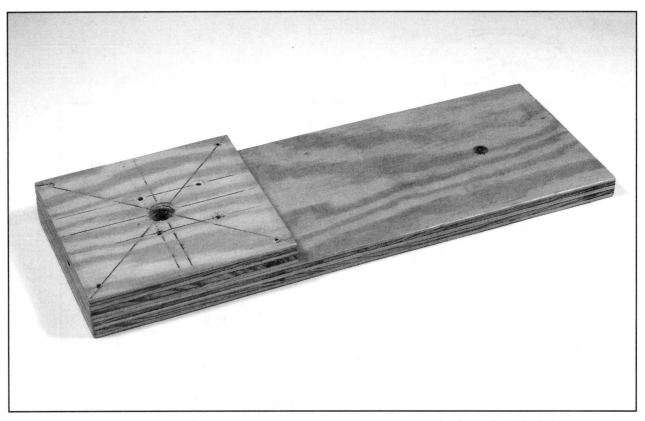

Figure 18-19. The plywood pieces for the bearing support and the pivot arm are glued together. The holes for mounting the chart drive motor and the pivot bolt are drilled.

specially designed to maintain the highly accurate speeds needed for things like clocks and industrial equipment timers. Because of the specialized applications in which they are used, they are *not* sold by electric motor shops, and you'll have to go to a Hurst distributor to buy one.

I'd originally hoped to include a list of Hurst distributors in this Chapter, but as the work on this book proceeded, I noticed that, with the exception of **W.W. Grainger,** Hurst's list of distributors kept changing. It would therefore be best for you to contact Hurst directly. Tell them where you live, and tell them that you need a PA-3202-014 6 rpm synchronous motor. Then ask them for the name and contact information of the nearest dealer who can sell you one. At the time of this writing, Hurst has a website on the Internet at **www.myhurst.com/hurstmfg**.

W.W. Grainger Inc. has so many retail outlets that you won't need an address. Just look in the white pages of your phone book, or go to their website at **www.grainger.com**. At the time of this writing, the Hurst PA-3202-014 synchronous motor is listed in the Grainger catalogue.

The chart drum assembly pivots on a wide plywood arm, so that it can be rotated into or away from the test stand's recording pen. The chart paper is a strip of 3" to 3-1/4"-wide, paper calculator tape. If you use a shorter cookie box, you'll need a narrower tape, but there are many widths available, and finding one that works shouldn't be a problem. Look for the tape at stationery stores and places like Office Depot.

Important note. *The drive shaft coupler was **not** easy to find.* After "striking out" at several hardware stores, I ended up again at W.W. Grainger. Grainger sells them in sizes to fit 1/4" dia. through 1" dia. shafts. The one that fits the Hurst motor is Grainger's stock #6L012. It is listed in the Grainger catalogue under "Rigid Steel Couplings". You can also buy these couplers from the dealers that sell gears and bearings. Look in *The Yellow Pages* under **BEARINGS**. What follows immediately are the instructions for putting together the chart recorder shown in the photos. *Figure 18-18* on **page 479** is an exploded view of the recorder's components, and the letters in the illustration refer to the capital letters in the materials list on **page 481**. To build a homemade chart recorder, refer to *Figure 18-18*, and proceed as follows.

Figure 18-20. A plastic box purchased at **Radio Shack** protects the motor's electrical connections.

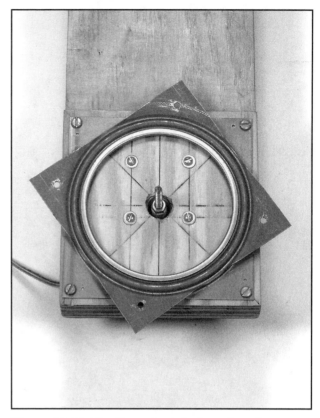

Figure 18-21. The "Lazy Susan" chart drum bearing is centered around the drive shaft, and mounted on the bearing support with 4 sheet metal screws.

MATERIALS LIST

- **A.** Chart drum: One 10-1/4" dia. X 4-1/4" high **Dak Imported Danish Butter Cookie** box (or similar)
- **B.** Chart drum bearing: One **Faultless** brand, 6" dia. "Lazy Susan" turntable bearing (or similar)
- **C.** Chart drum bearing support: One piece 3/4" CDX or better grade plywood, 6-1/2" square.
- **D.** Chart drive pivot arm: One piece 3/4" CDX or better grade plywood, 6-1/2" wide x 17-1/2" long.
- **E.** Pivot arm support: Two pieces 3/4" CDX or better grade plywood, 6-1/2" square.
- **F.** Pivot arm base: One piece 3/4" CDX or better grade plywood, 6-1/2" x 12".
- **G.** Chart drive motor: One **Hurst model PA, part number 3202-014**, 6 rpm synchronous motor.
- **H.** Four #8-32 x 2" R.H. bolts with washers.
- **J.** Drive shaft coupler: One 1/4" steel shaft coupler with two set screws. **Grainger's #6L012** (or similar).
- **K.** Chart drum drive shaft: One piece 1/4"x20 threaded steel rod 2" long with two nuts.
- **L.** Pivot arm bolt: One 1/2" x 3" hex head bolt with one nut and two washers.

NOT SHOWN in FIGURE 18-18. Four #10 x 1/2" F.H. sheet metal screws.

1. Cut the plywood pieces for the bearing support (*Figure 18-18-C*) and the pivot arm (*Figure 18-18-D*), and glue and clamp them together. Drill the hole pattern for mounting the chart drive motor, and drill the 1/2" dia. pivot bolt hole at the opposite end of the pivot arm (*Figure 18-19*). The screw holes for mounting the bearing are also visible in *Figure 18-19*.

2. Connect the chart drum drive shaft (*Figure 18-18-K*) and the drive shaft coupler (*Figure 18-18-J*) to the Hurst synchronous motor (*Figure 18-18-G*). Run the lower drive shaft nut down onto the shaft, and mount the motor to the underside of the pivot arm with the four #8-32 x 2" R.H. machine bolts and washers (*Figure 18-18-H*). Wire the motor as shown in Hurst's wiring diagram for 110 v. A.C. and *counterclockwise* rotation. *Figure 18-20* shows how I did it. I bought a small, plastic box at **Radio Shack** to protect the electrical connections and the motor's capacitor.

3. Mount the chart drum bearing (*Figure 18-18-B*) to the bearing support (*Figure 18-18-C*) with four #10 x 1/2" F.H. sheet metal screws (*Figure 18-21*), and center the bearing around the drive shaft as accurately as you can.

Figure 18-22. The upper drive shaft nut holds the chart drum firmly on the drive shaft. The positions of both nuts are adjusted to hold the drum gently down on the bearing.

Figure 18-23. The two pivot arm supports are glued and clamped to the pivot arm base, and the hole is drilled for the pivot arm bolt.

4. Drill a 1/4" dia. hole in the exact center of the bottom of the cookie box (**Figure 18-18-A**), *and locate it accurately.* If the hole is off-center, the cookie box-chart drum will wobble as it turns. Set the cookie box down on the bearing, so that the drive shaft extends up through the hole you just drilled. But *first* run the *lower* drive shaft nut up, so that it just touches the bottom of the box when the box is in place. Then screw on and tighten the *upper* drive shaft nut (**Figure 18-22**). You'll have to experiment with the exact positions of the two nuts until the cookie box rests firmly on the bearing, but *not* so tight that it jerks or binds when the motor turns.

5. Drill a 1/2" dia. hole through the center of the two pivot arm supports (**Figure 18-18-E**), and cut a 1-1/2" dia. hole in the center of the pivot arm base (**Figure 18-18-F**). This hole provides a recess for the head of the pivot arm bolt and one of its washers. Glue and clamp the two pivot arm supports to the pivot arm base (**Figure 18-23**). Then connect the plywood assembly you just made to the pivot arm with the pivot arm bolt (**Figure 18-18-L**), the pivot arm nut, and the two washers.

As you can see, I *don't* recommend that you install a switch in the chart drive circuit, and this is because, *for safety reasons,* you should be standing **at least 25 feet away** when you turn on the drive and fire a motor. I recommend, instead, that you plug the drive into a long extension cord, and plug that extension cord into one of the switched power strips designed for use with home computers.

Finally, because the test stand and the chart recorder are made of wood, spray the plywood parts with a few coats of lacquer or varnish before assembly, and store the test stand and the recorder in a warm, dry place when not in use.

● A Thrust-Time Curve

To demonstrate the chart recorder, I'll *again* use an experimental NV6-powered motor with a 3/4" i.d. casing, a 1/4" dia. nozzle throat, and a propellant mass (m_p) of 61.0 grams. To set up for the test, clamp the chart recorder in place on the test stand's base, and tape a strip of calculator tape tightly around the drum (**Figure 18-24**). Position the recorder so that the paper makes square contact with the pen when the chart drum is rotated into position. You'll find that spring clamps work nicely for holding the recorder in place, and for my *DAK-cookie-box* recording drum, the extra-wide 3" to 3-1/4" calculator tape works best.

Figure 18-24. The homemade test stand with an experimental motor and the homemade chart recorder. The chart paper has been taped in place, and the chart drum has been rotated into position against the pen.

1. Set up the test stand with an appropriate spring. Wrap masking tape around the front of the motor, and fit it *snugly* into the stand's motor holder. Then calibrate the stand with a known weight. As before, the weight should be close to the expected motor thrust, and for *this* test, 10 lbs. will again work nicely.

   With the motor in place, rotate the chart drum into position so that the pen touches the paper. Turn on the chart drive, and allow the pen to scribe a line around the full circumference of the drum (***Figure 18-25***). I call this the **zero line**, and it represents a thrust of **0 lbs.** Turn off the chart drive, and move the drum away from the pen. Find the bucket that you used for the previous thrust stand calibration (**pages 475 - 477**), and fill it with sand again until the total weight of the bucket *plus* the sand *plus* the hooks *plus* the chain equals 10.0 lbs. Hang the bucket from the eyebolt under the motor holder. Slide the chart drum into contact with the pen. Turn on the chart drive, and allow the pen to scribe a *second* line around the circumference of the drum (***Figure 18-25***). I call this the **calibration line**, and it represents a thrust of **10.0 lbs.**

2. Turn off the chart drive. Rotate the drum away from the pen, and remove the calibration weight (the sand-filled bucket, the hooks, and the chain). Insert a properly wired electric igniter into the rocket motor's nozzle. Rotate the chart drum back into place against the pen. Move back to a safe distance. Turn on the chart drive, fire the rocket motor, and wait for the thrust arm to come to rest. As the chart drum turns, the motion of the pen against the also-moving paper will generate an accurate thrust-time curve (***Figure 18-26***). Turn off the chart drive. Rotate the chart drum *away* from the pen, and cap the pen. Then remove the paper strip from the chart drum, and take it to your desk for analysis.

● **Chart Speed**

With the chart of a successful test in hand, you can obtain an amazing amount of information, but *first* you have to know how fast the chart is moving in inches-per-second. The Hurst model PA-3202-014 motor turns at 6 rpm, which works out to 1 revolution every 10 seconds. The diameter of the DAK cookie box is 10.25". To find the box's circumference, multiply its diameter by π (3.14). In *this* case, the answer is 32.185 inches, which you can round off to 32.2. To find the speed of the chart, divide the circumference of the box (32.2") by the time it takes to make one revolution (10 seconds). 32.2 / 10 = 3.22, so with *this* cookie box, the chart moves at a linear speed of 3.22 inches per second.

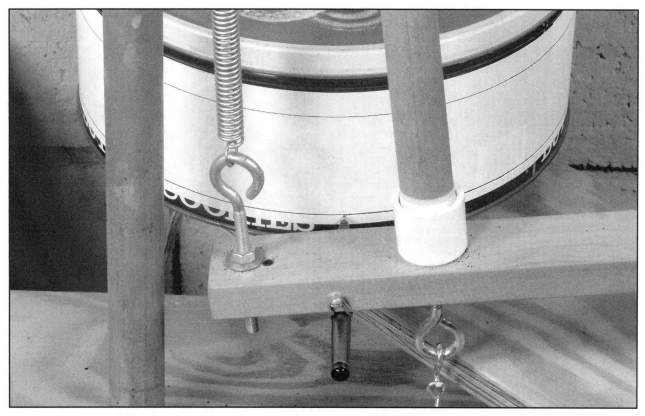

Figure 18-25. A zero line and a calibration line are scribed around the full circumference of the drum.

> **EXAMPLE 18-2:**
>
> *Let's suppose that your chart drive motor turns at a rate of 8 rpm., and the diameter of your cookie box is 9.625". What is the linear speed of the chart?*
>
> 8 revs. / min. = 8 revs. / 60 secs. = 1 rev. / 7.5 secs.
>
> 9.625" dia. x 3.14 (π) = 30.223" circumference, rounded to 30.2"
>
> 30.2" / 7.5 secs. = 4.027" / sec., rounded to 4.03" / sec.
>
> **The chart therefore moves at a linear speed of 4.03 inches per second.**

Synchronous motors are very accurate. If you use a **synchronous** motor, you'll know beforehand that the chart drum is turning at the exact rpm stated on the motor's label. If you use any other kind of a motor, you should double check its rate of rotation by measuring how many seconds it takes to make 50 revolutions. Then divide that number by 50. Let's assume, for example, that the label on your motor says that the rated speed is 8 rpm. To find its *exact* speed, time how long it takes to make 50 revolutions. If the answer is (for example) 365 seconds, dividing 365 by 50 yields a figure of 7.3. It therefore takes 7.3 seconds to make one revolution. To find its *exact* speed in "rpm", divide 60 seconds by 7.3. The answer is 8.22, so, though the *stated* speed is "8 rpm", the motor is *actually* turning at 8.22 rpm. Now apply the arithmetic illustrated above to determine how fast the chart is moving in inches-per-second, and you'll be ready to analyze your data.

● Maximum Thrust

With a ruler or an engineering scale divided in **tenths of an inch**, measure the distance from the zero line to the highest point on the thrust-time curve. I call this distance L_{max} (*Figure 18-27*). Then measure the distance from the zero line to the calibration line. I call this distance L_{cal} (*Figure 18-28*). To calculate the maximum thrust, (F_{max}), divide L_{max} by L_{cal}. Then multiply the result by the calibration weight (F_{cal}).

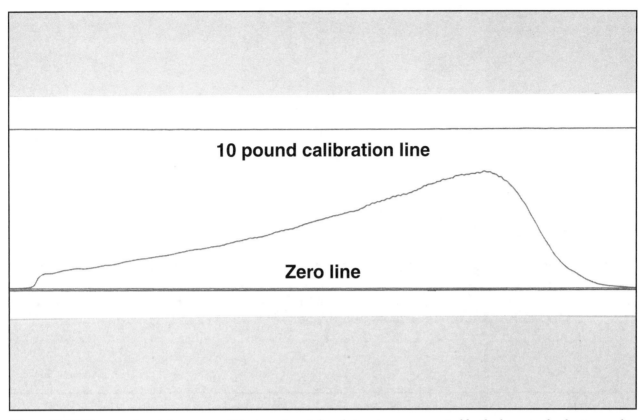

Figure 18-26. The raw thrust-time curve of the experimental rocket motor, as generated by the homemade chart recorder.

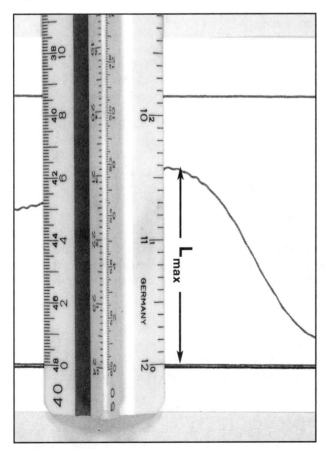 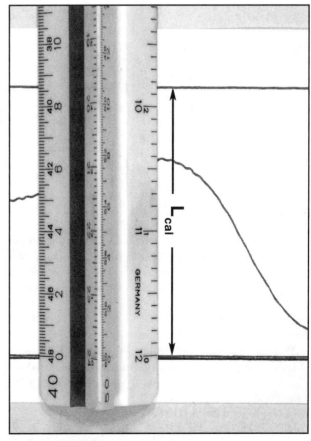

Figure 18-27. An engineering scale measures the distance from the chart's **zero line** to the **highest point on the thrust-time curve**. I call this distance L_{max}.

Figure 18-28. An engineering scale measures the distance from the chart's **zero line** to its **calibration line**. I call this distance L_{cal}.

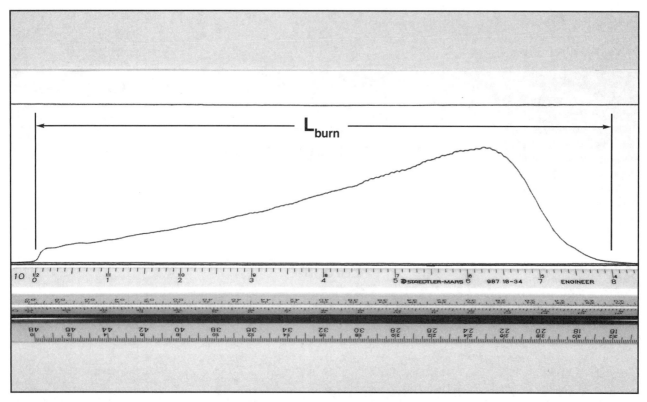

Figure 18-29. The total length of the thrust-time curve is measured from the point at which the pen first began to move, to the point at which it returned to the zero line. In this case the distance is 8.0".

> *EXAMPLE 18-3:*
>
> *As shown in Figure 18-27, the distance from the zero line to the highest point on the thrust-time curve is 1.58". As shown in Figure 18-28, the distance from the zero line to the calibration line is 2.15", and the calibration weight is 10.0 lbs. The motor's maximum thrust (F_{max}) is calculated as follows:*
>
> $$F_{max} = (L_{max} / L_{cal}) \times F_{cal}$$
>
> $$F_{max} = (1.58" / 2.15") \times 10.0 \text{ lbs.} = 7.35 \text{ lbs.}$$
>
> *The motor's maximum thrust is therefore 7.35 lbs.*

● Total Burn Time

With a ruler or an engineering scale graduated in **tenths of an inch**, measure the distance from the beginning of the thrust-time curve to the end of the thrust-time curve (*Figure 18-29*). This distance (L_{burn}) is measured from the point where the pen first began to move to the point where the pen returned to the zero line. To calculate the motor's total burn time (T), divide L_{burn} by the chart speed in inches-per-second.

> *EXAMPLE 18-4:*
>
> *As shown in Figure 18-29, the distance from the beginning of the thrust-time curve to the end of the thrust-time curve is 8.0". For this demonstration we're using the 10-1/4" dia. DAK cookie box and a Hurst 6 rpm motor, so as previously calculated, the chart moves at a speed of 3.22"/sec.*
>
> $$T = (L_{burn}) / \text{chart speed}$$
>
> $$T = 8.0" / 3.22" \text{ per sec.} = 2.48 \text{ secs.}$$
>
> *The motor's total burn time is therefore 2.48 secs.*

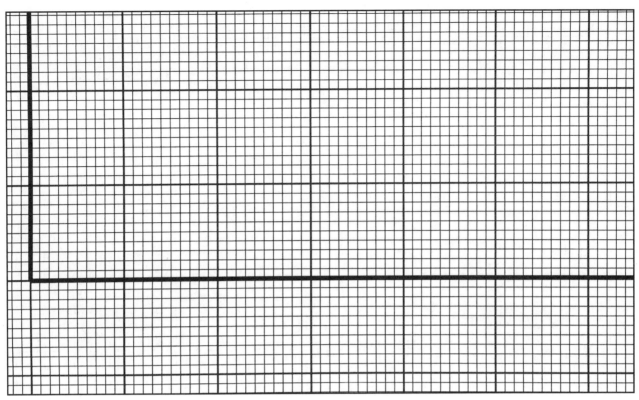

*Figure 18-30. A horizontal and a vertical axis are drawn on a sheet of quadrille ruled graph paper, so that they intersect as shown. The paper should be ruled in **tenths of an inch**.*

● **Total Impulse**

Finding a motor's total impulse is more time-consuming, but well worth the effort if you consider the importance of the data. To find the total impulse, proceed as follows:

> **MATERIALS LIST**
>
> 1. A sheet of tracing paper
> 2. A pen or a sharp pencil (An ultrafine-point "Sharpie" permanent marker works well.)
> 3. A piece of quadrille graph paper ruled in tenths of an inch (With a simple computer CAD program and a good printer, you can make it yourself.)

1. Draw a horizontal and a vertical axis on a sheet of quadrille-ruled graph paper, so that they intersect like the ones in *Figure 18-30*. The graph paper should be ruled in **tenths of an inch**. Then *carefully* trace the chart's thrust-time curve (*Figure 18-26*) onto a separate sheet of *tracing* paper, and be sure to include the zero line and the calibration line.

2. Tape the tracing *over* the graph paper, so that the thrust-time curve's zero line falls on the horizontal axis, and the beginning of the burn falls on the intersection of the two axes (*Figure 18-31*).

3. When viewed through the tracing, each little graph paper square takes on a thrust-time "impulse" value, and you find the motor's *total* impulse (I_t) by adding up all the squares between the zero line and the thrust-time curve. But *first* you have to know many pound-seconds of impulse each square represents.

4. To find a square's horizontal *time* value, count the number of squares between the beginning and the end of the thrust-time curve, and divide the total burn time by the number of squares. In this example the burn time is 2.48 seconds, and in *Figure 18-31* there are a total of 80 squares from beginning to end. 2.48 / 80 = 0.031. Therefore, each square along the *horizontal* axis has a *time* value of **0.031 seconds**.

5. To find a square's vertical *thrust* value, count the number of squares between the chart's zero line and its calibration line, and divide that number by the calibration weight. In this example the calibration weight is 10 lbs., and in *Figure 18-31* there are a total of 21.8 squares between the zero line and the calibration line. 10 / 21.8 = 0.459, rounded to 0.46. Therefore, each square along the *vertical* axis has a *thrust* value of **0.46 lbs**.

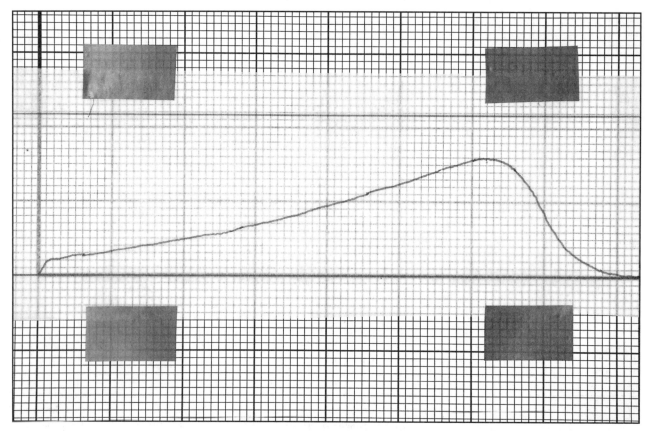

Figure 18-31. A tracing of the motor's thrust-time curve is taped over the graph paper, so that the zero line falls on the horizontal axis, and the beginning of the burn falls on the intersection of the two axes.

6. To find out how many lb.-secs. of impulse each square represents, multiply the square's vertical thrust value by its horizontal time value. In this example the answer is 0.46 lbs. x 0.031 secs. = 0.0143 lb.-secs. Therefore, each square represents **0.0143 lb.-secs. of impulse**.

7. To find the engine's *total* impulse (I_t), add up all the squares between the zero line and the thrust-time curve, and multiply that number by the impulse value of one square (**0.0143 lbs.-secs.**). Half squares are added up, and little corners of squares are tallied along with almost-whole ones to arrive at the final figure. In *Figure 18-32* I've marked them all off, and crossing off large blocks with multiple rows of 10 makes the counting easier. You'll also see a few vacant spaces. These vacant parts of squares were tallied with other parts of squares until a whole was obtained. One of the parts in the tally was then marked to represent the whole, and you can see that the grand total is 609.

Since each square represents 0.0143 lbs.-secs. of impulse, the total is 0.0143 x 609 = 8.71. The motor's total impulse is therefore 8.71 lb.-secs. Or translated into metrics (1 lb.-sec. = 4.45 Nt.-secs.), the motor's total impulse is 38.76 Nt.-secs.

● Average Thrust

To calculate the motor's average thrust in pounds, divide the total impulse (I_t) in lb.-secs. by the total burn time (**T**).

EXAMPLE 18-5:

In this example the total impulse is 8.71 lbs.-secs., and the burn time is 2.48 seconds. The average thrust is calculated as follows:

$$F_a = I_t / T$$

$$F_a = 8.71 \text{ lb.-secs.} / 2.48 \text{ secs.} = 3.51 \text{ lbs.}$$

To convert this value (3.51 lbs.) to Newtons, multiply it by 4.45. That is, 3.51 lbs. x 4.45 Nts./lb. = 15.6 Nts.

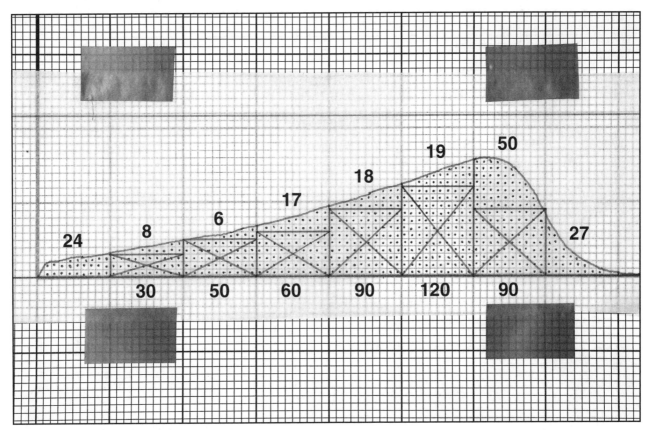

Figure 18-32. The motor's total impulse is found by adding up all the squares between the chart's zero line and the thrust-time curve. The numbers below the zero line are the totals for the crossed-out blocks, and the numbers above the thrust-time curve are the totals for the areas above or outside of the blocks. In this case the total is 609.

● Propellant Mass

The previous calclulations determined four of the motor's performance parameters, and now you're going to find its specific impulse. Specific impulse is important, because, like the gas mileage of a car, it gives you an actual number value for the motor's operating efficiency, an important thing to know when comparing it to other motors of similar design. To do the calculation, you'll have to know the motor's propellant mass (m_p), and the propellant mass is the exact amount of propellant that you loaded when you made the motor. **Important note**. The propellant mass includes *only* the motor's propellant. It does *not* include the time delay.

When you made the motor, you weighed out and mixed together specific amounts of propellant and baking soda, and their combined weight equaled the total amount of material in the mixing cup. To find the propellant mass, finish building the motor. Then dry and weigh what remains in the cup, and *subtract* that weight from the amount you originally mixed.

EXAMPLE 18-6:

In this example we'll assume that you started with 65.0 gms. of propellant, and added 9.75 gms. (15%) of baking soda.

Starting weight = 65.0 gms. propellant + 9.75 gms. (15%) baking soda = 74.75 gms.

Weight of the dried remains in the mixing cup = 13.75 gms.

74.75 gms. - 13.75 gms. = 61.0 gms.

Of the 74.75 gms. originally mixed, you actually loaded just 61.0 gms., so the motor's propellant mass (m_p) is 61.0 gms.

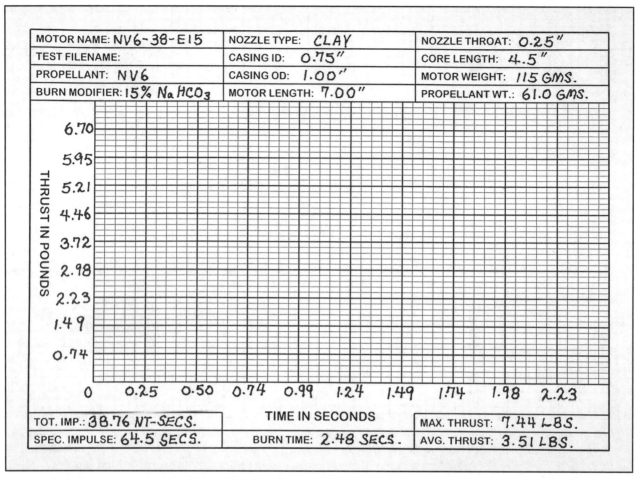

*Figure 18-33. To use the blank form on **page 491**, fill in the motor data. Then divide the motor's burn time and maximum thrust into 10 equal fractions. Round off the resultant figures to 2 decimal places, and write the appropriate numbers on the grid's horizontal and vertical axes. When the proper values are in place, **then** you can manually plot the motor's thrust-time curve.*

● Specific Impulse

To calculate the specific impulse (I_{sp}), divide the motor's total impulse (I_t) in pound-seconds by its propellant mass (m_p) in pounds. In the English system the motor's total impulse is **8.71 lb.-secs.** In the metric system its propellant mass (m_p) is **61.0 gms.** There are 453 grams in a pound, so in the English system the propellant mass is 61.0 / 453 = **0.135 lb.**

> *EXAMPLE 18-7:*
>
> *Specific impulse (I_{sp}) = total impulse (I_t) / propellant mass (m_p)*
>
> $$I_{sp} = I_t / m_p$$
>
> I_{sp} = **8.71 lb.-secs. / 0.135 lb. = about 64.5 lb.-secs. / lb. = 64.5 secs.**
>
> *Therefore, in this experimental motor, the NV6 propellant achieved a specific impulse of 64.5 seconds.*

With its I_t of 38.76 Nt.-secs. (8.71 lb.-secs.), and its F_a of 15.6 Nts. (3.51 lbs.), I'd call this motor an **NV6-38-E15**. Its F_{max} of 7.44 lbs. is more than 25% *below* the desired target figure of 10 lbs., so *in this case* I'd keep building, testing, and *reducing* the baking soda in 1% increments until I achieved an F_{max} of 9 to 10 lbs. Based on previous experience, I know that this would raise the I_t above 45 Nt.-secs., the propellant's I_{sp} above 70 secs., and push this motor's performance well into the "F" catagory.

On **page 491** is a blank motor data form with a blank grid for a thrust-time curve. You can make photocopies of this form, then use them to keep a record of your own motor tests. Vertically, each small rectangle represents 2% of the motor's maximum thrust (F_{max}), and horizontally, each small rectangle represents 2% of its burn time (**T**). The larger, darker rectangles represent 10% of those values.

MOTOR NAME:	NOZZLE TYPE:	NOZZLE THROAT:		MAX. THRUST:
TEST FILENAME:	CASING ID:	CORE LENGTH:		
PROPELLANT:	CASING OD:	MOTOR WEIGHT:		
BURN MODIFIER:	MOTOR LENGTH:	PROPELLANT WT.:		AVG. THRUST:

THRUST IN POUNDS / **TIME IN SECONDS**

| TOT. IMP.: | BURN TIME: |
| SPEC. IMPULSE: | |

To transfer a thrust-time curve to the grid, fill in the motor data. Then divide the motor's maximum thrust and its burn time into 10 parts, round the resultant figures to 2 decimal places, and write those values along the edges of the grid. **Figure 18-33** shows how you'd do it for the motor in the previous example. Though there's no room to write them, the number at the *top* of vertical axis is *assumed* to be the F_{max} of 7.35 lbs., and the number at the *right* end of the horizontal axis is *assumed* to be the **T** of 2.48 secs. Finally, using the same techniques used to calculate the maximum thrust and the burn time (**pages 484 - 486**), find the thrust at various time-points along your tracing of the thrust-time curve. Plot those points on the grid, and connect the dots with a pen. **Important note**. The more points you plot, the more accurate your thrust-time curve will appear.

● Equipment Limitations

Useful though it is, this homemade equipment has its limitations, the main one being its inability to record rapid-fire, short-duration changes in motor thrust. The thrust-time curves in **Chapters 11** through **15** show tiny fluctuations that don't appear on charts generated by the homemade equipment. "Inertia" is the tendency of an object to remain at rest or in a state of uniform motion until acted upon by an external force. During a rocket motor test, inertia expresses itself as a *resistance* to any change in the current state of motion (or rest) of the motor and the test stand parts. Large changes encounter greater resistance than small ones, so they take longer to occur.

When a rocket motor pushes against a load cell, a movement of *one ten thousandth of an inch* will appear as a visible feature in the thrust-time curve. Though the event may happen quickly, the motion within the load cell is so small that there's little resistance, so short-duration events are recorded. To record the same event with a mechanical system, the motor has to push the thrust arm and the pen a hundred times farther, and it takes longer to do it. The result is that short-duration changes in motor thrust begin and end faster than the system can respond, and these events are *not* recorded.

The second problem with the homemade system is that the motor's greater motion introduces a large amount of kinetic energy into the system's moving parts, and this energy has to be dissipated before the pen can come to rest. Large and *sudden* movements invest the pen with enough energy to make it overshoot its mark when rising, and recoil below the *actual* value on its return. The oil-filled oscillation damper eliminates most of this excess motion, but to eliminate *all* of it would require such extensive damping that all but the grossest features in the thrust-time curve would be lost.

The damper described in these instructions is a compromise. It preserves the most important details in the curve, and eliminates all but the most violent overshoot and recoil events. Overshoot and recoil are most prominent in motors with extremely short burn times, and also in warm weather, when the viscosity of the damping oil is lowered. The duct-taped, oil-filled margarine tub is homely and primitive, but it works quite well. If you want something more elegant, experiment with permanently-mounted metal cans, and adjust the diameter of the damping disk to fit.

Finally, though the spring-operated system is reasonably accurate, the locations of the zero and calibration lines will change with the exact weight of the motor being tested. And the force needed to stretch the spring a given distance will change a little with time, temperature, and other factors that tend to be unpredictable. Therefore, you should *always* recalibrate the stand (i.e. the bucket, the sand, etc.) before every test.

● Safety During Motor Testing

When testing rocket motors, you should always be conscious of the danger of setting a fire. Anticipating that the forces opposed to amateur rocketry might try to make an issue of this fact, I'll point out in advance that *commercially* made rocket motors often malfunction, and that *many* fires are started every year by people using commercialy made motors *in places where they shouldn't*. If you doubt what I'm saying, ask your state firemarshal. The problem is **not** related to who made the motor, but to whether it is used *responsibly*.

Acting *responsibly* means that you *only* test rocket motors in an area *completely* free of anything that might burn, including *trash, brush, dead wood, dry grass, paper or cardboard*, and *trees*. And even if you *don't* see anything flammable, before you conduct a test, *walk the area first to be sure*. If you find anything flammable within 200 feet of the test stand, remove the flammable material, wet it down with water, or move the test to a more suitable site.

To minimize the danger to yourself, **stand at least 25 feet away** when testing motors up through **E**, and **50 to 100 feet away** when testing anything larger. If, for any reason you find that you *must* be closer to the test, *protect yourself with the homemade shield shown in Figure 18-34*. It amounts to a sheet of 3/8" thick plywood, 4 feet wide by 5 feet high, hinged to a smaller sheet 4 feet wide by 3 feet high with lightweight chain supports. You can make the entire thing from a single 4' x 8' sheet of plywood. To use the shield, place it between you and the motor being tested, and crouch down behind it before you press the firing button. If the motor explodes, the shield will help protect you from sparks and flying debris. Of course an exploding motor can damage the homemade test equipment, so *only* use the equipment for gathering data. When testing a new propellant,

Figure 18-34. *If a motor explodes during a test, this simple plywood shield will help protect you from flying debris.*

Figure 18-35. *Fire danger during a rudimentary test can be minimized by placing the motor in a 4-inch diameter plastic pipe sunk below ground level. The motor holder is a disk of thick plywood with a hole bored to fit the motor. If the motor explodes, the pipe directs the debris upward, and horizontal spread is minimized.*

a new motor design, or a new type of casing, your *first* concern will be to eliminate "chuffing" (combustion instability), or to verify that the casing can withstand the pressure at which the motor operates. For these *simple* tests, you *don't* need a test stand or a chart recorder. A simple motor-holder will do.

To reduce the fire danger, sink a short length of 4" diameter PVC or ABS plastic pipe into the ground, and make the pipe go *deeper* than the longest motor you plan to test. Then mount the motor holder and the motor-to-be-tested *in* the pipe with the motor's nozzle an inch or two below ground level (***Figure 18-35***). The motor holder is a simple disk of thick plywood with a hole bored to fit the motor, and you can add or subtract sand to raise or lower the floor to the depth needed for the motor you are testing. For larger motors, you can stack several layers of plywood together. If the motor explodes, the pipe will direct the debris upward, and minimize its horizontal spread.

Important note. Though ***Figure 18-35*** is black & white, the grass in the photo is mostly-green, and *damp* from a morning rainstorm. ***Testing motors immediately after a rain*** can reduce or even ***eliminate*** the danger of setting a fire.

19. FLIGHT

A Basic Stick Rocket

To test a rocket motor *in-flight*, the *simplest* configuration is a **stick-rocket**. Be aware, however, that most states classify "stick-rockets" as *fireworks*. If fireworks are illegal, then stick-rockets will be illegal too. Also, because a stick-rocket has no parachute, it falls straight back to the ground when spent. Because it is still hot when it lands, it can set fire to anything flammable. Therefore, when flying these motors on sticks, remember that the largest has a range of *more than a mile*. Fly them *only* in an area where there's absolutely *nothing* that might catch fire, and check with local authorities before you proceed.

Aside from the legal and safety concerns, the first thing you should know about a stick-rocket (*Figure 19-1*) is that it will *not* fly as high as a rocket with an aerodynamic body. A stick-rocket has a blunt nose, and it's center of thrust is *not* aligned with its center of gravity. It therefore flies at a slightly cocked angle, and the resultant turbulence creates an excessive amount of aerodynamic drag that prevents the rocket from performing as well as it should. To streamline the rocket and alleviate *some* of the drag, you can fit the front of the motor with a paper nose cone.

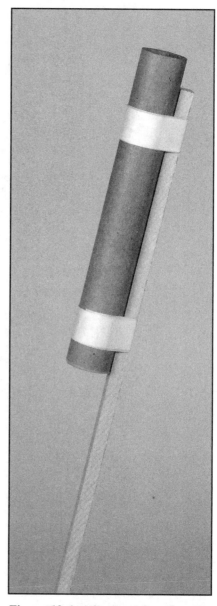

Figure 19-1. A basic stick rocket. The blunt nose and the off center stick generate excessive drag that degrades both speed and altitude.

MATERIALS LIST

1. A sheet of stiff paper (Like an old file folder.)
2. A strip of masking tape
3. Some white glue
4. A pair of scissors

1. From a sheet of stiff paper, cut a semicircle (half of a circle) about 4 times the outside diameter of the motor, and lay a strip of tape from the center to the outer edge (*Figure 19-2*).

2. Bend the semicircle into a cone. Stick it together with the tape, and make the diameter of the cone's base a little larger than the diameter of rocket motor (*Figure 19-3*).

3. Firmly press the cone down over the front of the motor, so that the motor's front edge marks the paper with a circular crease (*Figure 19-4*).

4. With the scissors, trim away the paper below the crease (*Figure 19-5*). Run a small bead of glue around the front of the motor (*Figure 19-6*), and press the cone firmly into place (*Figure 19-7*).

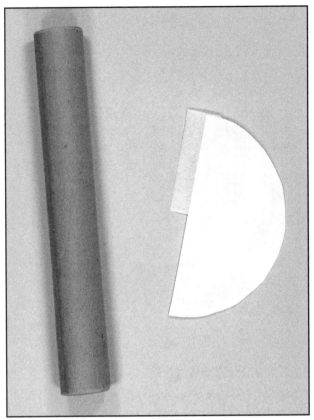

Figure 19-2. *A semicircle 4 times the outside diameter of the motor is cut from stiff paper, and a strip of tape is laid from the center to the outer edge.*

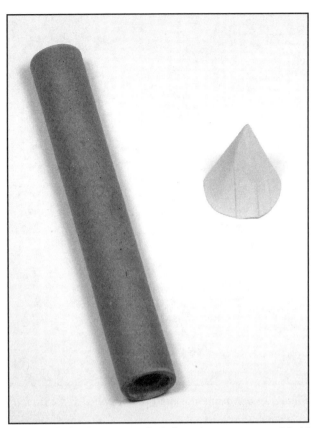

Figure 19-3. *The semicircle is bent into a cone, and held together with the tape. The base of the cone should be a little larger than the outside diameter of the motor.*

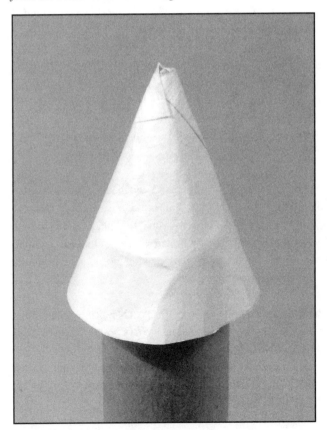

Figure 19-4. *The cone is pressed down over the front of the motor, creating a circular crease in the paper.*

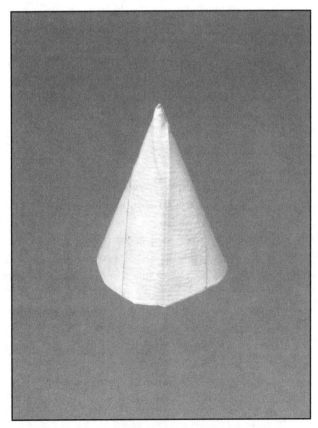

Figure 19-5. *The paper below the crease is trimmed away, leaving the smaller, finished cone.*

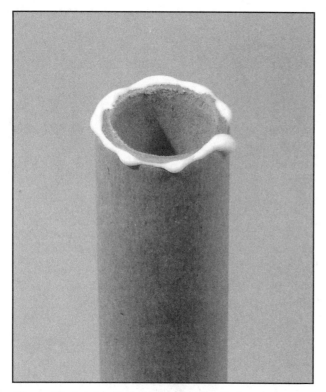

Figure 19-6. A bead of white glue is applied to the front of the motor.

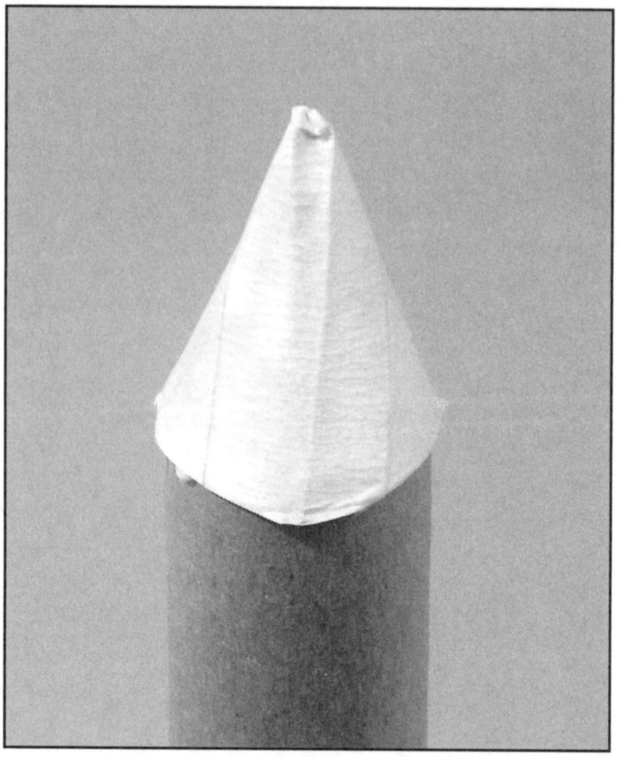

Figure 19-7. The finished nose cone is pressed into place.

Important note. Do *not* make a stick-rocket with a motor whose burn time is longer than 3 seconds. Most of the motors in this book have a low starting thrust. If the burn time exceeds 3 seconds, the longer period of low thrust can allow a stick-rocket to wander off course, and thereby increase the danger of flying downrange and setting a fire. *A motor with an **extended** burn time needs a properly balanced, aerodynamic body to establish and maintain **vertical** flight.*

When mounting a motor on the stick, it is important to mount it *firmly*, because the twisting forces generated during flight are powerful. If the stick is not firmly affixed, the motor will twist out of alignment, and the rocket will fly off-course. Also, the entire stick-motor assembly has to be properly balanced. If the stick is too long, the rocket will weathervane into the slightest wind. If the stick is too short, the rocket will fly in a gyrating spiral that generates drag and degrades performance. Finally, the stick has to be strong enough to withstand the rigors of flight at speeds up to 400 miles per hour. If the stick is too thin or too weak, *it will break*.

In the early 1980s I experimented with sticks made of pine and fir, but they broke too easily, so I changed to hardwoods like maple, mahogany, and alder. Hardwoods are expensive if you buy them at a lumber yard, but fortunately, the back roads of America are littered with a free supply. If you take a drive on the dirt roads outside of any city, you'll find places where people illegally dump trash, and that trash usually includes old, upholstered couches. These couches are made of cloth and padding stretched over hardwood frames. The wood is of poor cosmetic quality, and *not* good enough for making nice looking furniture. But it's *fine* for making rocket sticks, and one couch provides enough wood to make dozens.

Some couches have wider and thicker boards than others, so don't grab the first one you see. Turn each couch over, and look at the frame inside. With a little practice, you'll learn to spot the ones that are best for salvaging. When you take one of these couches, be a good citizen, and don't tear it apart at the side of the road. This leaves a mess for other people to clean up. The proper etiquette is to load it onto your truck, take it home, and knock it apart in the back yard. Then remove the nails, tacks, and staples, cut the long boards into the sticks you need with a table saw, and throw the leftovers in the trash.

When I did this many years ago, I used 1/2" square sticks for the 1" and 1-1/8" i.d. motors, 3/8" square sticks for the 3/4" i.d. motors, and 1/4" square sticks for the 1/2" i.d. motors. Some trial and error will be involved. If a stick breaks during a flight, make the next rocket with a *thicker* stick. To assemble a stick-rocket, and balance it for flight, proceed as follows:

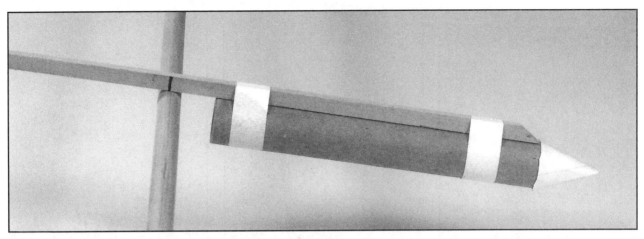

Figure 19-8. If the rocket tilts forward, the stick is too short.

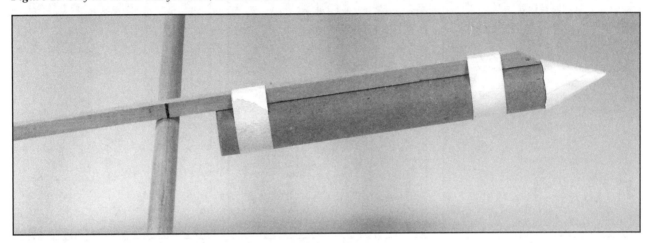

Figure 19-9. If the rocket tilts backward, the stick is too long.

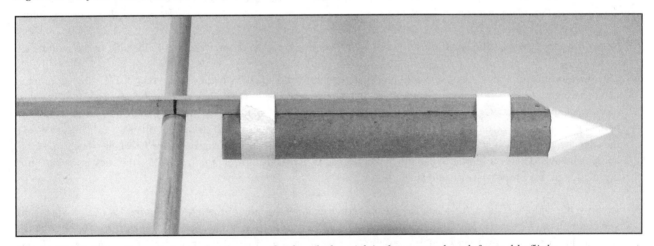

Figure 19-10. When the rocket rests level across the dowel, the stick is the correct length for stable flight.

MATERIALS LIST

1. A finished rocket motor with a paper nose cone
2. A stick of the proper size and length
3. Some masking tape
4. Some white glue

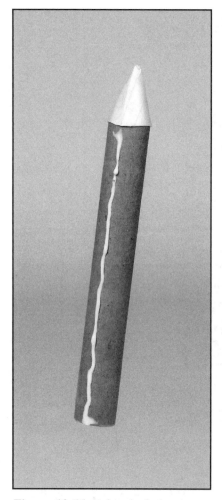 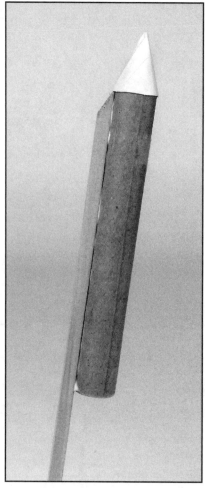 

Figure 19-11. A bead of glue is run lengthwise along the side of the motor.

Figure 19-12. The stick is pressed firmly into place.

Figure 19-13. Two rings of masking tape lock the stick firmly to the motor.

1. Select a stick of the size and length that you *think* you will need, and *temporarily* tape it to the side of the motor.

2. The *exact* location of a stick rocket's balance point can vary depending on the weight of the motor, the length of the motor, and the weight of the stick. But *most* of the stick rockets I've made seem to balance best **on the stick, about one motor diameter behind the motor's nozzle**. The motor in this example is 1" o.d., so I've made a mark on the stick exactly 1" behind the nozzle. *At this mark*, lay the rocket across a round, wooden dowel. If the rocket tilts forward (*Figure 19-8*), the stick is *too short*, and you'll have to use a longer stick. If the rocket tilts backward (*Figure 19-9*), the stick is *too long*, and you'll have to make it shorter. When the rocket rests *level* across the dowel (*Figure 19-10*), the stick is *probably* the correct length for proper balance and stable flight.

3. Remove the temporary tape, run a bead of glue lengthwise along the side of the motor (*Figure 19-11*), and press the stick into place (*Figure 19-12*). Notice in this photo how I've cut the front of the stick at an angle. This further reduces drag, and improves performance even more. Wrap the stick to the motor with two bands of masking tape *several layers thick* (*Figure 19-13*), and the rocket is finished.

● Launching a Single Stage Stick Rocket

You can make an inexpensive launcher from a scrap of schedule 40 PVC plastic pipe, a 2-foot length of steel rod, and two stainless steel hose clamps. The pipe should be longer than the rocket's stick. Cut the pipe and the rod to length, and clamp them together with the hose clamps (*Figure 19-14*). Then position the launcher at the angle you want, and push the rod into the ground (*Figure 19-15*). Insert the rocket's stick into the pipe (*Figure 19-16*), insert an electric igniter, hook up the firing system, retire to a safe distance, and fire the rocket. If the rocket flies straight and true, and stays on course, you correctly estimated the length of the stick. If it weathervanes into side-winds as it flies, make the stick on your next rocket *shorter*. If it gyrates or spirals as it climbs into the air, make the stick on your next rocket *longer*.

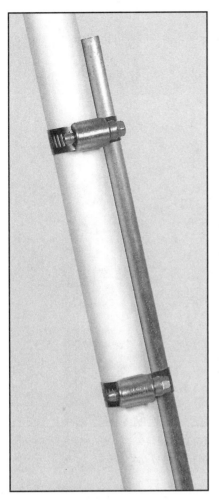

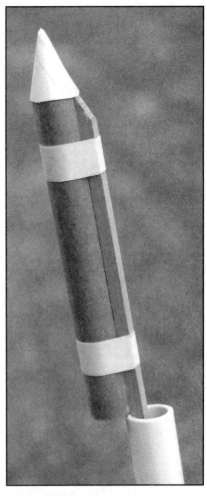

Figure 19-14. The pipe and the steel rod are clamped together.

Figure 19-15. The launch tube is stuck in the ground at the desired angle.

Figure 19-16. The rocket's stick fits down into the launch tube, while the motor rests on the tube's front edge.

A Two Stage Stick Rocket

Two stage rockets are interesting to fly, but making them work properly takes some practice. Because two-stage stick rockets and two-stage aerodynamic rockets stage in the same way, you can avoid risking an expensive, aerodynamic model by practicing with stick rockets until you're sure that you've got everything working right. A two stage rocket works best if the second stage motor is substantially smaller than the booster, so I'll use a **KG1-149-G104** and a **KG1-38-E33** in the following example. To make a two stage stick-rocket, proceed as follows:

MATERIALS LIST

1. A **KG1-149-G104** booster (a motor with a porthole but no time delay).
2. A **KG1-38-E33** "plugged" motor with a paper nose cone (a motor with a solid bulkhead and no time delay).
3. A soda straw and a small piece of cotton or tissue paper (i.e. "Kleenex")
4. A cardboard disk cut to fit snugly inside the front of the G104, with a center hole punched to hold the soda straw snugly in place.
5. A small amount of FF or FFF commercial black powder.
6. Two sticks of the correct size and weight to balance each motor properly for flight, and a short scrap of stick to shim the second stage stick away from the second stage motor.
7. Some white glue, some epoxy glue, and some masking tape.

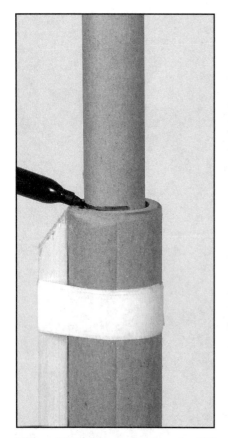 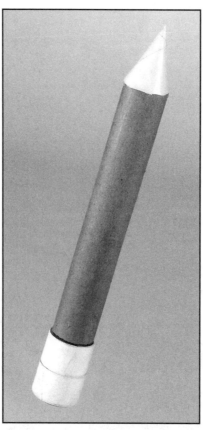 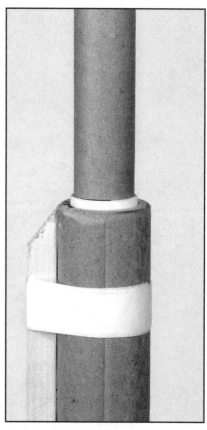

Figure 19-17. The point where the E33 exits the G104 is marked.

Figure 19-18. The nozzle-end of the E33 is built up with masking tape.

Figure 19-19. The E33 fits snugly into the front of the G104.

1. Mount the G104 on a stick of the correct size and length, and verify that the G104 is properly balanced for flight.

2. Insert the nozzle-end of the E33 into the front of the G104 so that it rests on the G104's forward bulkhead, and mark the point where it exits the front of the G104 (*Figure 19-17*). Then remove the E33, and prepare it for staging per the instructions on **pages 303 - 306**.

3. Starting at the mark you made in **step 2**, wrap masking tape around the nozzle-end of the E33 (*Figure 19-18*) until it fits snugly into the front of the G104 (*Figure 19-19*). Then *again* remove the E33 from the G104.

4. Mount the prepared E33 on a stick of the correct size and length, but when you do this, space the stick *away* from the E33 with a short piece of stick starting at the front of the tape you applied in **step 3** (*Figure 19-20*).

5. Prepare the G104 for staging per the instructions on **pages 301 - 303**. For clarity, I'll reiterate the process here. Cut or punch a disk of corrugated cardboard to fit inside the front of the G104, and punch a hole in the disk's center just big enough to fit the soda straw snugly. Press the disk into the front of the G104 until it rests firmly on the motor's forward bulkhead (*Figure 19-21*). Roughen one end of the soda straw with sandpaper. Insert the roughened end into the disk's hole so that the straw stands vertically out the front of the motor (*Figure 19-22*). Pour a 1/4" thick layer of epoxy on top of the disk around the straw's base (*Figure 19-23*), and let the epoxy harden.

6. Now plug the two rockets together. The soda straw (aka the *flame tube*) should extend almost to the front of the E33's core. If the straw is too long, cut it back until its end is about an inch behind the E33's igniter pellet when the two rockets are plugged together.

7. Separate the two rockets. Pour a small amount of loose commercial black powder into the straw (called a " flame tube"), and insert a small wad of tissue in on top of the powder. With a bamboo Bar-B-Que skewer, push the tissue *gently* down on top of the powder (*Figure 19-24*), and do *not* press it in tightly. Its only purpose is to keep the loose powder from flying forward when the G104 stops firing. Plug the two motors back together so that the E33's stick is on the side *opposite* the G104's stick (*Figure 19-25*), and the rocket is finished.

Figure 19-20. A short spacer-stick separates the E33 from its guide stick.

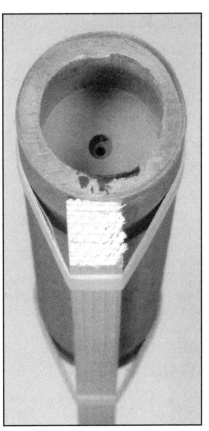

Figure 19-21. The cardboard disk rests firmly on the G104's forward bulkhead.

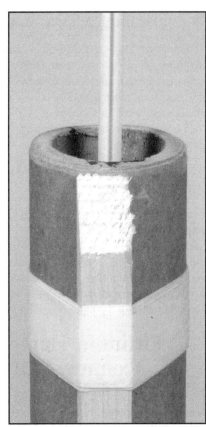

Figure 19-22. The soda straw extends vertically out the front of the motor.

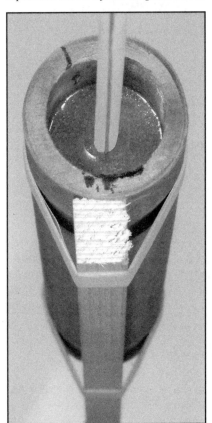

Figure 19-23. The cardboard disk is covered with a thick layer of epoxy.

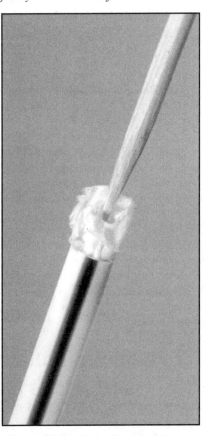

Figure 19-24. A small wad of tissue is inserted into the flame tube.

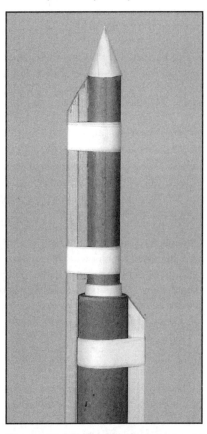

Figure 19-25. The E33's stick should be opposite the G104's stick.

● **Launching a Two Stage Stick Rocket**

To launch a two stage stick rocket (*Figure 19-26*), place the booster motor's stick (the *larger* stick) into the launch tube, insert an electric igniter, hook up the firing system, retire to a safe distance, and fire the rocket. If you've done everything right, the rocket will reach a modest altitude, at which point the booster will cease firing, and coast upward for a second or two. Then the upper stage will fire, separate from the booster, and continue on its way.

Practicing with a Time Delay

To practice with a time delay, make a properly balanced stick rocket with a time delay and a porthole. Then prepare it for parachute ejection per the instructions on **page 308**. Before gluing on the nose cone, fill the entire front of the motor *above* the ejection charge with dark blue or black tempera paint powder (For the tempera paint, look in *The Yellow Pages* under **ART SUPPLIES**). Then glue on the nose cone, and fire the rocket. When the motor stops firing, the rocket will coast upward for the duration of the time delay, at which point you'll hear a "pop" as the ejection charge blows out the paint powder, creating a small and visible cloud in the sky.

Fitting a Homemade Motor to a Commercially Made Rocket

Because of the process by which they are made, these motors are designed to fit loosely into standard sizes of schedule 40 water pipe. They will *not* be a perfect fit for most of the motor mounts used by the commercial rocket industry. Commercial mounts come in standard diameters of **24mm., 29mm., 38mm.,** and **54mm.** Your *homemade* motors have nominal diameters of **20mm., 25mm., 33mm., 38mm.,** and **50mm.** To use a homemade motor in a commercially-made rocket, you *shim* the motor to fit the commercial mount.

For a shimmed fit, a 20mm. motor can be made to fit a 24mm. mount. Other good matches are a 25mm. motor in a 29mm. mount, and a 33 mm. motor in a 38 mm. mount. A 38 mm. motor should fit *perfectly* in a 38mm. mount, and the giant 50 mm. I-65 should easily fit in a 54 mm. mount.

To shim a homemade motor, increase its diameter with 2 rings of masking tape (*Figure 19-27*), until it fits *snugly* in the commercial mount. Friction between the tape and the inside of the mount holds the motor in place. Then wrap an additional "stop ring" of tape around the motor's nozzle-end (*Figure 19-28*). This ring's *only* purpose is to keep the motor from sliding forward it fires. Depending on the size of the motor, it should be 3/8 to 3/4" wide.

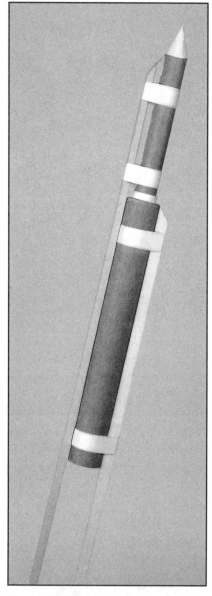

Figure 19-26. A two stage stick rocket ready for flight. Note that the booster motor's stick (the larger stick) is the one you insert into the launch tube.

Commercial rockets often keep the motor from sliding forward with a stop built into the *forward* end of the motor mount. But your homemade motor will almost *always* be longer than the motor for which the rocket was designed, and this stop will interfere with insertion of the homemade motor. To use the motor in the *commercial* rocket, you'll have to remove or bypass the manufacturer's retaining system. As you build a commercial rocket kit, pay close attention to the details of its motor mount. If the mount incorporates a forward stop, either *don't* install it, or cut it or sand it away. *Figure 19-29* shows a commercial rocket with a *homemade* motor mounted per the instructions above.

Aerodynamic Balance

To be aerodynamically stable and remain on course, a rocket must be properly balanced, and the correct balance is determined by the relationship between the rocket's **center of gravity** and its **center of wind pressure**. The **center of gravity**, abbreviated **"CG"**, is the point on the rocket's body where the *gravitational* forces to the front equal the *gravitational* forces to the rear. To find the center of gravity, load the rocket with *everything* it will contain at liftoff, including the motor, the payload, and the parachute. Then hang it from a loop of string. Slide the string back and forth until the rocket hangs level. Cut a small arrow from a piece of masking tape, label it "**CG**", and stick it to the body at the balance point (*Figure 19-30*).

Figure 19-27. In this photo, the diameter of a 20 mm. E21 is increased with masking tape until it fits snugly in the commercial 24 mm. motor mount.

Figure 19-28. An additional "stop ring" of narrow masking tape is built up around the motor's nozzle.

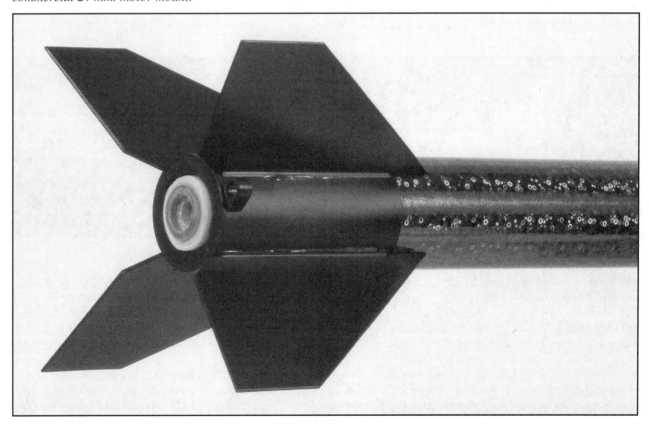

Figure 19-29. The motor is inserted into the motor mount up to the "stop ring". This particular rocket uses a twist-lock device to keep a commercial motor from falling out the rocket's tail. But the homemade motor is held in place by the masking tape with a **friction-fit**, so the twist-lock isn't needed.

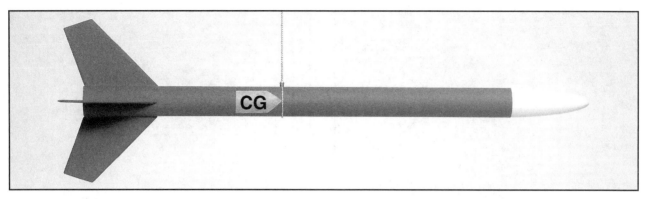

*Figure 19-30. The **center of gravity**, or "CG", is the point at which the rocket hangs level with the ground.*

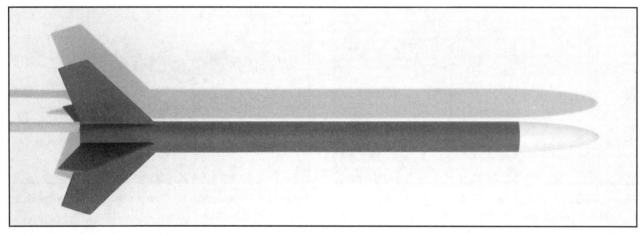

Figure 19-31. A single, bare light bulb casts a shadow of the rocket onto the cardboard, and the rocket is rotated until the shadow cast by its fins (top-to-bottom) is as wide as possible.

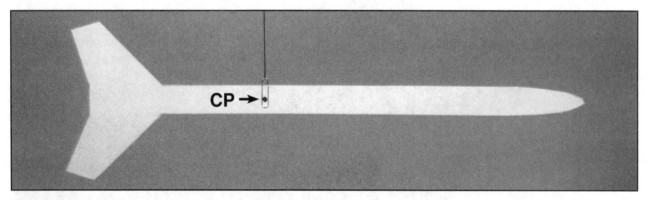

*Figure 19-32. The **center of wind pressure**, or **CP**, is the point where the rocket's silhouette hangs level.*

The **center of wind pressure**, or "CP", is the point on the rocket's body where the *aerodynamic* forces to the front equal the *aerodynamic* forces to the rear. In a simple and *practical* sense for *most* rockets, it is also the point where the front and rear forces exerted by *a rightangle side wind* are equal. Since a rightangle side wind encounters the rocket as a "straight-on" silhouette, you can *simulate* the relationship between the rocket and the side wind, and locate the **CP**, by cutting an exact silhouette of the rocket from stiff cardboard, and hanging it from a music wire hook with a paper clip.

To create the silhouette, tape the cardboard to a wall, and, with the motor removed, hold the rocket on a stick, parallel to, and as close to the cardboard as possible. Illuminate the rocket with a single, bare light bulb placed directly across the room, and as far away as possible. *Then rotate the rocket so that the shadow cast by its fins is as wide as possible* (***Figure 19-31***).

With a sharp pencil, trace the outline of the rocket's shadow onto the cardboard, and cut it out as accurately as you can. With a paper clip, hang this cut-silhouette from a music wire hook, and slide the clip back and forth until the silhouette hangs level. Then mark the location of the clip, and label it with the letters "**CP**" (***Figure 19-32***).

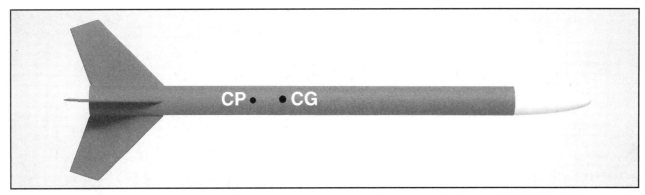

Figure 19-33. *For proper balance, the rocket's **CP** should be located about one body diameter **behind** the **CG**.*

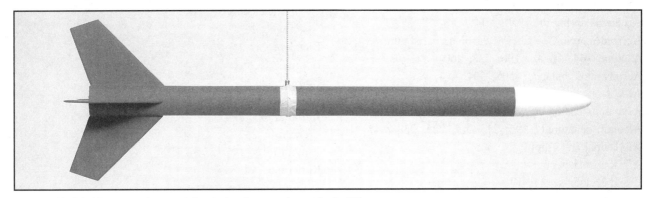

Figure 19-34. *The noose is taped firmly in place at the rocket's **CG**.*

Now hold the silhouette *next* to the rocket, and compare the positions of the two points. For proper balance, the **CP** should be located about one body diameter (called a "caliber") *behind* the **CG** (*Figure 19-33*). If it *isn't*, adjust the position of the **CG** by increasing or decreasing the weight of the rocket's nose. **Important note**. If the **CP** is located *more* than 1 caliber behind the **CG**, (2 or more calibers), the rocket will be "**overly stable**". In this condition, it will fly straight in *calm* air, but in windy weather it will tend to weathervane into any side winds it encounters on the way up. In practice, because the rocket's **CG** moves *forward* as its propellant is consumed, a starting position for the **CG** of 1 caliber *in front of* the **CP** (that's 1 body diameter *toward* the nosecone) will usually result in a good, stable flight.

As a final preflight test, *again* load the rocket with everything it will contain at lift-off. Make a noose at the end of a 6 foot to 8 foot piece of string. Slip the string over the rocket's body, and tape the noose *firmly* in place at the **CG** (*Figure 19-34*). Then swing the rocket on the end of the string around your head. If it flies *forward* into the wind like a weathervane, it is properly balanced for flight. If it flies at an angle, flies sideways, or flies backward, add weight to the nose, and readjust the position of the string until it hangs level again. Then repeat this "swing-test", and keep adjusting the weight of the nose until it flies *forward*, and the problem is corrected.

Flight Safety

Never light a rocket motor with a fuse. *Always* use an electric igniter. When launching rockets powered by motors up through **E** in size, *stand at lease 25 feet away from the launch pad*. When launching rockets powered by **F** through **I** motors, *stand 50 to 100 feet away*. If a rocket flies off course, it can travel downrange anywhere from a few dozen feet to a distance approximately equal to its expected maximum altitude. Therefore make sure that the launch site is clear of all flammable materials for a radius equal to the rocket's expected altitude. For example, if the rocket is capable of reaching an altitude of 2,500 feet, make sure that there's nothing flammable within 2,500 feet of the launch pad.

The lightweight "model rocket" kits sold at hobby and toy shops will handle motors up through **E**, but they are *not* sturdy enough to withstand the high speeds and accelerations produced by the larger motors. For a motor *larger* than an **E**, you need a "high power" rocket kit. *Some* hobby and toy shops sell "high power" kits, but because they are more expensive, the selection is limited, and you might have better luck dealing directly with the manufacturers. Most of them have websites on the Internet, and a search on the phrase, "high power rocket", should bring up dozens. You'll find general information about high power rocketry on the Tripoli website at **www.tripoli.org**, and for motor-making info & supplies, look over **The Teleflite Corporation** website at **www.teleflite.com**, or **www.amateur-rocketry.com**. At Teleflite's website you can view Teleflite's catalogue, and place orders with a credit card over a secured server. You'll also find answers to frequently asked questions.

Index

A

A-2 rocket (German) 8
A-3 rocket (German) 8
A-4 rocket (German) 8–9
A-5 rocket (German) 8
Accidents during the 1950s 11
Accroides gum. *See* Chemicals: binders: red gum
Acetone 44, 46–48, 196, 238, 460
Aerodynamic balance 500, 504
AIAA 10
Aircraft extension bit 89
Alcohol, denatured 196, 213–215, 225, 238, 447
Ali, Haidar and Tippu 2
Amateur rocketry
   and accidents 11, 19
   campaign against 19
   definition 9
   future of 19
American Fireworks News 175, 195, 211
American Institute of Aeronautics and Astronautics. *See* AIAA
Ammonium nitrate 181–182
Anthracite 188, 206–207, 234
   Anthracite Carbon Filter Media Company 188
   milled 206–207
Apogee 25
Arch punch 287, 301
Average thrust 21, 488

B

Baking soda 195, 237–238, 292
   effect on motor performance 292
Balances. *See* Scales
Ball mills 42–46
Becker, Walter, Colonel 8
Binder solvent, adding 238
Binders 194–195
Black powder
   and homemade bombs 18
   commercial 459
   commercial manufacture 40–41
   definition and uses 12–13
   disposal 16
   piezoelectric effect 15
   safe handling 14
   sensitivity
      friction 14
      impact 13
      spark 14
      static electricity 14
   storage 16
Blenders 41
Bombs 18
Booster 34
Brass 44
Burn rate
   control 237
   modifiers 195, 237–238
Burn time 21

C

Caraustar (formerly Star Paper Tube) 212
Carbon black 191, 209, 234
Casing retainers 88, 299
   improvised 65–66
   slipped 299
Casings. *See* Motor casings
Caustic potash. *See* Chemicals: potassium hydroxide
Caustic soda. *See* Chemicals: sodium hydroxide
Center of gravity 504–507
Center of wind pressure 504–507
Centering rings 59
CG. *See* Center of gravity
Charcoal 188–191, 207–209, 234
   activated 191
   airfloat 190, 209, 234
   garden 190, 208, 234
      milled 208
   mesquite 189, 208, 234
      milled 207
charcoal 12
Chart speed determination 483–484
Chemcentral Corp. 195
Chemical grades 174
Chemicals
   binders
      goma laca shellac 194–195, 234
      red gum 194, 234
      vinsol resin 195, 234

burn rate modifiers   195, 237–238
   baking soda   195, 292
   baking soda: effect on motor performance   292
chemical grades   174
drying   200
fuels   187–192, 206–209
   anthracite   188, 206, 234
   anthracite, milled   206
   carbon black   191, 209, 234
   charcoal   188–191, 207–209, 234
   charcoal, activated   191
   charcoal, airfloat   190, 209, 234
   charcoal, garden   190, 208, 234
   charcoal, garden, milled   208
   charcoal, mesquite   189, 207, 234
   charcoal, mesquite, milled   207
   sugar, powdered   192, 209
oxidizers   174–187, 200–206
   ammonium nitrate   181–182
   potassium nitrate   175–180, 234
   potassium nitrate, extracted   176–180
   potassium nitrate, fertilizer   175
   potassium nitrate, from chemical dealer   175
   potassium nitrate, milled   200–201
   potassium nitrate, precipitated   202–206
   potassium nitrate, stump remover   175
   sodium nitrate   181–187, 200–201, 234
   Sodium nitrate, extracted   185–187
   sodium nitrate, homemade   181–185
   sodium nitrate, milled   200–201
silica gel   454–455
sodium hydroxide   182–183
solvents   196–197
   acetone   44, 46–48, 196, 238, 460
   alcohol, denatured   196, 213–215, 225, 238, 447
   evaporation rates   196–197
sulfur   12, 192–193, 209
   dusting   193
   flowers   193
   powdered   193
   soil   193
Chile saltpeter. *See* Chemicals: oxidizers: sodium nitrate
Clay, powdered   197–198, 209
Coal crusher, homemade   86
Colander   52
Confectioner's sugar. *See* Chemicals: fuels: sugar, powdered
Congreve, William   2
Core length   25, 292
   effect on motor performance   292
Core spindles   89
   improvised   72
   tilted   296
CP. *See* Center of wind pressure
Cylindrical core burner   24

D

De Laval, Carl Gustav   20–21
De Laval nozzles   20–21, 319–322
Definitions
   amateur rocketry   9
   apogee   25
   average thrust   21
   booster   34
   burn time   21
   casing i.d.   22
   casing o.d.   22
   center of gravity   504–507
   center of wind pressure   504–507
   core length   25
   cylindrical core burner   24
   end burner   23
   first stage   34
   flame tube   34
   fuel rich   247
   grain geometry   23
   model rocketry   9
   motor casing   20
   motor length   22
   motor weight   22
   nozzle   20
   nozzle, De Laval   20–21
   nozzle throat   22
   propellant   20
   propellant grain   20
   propellant mass   21
   second stage   34
   single stage rocket   29
   specific impulse   21
   stoichiometric mixture   245
   thrust   21
   thrust-time curve   22
   total impulse   21
   two stage rocket   34
Dexol gopher gassers   177–180
Dornberger, Walter, Captain   8
Drying (chemicals)   200
Drying (rocket motors)
   air   290, 454–455
   vacuum   290–291

E

End burner   23, 310
Ertco   449
Eveready Thermometer Co.. *See* Ertco

F

Federov, Nikolai   3
Firesand. *See* Materials: nozzle clay components: grog
First stage   34
Flame tube   34, 502
Flight safety   507

Fork test   239, 243
French fry basket   52
Fuel rich   247
Fuels   187–192, 206–209

G

Gardner Spring Co.   470
German rocket program   6–9
Glue
   aliphatic   214, 232, 496, 499
   contact cement   232
   dextrin   213
   silicate   213
Goddard, Robert   4, 6
Goma laca shellac   194–195, 234
Gopher gassers, Dexol   177–180
Grain geometry   23
Grog   198, 209

H

Hale, William   2
Heat gun   460
Heat shrink tubing. *See* Tubing, heat shrink
Hercules Inc.   195
History
   Ali, Haidar and Tippu   2
   Army Signal Corps   4
   Becker, Walter, Colonel   8
   Congreve, William   2
   De Laval, Carl Gustav   20
   Dornberger, Walter, Captain   8
   Federov, Nikolai   3
   German rocket program   6–9
   German Rockets
      A-2   8
      A-3   8
      A-4   8
      A-5   8
      Kegelduse   7, 8
      Repulsor   8
      V-2   9
   Goddard, Robert   4, 6
      Clark University   4
      Guggenheim Foundation   5
      Navy work   5
      Roswell, New Mexico   5
      Smithsonian Institution   4
   Hale, William   2
   Kai-Fung-Fu, battle at   1
   Kummersdorf Proving Ground   8
   Lang, Fritz   7
   Lindbergh, Charles   5
   Nebel, Rudolph   7
   Oberth, Hermann   6–9
      work at UFA film studio   7
      work with the VfR   7

   Paulet, Pedro   5
   Peenemunde   8
   Scherschevsky, Alexander   5, 7
   Tsiolkovskii, Konstantin   3, 6
   Valier, Max   7
   Versailles Treaty   8
   VfR   7, 8
   von Braun, Wernher   8
   von Harbou, Thea   7
   von Opel, Fritz   7
Homemade centigram balance   40
Hurst Manufacturing   478–480
Hydrometer cylinders   450
Hydrometers   448–449
   proof scale   449
   tralle scale   449

I

Igniters, electric   456–462
   high heat   459–460
   mini   460–462
   paper match   456–457
Isp. *See* Definitions: specific impulse

K

Kai-Fung-Fu, battle at   1
Kegelduse rocket (Oberth - German)   7, 8
Kummersdorf Proving Ground   8

L

Lampblack. *See* Chemicals: fuels: carbon black
Landing gear wire. *See* Materials: wire: music
Lang, Fritz   7
Lindbergh, Charles   5
Load cell   468
Lortone Inc.   43, 235
Lye. *See* Chemicals: sodium hydroxide

M

Match, electric. *See* Igniters, electric
Materials
   glue
      aliphatic   214, 232, 496, 499
      contact cement   232
      dextrin   213
      silicate   213
   grease, vacuum, silicone   84
   metals   66
      brass   44
   nozzle clay components
      clay, powdered   197–198, 209
      grog   198, 209
      paraffin wax   209
      sand   198
   nylon   73–74
      lathe tool for cutting nylon   73

paper
  converters 222
  dealers 222
  how to buy 221–225
  kraft, 60 lb. 221–222
  liner board 222–223
  mills 222
  red rosin 224
  shopping bags 214
  tag board 224
pipe 60
  plastic 65
  steel 60, 88
polyethylene 51
wire
  hookup 460
  magnet 460
  music 287
  nichrome 460
  stereo speaker 456, 462
  thermostat-doorbell 456

Maximum thrust 484–486
Mixing cups 51
Model rocketry 9
Monoject syringes 48
Mortar & pestal 41
Motor building: original 1983 method 446–455
  casing preparation 447–451
  drying 454–455
    silica gel 454–455
  hydrometer cylinders 450
  hydrometers 448–449
    proof scale 449
    tralle scale 449
  motor building process 452–454
  propellant formulas 447
  thermometers 449
Motor casings 20
  drying 232
  glue coating 232–233
  homemade, large 225–231
  homemade, small 213–220
  telescoping 231
Motor classification 34–35
Motor length 22
Motor weight 22
Multistage ignition 301–306
Music wire 287

N

NAR 9, 10
National Association of Rocketry. *See* NAR
Nebel, Rudolph 7
New England Paper Tube 212
Newark Paperboard Products 212
Nozzle clay components 197–199, 209
Nozzle molds 88

Nozzle throat 22
Nozzles
  clay 20
  De Laval 20–21, 319–320
Nylon 73–74

O

Oberth, Hermann 6–9
Oxidizers 174–187

P

Pacific Fabric Reels 212
Paper 221–225
  converters 222
  dealers 222
  how to buy 221–225
  kraft, 60 lb. 221–222
  kraft liner board 222–223
  mills 222
  red rosin 224
  shopping bags 214
  tag board 224
Parachute ejection 308
Paraffin wax 209
Paulet, Pedro 5
Peenemunde 8
Pellet test 238, 240–245, 446
Piano wire. *See* Materials: wire: music
Pipe
  plastic 65
  steel 60, 88
Polyethylene 51
Potassium nitrate 175–184, 200–205, 234
  extracted 176–181
  fertilizer 175
  milled 200–201
  precipitated 201–206
  stump remover 175
Powder mills 235
  maintenance 45
  setup 44–45
Proof scale (hydrometers) 449
Propellant
  definition 20
  development process 245–251
  formulas 251
    from original 1983 book 447
    KG1 251, 328
    KG3 348
    KG6 311
    KS 252
    NG6 382
    NV6 418
  grain 20
  mixing procedure 235–237
  naming system 234
Propellant mass 21, 489

R

Radio Shack 481
Raketenflugplatz 8
Reaction Research Society 12
Red gum 194, 234
Reference bar 69
Reference plate 69
Repulsor rocket (German) 8
Rock tumblers 42–43
Rocket
   single stage 29–34
   stick 496–504
   two stage 34, 501–504
Rocket kits, commercial 504
   adapting a homemade motor 504
Rocket motor designs 330–445
   KG1 motors
     KG1-6-C6 330-331
     KG1-12-D18 332-333
     KG1-18-D12 334-335
     KG1-38-E33 336-337
     KG1-44-F25 338-339
     KG1-97-G60 340-341
     KG1-71-F40 342-343
     KG1-149-G104 344-345
     KG1-193-H144 346-347
   KG3 motors
     KG3-6-C8 350-351
     KG3-12-D17 352-353
     KG3-14-D12 354-355
     KG3-26-E31 356-357
     KG3-37-E26 358-359
     KG3-78-F55 360-361
     KG3-63-F41 362-363
     KG3-116-G102 364-365
     KG3-140-G139 366-367
   KS motors
     KS-6-C7 370-371
     KS-14-D20 372-373
     KS-14-D10 374-375
     KS-30-E30 376-377
     KS-38-E22 378-379
     KS-50-F25 380-381
   NG6 motors
     NG6-9-C8 384-385
     NG6-10-D7 386-387
     NG6-18-D23 388-389
     NG6-21-E23 390-391
     NG6-29-E17 392-393
     NG6-41-F21 394-395
     NG6-56-F41 396-397
     NG6-71-F42 398-399
     NG6-71-F32 400-401
     NG6-87-G30 402-403
     NG6-134-G78 404-405
     NG6-155-G88 406-407
     NG6-94-G36 408-409
     NG6-129-G37 410-411
     NG6-191-H100 412-413
     NG6-313-H119 414-415
     NG6-227-H137 416-417
   NV6 motors
     NV6-15-D8 420-421
     NV6-17-D8 422-423
     NV6-27-E21 424-425
     NV6-49-F20 426-427
     NV6-61-F20 428-429
     NV6-83-G44 430-431
     NV6-102-G32 432-433
     NV6-142-G32 434-435
     NV6-217-H75 436-437
     NV6-158-G43 438-439
     NV6-193-H47 440-441
     NV6-340-I100 442-443
     NV6-417-I65 444-445
Rocket motors
  and terrorism 18
  basic function 20
  sale of 18
  storage and handling 16
RRS. *See* Reaction Research Society

S

Safety record, Teleflite 18
Saltpeter. *See* Chemicals: oxidizers: potassium nitrate
Sand 198
Scales 40
   balance, centigram, homemade 40
   balance, Harvard-Trip 40
   balance, triple beam 37–39
   electronic, digital 40
   reloading 40
Scherschevsky, Alexander 5, 7
Second stage 34, 303
Shellac thinner. *See* Chemicals: solvents: alcohol, denatured
Sherwood Medical (Monoject syringes) 48
Sieves, homemade 85–86
Silica gel 454–455
Sine bar 69
Smart & Final Iris 52
Sodium bicarbonate. *See* Chemicals: burn rate modifiers: baking soda
Sodium hydroxide 182–185
Sodium nitrate 181–187, 200–201, 234
   homemade 181–185
   milled 200–201
sodium nitrate 185
sodium nitrate, extracted 185, 185–186
Solvent evaporation, rapid 299
Solvent requirement 238
Solvents 196–197
Sonoco Products Co. 212

Sources
  chart drive motors
    Hurst Manufacturing   478–480
  chemicals
    Anthracite Carbon Filter Media Company   188
    Chemcentral Corp.   195
    Hercules Inc.   195
    Teleflite Corporation   191, 194, 195
    Vicksburg Chemical Co.   175
  convolute cardboard tubes
    Caraustar (formerly Star Paper Tube)   212
    New England Paper Tube   212
    Newark Paperboard Products   212
    Pacific Fabric Reels   212
    Sonoco Products Co.   212
    Teleflite Corporation   211
    Thames River Tube Co.   212
  electronics
    Radio Shack   481
  miscellaneous
    Ertco   449
    Gardner Spring Co.   470
    Lortone Inc.   43
    Sherwood Medical (Monoject syringes)   48
    Smart & Final Iris   52
  nichrome wire
    Teleflite Corporation   460
  reading & reference material
    American Fireworks News   175, 195, 211
    Teleflite Corporation   19
Specific impulse   21, 490
Speedbor   73
Spoon sizes for propellant loading   283–284
Stoichiometric mixture   245
Strain gauge   469
Stump remover   175, 176–177
Sugar, powdered   192, 209
Sulfur   192–193, 209
  dusting   193
  powdered   193
  soil   193
Suppliers. *See* Sources
Syringes   48–50

T

Tamps   72–81, 89
Taper, convergent   22
Taper, divergent   22, 54
Teleflite Corporation   19, 40, 191, 194, 195, 211, 460
Telescoping   231
Terrorism   18
Test equipment
  electronic
    load cell   469
    strain gauge   469
  homemade   470–490
    chart recorder   478–483
    equipment Limitations   492
    safety during motor testing   492
    test stand   470–475
Test equipment use
  average thrust   488
  chart speed   483–484
  Maximum thrust   475–477, 484–486
  Specific impulse   490
  Thrust-time curve   482–483
  Total burn time   486
  Total impulse   487–488
Thames River Tube Co.   212
The Rocket Book   10
Thermometers   449
Thrust
  average   21
  definition   21
  maximum   484–486
Thrust-time curve   22
Time delay burn rate test   293–296
Time delays   25
  practicing with stick rockets   502
Tool Drawings. *See* Drawings, tool
Tools
  dedicating   301
  general
    aircraft extension bits   89
    blenders   41
    heat guns   460
    hydrometer cylinders   450
    hydrometers   448–449
    mortar & pestal   41
    reference bar   69
    reference plate   69
    sine bar   69
    Speedbors   73
    thermometers   449
    vacuum chamber, homemade   81–84
    vacuum pump maintenance   291
    vacuum pumps   46–47
    vacuum pumps, how to buy   46–47
    X-acto knife   287
  propellant preparation
    ball mills   42–46
    coal crusher, homemade   86
    colanders   52
    french fry baskets   52
    mixing cups   51
    powder mill maintenance   45
    powder mills   235
    rock tumblers   42–43
    sieves, homemade   85–86
    syringes   48–50
    syringes, how to use   48–50

rocket motor tooling
    arch punches 287
    casing retainers 60–66, 88
    casing retainers, improvised 65–66
    centering rings 59
    core spindles 89
    core spindles, improvised 72
    nozzle molds 88
    porthole wires 288
    tamps 72–81, 89
    tamps, danger of metal tamps 73
scales
    balance, centigram, homemade 40
    balance, Harvard-Trip 40
    balance, triple beam 37–39
    electronic, digital 40
    reloading 40
Total burn time 486
Total impulse 21, 487–488
Tralle scale (hydrometers) 449
Tripoli Rocketry Association 12
Tsiolkovskii, Konstantin 3, 6
Tubes, cardboard
  convolute 211–219, 225–233
    drying 232
    glue coating 232–233
    homemade, large 225–231
    homemade, small 213–220
    telescoping 231
  spiral 211
Tubing, heat shrink 460

V

V-2 rocket (German) 9
Vacuum chamber, homemade 81–84
Vacuum drying 290–291
Vacuum grease, silicone 84
Vacuum pumps 46–47, 290–292
  how to buy 46–47
  maintenance 291
Valier, Max 7
Verein fur Raumschiffahrt. *See* History: VfR
Versailles Treaty 8
VfR 7, 8
Vicksburg Chemical Co. 175
Vinsol resin 195, 234
von Braun, Wernher 8
von Harbou, Thea 7
von Opel, Fritz 7

W

Wan Hoo 1
Wire
  hookup 460
  magnet 460
  music 287
  nichrome 460
  stereo speaker 456, 462
  thermostat-doorbell 456

X

X-acto knife 287

Y

Yacca gum. *See* Chemicals: binders: red gum